364
Topics in Current Chemistry

Aims and Scope

The series Topics in Current Chemistry presents critical reviews of the present and future trends in modern chemical research. The scope of coverage includes all areas of chemical science including the interfaces with related disciplines such as biology, medicine and materials science.

The goal of each thematic volume is to give the non-specialist reader, whether at the university or in industry, a comprehensive overview of an area where new insights are emerging that are of interest to larger scientific audience.

Thus each review within the volume critically surveys one aspect of that topic and places it within the context of the volume as a whole. The most significant developments of the last 5 to 10 years should be presented. A description of the laboratory procedures involved is often useful to the reader. The coverage should not be exhaustive in data, but should rather be conceptual, concentrating on the methodological thinking that will allow the non-specialist reader to understand the information presented.

Discussion of possible future research directions in the area is welcome.

Review articles for the individual volumes are invited by the volume editors.

Readership: research chemists at universities or in industry, graduate students.

More information about this series at
http://www.springer.com/series/128

Anouk M. Rijs • Jos Oomens
Editors

Gas-Phase IR Spectroscopy and Structure of Biological Molecules

With contributions by

J.L. Alonso · O.V. Boyarkin · P. Çarçabal · E.J. Cocinero · R.C. Dunbar · M.-P. Gaigeot · E. Gloaguen · J.C. López · M. Mons · J. Oomens · A.L. Patrick · N.C. Polfer · A.M. Rijs · T.R. Rizzo · R. Spezia · M.S. de Vries

Springer

Editors
Anouk M. Rijs
Molecular and Biophysics
FELIX Laboratory
Institute for Molecules and Materials
Radboud University
Nijmegen
The Netherlands

Jos Oomens
Molecular Structure and Dynamics
FELIX Laboratory
Institute for Molecules and Materials
Radboud University
Nijmegen
The Netherlands

ISSN 0340-1022 ISSN 1436-5049 (electronic)
Topics in Current Chemistry
ISBN 978-3-319-19203-1 ISBN 978-3-319-19204-8 (eBook)
DOI 10.1007/978-3-319-19204-8

Library of Congress Control Number: 2015942232

Springer Cham Heidelberg New York Dordrecht London

Printed on acid-free paper

Springer International Publishing AG Switzerland is part of Springer Science+Business Media (www.springer.com)

Preface

While obviously not the natural environment of biological molecules, the gas phase provides the ultimate degree of isolation. It is a reductionist's paradise, where the subject under study can be investigated in its purest form. Under these conditions, inter- and intramolecular interactions can be scrutinized in the highest possible detail, suspending any inhomogeneities induced by perturbations from the surroundings, and enabling a true comparison with results from quantum-chemical calculations at the highest levels of sophistication. At the same time, the influence of (aqueous) solution can be accurately tracked by non-covalent attachment of individual water molecules.

Spectroscopy, in the various frequency ranges of the electromagnetic spectrum, offers the most direct experimental probe into the quantum-mechanical nature, and hence into the molecular structure, of these dilute isolated biomolecules. Laser-based spectroscopy of jet-cooled gaseous biomolecules has a long and rich history, but has particularly seen impressive developments over the past decade, e.g. with the introduction of laser desorption methods, new infrared laser sources, and sophisticated multi-resonance excitation and detection schemes. The conformation selectivity provided by many of these methods is perhaps one of their foremost assets in the study of biological molecules. The fine interplay between many (non-covalent) interactions, ultimately determining the folding structure of the molecule and hence its biological functioning, can thus be unraveled.

Studies on the fundamental properties of biological molecules have also brought the fields of ion chemistry and spectroscopy closer together over the past decade. With electrospray ionization tandem mass spectrometry, ion chemists have been able to manipulate and characterize biological molecules in the gas phase since the late 1980s. The past decade has seen particularly rapid developments in the application of laser-based spectroscopy to mass-selected molecular ions, with the aim of determining the structure of charged biomolecules. The use of widely tunable infrared free electron lasers and cryogenic ion trapping devices, as well as the combination of ion-mobility conformer separation with ion spectroscopy, have led to rapid progress in this field.

In this book, we have tried to capture some of the excitement of these recent developments in the spectroscopy of gas-phase biological molecules, admittedly without having the slightest illusion of our coverage being anywhere near complete.

The chapter "IR Spectroscopic Techniques to Study Isolated Biomolecules" gives an overview of some of the most common experimental practices currently in use to characterize the structure of isolated biomolecules by infrared spectroscopy. We address especially two main categories of experimental approaches: conformation-selective infrared spectroscopy of jet-cooled neutral species and infrared (multiple-photon) dissociation spectroscopy of mass-selected ionized biomolecules. Molecular beam laser spectroscopy methods form the experimental basis for the topics covered in the sixth to eighth chapters. Mass spectrometry-based ion spectroscopy provided the experimental data for the studies reviewed in fourth and fifth chapters (and seventh inpart).

The use of gas-phase biomolecule spectroscopy for structural characterization relies strongly on the interplay between experiment and theory. Structural properties can usually only be extracted from experimental spectra with the use of high-level quantum-chemical calculations. The chapter "Theoretical Methods for Vibrational Spectroscopy and Collision Induced Dissociation in the Gas Phase" reviews some recent advances in the theoretical methods applied to predict vibrational spectra, including molecular-dynamics-based methods to model photo-dissociation spectra and DFT-based molecular dynamics to predict spectra in the far-infrared region. The use of trajectory calculations on a semi-empirical potential is investigated as an alternative to transition-state calculations for the modeling of collision-induced dissociation of protonated peptides.

The introduction of cryogenic ion traps with optical access has opened up an entire range of new approaches for ion spectroscopy, both in the UV/vis and IR wavelength regimes. These methods enable the use of conformation-selective ion spectroscopy either by multiple-resonant laser schemes or by combination with ion mobility. The push towards spectroscopy of larger biological molecules is largely carried by these technologies. The state of the art in cryogenic ion spectroscopy is reviewed in the chapter "Cryogenic Methods for the Spectroscopy of Large, Biomolecular Ions".

The chapters "Peptide Fragmentation Products in Mass Spectrometry Probed by Infrared Spectroscopy" and "Spectroscopy of Metal-Ion Complexes with Peptide-Related Ligands" deal with the application of infrared ion spectroscopy to amino acids and peptides as protonated species and as ligands in metal-ion coordination complexes. Chapter 4 discusses how ion spectroscopy has been instrumental in gaining a deeper insight into the dissociation chemistry of protonated (and deprotonated) peptides. Despite their widespread use in peptide and protein sequencing in proteomics studies, the actual pathways of the dissociation reactions have been under much debate. Using the MS^n capabilities of modern mass spectrometers in combination with ion spectroscopy reveals the molecular structures of peptide fragments, yielding key information on the reaction pathways followed. The fifth chapter addresses the coordination motifs of amino acids and peptides to metal ions, which are usually impossible to resolve from MS data alone. Infrared spectroscopy of the ion-ligand complexes reveals the isomeric motifs adopted by the ligands and provides further information on the metal binding anchors and the conformeric preferences.

Applications of spectroscopy to the main classes of biological molecules and in particular their oligomeric building blocks – nucleobases, carbohydrates, and peptides – in their neutral forms are reviewed in the sixth to ninth chapters. These studies employ either conformer-selective IR–UV double resonance spectroscopy or microwave spectroscopy. The structural characterization of isolated neutral amino acids and peptides, reviewing the major secondary structure elements, is discussed in the chapter "Isolated Neutral Peptides". This chapter also briefly reviews density functional theory calculations used to interpret the conformation-selective infrared spectra. The chapter "Gas-Phase IR Spectroscopy of Nucleobases" gives an overview of IR spectroscopy applied to identify the structures of nucleobases, nucleosides, and nucleotides. The tautomeric preferences of monomeric units as well as the hydrogen-bonding interactions stabilizing base pairs are some of the important structural elements which can be deciphered from the spectral data. Furthermore, the UV spectra convey information on the excited state dynamics of isolated nucleobases, base pairs, and their clusters with water. Spectroscopic studies on ionized building blocks of DNA and RNA are also reviewed in the seventh chapter. In the chapter "Carbohydrates", the extensive body of gas-phase spectroscopic studies on carbohydrates employing both infrared and microwave methods is reviewed. The high conformational flexibility and structural diversity of this class of species are countered with the high resolving power of molecular beam spectroscopy. The structural preferences and the interactions with water and peptides or peptide mimics are some of the issues addressed in this chapter.

Finally, the chapter "Microwave Spectroscopy of Biomolecular Building Blocks" reviews the use of microwave spectroscopy as a structural probe of jet-cooled biological molecules including amino acids, carbohydrates, and nucleobases, revealing rotational constants, nuclear quadrupole coupling interactions, and dipole moments, derived from their pure rotational spectra.

It has been a pleasure to compile this volume and work with so many experts in the field. We hope this book will serve as a benchmark of the state of the art in 2014 and as an inspiration for current and future investigators of gas-phase biomolecule spectroscopy.

Nijmegen, The Netherlands
Spring 2015

Anouk Rijs
Jos Oomens

Contents

Top Curr Chem (2015) 364: 1–42
DOI: 10.1007/128_2014_621

Published online: 27 March 2015

IR Spectroscopic Techniques to Study Isolated Biomolecules

Anouk M. Rijs and Jos Oomens

Abstract The combination of mass spectrometry, infrared action spectroscopy and quantum-chemical calculations provides a variety of approaches to the study of the structure of biologically relevant molecules in vacuo. This chapter reviews some of the experimental methods that are currently in use, which can roughly be divided into two main categories: (1) low-temperature neutral molecules in a molecular beam environment, which can be investigated in a conformationally selective manner by the application of double-resonance laser spectroscopy and (2) ionized species which can conveniently be manipulated and selected by mass spectrometric methods and which can be investigated spectroscopically by wavelength-dependent photo-dissociation. Both approaches rely on the application of infrared tunable laser spectroscopy and the laser sources most commonly used in current studies are briefly reviewed in Sect. 3. Along with quantum-chemical calculations, reviewed in Chapter 3 of this book (Gaigeot and Spezia, Top Curr Chem doi:10.1007/128_2014_620), the experimental IR spectra reveal a wealth of information on the structural properties of the biological species.

Keywords Biomolecules · Conformation selective · Free electron laser · FTICR · Ion trap · IRMPD · IR–UV ion dip spectroscopy · Laser desorption · Molecular beam · OPO · Tandem mass spectrometry

---

A.M. Rijs (✉)
Radboud University Nijmegen, Institute for Molecules and Materials, FELIX Laboratory, Toernooiveld 7c, 6525 ED Nijmegen, The Netherlands
e-mail: a.rijs@science.ru.nl

J. Oomens
Radboud University Nijmegen, Institute for Molecules and Materials, FELIX Laboratory, Toernooiveld 7c, 6525 ED Nijmegen, The Netherlands

Van 't Hoff Institute for Molecular Sciences, University of Amsterdam, Science Park 904, 1098 XH Amsterdam, The Netherlands
e-mail: j.oomens@science.ru.nl

Contents

1 Conformation-Selective IR Spectroscopy of Neutral Biomolecules

1.1 Isolated Biomolecules and Their Interactions

The selectivity and function of biologically active molecules is closely intertwined with their three-dimensional structure, which in turn depends on the skeletal shape resulting from covalent and non-covalent interactions, such as hydrogen bonds between neighbouring parts of the biomolecule. In addition to function, structure also controls molecular transport, molecular recognition and selective binding of ligands, such as adenosine triphosphate (ATP) association at the active site of ATPases to initiate enzymatic activity. A delicate balance between intra- and intermolecular hydrogen-bonding forces and hydrophobic interactions controls the resulting biomolecular conformation, and is key to understanding the function of biological molecules. Not surprisingly, much effort has been devoted to the elucidation of biomolecular structure and the associated interactions. Well-known techniques such as X-ray crystallography, nuclear magnetic resonance (NMR), single molecule spectroscopy, mass spectrometry and computational simulations have been employed to clarify the structure-to-function relation of biomolecules. In addition, molecular spectroscopic techniques have been developed to study biological molecules under isolated conditions, removing inhomogeneous line broadening induced by environmental influences. Two parallel pathways, both using molecular spectroscopy to probe either neutral or ionic isolated species, have led to the development of state-of-the-art techniques, which are described in Sects. 1 and 2 of this chapter, respectively.

To employ the rich toolbox of gas-phase optical spectroscopy techniques in the study of this class of molecules, the biomolecules need to be brought into the gas phase and cooled to their lowest energy conformations. Pioneering experiments on the amino acid tryptophan were performed by Levy et al. using thermospray evaporation combined with a supersonic beam [1]. This allowed them to record the electronic spectrum of tryptophan and tryptophan analogs under the cold and isolated conditions of a supersonic molecular beam. For the first time, the existence of different conformations was inferred from the power dependence of the signals resulting from electronic transitions [1–5]. Later, a more elegant version of conformation selection was used, IR–UV double resonance spectroscopy [6–8], to identify spectral features originating from distinct conformers [9–11]. Its full potential was shown when it was combined with quantum-chemical calculations to assign vibrational bands and reveal conformational structures [11]. Since then, this method and a number of its variants have been applied to numerous biomolecular systems. In Sect. 1 of this chapter we focus on the methods currently employed to obtain these IR spectra of mass-selected, conformation-selected neutral biomolecules in the gas phase.

1.2 Neutral Biomolecules in the Gas Phase

As virtually all biological molecules possess low vapour pressures gas-phase molecular spectroscopy methods to investigate isolated neutral biomolecules require volatilization methods other than thermal evaporation. The combination of laser desorption with a supersonic molecular beam expansion together with the selectivity of IR and UV double resonance methods opened up the possibility of characterizing isolated, neutral biomolecules and their clusters with the biological environment.

1.2.1 Transfer into the Gas Phase: Laser Desorption

The simplest method to transfer molecules as neutrals into the gas phase is by thermal heating. Here, a temperature-controlled oven is located either before or after the nozzle in the pre-expansion region. This method has been used for a limited number of small biomolecules including some nucleobases [12–15], amino acids [1–5], anaesthetics [16–23] and several neurotransmitters [24–26]. However, this method often requires very high temperatures to achieve significant vapour pressures, resulting in impractical experimental conditions and unacceptable high levels of thermal decomposition.

Thermal degradation can be avoided by instant heating of the biomolecule of interest using laser desorption. Laser desorption in combination with supersonic expansion has been widely used to bring intact biomolecules into the gas phase [27]. The non-volatile molecules are deposited on a sample bar made of a material that is believed to assist the desorption process. Various matrices, such as activated carbon, fritted glass [28, 29], polyethylene [30] or graphite [31–35], have been used, although matrix-free desorption (bare molecules) has been performed as well

[36]. Besides choice of sample bar material, variations in sample preparation (deposited as a thin layer, doped, pre-mixed), sample bar shape (rods, pressed discs, flat stages) and the choice of desorption laser (pulse length, wavelength, fluence) result in slight modifications in the desorption mechanism. For the experiments discussed in Chaps. [37–39] of this book, molecules are desorbed from a sample bar of ultrafine grain graphite with a low intensity pulsed Nd:YAG laser (fundamental or frequency doubled) with output energies of about 1–2 mJ.

Various mechanisms have been proposed to explain the process of laser desorption [40–44]. Here, we focus on the mechanism where the matrix plays a passive role as energy transmitter, i.e. laser-induced thermal desorption (LITD) by indirect heating of the graphite substrate to bring neutral non-volatile molecules into the gas phase. When the desorption laser hits the surface of the graphite sample bar, instant, fast and extreme heating takes place. The main difference between laser desorption and thermal heating lies in the heating rate of the substrate. Rates for resistive heating are of the order of 10^0–10^2 K s^{-1}, while for laser desorption they are of the order of 10^{10}–10^{12} K s^{-1} (about 1,000 K in a 1-ns laser pulse). Two processes now compete: the molecules either desorb from the surface and enter the gas phase intact or they react and fragment [45]. Although the reaction/fragmentation energy barrier might be lower than the barrier for intact desorption, it was shown that high heating rates allow the energetically unfavourable desorption process to occur on account of entropic reasons [44]. It should be noted that higher heating rates are not always better. After an optimum is reached, higher laser fluences result in a high amount of desorbed sample molecules, which may disturb the supersonic expansion, the cooling conditions and even complete molecular clusters may be ablated from the sample substrate.

The major disadvantage of laser desorption lies in shot-to-shot fluctuations in the amount of sample molecules injected into the supersonic expansion. A small fluctuation in desorption laser power can cause larger fluctuations in the amount of desorbed molecules [27]. Variations can be minimized by maintaining a constant laser fluence and by creating a homogeneous sample. The stability and strength of the signal (the amount of evaporated sample molecules) can be improved by mixing the sample bar with carbon black powder [46], forming a more homogeneous sample.

Figure 1 presents a schematic overview of a typical molecular beam time-of-flight mass spectrometer equipped with a laser desorption source. In the studies presented in this book, the sample bar is made from graphite. Accurate positioning of the sample bar with respect to the nozzle is required for optimal performance. It is typically mounted on a double translation stage (Fig. 1). The vertical travel (x-direction) with a typical accuracy better than 0.01 mm allows for optimal cooling with minimal distortion of the molecular beam expansion. The sample bar is typically positioned about 0.1 mm below the aperture of the pulsed molecular beam valve. Travel in the horizontal direction (y-axis) of 50 mm (length of the sample bar) with a position accuracy of about 0.1 mm ensures desorption of fresh sample at every laser shot. Both positioning options can be controlled under operating conditions. Finally, the distance along the molecular beam (z-axis)

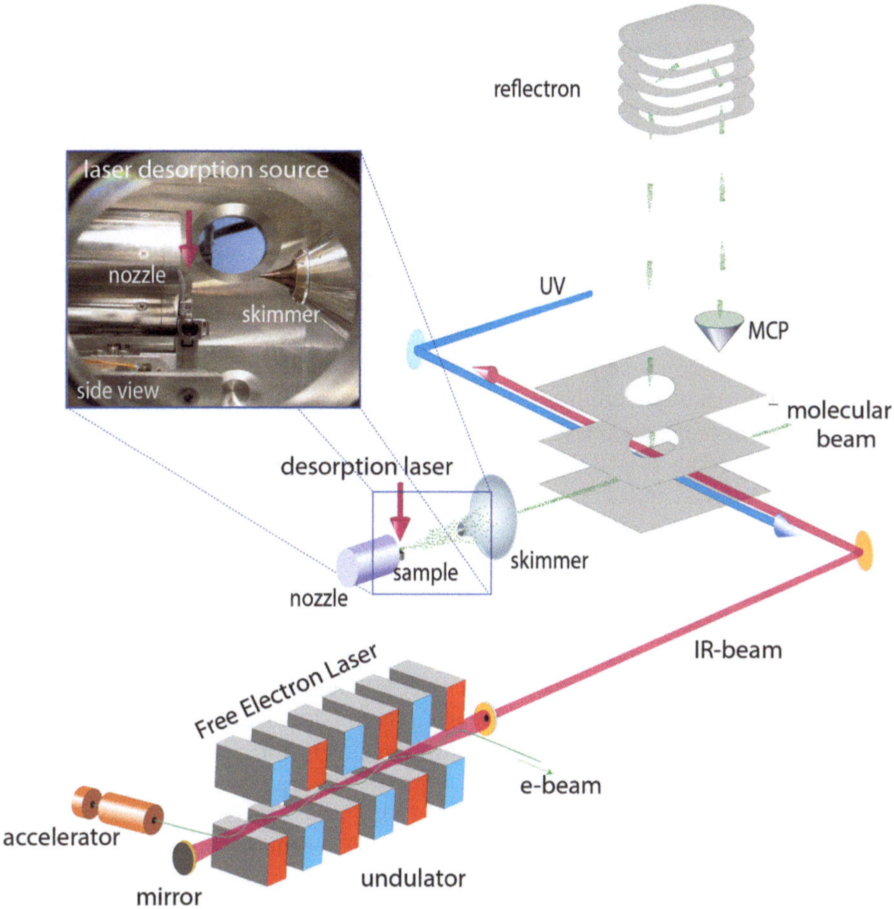

Fig. 1 Schematic representation of the molecular beam set-up equipped with a laser desorption source and a reflectron Time-of-Flight mass spectrometer used for IR–UV double resonance experiments at the FELIX Facility. The *inset* shows details of a typical laser desorption source

between the sample bar and nozzle can be adjusted when necessary, but should be kept as close as possible for optimal cooling. The shape of the sample bar (knife shaped) as well as its distance to the nozzle and the molecular beam axis is chosen to influence the molecular beam expansion as little as possible while obtaining as much signal as possible.

Desorption of molecules ranging in size from single amino acids to systems with molecular weights up to 2,000 Da has been demonstrated, where a monolayer or less is desorbed by the heat generated at the surface. However, some species are difficult to bring into the gas phase without inducing fragmentation. For example, arginine-containing peptides tend to undergo substantial fragmentation [47, 48]. In addition, nucleosides show varying behaviour; guanosine does not fragment at all, but adenosines may show significant fragmentation [49, 50].

In most experiments, argon at backing pressures ranging from 1 to 8 bar is used as a seed gas, although other noble gases have been used as well. For example, for molecules with a molecular weight close to 2,000 Da, xenon has been found to improve the supersonic cooling of the sample [51].

1.2.2 Molecular Beam Techniques: Supersonic Cooling

The laser-desorbed molecules are injected into the collision zone of a pulsed supersonic expansion where they are entrained in the carrier gas pulse and internally cooled by multiple collisions with the carrier gas atoms. When the molecules are desorbed into the gas phase, it is assumed that the different conformations are populated according to a Boltzmann distribution at a temperature of about 600–800 K. An important question is to what extent this initial distribution is retained in the supersonic cooling process.

Initially, collisions between the molecules and seed gas atoms accelerate the sample molecules in the same direction as the seed gas atoms. This translational cooling takes place close to the orifice and results in a translational temperature (velocity spread in the direction of the molecular beam) similar to that of the seed gas atoms. Once the velocity of the biomolecules and argon atoms are equilibrated, only low energy collisions occur, decreasing the rotational and vibrational temperatures of the sample molecules via energy transfer to the gas bath [52, 53]. The efficient cooling of the translational and rotational degrees of freedom leads to translational temperatures smaller than 2 K [54, 55] and rotational temperatures of about 1–10 K [27, 53, 55]. Vibrational cooling is less efficient because of a smaller vibrational depopulation cross section compared to the cross section of rotational depopulation. Therefore, the vibrational temperature may not fully equilibrate, providing temperatures ranging from 15 to 50 K. Such higher temperatures are not necessarily a problem for the experiments presented here, because even at these temperatures most vibrationally excited states are not significantly populated according to a Boltzmann distribution. Additionally, low-frequency vibrations have a larger depopulation cross section and are therefore more efficiently cooled. The vibrational temperature can be determined experimentally by comparing the intensities ($I_{0\text{–}0}$ and $I_{1\text{–}0}$) of the $v''=0 \rightarrow v'=0$ origin and the $v''=1 \rightarrow v'=0$ hot band of an electronic transition using the relation (assuming that the Franck-Condon factor for both transitions is comparable)

$$\frac{I_{0-0}}{I_{1-0}} = e^{\left(\frac{-\Delta E}{kT}\right)}, \tag{1}$$

where ΔE is the vibrational energy of the $v''=1$ state. For example, in the case of laser-desorbed Z-Glu-OH, a rather flexible molecule, a vibrational temperature of about 15 K can be obtained [47]. This cooling can be as efficient as in conventional molecular beam seeding experiments based on thermal evaporation provided that the desorption takes place within a few nozzle diameters from the nozzle.

The biomolecules under investigation, even single amino acids, have many flexible coordinates and can thus potentially adopt many low energy conformations. Their potential energy surfaces are very complex and contain a large number of local minima and many transition states between them. Zwier and co-workers showed that a relatively small molecule such as *N*-acetyl tryptophan methyl amide (NATMA) already contains 164 minima, which are connected by 714 transition states [56]. The question remains as to which conformers (local and global minima on the potential energy surface) are present in our cooled molecular beam and how this ensemble represents the distribution before cooling? When the barriers between the conformers on the potential energy surface are higher than the average collision energies, the relative populations present at high temperatures (determined by the Gibbs free energy at 300 K) are frozen into the low temperature environment of the molecular beam. However, if the barriers between the conformers are lower than the collision energy, conformational interconversion can occur and relaxation brings the molecule down into its lowest energy conformers. This has been summarized by the rule of thumb proposed by Simons [57, 58]. When the barrier is higher than 800 cm^{-1} (~2 kcal/mol), the initial Boltzmann distribution remains unchanged; however, for lower barriers, conformational isomerization occurs, resulting in a conformational distribution reflecting the lower temperature. Conformational isomerization therefore depends on the carrier gas, where argon and krypton are more efficient in inducing interconversion than helium.

Supersonic expansion cooling in the study of biomolecular systems thus results in a significant depopulation of excited rotational and vibrational levels, which decreases the spectral congestion dramatically. This allows one to characterize individual conformers – and their complexes – using double resonance spectroscopic techniques. These methods are described in Sects. 1.3.4–1.3.7.

1.2.3 Non-covalent Complexes: The Biological Environment in the Gas Phase

In their natural environment, the structure and conformational preferences of biomolecules are (fine-)tuned by interactions with their environment. The advantage of the gas-phase approach is that it allows one to study the systems under high-resolution spectroscopic conditions, and, moreover, to generate non-covalent complexes to mimic the interactions in a biological environment, for example by clustering with ligands, metal ions and/or water molecules. This enables one to assess the influence of these bio-environments in a precisely controlled manner. For example, micro-hydration, i.e. studying the properties of clusters with one or more individual water molecules added, is even more appealing when realizing that most of the basic chemical processes in proteins occur at a well-defined, protected site shielded by the protein environment (hydrophobic pocket) [59–61]. Hence, the hydrated complexes generated in the gas phase can be realistic mimics, because only one or a few water molecules are typically present in these hydrophobic pockets and play a role in stabilizing the active sites of enzymes. These water complexes can be produced in the molecular beam by co-expanding water-vapour with the carrier gas [62–72], while the metal complexes or ligand complexes are formed by co-desorbing them together with the sample molecules [25].

1.2.4 Time-of-Flight Detection

Several methods can be applied to probe sample molecules spectroscopically in the supersonic expansion, such as laser-induced fluorescence (LIF) and resonance-enhanced multi-photon ionization (REMPI). The latter, frequently used in the experiments presented in this book, is more sensitive, can be applied more generally and, moreover, provides mass selectivity. The neutral biomolecules of interest are most commonly ionized via a two-photon resonant process (see Sect. 1.3.3) and detected in a time-of-flight mass spectrometer (TOF-MS). In a TOF-MS, ions generated at a well-defined position and time in a uniform electric field are accelerated to the same kinetic energy, so that ions of different mass-to-charge ratio reach different velocities. The flight time (t_{flight}) through a field-free region of length L is measured, so that the mass-to-charge ratio m/z of the ion can be determined from

$$t_{\text{flight}} = \frac{L}{v} = L\sqrt{\frac{m}{2z(sE_s + dE_d)}}, \tag{2}$$

where E_s and E_d are the electric potential differences between repeller and extractor and between extractor and ground plate, respectively, over which the ions are accelerated. The distance from the ionization position (molecular beam-laser interaction) to the extractor plate (ideally half way between the repeller and extractor plates) is given by s, while d represent the distance between the extractor and ground plate. Virtually all TOF-MS systems feature a Wiley–McLaren configuration, in which the acceleration zone is divided into two regions of different electric field strength [73]. In this geometry, tuning of the voltages applied to repeller and extractor plates can be used to place the focal plane of the mass analyzer at the surface of the detector located at the far end of the field-free zone. In addition to Wiley-McLaren space focussing, a reflectron can be used to enhance the mass resolution. Here, a series of cylindrically shaped electrodes with gradually increasing DC bias voltages placed at the end of the flight tube acts as an ion mirror. The reflectron not only increases the flight path but also corrects for kinetic energy spread of the ion package.

1.3 Spectroscopic Probes

Of all the numerous spectroscopic methods to probe molecular conformation (including NMR, circular dichroism, Raman spectroscopy, two-dimensional IR spectroscopy, etc.), UV or IR action spectroscopy is most suited for the application to gas-phase molecules, either neutral or charged. Optical spectroscopy in the microwave, infrared or UV/vis spectral range can provide detailed structural information, especially in combination with molecular beam techniques or other cryogenic methods (low internal temperatures), resulting in highly resolved spectroscopic signatures [73].

1.3.1 Action Spectroscopy to Probe Isolated, Neutral Molecules

Because IR absorption measurements discussed in this book are all performed in optically extremely dilute samples, such as those occurring in molecular beams or tandem mass spectrometers, *action spectroscopy* is employed instead of *direct absorption* IR spectroscopy. In conventional *direct absorption* IR spectroscopy, the attenuation of the light beam transmitted through the sample is measured as a function of the frequency of the IR radiation. By measuring the IR laser beam intensity before and after the sample, the absorbance as a function of the frequency of the IR photons is obtained according to the Lambert–Beer law:

$$I(\nu) = I_0 e^{-\sigma(\nu)Ln}, \tag{3}$$

where $I(\nu)$ and I_0 are the intensities of transmitted and incident light beams as a function of the frequency ν, $\sigma(\nu)$ is the absorption cross section, L the path length and n the number density of the sample, assuming that all molecules are initially in the ground vibrational state. The product of number density and optical path length occurring in the exponent of Beer's law is typically so tiny in molecular beams or ion traps that the difference between $I(\nu)$ and I_0 is not measurable. In contrast, *action spectroscopy* methods make use of the *effect* of IR radiation on the molecules or ions, monitoring, for example, a change in ionization yield, quantum state, fluorescence yield, fragmentation or electron detachment. The fraction of molecules affected can be expressed as

$$N(\nu) = N_0 e^{-\sigma(\nu)\Phi(\nu)}, \tag{4}$$

where $\Phi(\nu)$ is the photon fluence experienced by the sample. Using typical values for absorption cross sections and laser intensities, it is seen that the argument of the exponent assumes values of the order of about 1 for photon fluxes typical of pulsed laser systems.

The types of action spectroscopy presented in this book include IR–UV ion dip spectroscopy for neutral species and Infrared Multiple Photon Dissociation spectroscopy for ionized species. IRMPD spectroscopy is discussed in more detail in Sect. 2.3.3 of this chapter. In IR–UV ion dip spectroscopy on neutral molecules in a molecular beam expansion, the IR and UV transitions share a common ground state; removal of population from this ground state by IR excitation is probed as a decrease in the UV induced ionization signal. Combined with a resonant ionization scheme, this technique allows one to record conformation specific IR spectra, as discussed in Sects. 1.3.5–1.3.7.

1.3.2 Electronic Spectroscopy

UV spectroscopy in molecular beams involves either laser-induced fluorescence (LIF) or resonance-enhanced multiphoton ionization (REMPI) methods. The latter method has the advantage that the resulting ionized molecules can be mass-analysed in a TOF mass spectrometer. The application of either REMPI or LIF spectroscopy requires the molecule of interest to incorporate a UV chromophore, such as an aromatic moiety. The DNA and RNA bases adenine, guanine, cytosine, thymine and uracil are aromatic molecules with well-known UV absorptions. Of the 20 naturally occurring proteinogenic amino acids, 3 – phenylalanine, tyrosine and tryptophan – feature an aromatic side chain, as do many neurotransmitter molecules. To study molecules that lack a UV chromophore, such as peptides without Trp, Tyr and Phe residues and carbohydrates, a UV chromophore needs to be chemically attached [47, 48, 74].

UV spectroscopy is a sensitive probe for the environment of the UV chromophore, resulting in slightly different UV transition frequencies for each conformation. Shifts in the UV absorption band induced by the peptide environment and the intramolecular interactions are typically $<100\ \text{cm}^{-1}$ for UV transitions around $37{,}000\ \text{cm}^{-1}$. The combination of a narrow-band UV laser source (e.g. a frequency-doubled pulsed dye laser) and a collisionless low-temperature environment (molecular beam) minimizes line broadening, allowing UV transitions of different conformers to be resolved.

Additionally, the UV spectra might reveal the conformational abundances present in the molecular beam expansion. However, this information ought to be interpreted with care, since specific detection schemes may introduce a quantitative bias. For REMPI detection, the photo-ionization efficiency can be conformer-dependent [75], decreasing for instance in the presence of NH-π intramolecular interactions [76]. When employing fluorescence detection, strongly fluorescent conformers may be overestimated, while poorly fluorescent conformers are underestimated.

1.3.3 REMPI Spectroscopy

Resonance Enhanced MultiPhoton Ionization (REMPI) is an effective approach to ionize molecules and study their electronic structure. In the studies presented in this

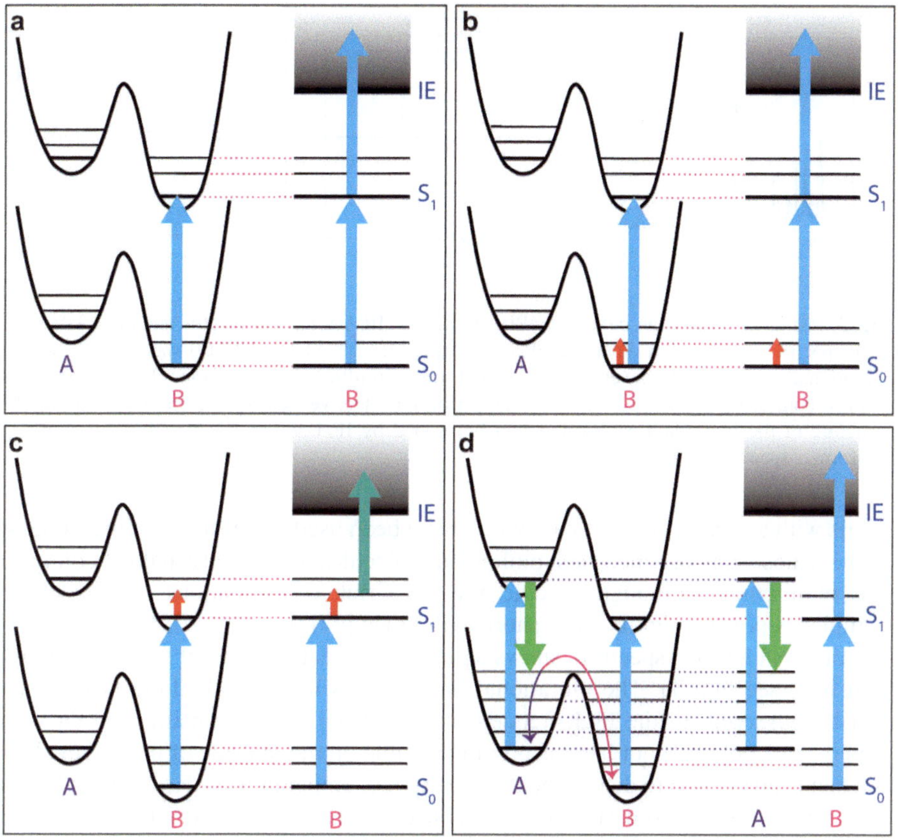

Fig. 2 Schematic representation of: (**a**) a one-colour two-photon REMPI scheme; (**b**) the modus operandi of IR–UV ion-dip spectroscopy; (**c**) IR probing scheme for the excited state using UV–IR–UV spectroscopy; (**d**) a general scheme for stimulated emission pumping for conformer A and B

book, the molecules of interest are usually resonantly ionized using two photons of the same frequency via a one-colour R2PI or (1 + 1) REMPI process.

In one-color R2PI, a molecule is ionized via an intermediate electronically excited state by two photons of the same energy, originating from a single laser (see Fig. 2a). In most cases, the energy difference between the excited state and the ionization continuum is sufficiently small to allow for the use of a single frequency for the excitation and ionization steps. However, for molecules in which the excited state energy is smaller than half the ionization energy, two-color R2PI must be employed. Using two independent lasers, the photon energy of the second laser can be chosen freely so that one can make sure that the ionization limit is reached.

One of the requirements for recording REMPI spectra with sharp resonances is that the excited electronic state should be sufficiently long-lived to allow for the absorption of the second photon. For the work presented here, pulsed nanosecond

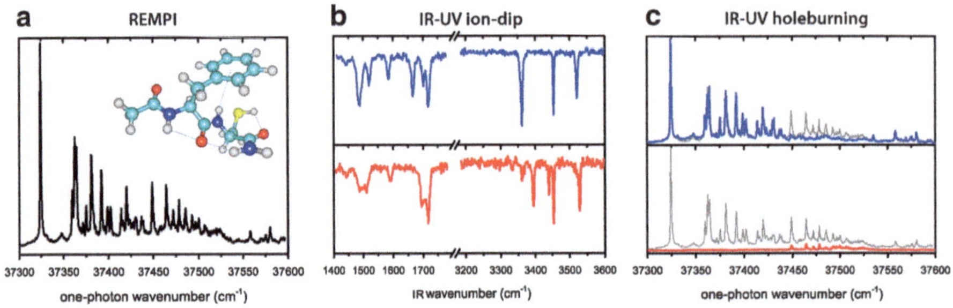

Fig. 3 **a**) mass selected one-color REMPI spectrum, **b**) IR-UV ion dip spectra, and **c**) IR-UV hole-burn spectra of the capped dipeptide PheCys adapted from [78] The REMPI spectrum shows the vibrational progression of the electronic excited state of multiple conformations. The contributions of these conformers are visualized using the conformer selective IR-UV hole-burning method. The structure of the individual conformers can be probed by IR-UV ion-dip spectroscopy, yielding conformer specific ground-state IR spectra.

lasers with a narrow spectral bandwidth have been used, requiring lifetimes at least in the nanosecond regime. Most biological molecules under study in this book fulfil this requirement, although specific Watson–Crick conformations of nucleobase complexes possess significantly shorter lifetimes [77].

A typical REMPI spectrum shows a plethora of peaks besides the $S_1 \leftarrow S_0$ origin transition see Fig. 3a. The vibrational progressions in the excited electronic state can be observed on the high-frequency side of the origin transition. Even with supersonic cooling, a hot band is possibly visible on the red side of the origin. This hot band can be used to determine the vibrational temperature as described in (1). In addition to this spectral structure, several conformers may coexist for a typical biomolecule. These conformers have slightly different $S_1 \leftarrow S_0$ transition energies, which therefore appear as separate peaks in the REMPI spectrum. This can result in a severe complexity of the REMPI spectrum because each conformer can display its own vibrational progression, resulting in many closely spaced peaks (see Fig. 3a). However, because every conformer has its unique $S_1 \leftarrow S_0$ transition frequency, each gas-phase conformation can be addressed individually by applying one of the double-resonance methods described below.

1.3.4 Double-Resonance Methods

Conformer distinction is achieved via double-resonance methods, which are generally based on addressing two transitions sharing a common lower state. In such a situation, the depopulation of the ground state induced with one laser can be probed with the second laser in a REMPI or LIF process. The depopulation can be induced either by UV or IR pumping.

UV–UV double-resonance spectroscopy can be used to determine which peaks in the REMPI spectrum originate from the same conformation. The first laser pulse, the so-called *burn* laser pulse, resonantly depletes the ground state. The second laser (*probe*), which is fired about 100–400 ns later, probes the effect induced by the burn laser as a decrease in ion or fluorescence signal. UV–UV double-resonance methods are commonly distinguished as either "hole-burning" or "ion-dip" spectroscopy, depending on whether the probe or the burn laser is tuned in frequency. In ion-dip spectroscopy, the burn laser frequency is tuned while that of the probe laser is fixed. This results in a burn laser-induced dip in the ion signal whenever its frequency is resonant with a transition of the conformer selected by the probe laser.

In hole-burning spectroscopy, on the other hand, the burn laser frequency is fixed while the probe laser frequency is tuned. The pump laser depletes the ground state of a single conformation, so that the probe laser records an excitation spectrum of all conformers except the one selected by the burn laser. It should be noted that in both cases the hot bands are not affected, because only the ground state is depopulated. In addition, for both methods to be effective, the burn laser pulse energy should be sufficient to induce a substantial ground state depletion.

1.3.5 Probing the Structure of the Electronic Ground State: IR–UV Ion Dip Spectroscopy

One of the most powerful double-resonance methods for conformational assignment is IR–UV spectroscopy. When using an IR laser instead of a UV laser as the pump source, a conformation specific IR spectrum can be obtained. This IR–UV ion-dip technique (see Fig. 2b), first developed by Lee and coworkers in the 1980s, is very similar to UV–UV depletion spectroscopy [6]. For IR–UV ion-dip spectroscopy, the UV-laser (probe laser) is parked on a transition in the REMPI spectrum of one specific conformer to generate a constant ion signal. The IR laser, which interacts with the molecules about 100–400 ns prior to the UV pulse, is scanned over the frequency range of interest. Whenever the IR laser frequency is resonant with a vibrational transition of the conformation probed by the UV laser, the ground state of that conformation is depopulated and the ion current decreases, resulting in a conformation-selective IR ion-dip spectrum (see Fig. 3b). By selecting different UV probe wavelengths, the IR spectra of all conformations present can be recorded separately.

In a similar manner, by fixing the IR laser frequency on a vibrational transition and scanning the UV probe frequency, an IR–UV hole-burning spectrum can be recorded. The IR laser depopulates the ground state of the selected conformation and in the UV excitation spectrum all peaks belonging to that conformation disappear (or are strongly reduced); see Fig. 3c. An IR–UV hole-burn spectrum provides the same information as a UV–UV hole-burn spectrum, although it is often more likely that different conformers possess overlapping IR absorption bands than overlapping UV absorption bands; the IR wavelength should therefore be selected with care.

The observed IR induced depletion of the ion signal often exceeds 50%, and even reaches dips up to 100%, especially when employing powerful free electron lasers. This indicates that the transitions studied cannot be considered as a simple two-level system [79]. Depletions greater than 50% are believed to result from the vibrational energy either being redistributed over background vibrational states by IVR or inducing (undetectable) fragmentation of the molecule.

The conformer selectivity in IR–UV double resonance methods relies on the excitation bands of different conformers being resolvable in the UV spectrum. For unresolved UV spectra, indicating either short excited state lifetimes (<1 ns) or strong conformational heterogeneity, the IR–UV ion dip scheme can be modified by introducing either a femtosecond ionization laser (IR femtosecond multiphoton ionization) [80, 81] or an additional IR step (IR–IR–UV spectroscopy) [21, 82–84].

1.3.6 Probing the Structure of the Electronic Ground State: IR–VUV Spectroscopy

The above excitation schemes all require a UV chromophore in the molecule and a sufficiently long-lived resonant intermediate electronic state. Nevertheless, one could employ schemes that circumvent the use of R2PI detection by using VUV single photon ionization. Although VUV ionization is a more general ionization method, the disadvantage of the method is that conformer selection is compromised. The frequently used 118 nm VUV light, obtained via the tripling of the third harmonic of a Nd:YAG laser in xenon, is capable of producing parent ions of various molecules with ionization energies up to 10.5 eV [85–87]. This can, for instance, be used to ionize directly all amino acids.

Employing IR–VUV action spectroscopy, absorption of an IR photon prior to VUV ionization can be made to influence the detected mass-selective ion signal in two ways. One option is to choose a VUV photon energy that is just not sufficient to ionize the molecule; absorption of an IR photon in addition to the VUV photon provides just enough energy to reach the ionization threshold, leading to ionization of the neutral molecule. Alternatively, using a VUV photon energy sufficient to ionize the molecule, the absorption of an IR photon may open or enhance photodissociation channels upon ionization. As a result, by monitoring the mass-selective ion intensities while scanning the infrared wavelength, an IR–VUV double-resonance spectrum can be acquired [86, 88, 89].

1.3.7 Far-IR Action Spectroscopy

The use of free electron lasers (see Sect. 3.2) offers the possibility to explore the low frequency vibrational motions in biomolecules. Until recently this far-IR region ($<500\ cm^{-1}$) has hardly been considered as a structural probe to characterize biomolecules [68, 90]. Including the far-IR spectrum in a structural analysis has several advantages. Mid-IR vibrations, such as the NH stretch, C=O stretch and NH bend vibrations, are typically strongly localized modes providing information on the local hydrogen bond and electrostatic environment. In contrast, the soft vibrations in the far-IR range are typically delocalized over large parts of the peptide and are hence expected to be highly sensitive to secondary structure motifs [91]. Furthermore, even in the environment of a supersonic molecular beam expansion, the IR spectra of larger peptides show a high degree of spectral congestion in the mid-IR region, possibly to an extent where only families of structures can be identified rather than a single conformation [48, 51, 92–94]. In such cases, the far-IR spectrum often still shows a wealth of well-resolved absorption bands allowing for a detailed structural assignment, provided that reliable theoretical predictions are available.

Deducing structural information from experimental IR spectra is usually accomplished through comparison with predicted IR spectra from quantum-chemical calculations. The vibrational frequencies of localized modes such as CO stretches and NH bends in the mid-IR range can be computed fairly reliably within the harmonic approximation, ignoring the intrinsic anharmonic character of the vibrations. An empirical uniform scaling factor is usually applied to correct for this deficiency [95, 96]. In the far-IR spectral range, such a comparison between experiment and theory is often complicated by the generally more substantial anharmonicity of the vibrations, thereby limiting the use of this range of the IR spectrum for conformational elucidation [97]. Recent developments showed that Born–Oppenheimer Molecular Dynamics (BOMD) techniques as presented in [98] are a promising tool for the detailed structural characterization of jet-cooled peptides [99]. BOMD simulations naturally include the effects of anharmonic vibrational potentials [100] and have been applied successfully to reveal the far-IR signatures of specific secondary motifs and hydrogen-bond stretching vibrations in capped peptides [99].

1.3.8 IR Probing of Electronically Excited States

The combination of IR and UV spectroscopies can also be applied to the study of vibrational structure of electronically excited states, although technical challenges have limited the broad application of these methods as compared to ground-state IR absorption studies [84, 101–105]. As compared to ground state IR–UV ion dip spectroscopy, a two-color R2PI scheme must be chosen, so that the IR photon can be timed after the first UV excitation pulse, but before the second UV ionization pulse (see Fig. 2c). The IR laser now excites vibrations of the electronically excited S_1 state, resulting, for example, in a depopulated vibrational ground level of the S_1 state and therefore in a depletion of the ion signal. Obviously, timing between the various laser pulses is more critical than for ground state IR–UV ion dip spectroscopy. The time delay between the first UV laser pulse (excitation from S_0 to S_1) and the IR laser should be large, minimizing temporal overlap between the two pulses, so that contributions from vibrations in the S_0 ground state are avoided. However, the excited state lifetime of biomolecules of interest is typically only a few nanoseconds (or less), and the nanosecond pulse durations of the commonly used UV and IR lasers lead to residuals of S_0 vibrations in the excited state IR spectrum. In this respect, despite their optical brightness, FEL-generated IR pulses are less suitable for the study of electronically excited states on account of their μs temporal structure. As for ground state IR–UV ion dip spectroscopy, the depletion in the electronically excited state caused by IR absorption can also be probed via LIF. LIF detection should be equally applicable to ground state and excited states, but has more limited application as strongly fluorescent molecules are required [102, 104, 105].

Although the opportunity to characterize the electronically excited state is very valuable, both the excited-state lifetime constraint and intersystem crossing effects have resulted in a limited, but promising, application of this method to biomolecular systems [84, 103, 106, 107]. In addition to the UV–IR–UV scheme described above, various other multiple-resonance IR/UV schemes have been developed to study selectively the IR absorption of the electronically excited state. These have been summarized by Gerhards and Schwing [108], showing interesting possibilities for the future.

1.3.9 Spectroscopic Methods to Probe Conformational Dynamics

The flexible character of biomolecules results in a complex potential energy surface (PES) with multiple stable, low-energy conformations. The double resonance techniques described above allow us to probe and to characterize these low-energy conformations. Zwier et al. have developed methods to induce and analyse conformational isomerization between these low-energy conformers [56, 109, 110]. These techniques, hole-filling spectroscopy (HF) and IR-induced population transfer (PTS), have in common that a specific vibration is selectively excited early in the expansion. If the IR energy lies above the isomerization barrier, isomerization can occur. Subsequent cooling in the molecular beam expansion results in a population redistribution over the conformers, which can be probed by either laser-induced fluorescence or IR–UV ion-dip spectroscopy. The difference between HF and PTS spectroscopy lies mainly in the choice as to which of the two lasers is set to a fixed frequency and which is tuned. For hole-filling spectroscopy the IR excitation laser frequency is fixed, while for IR-induced population transfer the IR laser is tuned while the probe laser frequency is fixed, monitoring the changes in the population of that specific conformer. It is remarkable that, for the molecules under study, no new conformers were formed, but IR excitation only resulted in repopulation of existing conformers. However, this method provides a unique way to map connections between the various low energy conformers [56, 109, 110].

To date, only IR photon energies in the H-stretching range have been used in IR-induced population transfer experiments, which for most systems results in excitation far above the isomerization barrier, impeding an experimental determination of the isomerization barriers directly. Zwier and coworkers presented an elegant alternative, where stimulated emission pumping (SEP) is combined with HF or PTS methods, see Fig. 2d. This combination allows one to probe the barrier to conformational isomerization [56]. The method consists of two steps: a pump-dump SEP followed by a probe laser interrogation to determine the new conformational distribution. In the early part of the molecular beam expansion, SEP prepares the molecules in a specific vibrationally excited state at an energy corresponding to the difference between the pump and dump photon energies. Conformational isomerization can occur when the energy is sufficient for the molecule to cross an isomerization barrier. In the remaining part of the molecular beam, the molecules relax into their new conformational distribution, which is probed using either LIF or REMPI methods. By tuning the dump laser frequency while keeping that of the pump and probe laser fixed (SEP-PTS method), different initially populated vibrational states can be selected so that barrier heights can be determined.

In conclusion, laser desorption in combination with molecular beam high-resolution spectroscopic methods has enabled the structural characterization of gas-phase biomolecules of significant size. Moreover, IR–UV ion dip spectroscopy offers information about the number and types of conformers present. Applications to diverse classes of biomolecular systems such as peptides, carbohydrates and nucleobases, are described in the various chapters of this book.

2 IR Spectroscopy of Ionized Biological Molecules

2.1 Combining IR Spectroscopy with Tandem Mass Spectrometry

With the advent of soft-ionization methods, mass spectrometry (MS) has become an important analytical tool in biochemistry. Transferring intact biological molecules, ranging in size from simple amino acids to entire protein assemblies, from their condensed-phase environment into the gas phase has become standard practice using electrospray ionization (ESI), matrix-assisted laser desorption ionization (MALDI) and several of their descendants. These methods generally produce protonated (or deprotonated) species, which can be manipulated and analysed using a variety of mass-spectrometric methods, including time-of-flight (TOF), ion trap, triple-quadrupole and Fourier transform mass spectrometers. To the *mass spectrometrist*, an IR spectrum of the mass-selected ion can provide detailed structural information unavailable from more conventional MS/MS methodologies. To the *spectroscopist*, the MS toolbox provides a variety of methods to vaporize, ionize and manipulate gas-phase biological molecules as alternatives to the (laser-desorption) molecular beam methods described in the previous section. The challenge is thus to couple tandem mass spectrometry and optical spectroscopy so that IR spectra of mass-selected molecular ions can be recorded.

Although the IR spectroscopy of gaseous molecular ions has a long history, its application to molecules of biological interest is fairly recent. The foremost impediment to obtaining IR absorption spectra of gaseous ionized species is their mutual Coulombic repulsion, limiting obtainable ion densities to values typically far too low to induce an observable attenuation of the incident light beam in a typical direct absorption set up. Comparing ballpark figures for the typical ion densities in an (ion trap) mass spectrometer of up to 10^6 cm^{-3} with the typical pressure inside a gas cell used in an IR spectrometer of around 1 mbar, corresponding to about 10^{16} cm^{-3}, the mismatch in densities is roughly estimated to be in the range of 10 orders of magnitude!

Before embarking on a discussion of the various methods used to address the above challenge – recording an IR spectrum of ionized gaseous species – we note that numerous alternative methods are available to provide information on the vibrational structure of ionized species. Conventional IR spectrometers can be applied to condensed-phase samples containing (solvated) ionic species. To minimize interactions with the condensed-phase environment, and hence its effect on the spectra, ions can be isolated in cryogenic rare-gas matrices. In all cases, though, the sample has to maintain overall charge neutrality, so that the sample necessarily contains additional species other than the ion under study, which may have an (unknown) influence on the spectrum to be recorded. Vibronically resolved UV/vis spectra contain vibrational information and can be obtained via molecular beam photo-ionization spectroscopy schemes related to those discussed in the previous section of this chapter. In particular, mass-analysed threshold ionization (MATI) and zero-kinetic energy (ZEKE) photo-electron spectroscopy have been applied to reveal spectroscopic information on the ion. UV photo-dissociation or UV induced fluorescence spectroscopy on samples of trapped ions can reveal vibrational information. Photo-electron spectroscopy (PES) has been used extensively to obtain information on the vibrational structure of anionic species. These methods are beyond the scope of the current overview, where we focus on methods used to obtain IR spectra of mass-selected, gaseous molecular ions.

2.2 *Direct Absorption*

Early ion spectroscopy studies circumvented the density problem in several ways. First of all, massive ionization methods such as, in particular, electron ionization (EI) were mostly applied [111], limiting the studies to relatively simple molecular ions. Discharge sources generating plasmas containing not only the positively charged molecular ions but also free electrons can evidently reach much higher ion densities, although mass selection is not possible [112–114]. Discharge and EI ionization combined with a slit jet expansion [111] are currently among the most popular instruments to record direct absorption spectra of molecular ions. As an alternative, or in addition, the overlap between the ionic sample and the optical path can be maximized to increase the sensitivity in direct-absorption experiments, which is typically achieved either by a long, straight ion path (such as in a linear ion guide) or a long optical path (such as in multipass or cavity-enhanced light beam arrangements [111, 114]). In combination with sophisticated modulation schemes, direct absorption spectra of various molecular ions, particularly of astrophysical interest, have been reported [112, 113, 115].

2.3 *Action Spectroscopy*

While the light attenuation induced by a sample of mass-selected molecular ions is hard, or usually even impossible, to detect, dissociation of the ion is routinely detected in a tandem mass spectrometer. Hence, dissociation induced by wavelength-selective absorption of IR radiation can be used to reconstruct a "surrogate IR spectrum" from a series of mass spectra recorded with the radiation of a tunable laser set to different wavelengths. A difficulty with this strategy, however, lies in the low energy of an IR photon, typically much lower than the thresholds to dissociation. Hence, for an IR photo-dissociation spectrum to be recorded, the system under study must either have a very low barrier to dissociation (i.e. be weakly bound with bond energies not exceeding one or a few tens of kJ/mol) or the laser source must be sufficiently powerful to induce the absorption of multiple photons by a single ion. Experimental methods based on either of these two strategies have developed into the most commonly applied tools in IR ion spectroscopy to date, *messenger spectroscopy* and *multiple-photon dissociation spectroscopy*.

2.3.1 Messenger Spectroscopy

In the 1980s, Y.T. Lee and co-workers developed the method of messenger spectroscopy [116] based on molecular beam methods described in the previous section of this chapter. Molecules seeded in a supersonic expansion were ionized by electron ionization using a filament ionizer mounted immediately after the nozzle or by a corona discharge implemented in the nozzle. Because of the low temperatures of the jet expansion, the small molecular ions or clusters formed weakly (ion – induced dipole) bound clusters with co-expanded H_2. Carefully avoiding collisional heating, these clusters were injected into a sector magnet for mass selection and stored in a linear radio-frequency (rf) ion trap. While in the trap, the ion cloud was intersected with the tunable IR beam from an OPO and the degree of IR induced detachment of the H_2-messenger was determined subsequently in a quadrupole mass analyser.

The objective of these experiments was to obtain an IR spectrum of the ionic species under study, and hence the influence of the messenger on the IR spectrum should be minimal. In Lee's words, "like a spy, the messenger's role is to gather information as unobtrusively as possible". Having only weak interaction with the ion under study, no intrinsic IR absorption and producing an easily observable shift in the mass spectrum upon detachment, rare gas atoms became popular as messenger [117–120].

Various groups have now implemented messenger spectroscopy in different forms using a variation of MS hardware. Instruments based on triple-quadrupole (-like) geometries were developed [117, 120–122] (see Fig. 4) and have, among others, led to the successful application of messenger spectroscopy in cryogenic ion

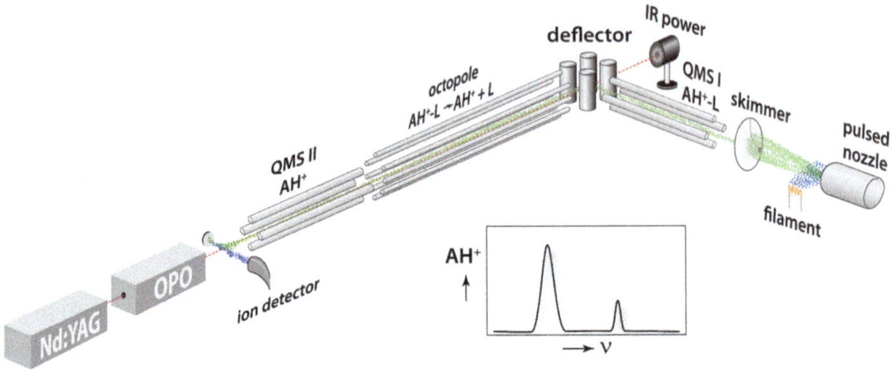

Fig. 4 Typical tandem MS set-up for messenger spectroscopy based on quadrupole mass selector and analyser tuned to the *m/z* values of the weakly bound ion-ligand complex (AH^+-L) and that of the bare ion (AH^+), respectively. The ligand is usually a small molecule (e.g. H_2) or a rare gas atom. Figure adapted from Dopfer [117]

traps. Time-of-flight (TOF) instruments have been employed, where the turning point of a reflectron can be used as laser interaction zone; using an ion gate, the two legs of the reflectron can then be used as mass selector and mass analyser [118, 123]. Especially for the study of ionic clusters containing one or more metal ions, laser ablation sources [123] are an attractive alternative to EI or discharge sources. Resonant UV photo-ionization (as described in Sect. 1.3.3) provides much control over the ionization process and has been implemented in particular in combination with TOF MS [124].

While instrumental in the development of infrared ion spectroscopy, relatively few studies have employed the messenger method to study molecular ions of true biological interest, apart from relatively small model systems [119, 122]. To generate biomolecular ions efficiently, most (commercial) MS platforms are equipped with ESI or MALDI ion sources, in which the formation of weakly-bound clusters is usually not observed. Particularly nano-spray sources are well-known to produce hydrated ions, which can be investigated spectroscopically using the attached H_2O-molecule as the messenger [121]. Although the H_2O-tag can clearly not be considered an “unobtrusive observer”, such studies have revealed a wealth of fundamental information on the structural rearrangement upon solvation [125]. Cryogenic cooling of the stored ions can be used to attach more weakly bound, “true” messengers to the ion of interest for subsequent IR spectroscopic interrogation [126–128].

Although not formally regarded as messenger spectroscopy, IR spectroscopy in helium nanodroplets is closely related. Here, ions are embedded in superfluid clusters typically consisting of 10^3–10^5 (or more) He atoms produced using molecular beam techniques. These droplets provide a cryogenic (0.37 K), virtually non-interacting matrix and have been used for (IR) spectroscopy of neutral (bio-) molecules and clusters since the 1990s [129–131]. Resonant IR absorption of the ion leads to its ejection of the cluster, which is monitored by a mass spectrometer

[132]. Combination of an He droplet source with a tandem mass spectrometer provides opportunities for capturing ionized biomolecular systems in these droplets [133] and to record their spectra at very low temperature [134].

2.3.2 Cryogenic Ion Trap Spectroscopy

Much of the current work on ion spectroscopy in cold traps finds its origin in the studies on cryogenic multipole ion traps by Gerlich [135]. The development of these traps was largely driven by the quest for accurate ion–molecule reaction rates at low temperatures which could serve as input for models of interstellar chemistry. Ion temperatures in the range of a few K are achieved by mounting an rf ion trap on the cold head of a closed-cycle He refrigerator. Ions stored in the trap are cooled sympathetically with a pulse of He gas. To avoid rf heating of the ions, Gerlich introduced linear multipole ion traps (the 22-pole trap had become popular) which feature much flatter potentials near the axis of the trap as compared to more conventional quadrupole, hexapole or octopole linear ion traps [135]. The cold trap is usually implemented in a tandem mass spectrometer, e.g. by placing it between two quadrupoles for mass selection and mass analysis.

Various action spectroscopy schemes have emerged from combining cold ion traps with tunable UV/vis and IR lasers. Rizzo and co-workers developed a saturation spectroscopy scheme based on resonant UV photo-dissociation probing and IR pumping. One of the most interesting features of this method is its conformer selectivity, which is achieved in a fashion analogous to the molecular beam methods described in the previous section and which makes this method particularly useful for the study of biological molecules. Chapter 2 of this Volume (Rizzo and Boyarkin [136]) is dedicated to this method. Schlemmer and co-workers introduced IR laser-induced reactions (LIR) as a tool to obtain action spectra [137]. A slightly endothermic ion-molecule reaction, such as proton transfer between two bases with slightly different proton affinities, is induced by the resonant absorption of a single IR photon. Because the internal energy distribution at low temperature is much narrower than the energy of an IR photon, a high contrast is observed between on- and off-resonance irradiation, even down to frequencies in the THz range, expanding action spectroscopy methods to pure rotational spectra [138]. Combination of this method with a frequency comb-stabilized cw OPO allowed Schlemmer and co-workers to observe rovibrational transitions of the CH_5^+ ion with a FWHM bandwidth of 70 MHz (0.0023 cm^{-1}) [139]. The cryogenic environment of the trap allows for the formation of weakly bound clusters so that messenger spectroscopy can be applied, as utilized for instance by the groups of Asmis [126, 127] and Johnson [128]. Interestingly, some of these experiments returned to using H_2 as the tag [128, 140] (to avoid condensation on the 4 K parts of the trap), as in the original messenger spectroscopy experiments of Y.T. Lee. Also here, sophisticated multi-laser hole-burning schemes have been applied to obtain (partially) conformer-selective spectra [140].

2.3.3 IRMPD Spectroscopy

Messenger spectroscopy, as developed mainly by spectroscopists, relies on the ability to produce weakly bound species and hence requires the production of ions with low (sub-thermal) internal energies, which is commonly achieved using cryogenic or supersonic expansion cooling. However, conditions prevalent in common (commercial) tandem mass spectrometers generally do not promote the formation of such weakly bound complexes. Ion spectroscopy strategies developed within the MS community have therefore mainly followed a different route, where dissociation is induced by the absorption of multiple (tens to hundreds) IR photons. Ion storage mass spectrometers have been particularly useful because the ions can be irradiated for a user-defined time and slow dissociation processes are observable.

Because of the low collision rate in the high vacuum environment of a Fourier transform mass spectrometer (FTMS), vibrationally excited molecular ions cool predominantly by IR fluorescence. For typical IR transition dipole moments and frequencies in the mid-IR, spontaneous emission is expected to occur at a rate in the range of 1–100 s^{-1}. To energize an ion efficiently using IR multiple-photon excitation (MPE), the rate of photon absorption – the product of absorption cross section and photon flux – should exceed the emission rate. From such a back-of-an-envelope estimate, one finds that radiation sources producing several Watts/cm^2 are required to induce efficient dissociation [141]. Note that the demands on laser power may further increase because of the limited residence time of the ions in the laser field, collisional deactivation in traps at higher pressures, limited spectral overlap between molecular absorption and laser emission profiles, etc.

On account of its high output power and its wavelength tunability (albeit only between a few dozen discrete emission lines in the 9–11-μm spectral range), CO_2 lasers have been employed by several groups since the late 1970s to obtain IR spectral information on mass-selected molecular ions in an FTMS [143–147]. These studies may be regarded as the IR analogue of the UV photo-dissociation spectroscopic studies on mass-selected ions in an FTICR MS pioneered by Brauman, Dunbar and others since the late 1960s [148, 149].

While the practical application of IRMPD spectroscopy to mass-selected molecular ions had thus been shown, its widespread use as a structural tool in ion chemistry was impeded by the limited tunability of the CO_2 laser and the absence of other useful laser sources featuring a high power and wide tunability across the IR spectrum. The interest in IRMPD spectroscopy of gaseous ions revived around the turn of the millennium, when IR free electron lasers (FELs) as well as novel high pulse-energy OPOs were coupled with ion storage tandem mass spectrometers.

In the earliest applications of FEL-based ion spectroscopy, relatively simple tandem mass spectrometers were used. At the FELIX facility, an rf quadrupole ion trap was used, where the ring electrode had been modified to provide access for the FEL beam to the stored ion cloud [150]. A short TOF MS was used for mass analysis, while ions were generated in the volume of the trap by non-resonant UV photo-ionization of aromatic molecules evaporated from a small oven. At the CLIO

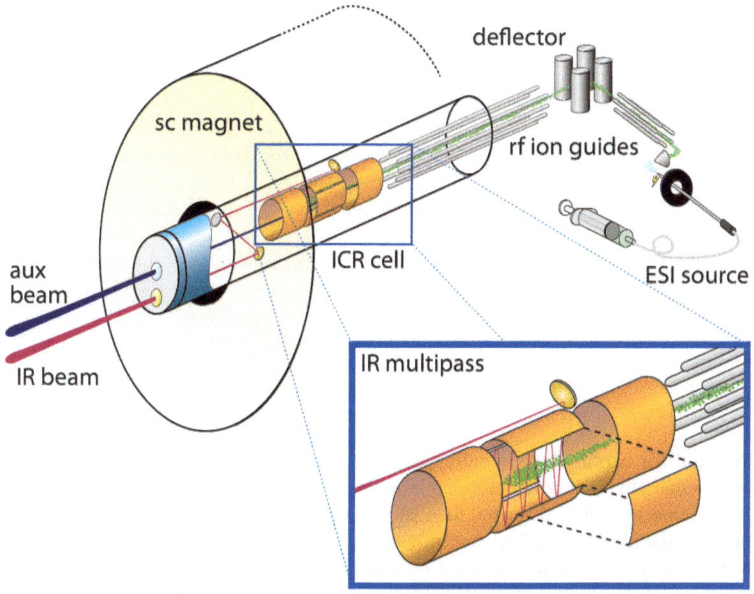

Fig. 5 Fourier-Transform Ion Cyclotron Resonance (FTICR) mass spectrometer with electrospray ionization (ESI) source used for IRMPD ion spectroscopy at the FELIX facility. The hexapole ion guide is also used to accumulate ions from the ESI source before being pulse-injected into the ICR cell. The inside of the excite electrodes of the ICR cell are polished so that they act as a multipass reflection cell for the IR beam from FELIX (or an OPO) achieving approximately 10 passes through the ion cloud. Adapted from [142]

facility, a compact 1.25-T permanent magnet based FTICR MS (nicknamed MICRA) was developed and coupled to the IR beamline. Ions were generated internally either by EI ionization of vapours introduced through pulsed valves [151] or by metal ion attachment to the vapour using laser ablation of a metal target [152]. Although this instrument provided the full suite of ion manipulation methods common to FTICR mass spectrometers, the step towards application to ions of true bio-molecular interest came with the introduction of ESI-equipped tandem mass spectrometers at these FEL facilities.

At the FELIX facility, a homebuilt FTICR-MS based on a 4.7-T actively-shielded magnet was installed, which initially had an internal filament for EI generation of ions [153]. The instrument was later extended with an external ESI source (Waters Z-Spray) and an rf octopole ion guide to transport the ions from the source to the ICR cell [142] (Fig. 5). As an interesting novelty of this FTICR-MS, the DC-bias of the octopole ion guide is switchable, so that the kinetic energy of the ions inside the guide can be re-referenced [154]. This is applied to reduce the kinetic energy of the ions for efficient trapping in the ICR cell without the use of a gas pulse, which avoids collisional heating of the ions. The method is explained in Fig. 6.

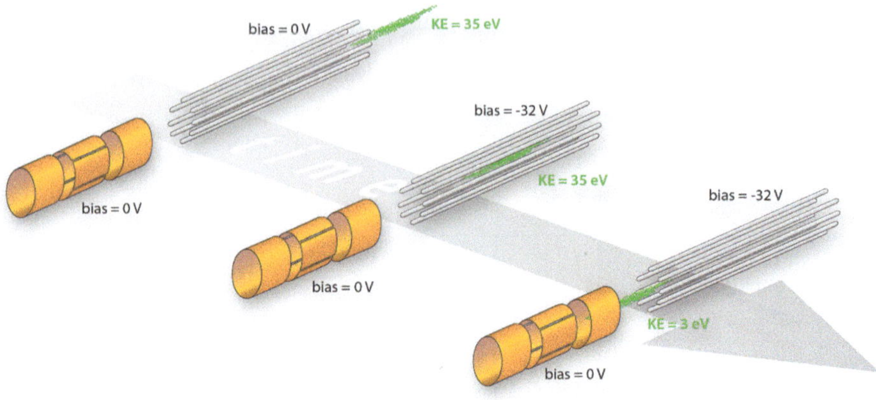

Fig. 6 Method for capturing externally generated ions in the ICR cell. A bunch of positively charged ions enters the octopole rf ion guide with a kinetic energy of around 35 eV. While in the guide, the DC bias is switched from 0 to −32 V. Nonetheless, the ions keep their original kinetic energy because the bias switch does not induce any potential gradient along the guide axis. Upon exiting the guide, the ions encounter the ICR cell at ground potential and slow down to a low kinetic energy (3 eV in this example), so that they can be efficiently captured in the cell. Figure adapted from Polfer et al. [154]

To enhance the overlap of the IR beam with the trapped ion cloud, the copper excite-electrodes of the ICR cell are polished and serve as a multipass reflection cell for the laser beam; the curvature of the electrodes naturally refocuses the beam (in one direction only) on each pass. The off-axis geometry of the FEL beam has the additional advantage that a second laser beam can access the ion cloud on-axis. A CO_2-laser can, for instance, be used in addition to the FEL to induce dissociation, so that product ions can be mass-isolated and spectroscopically investigated. Moreover, a CO_2-laser can be employed to enhance the fragmentation yield.

At the CLIO FEL facility, a 7-T Bruker Apex Qe FTICR-MS was coupled to the IR beamline [155]. The FEL beam, mildly focused using a 2-m focal length mirror, enters the instrument through an IR window on the backside of the instrument which is normally used for CO_2-laser activation of the ions. The beam passes through a hollow cathode which can be used to induce electron capture dissociation (ECD) and then enters the Infinity ICR cell on-axis through a 6-mm hole. The instrument provides extensive MS^n capabilities, including ECD, which provides ample opportunities for the spectroscopic characterization of the structure of reaction products in ion chemistry [156]. A simpler quadrupole ion trap mass spectrometer (Bruker Esquire) has now also been coupled to the CLIO beamline, which reduces the time necessary to record an ion spectrum and allows for ion structure characterization of a range of molecular ions of interest in bio- and other branches of chemistry [157]. Finally, it is worth mentioning that the IR FEL at the Tokyo University of Science has also been coupled to an FTICR MS (Bruker BioAPEX III with a 4.7-T magnet) to obtain IRMPD spectra of ionized peptides and oligosaccharides, although relatively few studies have been reported

[158]. Furthermore, the new IR FEL at the Fritz Haber Institute in Berlin has very recently been applied to obtain IR ion spectra [159].

Along with the use of IR FELs for IRMPD ion spectroscopy, the rapid developments in table-top IR laser sources based on nonlinear frequency downconversion have had a profound influence on the recent revival of IR ion spectroscopy. Sources based on difference frequency generation (DFG) as well as those based on optical parametric oscillators (OPO) have been used to investigate the spectral range around 3 μm covering the hydrogen stretching frequencies (−OH, −NH, −CH). Williams and co-workers coupled a LaserVision OPO to a 2.75-T ESI-FTICR-MS to obtain IR spectra of a wide range of bio-molecular complexes including alkali metal ion complexed amino acids [160] and selectively hydrated protonated amino acids [125]. This source is now also in use for IR ion spectroscopy in several other labs [118, 122, 127, 128, 155, 161, 162]. Other pulsed [163, 164] as well as cw [165] infrared OPO sources have also been coupled to FTICR and ion trap [162, 165] mass spectrometers and used in ion spectroscopy studies.

2.3.4 Multiple-Photon Dissociation Mechanism

In sharp contrast to conventional spectroscopic methods based on direct one-photon absorption, IRMPD spectroscopy relies on the sequential absorption of a large number of IR photons. This excitation mechanism leaves an imprint on the observed IR spectrum in the sense that vibrational bands are typically broadened, red-shifted and affected in relative intensity to some extent. While the intramolecular processes underlying these spectral modifications have been addressed and qualitatively modelled in a large number of studies [166–172], it is often hard to predict quantitatively an IRMPD spectrum because the required molecular parameters, in particular the anharmonic couplings between vibrational normal modes at high internal energies, are usually unknown and cannot be calculated accurately using current quantum-chemical methods. In practice, most experimental IRMPD spectra are therefore analysed on the basis of computed linear absorption spectra, which usually provide a reasonable approximation to the IRMPD spectrum.

Molecular vibrational potentials inherently deviate from being harmonic so that the energy spacings between adjacent bound states in the potential are not equidistant. In general, the energy spacing decreases with increasing vibrational quantum number v_i, which prevents the molecule from climbing the ν_i vibrational energy ladder by absorbing multiple IR photons from a monochromatic laser source. This effect is commonly referred to as the *anharmonicity bottleneck* [167]. On the other hand, however, anharmonic terms in the potential introduce couplings between the normal modes that are truly orthogonal only in the harmonic approximation. Because of these couplings, energy can flow from one vibrational mode into another, so that an IR excited vibrational level can be deactivated by energy dissipation into other modes. Note here that the set of iso-energetic background states consists not only of other fundamental vibrations, but predominantly of combination modes. Once the fundamental vibrational level is deactivated, another photon can be absorbed on the same $v_i = 1 \leftarrow 0$ transition. This process can repeat itself many times, thereby progressively energizing the molecule.

The efficiency of the multiple-photon excitation (MPE) process depends crucially upon the rate at which the excited vibrational level is deactivated by intramolecular vibrational redistribution (IVR). Because the density of vibrational combination states at energies relevant in the MPE process is large even for relatively small molecules, the process is best described by Fermi's Golden Rule, which does not consider individual states but rather relates the IVR rate to the density of states (DOS, ρ) and the average coupling matrix element squared $<W>^2$. It is believed that there are generally just a few states that couple particularly strongly to the bright state and vibrational energy is first channelled into these so-called *tier-1* states [173, 174]; from there, energy is diffused into a second tier containing more states, and so on. According to Fermi's Golden Rule, an increasing DOS does not necessarily increase the IVR rate, because the average coupling strength may drop as a consequence of adding states far outside the first tiers [175]. Nonetheless, the increasing phase-space enhances the degree of irreversibility of the process. In other words, a higher vibrational entropy enhances the unidirectionality of the flow of energy, away from the bright state, and thereby enhances the MPE efficiency.

Even though absorption takes place on the same $v_i = 1 \leftarrow 0$ transition over and over again, the centre frequency of the ν_i absorption band undergoes a slight shift as the internal energy of the molecule rises. Excitation in other vibrational degrees of freedom influences ν_i through the anharmonic couplings. However, because the excitation in each individual vibrational coordinate remains low, the frequency shift is small compared to that induced by high vibrational excitation in one coordinate (because the deviation from harmonic behaviour is proportional to $(v_i + ½)^2$ and higher order terms). The process is sketched in Fig. 7.

As a consequence of the IVR-mediated nature of the multiple-photon excitation process, the vibrational excitation is randomized as the dissociation threshold is approached. Hence, the molecule has no memory of the vibrational coordinate that was originally excited. Dissociation therefore occurs statistically and can be modelled using the Arrhenius equation or phase-space theories. Mode-selective dissociation is normally not observed.

An issue which has been under discussion is whether or not the internal energy distribution generated by MPE resembles a thermal distribution at elevated temperatures. Modelling by Y.T. Lee and co-workers showed that an MPE generated distribution induced by a pulsed laser can be significantly different from a thermal distribution [169]. In particular, the energy distribution can be much narrower than a Boltzmann distribution at the same average internal energy [169, 176]. In contrast, we note here that excitation induced by a cw (CO_2-)laser gives rise to slow heating of the ion, generating an internal energy distribution very close to thermal if concurrent deactivation by spontaneous emission is taken into account, as followed from steady-state models by Dunbar and others [141, 177, 178]. Tandem mass

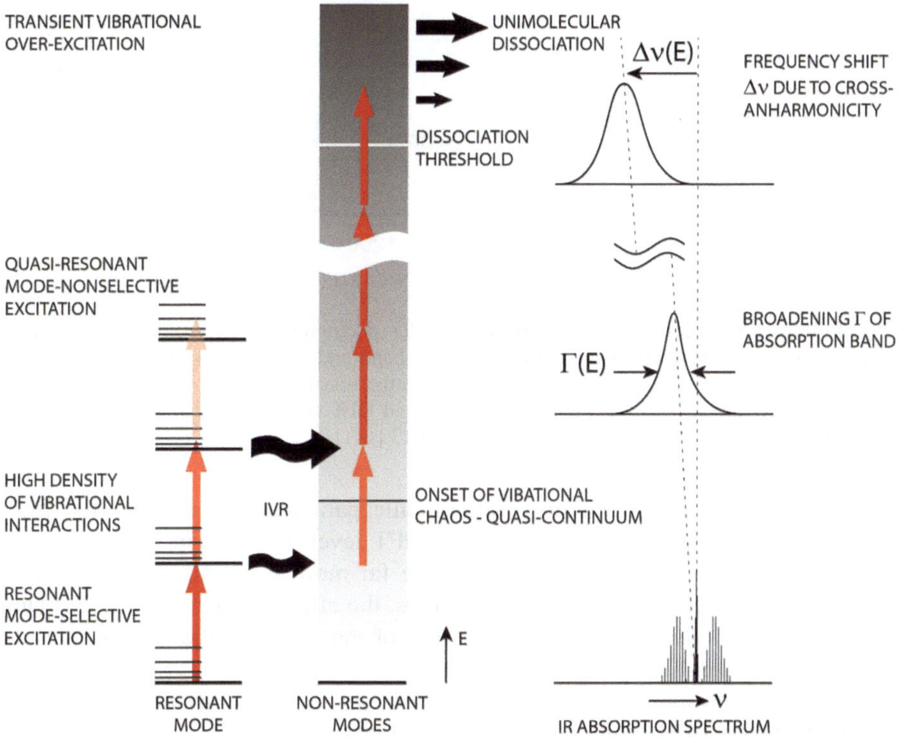

Fig. 7 Schematic overview of the IRMPD process. Adapted from Makarov et al. [168]

spectra obtained via CID and IRMPD are therefore typically very similar (commercial mass spectrometers are equipped with cw CO_2-lasers for IRMPD).

Of particular interest in the current context is the effect of anharmonicity on the MPE process and eventually on the IRMPD spectrum. One model [169, 170, 172, 176, 179], including the anharmonic shifting and broadening of absorption bands as a function of internal energy, considers a set of coupled rate equations describing the instantaneous population $N_m(t)$ of levels m with energy $E_m = mh\nu_L$, with ν_L the laser frequency. The 'levels' here do not strictly correspond to vibrational eigenstates, but are rather characterized in terms of their degeneracy, i.e. the density of states $\rho(E_m)$. The frequency dependence of the absorption cross sections $\sigma(\nu, E_m)$ coupling adjacent m-levels is assumed to vary with internal energy, so that anharmonic shifting and broadening can be taken into account. Numerical integration of the set of differential equations over the duration of the IR pulse then yields an internal energy distribution for each laser frequency, from which the dissociation yield can be calculated using an Arrhenius-like relation. A different approach involves a Monte Carlo evaluation of trajectories describing the time-evolution of the internal state of a molecule, accounting for the internal energy-dependent rates for absorption, stimulated and spontaneous emission and dissociation [171]. In this particular study [171], dissociation was implemented using phase-space theory

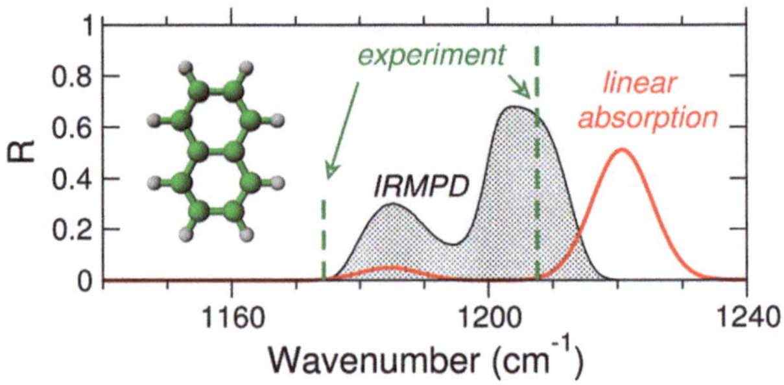

Fig. 8 Modelling (*black*) of the experimental IRMPD spectrum [150] (*green*) of the naphthalene radical cation. The redshift of the main absorption band and the intensity enhancement of the minor band in the IRMPD spectrum as compared to the linear spectrum (*red*) are satisfactorily modelled based upon the model of Parneix et al. and DFT computed anharmonic parameters. Reproduced from Parneix et al., J Phys Chem A, 2013 [171]

rather than Arrhenius modelling. Anharmonic parameters were extracted from anharmonic frequency calculations at the DFT level, rather than being assumed or empirically determined. Both models are far more sophisticated than can be discussed here and incorporate, among others, the effects of the initial Boltzmann population distribution, the finite bandwidth of the laser and the spatial intensity profile of the laser beam.

These models can reasonably describe the typical features of experimental IRMPD spectra such as the observed bandwidths (typically >20 cm^{-1}) and frequency shifts (typically a few %). More system-specific effects are also evident from such modelling. For instance, shifting of the absorption profile during MPE (see Fig. 7) may cause nearby absorptions in the IR spectrum to become resonant during the excitation process, which can induce a second burst of absorption and lead to severe deviations from relative band intensities in a linear absorption spectrum [171, 179]. An example is shown in Fig. 8, where a weak absorption in the linear IR spectrum of the naphthalene radical cation is amplified in the IRMPD spectrum because of the nearby strong absorption band, as correctly modelled by Parneix et al. [171].

The use of IR action spectroscopy in the structural characterization of gas-phase (bio-)molecular ions has become well established within the fields of ion chemistry and mass spectrometry. It has been the subject of various recent reviews [142, 164, 172, 180–184] and has been applied to extract structural information from ions ranging in size from simple amino acids to entire proteins [185–187]. Applications to diverse classes of biomolecular ions are described in various chapters of this book.

3 Tunable IR Laser Sources

The application of action spectroscopy methods to obtain IR spectra of gas-phase biological molecules generally relies on the use of tunable IR laser sources. A brief overview of the various IR laser systems available is presented here.

3.1 *Gas Discharge Lasers*

A CO_2 laser operates on the emission bands between vibrational combination states generating emission on discrete rovibrational transitions in the $\nu_1 \leftarrow \nu_3$ and $2\nu_2 \leftarrow \nu_3$ bands, centred around 10.6 and 9.6 μm, respectively. Population inversion is achieved by collisional energy transfer from plasma-excited N_2 to CO_2, usually in a mixture with He. A particular rovibrational emission line can be selected using a rotatable diffraction grating incorporated in the laser cavity. CO_2 lasers can achieve very high continuous-wave (cw) power levels of up to 100 W from commercially available systems. In addition, CO_2 lasers are robust, narrow-bandwidth and low-cost systems well able to induce IRMPD, but a disadvantage is clearly its limited tunability. It should be noted that fixed-frequency CO_2 lasers are used routinely in commercial MS platforms to induce dissociation as an alternative to CID.

Based on a similar principle, carbon monoxide lasers emit on individual rotational lines of the $v = 0 \leftarrow 1$ band of CO in the range of about 1,400–2,000 cm^{-1}. The design of this laser is far more complicated than that of a CO_2 laser, because liquid-nitrogen cooling of the laser gas mixture is required and the gain is lower, resulting in much larger systems with higher operational cost. We are aware of only very few studies which have used this laser for IR ion spectroscopy [188].

3.2 *Free Electron Laser (FEL)*

In an FEL, the lasing medium is formed by electrons that are unbound to an atomic nucleus. As these free electrons have no self-absorption, FELs can be used to generate radiation over a very wide range of the electromagnetic spectrum. An FEL operates on the principle that accelerated electrons produce synchrotron radiation. In practice, relativistic electrons are injected into a so-called wiggler (or undulator), a periodic magnetic structure as depicted in Fig. 9, where the radiation produced at each turning point in the sinusoidal trajectory is captured in a laser cavity. Freshly injected electron bunches interact with light pulses circulating in the cavity, creating gain on each successive pass. Because of their curved trajectory, the electrons travel a longer distance than the light beam, and hence coherent radiation is generated only at those wavelengths for which the pathlength difference over one undulator period (Λ_u) corresponds to one (or an integer) optical

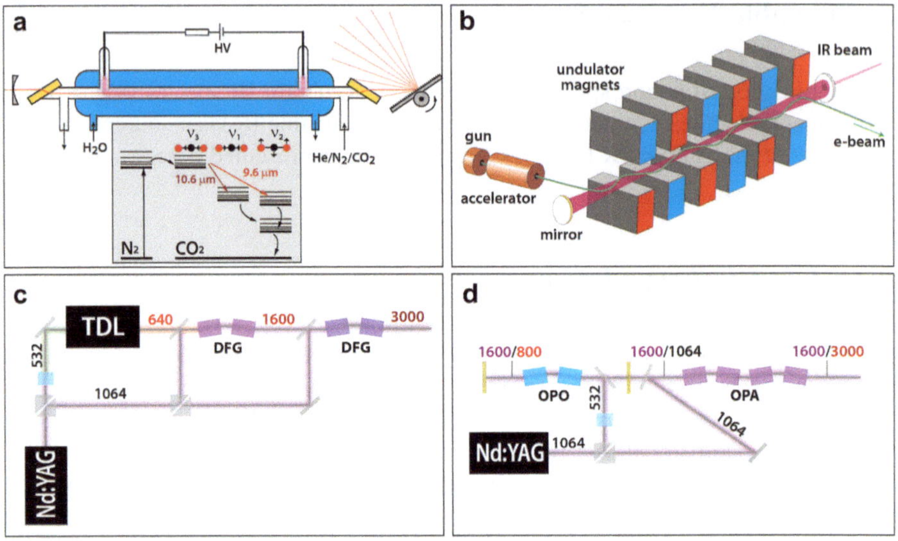

Fig. 9 Schematic overview of tunable IR laser sources. (**a**) CO_2-laser. Gas discharge, water cooling jacket, Brewster-angle ZnSe windows, rotatable grating and outcoupling mirror are shown. Discharge induced vibrational excitation of N_2 is collisionally quenched by the near-resonant ν_3 vibration in CO_2; radiative decay to the ν_1 and $2\nu_2$ levels leads to laser emission. (**b**) Free electron laser. The gap between the two arrays of magnets can be adjusted to change the magnetic field strength, providing a convenient way to tune continuously the lasing frequency. It should be noted that, in reality, the accelerator is much larger; the overall length of an IR FEL is of the order of 20 m. (**c**) Difference frequency generation. Two stages of DFG with the 1,064 nm Nd: YAG fundamental are used to down-convert the radiation from a tunable dye laser (TDL) operating on a red laser dye into the 3 μm range. Wavelengths are indicated in nm, where *numbers in color* represent tunable wavelengths. (**d**) Optical parametric oscillator. A schematic outline of the LaserVision OPO/OPA system is depicted. Tunable wavelengths (signal/idler in nm) are in *color*. The idler beam generated in the oscillator (OPO) serves as the signal beam injection seeder for the amplifier (OPA) stage

wavelength. Taking into account the relativistic length contraction in the frame of the moving electrons and the relativistic Doppler shift of the wavelength observed in the lab frame, one arrives at the FEL resonance condition

$$\lambda_{\mathrm{obs}} = \Lambda_{\mathrm{u}} \frac{1 + K^2}{2\gamma^2} \quad \text{with } \gamma = \frac{1}{\sqrt{1 - \frac{v^2}{c^2}}}, \tag{5}$$

where γ is the relativistic Lorentz factor and K is a dimensionless parameter scaling with the magnetic field strength and representing the deviation of the electron path from the straight path of the light beam. From this equation it is seen that wavelength tuning can be achieved by changing the relativistic energy of the electrons (γ) or the magnetic field strength (K). The undulator is usually made of an array of permanent magnets so that Λ_u is normally fixed.

Table 1 Selected specifications of the CLIO and FELIX FELs [189, 190]

	CLIO	FELIX laboratory	
		FELIX	FLARE
Beam energy (MeV)	20–50	15–45	10–15
Wavelength (μm)	3–150	2.7–150	100–1,500
Macropulse energy (mJ)	≤60	≤100	≤100
M-pulse rep rate (Hz)	≤25	≤10	≤10
Bandwidth (%)	0.2–10	0.2–5	<0.5–2.0

In the FELs at the CLIO and FELIX facilities, which have thus far been applied for IR action spectroscopy more than any other FEL worldwide, wavelength tuning during an IR scan is achieved by changing the gap between the two arrays of magnets, changing the magnetic field strength. For a fixed setting of the electron beam energy, this allows one to tune the wavelength over a factor of about three.

The temporal structure of FEL radiation typically consists of μs-long macropulses, which are made up of a train of ps-long micropulses at MHz to GHz repetition rates. Other parameters of the CLIO and FELIX FELs are collected in Table 1 [189, 190]. At the FELIX facility, an FEL for wavelengths between 100 μm and 1.5 mm has been constructed (FLARE). Other tunable infrared FELs in use for action spectroscopy applications are located at the Tokyo University of Science and the Fritz-Haber Institute in Berlin; their specifications are comparable to those listed for FELIX and CLIO in Table 1.

3.3 *Difference Frequency Generation (DFG)*

DFG is a nonlinear optical technique in which tunable infrared radiation is generated by mixing the radiation from a fixed-frequency pump laser with that from a tunable laser source in a birefringent crystal. Typically, the fixed-frequency source is a Nd:YAG laser, while the tunable radiation is generated by a dye laser (often pumped with the same Nd:YAG).

Energy conservation requires that the generated difference frequency $\nu_{DFG} = \nu_{pump} - \nu_{tune}$. However, momentum conservation requires that a similar relation holds for the wave vectors $\mathbf{k}_{DFG} = \mathbf{k}_{pump} - \mathbf{k}_{tune}$, where $\mathbf{k} = 2\pi/\lambda = 2\pi n(\nu)\nu/c$ with $n(\nu)$ the index of refraction. Because of the frequency dependence of $n(\nu)$, both requirements can only be satisfied simultaneously in birefringent materials having different indices of refraction depending on the propagation direction and polarization of the light with respect to the lattice orientation of the crystal. Crystals with an anisotropic structure are usually birefringent and are mostly used in nonlinear optics. By tuning the angle of incidence into the crystal, two beams of different frequency and with perpendicular polarization can be made to travel through the crystal with equal phase velocities, hence the term ‘phase matching’. Scanning the wavelength of the infrared DFG radiation thus requires synchronous tuning of the dye laser and the angle of the birefringent crystal, which can be achieved with an auto-tracker.

To produce tunable radiation in the 3-μm wavelength range, various mixing schemes have been devised. Mixing the 1,064 nm Nd:YAG fundamental with the radiation from a tunable dye laser operating on a common red dye (such as DCM) requires two DFG stages to reach the desired wavelength range [121], as schematically indicated in Fig. 9c. The choice of nonlinear crystal depends crucially on the requirement that the material needs to be transparent at all frequencies ν_{DFG}, ν_{pump} and ν_{tune}. Materials typically used in generation of IR radiation around wavelengths of 3 μm include potassium titanyl phosphate (KTP), potassium titanyl arsenate (KTA) and lithium niobate ($LiNbO_3$).

3.4 Optical Parametric Oscillation (OPO)

As in DFG, an OPO makes use of nonlinear frequency mixing to generate tunable infrared radiation. Here, a single pump laser photon is frequency downconverted into two new photons known as the *signal* and *idler* waves, so that energy conservation gives $\nu_{pump} = \nu_{signal} + \nu_{idler}$. As in DFG and other applications involving nonlinear optics, phase matching is also required here and can again be achieved using birefringent crystals and angle tuning.

The OPO source marketed by LaserVision has been particularly influential in gas-phase (bio-)molecule spectroscopy. At wavelengths in the 3-μm range, this source reaches pulse energies exceeding 10 mJ in an approximately 6-ns pulse, which has been shown in many experiments to be sufficient to induce multiple-photon excitation and dissociation. The 1,064-nm output of a pulsed Nd:YAG laser is frequency-doubled (532 nm) and injected into an angle-tuned OPO oscillator, where tunable signal and idler beams at wavelengths around 800 and 1,600 nm, respectively, are generated. The 1,600-nm beam is seeded into an optical parametric amplifier (OPA) stage along with a 1,064-nm pump beam (split off from the Nd:YAG before second harmonic generation). Amplification of the 1,600-nm beam by optical parametric frequency downconversion of the pump beam generates a second (idler) photon in the 3-μm range. Synchronous angle tuning of the nonlinear crystals (KTA) in the OPO and OPA stages allows one to scan the wavelength of the nominally 3-μm beam roughly between 2,200 and 4,500 cm^{-1}. An additional nonlinear crystal ($AgGaSe_2$) can be used to induce DFG between the signal and idler beams producing tunable radiation in the 600–2,500-cm^{-1} range, though at pulse energies below 1 mJ. While too low for IRMPD, such energies have been shown to be sufficient for messenger atom and saturation spectroscopy.

Even if the phase velocities are matched, there is generally a gradual walk-off of the different beams as they travel through the crystal. Increasing the frequency conversion efficiency by employing longer crystals is therefore generally not useful beyond about 1 cm. This has long hampered the development of nonlinear optics using cw sources, which have much lower instantaneous output powers as compared to Q-switched laser sources. This situation changed with the development of periodically poled nonlinear crystals, which are microstructured materials with alternating domains where the optical axis points in opposite directions [191]. This avoids beam walk-off in these materials so that crystals up to several cm in length induce efficient frequency conversion. Moreover, in these materials true phase-matching is replaced by quasi-phase matching (QPM):

$$\frac{n_{\text{pump}}}{\lambda_{\text{pump}}} = \frac{n_{\text{signal}}}{\lambda_{\text{signal}}} + \frac{n_{\text{idler}}}{\lambda_{\text{idler}}} + \frac{1}{\Lambda}, \tag{6}$$

where Λ is the period of the poling of the optical axis in the material. Changing this period can be used to tune the signal and idler frequencies. A popular material for cw OPOs generating idler wavelengths around 3 μm is periodically poled lithium niobate (PPLN). Tunable OPOs with cw output powers exceeding 1 W are now commercially available from various vendors; depending on the oscillator geometry and wavelength-selective elements in the cavity, the bandwidths of these sources vary from a few cm^{-1} down to the MHz range.

Acknowledgements The material described in this chapter that refers to our own work would not have been possible without the expert and dedicated contributions of numerous graduate students, postdocs, collaborators and staff of the FELIX Laboratory, whom we would hereby like to gratefully acknowledge. We thank in particular Dr. G. Berden for his helpful comments and careful proofreading of the manuscript.

References

1. Rizzo TR, Park YD, Levy DH (1985) A molecular-beam of tryptophan. J Am Chem Soc 107 (1):277–278
2. Rizzo TR, Park YD, Levy DH (1986) Dispersed fluorescence of jet-cooled tryptophan – excited-state conformers and intramolecular exciplex formation. J Chem Phys 85 (12):6945–6951
3. Rizzo TR, Park YD, Peteanu L, Levy DH (1985) Electronic-spectrum of the amino-acid tryptophan cooled in a supersonic molecular-beam. J Chem Phys 83(9):4819–4820
4. Rizzo TR, Park YD, Peteanu LA, Levy DH (1986) The electronic-spectrum of the amino-acid tryptophan in the gas-phase. J Chem Phys 84(5):2534–2541
5. Park YD, Rizzo TR, Peteanu LA, Levy DH (1986) Electronic spectroscopy of tryptophan analogs in supersonic jets – 3-indole acetic-acid, 3-indole propionic-acid, tryptamine, and *N*-acetyl tryptophan ethyl-ester. J Chem Phys 84(12):6539–6549
6. Page RH, Shen YR, Lee YT (1988) Local modes of benzene and benzene dimer, studied by infrared-ultraviolet double-resonance in a supersonic beam. J Chem Phys 88(8):4621–4636
7. Ambartzumian RV, Letokhov VS, Makarov GN, Puretzki AA (1972) Direct measurement of population on molecular vibrational levels excited by laser radiation. Chem Phys Lett 16 (2):252–254

8. Lipert RJ, Colson SD (1989) Persistent spectral hole burning of molecular clusters in a supersonic jet. J Phys Chem 93(10):3894–3896
9. Stanley RJ, Castleman AW (1991) Cluster ion dip spectroscopy of hydrogen-bonded phenol $(H_2O)_n$ clusters, n = 0–4. J Chem Phys 94(12):7744–7756
10. Tanabe S, Ebata T, Fujii M, Mikami N (1993) OH stretching vibrations of phenol-$(H_2O)_n$(n = 1–3) complexes observed by IR–UV double-resonance spectroscopy. Chem Phys Lett 215(4):347–352
11. Pribble RN, Zwier TS (1994) Size-specific infrared-spectra of benzene-$(H_2O)_n$ clusters (n = 1 through 7) – evidence for noncyclic $(H_2O)_n$ structures. Science 265(5168):75–79
12. Colarusso P, Zhang KQ, Guo BJ, Bernath PF (1997) The infrared spectra of uracil, thymine, and adenine in the gas phase. Chem Phys Lett 269(1–2):39–48
13. Kim SK, Lee W, Herschbach DR (1996) Cluster beam chemistry: hydration of nucleic acid bases; ionization potentials of hydrated adenine and thymine. J Phys Chem 100 (19):7933–7937
14. Lin J, Yu C, Peng S, Akiyama I, Li K, Lee LK, Lebreton PR (1980) Ultraviolet photoelectron studies of the ground-state electronic-structure and gas-phase tautomerism of purine and adenine. J Am Chem Soc 102(14):4627–4631
15. Luhrs DC, Viallon J, Fischer I (2001) Excited state spectroscopy and dynamics of isolated adenine and 9-methyladenine. Phys Chem Chem Phys 3(10):1827–1831
16. Leon I, Cocinero EJ, Lesarri A, Castano F, Fernandez JA (2012) A spectroscopic approach to the solvation of anesthetics in jets: propofol$(H_2O)_n$, n = 4–6. J Phys Chem A 116 (36):8934–8941
17. Leon I, Cocinero EJ, Millan J, Jaeqx S, Rijs AM, Lesarri A, Castano F, Fernandez JA (2012) Exploring microsolvation of the anesthetic propofol. Phys Chem Chem Phys 14 (13):4398–4409
18. Leon I, Cocinero EJ, Millan J, Rijs AM, Usabiaga I, Lesarri A, Castano F, Fernandez JA (2012) A combined spectroscopic and theoretical study of propofol center dot $(H_2O)_3$. J Chem Phys 137(7)
19. Leon I, Cocinero EJ, Rijs AM, Millan J, Alonso E, Lesarri A, Fernandez JA (2013) Formation of water polyhedrons in propofol-water clusters. Phys Chem Chem Phys 15(2):568–575
20. Leon I, Millan J, Castano F, Fernandez JA (2012) A spectroscopic and computational study of propofol dimers and their hydrated clusters. ChemPhysChem 13(17):3819–3826
21. Leon I, Millan J, Cocinero EJ, Lesarri A, Castano F, Fernandez JA (2012) Mimicking anaesthetic-receptor interaction: a combined spectroscopic and computational study of propofol···phenol. Phys Chem Chem Phys 14(25):8956–8963
22. Leon I, Millan J, Cocinero EJ, Lesarri A, Fernandez JA (2013) Magic numbers in the solvation of the propofol dimer. ChemPhysChem 14(8):1558–1562
23. Leon I, Millan J, Cocinero EJ, Lesarri A, Fernandez JA (2013) Transition from planar to nonplanar hydrogen bond networks in the solvation of aromatic dimers: propofol(2)-(H2O) (2–4). J Phys Chem A 117(16):3396–3404
24. Aguado E, Leon I, Cocinero EJ, Lesarri A, Fernandez JA, Castano F (2009) Molecular recognition in the gas phase: benzocaine-phenol as a model of anaesthetic-receptor interaction. Phys Chem Chem Phys 11(48):11608–11616
25. Aguado E, Leon I, Millan J, Cocinero EJ, Jaeqx S, Rijs AM, Lesarri A, Fernandez JA (2013) Unraveling the benzocaine-receptor interaction at molecular level using mass-resolved spectroscopy. J Phys Chem B 117(43):13472–13480
26. Leon I, Aguado E, Lesarri A, Fernandez JA, Castano F (2009) Combined experimental and theoretical study of the benzocaine/Ar van der Waals system in supersonic expansions. J Phys Chem A 113(6):982–988
27. Meijer G, Devries MS, Hunziker HE, Wendt HR (1990) Laser desorption jet-cooling of organic-molecules – cooling characteristics and detection sensitivity. Appl Phys B-Photophys Laser Chem 51(6):395–403
28. Meijer G, Berden G, Meerts WL, Hunziker HE, Devries MS, Wendt HR (1992) Spectroscopy on triphenylamine and its van der Waals complexes. Chem Phys 163(2):209–222

29. Bescos B, Orea JM, Urena AG (2000) Laser desorption dynamics of carbendazim on a glass surface. Laser Chem 18(4):219–237
30. Lindner J, Grotemeyer J, Schlag EW (1990) Biomolecules in the gas-phase. 8. Applications of multiphoton ionization mass-spectrometry – small protected nucleosides and nucleotides. Int J Mass Spectrom Ion Process 100:267–285
31. Handschuh M, Nettesheim S, Zenobi R (1999) Is infrared laser-induced desorption a thermal process? The case of aniline. J Phys Chem B 103(10):1719–1726
32. de Vries MS, Hobza P (2007) Gas-phase spectroscopy of biomolecular building blocks. Annu Rev Phys Chem 58:585–612
33. Weinkauf R, Schermann JP, de Vries MS, Kleinermanns K (2002) Molecular physics of building blocks of life under isolated or defined conditions. Eur Phys J D 20(3):309–316
34. Stanca-Kaposta EC, Simons JP (2011) High-resolution infrared-ultraviolet (IR–UV) double-resonance spectroscopy of biological molecules. In: Quack M, Merkt F (eds) Handbook of high-resolution spectroscopy. Wiley, Chichester, pp 1911–1941
35. Chin W, Piuzzi F, Dimicoli I, Mons M (2006) Probing the competition between secondary structures and local preferences in gas phase isolated peptide backbones. Phys Chem Chem Phys 8(9):1033–1048
36. Saigusa H, Tomioka A, Katayama T, Iwase E (2006) A matrix-free laser desorption method for production of nucleobase clusters and their hydrates. Chem Phys Lett 418(1–3):119–125
37. Gloaguen E, Mons M (2014) Isolated neutral peptides. Top Curr Chem. doi:10.1007/128_2014_580
38. de Vries MS (2014) Gas-phase IR spectroscopy of nucleobases. Top Curr Chem. doi:10.1007/128_2014_577
39. Cocinero EJ, Çarçabal P (2014) Carbohydrates. Top Curr Chem. doi:10.1007/128_2014_596
40. Voumard P, Zenobi R (1995) Laser-induced thermal-desorption of aniline from silica surfaces. J Chem Phys 103(15):6795–6805
41. Zare RN, Levine RD (1987) Mechanism for bond-selective processes in laser desorption. Chem Phys Lett 136(6):593–599
42. Zenobi R, Hahn JH, Zare RN (1988) Surface-temperature measurement of dielectric materials heated by pulsed laser-radiation. Chem Phys Lett 150(5):361–365
43. Li YZ, McIver RT, Hemminger JC (1990) Experimental-determination of thermal and nonthermal mechanisms for laser desorption from thin metal-films. J Chem Phys 93 (7):4719–4723
44. Hall RB (1987) Pulsed-laser-induced desorption studies of the kinetics of surface-reactions. J Phys Chem 91(5):1007–1015
45. Levis RJ (1994) Laser-desorption and ejection of biomolecules from the condensed-phase into the gas-phase. Annu Rev Phys Chem 45:483–518
46. Woodward JR, Watanabe H, Ishiuchi S-I, Fujii M (2012) A two-color tunable infrared/ vacuum ultraviolet spectrometer for high-resolution spectroscopy of molecules in molecular beams. Rev Sci Instrum 83(1)
47. Jaeqx S, Du WN, Meijer EJ, Oomens J, Rijs AM (2013) Conformational study of Z-Glu-OH and Z-Arg-OH: dispersion interactions versus conventional hydrogen bonding. J Phys Chem A 117(6):1216–1227
48. Jaeqx S, Oomens J, Rijs AM (2013) Gas-phase salt bridge interactions between glutamic acid and arginine. Phys Chem Chem Phys 15(38):16341–16352
49. Nir E, de Vries MS (2002) Fragmentation of laser-desorbed 9-substituted adenines. Int J Mass Spectrom 219(1):133–138
50. van Zundert GCP, Jaeqx S, Berden G, Bakker JM, Kleinermanns K, Oomens J, Rijs AM (2011) IR spectroscopy of isolated neutral and protonated adenine and 9-methyladenine. ChemPhysChem 12(10):1921–1927
51. Rijs AM, Kabelac M, Abo-Riziq A, Hobza P, de Vries MS (2011) Isolated gramicidin peptides probed by IR spectroscopy. ChemPhysChem 12(10):1816–1821
52. Johnston MV (1984) Supersonic jet expansions in analytical spectroscopy. TrAC Trends Anal Chem 3(2):58–61

53. Miller TA (1984) Chemistry and chemical intermediates in supersonic free jet expansions. Science 223(4636):545–553
54. Levy DH (1980) Laser spectroscopy of cold gas-phase molecules. Annu Rev Phys Chem 31:197–225
55. Smalley RE, Wharton L, Levy DH (1977) Molecular optical spectroscopy with supersonic beams and jets. Acc Chem Res 10(4):139–145
56. Zwier TS (2006) Laser probes of conformational isomerization in flexible molecules and complexes. J Phys Chem A 110(12):4133–4150
57. Schermann JP (2008) Spectroscopy and modeling of biomolecular building blocks. Elsevier, Amsterdam
58. Snoek LC, Van Mourik T, Simons JP (2003) Neurotransmitters in the gas phase: a computational and spectroscopic study of noradrenaline. Mol Phys 101(9):1239–1248
59. Dittrich M, Schulten K (2005) Zooming in on ATP hydrolysis in F-1. J Bioenerg Biomembr 37(6):441–444
60. Wyttenbach T, Bowers MT (2009) Hydration of biomolecules. Chem Phys Lett 480 (1–3):1–16
61. Zimmer M (2002) Green fluorescent protein (GFP): applications, structure, and related photophysical behavior. Chem Rev 102(3):759–781
62. Blom MN, Compagnon I, Polfer NC, von Helden G, Meijer G, Suhai S, Paizs B, Oomens J (2007) Stepwise solvation of an amino acid: the appearance of zwitterionic structures. J Phys Chem A 111(31):7309–7316
63. Teh CK, Sipior J, Sulkes M (1989) Spectroscopy of tryptophan in supersonic expansions - addition of solvent molecules. J Phys Chem 93(14):5393–5400
64. Watanabe T, Ebata T, Tanabe S, Mikami N (1996) Size-selected vibrational spectra of phenol-$(H_2O)_n$ (n = 1–4) clusters observed by IR–UV double resonance and stimulated Raman-UV double resonance spectroscopies. J Chem Phys 105(2):408–419
65. Zwier TS (1996) The spectroscopy of solvation in hydrogen-bonded aromatic clusters. Annu Rev Phys Chem 47:205–241
66. Abo-Riziq A, Crews B, Grace L, de Vries MS (2005) Microhydration of guanine base pairs. J Am Chem Soc 127(8):2374–2375
67. Carcabal P, Jockusch RA, Hunig I, Snoek LC, Kroemer RT, Davis BG, Gamblin DP, Compagnon I, Oomens J, Simons JP (2005) Hydrogen bonding and cooperativity in isolated and hydrated sugars: mannose, galactose, glucose, and lactose. J Am Chem Soc 127 (32):11414–11425
68. Carcabal P, Kroemer RT, Snoek LC, Simons JP, Bakker JM, Compagnon I, Meijer G, von Helden G (2004) Hydrated complexes of tryptophan: ion dip infrared spectroscopy in the 'molecular fingerprint' region, 100–2000 cm^{-1}. Phys Chem Chem Phys 6(19):4546–4552
69. Carney JR, Hagemeister FC, Zwier TS (1998) Hydrogen-bonding topologies of indole-(water)(n) clusters from resonant ion-dip infrared spectroscopy. J Chem Phys 108 (9):3379–3382
70. Diken EG, Headrick JM, Johnson MA (2005) Photoelectron spectroscopy of the glycine center dot(H2O)(1,2) (−) clusters: sequential hydration shifts and observation of isomers. J Chem Phys 122(22)
71. Ebata T, Hashimoto T, Ito T, Inokuchi Y, Altunsu F, Brutschy B, Tarakeshwar P (2006) Hydration profiles of aromatic amino acids: conformations and vibrations of L-phenylalanine-(H2O)(n) clusters. Phys Chem Chem Phys 8(41):4783–4791
72. Lee KT, Sung J, Lee KJ, Kim SK, Park YD (2002) Resonant two-photon ionization study of jet-cooled amino acid: L-phenylalanine and its monohydrated complex. J Chem Phys 116 (19):8251–8254
73. Wiley WC, McLaren IH (1955) Time-of-flight mass spectrometer with improved resolution. Rev Sci Instrum 26(12):1150–1157
74. Simons JP, Jockusch RA, Carcabal P, Hung I, Kroemer RT, Macleod NA, Snoek LC (2005) Sugars in the gas phase. Spectroscopy, conformation, hydration, co-operativity and selectivity. Int Rev Phys Chem 24(3–4):489–531

75. Buchanan EG, James WH III, Gutberlet A, Dean JC, Guo L, Gellman SH, Zwier TS (2011) Single-conformation spectroscopy and population analysis of model gamma-peptides: new tests of amide stacking. Faraday Discuss 150:209–226
76. Gloaguen E, Pagliarulo F, Brenner V, Chin W, Piuzzi F, Tardivel B, Mons M (2007) Intramolecular recognition in a jet-cooled short peptide chain: gamma-turn helicity probed by a neighbouring residue. Phys Chem Chem Phys 9(32):4491–4497
77. Kleinermanns K, Nachtigallova D, de Vries MS (2013) Excited state dynamics of DNA bases. Int Rev Phys Chem 32(2):308–342
78. Yan B, Jaeqx S, van der Zande WJ, Rijs AM (2014) A conformation-selective IR–UV study of the dipeptides Ac-Phe-Ser-NH_2 and Ac-Phe-Cys-NH_2: probing the SHO and OHO hydrogen bond interactions. Phys Chem Chem Phys 16(22):10770–10778
79. Rijs AM, Compagnon I, Oomens J, Hannam JS, Leigh DA, Buma WJ (2009) Stiff, and sticky in the right places: binding interactions in isolated mechanically interlocked molecules probed by mid-infrared spectroscopy. J Am Chem Soc 131(7):2428–2429
80. Nosenko Y, Kunitski M, Stark T, Goebel M, Tarakeshwar P, Brutschy B (2013) Vibrational signatures of Watson-Crick base pairing in adenine-thymine mimics. Phys Chem Chem Phys 15(27):11520–11530
81. Nosenko Y, Kunitski M, Thummel RP, Kyrychenko A, Herbich J, Waluk J, Riehn C, Brutschy B (2006) Detection and structural characterization of clusters with ultrashort-lived electronically excited states: IR absorption detected by femtosecond multiphoton ionization. J Am Chem Soc 128(31):10000–10001
82. Shubert VA, Mueller CW, Zwier TS (2009) Water's role in reshaping a macrocycle's binding pocket: infrared and ultraviolet spectroscopy of benzo-15-crown-5-$(H_2O)_n$ and 4′-aminobenzo-15-crown-5-$(H_2O)_n$, $n = 1, 2$. J Phys Chem A 113(28):8067–8079
83. Shubert VA, Zwier TS (2007) IR–IR–UV hole-burning: conformation specific IR spectra in the face of UV spectral overlap. J Phys Chem A 111(51):13283–13286
84. Weiler M, Bartl K, Gerhards M (2012) Infrared/ultraviolet quadruple resonance spectroscopy to investigate structures of electronically excited states. J Chem Phys 136(11)
85. Fielicke A, Lyon JT, Haertelt M, Meijer G, Claes P, de Haeck J, Lievens P (2009) Vibrational spectroscopy of neutral silicon clusters via far-IR–VUV two color ionization. J Chem Phys 131(17)
86. Hu YJ, Guan JW, Bernstein ER (2013) Mass-selected IR–VUV (118 nm) spectroscopic studies of radicals, aliphatic molecules, and their clusters. Mass Spectrom Rev 32(6):484–501
87. Ng C-Y (2014) State-to-state spectroscopy and dynamics of ions and neutrals by photoionization and photoelectron methods. Annu Rev Phys Chem 65:197–224
88. Hu Y, Bernstein ER (2008) Vibrational and photoionization spectroscopy of biomolecules: aliphatic amino acid structures. J Chem Phys 128(16)
89. Hu Y, Bernstein ER (2009) Vibrational and photoionization spectroscopy of neutral valine clusters. J Phys Chem A 113(30):8454–8461
90. Bakker JM, Aleese LM, Meijer G, von Helden G (2003) Fingerprint IR spectroscopy to probe amino acid conformations in the gas phase. Phys Rev Lett 91(20)
91. Plusquellic DF, Siegrist K, Heilweil EJ, Esenturk O (2007) Applications of terahertz spectroscopy in biosystems. ChemPhysChem 8(17):2412–2431
92. Rijs AM, Kay ER, Leigh DA, Buma WJ (2011) IR spectroscopy on jet-cooled isolated two-station rotaxanes. J Phys Chem A 115(34):9669–9675
93. Rijs AM, Ohanessian G, Oomens J, Meijer G, von Helden G, Compagnon I (2010) Internal proton transfer leading to stable zwitterionic structures in a neutral isolated peptide. Angew. Chem. Int. Ed. 49(13):2332–2335
94. Rijs AM, Saendig N, Blom MN, Oomens J, Hannam JS, Leigh DA, Zerbetto F, Buma WJ (2010) Controlled hydrogen-bond breaking in a rotaxane by discrete solvation. Angew Chemi Int Ed 49(23):3896–3900
95. Merrick JP, Moran D, Radom L (2007) An evaluation of harmonic vibrational frequency scale factors. J Phys Chem A 111(45):11683–11700

96. Scott AP, Radom L (1996) Harmonic vibrational frequencies: an evaluation of Hartree-Fock, Moller-Plesset, quadratic configuration interaction, density functional theory, and semiempirical scale factors. J Phys Chem 100(41):16502–16513
97. Cirtog M, Rijs AM, Loquais Y, Brenner V, Tardivel B, Gloaguen E, Mons M (2012) Far/mid-infrared signatures of solvent solute interactions in a microhydrated model peptide chain. J Phys Chem Lett 3(22):3307–3311
98. Gaigeot M-P, Spezia R (2014) Theoretical methods for vibrational spectroscopy and collision induced dissociation in the gas phase. Top Curr Chem. doi:10.1007/128_2014_620
99. Jaeqx S, Oomens J, Cimas A, Gaigeot M-P, Rijs AM (2014) Gas-phase peptide structures unraveled by far-IR spectroscopy: combining IR–UV ion-dip experiments with Born-Oppenheimer molecular dynamics simulations. Angew Chem Int Ed Engl 53(14):3663–3666
100. Gaigeot M-P (2010) Theoretical spectroscopy of floppy peptides at room temperature. A DFTMD perspective: gas and aqueous phase. Phys Chem Chem Phys 12(14):3336–3359
101. Chen Y, Topp MR (2002) Infrared-optical double-resonance measurements of hydrogen-bonding interactions in clusters involving aminophthalimides. Chem Phys 283(1–2):249–268
102. Ebata T, Mizuochi N, Watanabe T, Mikami N (1996) OH stretching vibrations of phenol-$(H_2O)_1$ and phenol-$(H_2O)_3$ in the S-1 state. J Phys Chem 100(2):546–550
103. Fricke H, Bartl K, Funk A, Gerlach A, Gerhards M (2008) Proton/hydrogen-transfer coordinate of 2,5-dihydroxybenzoic acid investigated in a supersonic beam: combined IR/UV spectroscopy in the S_0, S_1, and D_0 states. ChemPhysChem 9(17):2592–2600
104. Southern CA, Levy DH, Florio GM, Longarte A, Zwier TS (2003) Electronic and infrared spectroscopy of anthranilic acid in a supersonic jet. J Phys Chem A 107(20):4032–4040
105. Walther T, Bitto H, Minton TK, Huber JR (1994) UV–IR double-resonance spectroscopy of jet-cooled propynal detected by the fluorescence dip method. Chem Phys Lett 231(1):64–69
106. Bartl K, Funk A, Gerhards M (2009) Structure of isolated xanthone in the T-1 state obtained via combined UV/IR spectroscopy. ChemPhysChem 10(11):1882–1886
107. Bartl K, Funk A, Schwing K, Fricke H, Kock G, Martin HD, Gerhards M (2009) IR spectroscopy applied subsequent to a proton transfer reaction in the excited state of isolated 3-hydroxyflavone and 2-(2-naphthyl)-3-hydroxychromone. Phys Chem Chem Phys 11 (8):1173–1179
108. Schwing K, Gerhards M (2014) Combined infrared/ultraviolet spectroscopy in molecular beam experiments. Bunsen-Magazin 16(3)
109. Clarkson JR, Dian BC, Moriggi L, DeFusco A, McCarthy V, Jordan KD, Zwier TS (2005) Direct measurement of the energy thresholds to conformational isomerization in tryptamine: experiment and theory. J Chem Phys 122(21)
110. Dian BC, Clarkson JR, Zwier TS (2004) Direct measurement of energy thresholds to conformational isomerization in tryptamine. Science 303(5661):1169–1173
111. Linnartz H, Verdes D, Speck T (2000) High resolution infrared direct absorption spectroscopy of ionic complexes. Rev Sci Instrum 71:1811–1815
112. Gudeman CS, Begemann MH, Pfaff J, Saykally RJ (1983) Velocity-modulated infrared laser spectroscopy of molecular ions: the ν_1 band of HCO^+. Phys Rev Lett 50:727–731
113. Crabtree KN, Hodges JN, Siller BM, Perry AJ, Kelly JE, Jenkins PA II, McCall BJ (2012) Sub-Doppler mid-infrared spectroscopy of molecular ions. Chem Phys Lett 551:1–6
114. Oka T (1980) Observation of the infrared spectrum of H_3^+. Phys Rev Lett 45:531–534
115. Berden G, Engeln R (2009) Cavity ring-down spectroscopy of molecular transients of astrophysical interest. Wiley, Chichester
116. Lisy JM (2006) Infrared studies of ionic clusters: the influence of Yuan T. Lee. J Chem Phys 125:132302
117. Dopfer O (2003) Spectroscopic and theoretical studies of $CH3^+$-Rg_n clusters (Rg = He, Ne, Ar): from weak intermolecular forces to chemical reaction mechanisms. Int Rev Phys Chem 22:437–495
118. Duncan MA (2012) Infrared laser spectroscopy of mass-selected carbocations. J Phys Chem A 116:11477–11491

119. Nicely AL, Miller DJ, Lisy JM (2009) Charge and temperature dependence of biomolecule conformations: K^+tryptamine$(H_2O)_{n=0-1}Ar_{m=0-1}$ cluster ions. J Am Chem Soc 131:6314–6315
120. Bieske EJ, Dopfer O (2000) High-resolution spectroscopy of cluster ions. Chem Rev 100:3963–3998
121. Kamariotis A, Boyarkin O, Mercier SR, Beck RD, Bush MF, Williams ER (2006) Infrared spectroscopy of hydrated amino acids in the gas phase: protonated and lithiated valine. J Am Chem Soc 128:905–916
122. Miller DJ, Lisy JM (2007) Modeling competitive interactions in proteins: vibrational spectroscopy of M^+(n-methylacetamide)$_1$$(H_2O)$n=0–3, M=Na and K, in the 3 μm region. J Phys Chem A 111:12409–12416
123. Duncan MA (2012) Invited review article: laser vaporization cluster sources. Rev Sci Instrum 83:041101
124. Piest H, von Helden G, Meijer G (1999) Infrared spectroscopy of jet-cooled neutral and ionized aniline-Ar. J Chem Phys 110:2010–2015
125. Prell JS, Chang TM, O'Brien JT, Williams ER (2010) Hydration isomers of protonated phenylalanine and derivatives: relative stabilities from infrared photodissociation. J Am Chem Soc 132:7811–7819
126. Asmis KR, Neumark DM (2012) Vibrational spectroscopy of microhydrated conjugate base anions. Acc Chem Res 45:43–52
127. Goebbert DJ, Wende T, Bergmann R, Meijer G, Asmis KR (2009) Messenger-tagging electrosprayed ions: vibrational spectroscopy of suberate dianions. J Phys Chem A 113:5874–5880
128. Garand E, Fournier JA, Kamrath Z, Schley ND, Crabtree RH, Johnson MA (2012) Characterization of an activated iridium water splitting catalyst using infrared photodissociation of H_2 tagged ions. Phys Chem Chem Phys 14:10109–10113
129. Goyal S, Schutt DL, Scoles G (1992) Vibrational spectroscopy of sulfur hexafluoride attached to helium clusters. Phys Rev Lett 69:933–936
130. Hartmann M, Miller RE, Toennies JP, Vilisov AF (1996) High-resolution molecular spectroscopy of van der Waals clusters in liquid helium droplets. Science 272:1631–1634
131. Lindinger A, Toennies JP, Vilisov AF (1999) High resolution vibronic spectra of the amino acids tryptophan and tyrosine in 0.38 K cold helium droplets. J Chem Phys 110:1429–1436
132. Smolarek S, Brauer NB, Buma WJ, Drabbels M (2010) IR spectroscopy of molecular ions by nonthermal ion ejection from helium nanodroplets. J Am Chem Soc 132:14086–14091
133. Bierau F, Kupser P, Meijer G, Von Helden G (2010) Catching proteins in liquid helium droplets. Phys Rev Lett 105:133402
134. Filsinger F, Ahn D-S, Meijer G, Von Helden G (2012) Photoexcitation of mass/charge selected hemin$^+$, caught in helium nanodroplets. Phys Chem Chem Phys 14:13370–13377
135. Gerlich D (1992) Inhomogeneous RF fields: a versatile tool for the study of processes with slow ions. In: Ng C-Y, Baer M (eds) State-selected and state-to-state ion-molecule reaction dynamics. Part 1: experiment, vol LXXXII. Wiley, New York
136. Rizzo TR, Boyarkin OV (2014) Cryogenic methods for the spectroscopy of large, biomolecular ions. Top Curr Chem. doi:10.1007/128_2014_579
137. Asvany O, Giesen T, Redlich B, Schlemmer S (2005) Experimental determination of the ν_5 cis-bending vibrational frequency and Renner-Teller structure in ground state $(X^2\Pi_u)$ $C_2H_2^+$ using laser induced reactions. Phys Rev Lett 94:073001
138. Asvany O, Ricken O, Muller HSP, Wiedner MC, Giesen TF, Schlemmer S (2008) High-resolution rotational spectroscopy in a cold ion trap: H_2D^+ and D_2H^+. Phys Rev Lett 100:233004
139. Asvany O, Krieg J, Schlemmer S (2012) Frequency comb assisted mid-infrared spectroscopy of cold molecular ions. Rev Sci Instrum 83:093110
140. Heine N, Fagiani MR, Rossi M, Wende T, Berden G, Blum V, Asmis KR (2013) Isomer-selective detection of hydrogen-bond vibrations in the protonated water hexamer. J Am Chem Soc 135:8266–8273

141. Woodin RL, Bomse DS, Beauchamp JL (1978) Multiphoton dissociation of molecules with low power continuous wave infrared laser radiation. J Am Chem Soc 100:3248–3250
142. Polfer NC, Oomens J (2007) Reaction products in mass spectrometry elucidated with infrared spectroscopy. Phys Chem Chem Phys 9:3804–3817
143. Bomse DS, Woodin RL, Beauchamp JL (1978) Molecular activation with low-intensity CW infrared laser radiation. Multiphoton dissociation of ions derived from diethyl ether. J Am Chem Soc 100:5503–5512
144. King DS, Stephenson JC (1978) Intrinsic isotopic selectivity factors: CO_2 TEA laser photolysis of CF_2Cl_2. J Am Chem Soc 100:7151–7155
145. Gaumann T, Riveros JM, Zhu Z (1990) The infrared multiphoton-dissociation spectra of bromopropene isomeric cations. Helv Chim Acta 73:1215–1218
146. Shin SK, Beauchamp JL (1990) Infrared multiphoton dissociation spectrum of $CF_3Mn(CO)_3(NO)^-$. J Am Chem Soc 112:2066–2069
147. Peiris DM, Cheeseman MA, Ramanathan R, Eyler JR (1993) Infrared multiple photon dissociation spectra of gaseous ions. J Phys Chem 97:7839–7843
148. Brauman JI, Smith KC (1969) Photodetachment energies of negative ions by ion cyclotron resonance spectroscopy. Electron affinities of neutral radicals. J Am Chem Soc 91:7778–7780
149. Dunbar RC (1971) Photodissociation of the CH_3Cl^+ and N_2O^+ cations. J Am Chem Soc 93:4354–4358
150. Oomens J, van Roij AJA, Meijer G, Von Helden G (2000) Gas-phase infrared photodissociation spectroscopy of cationic polyaromatic hydrocarbons. Astrophys J 542:404–410
151. Lemaire J, Boissel P, Heninger G, Mauclaire G, Bellec G, Mestdagh H, Simon A, LeCaer S, Ortega JM, Glotin F, Maitre P (2002) Gas phase infrared spectroscopy of selectively prepared ions. Phys Rev Lett 89:273002
152. Reinhard BM, Lagutschenkov A, Lemaire J, Maitre P, Boissel P, Niedner-Schatteburg G (2004) Reductive nitrile coupling in niobium-acetonitrile complexes probed by free electron laser IR multiphoton dissociation spectroscopy. J Phys Chem A 108:3350–3355
153. Valle JJ, Eyler JR, Oomens J, Moore DT, van der Meer AFG, von Helden G, Meijer G, Hendrickson CL, Marshall AG, Blakney G (2005) Free electron laser-Fourier transform ion cyclotron resonance mass spectrometry facility for obtaining infrared multiphoton dissociation spectra of gaseous ions. Rev Sci Instrum 76:023103
154. Polfer NC, Oomens J, Moore DT, von Helden G, Meijer G, Dunbar RC (2006) Infrared spectroscopy of phenylalanine Ag(I) and Zn(II) complexes in the gas phase. J Am Chem Soc 128:517–525
155. Bakker JM, Besson T, Lemaire J, Scuderi D, Maitre P (2007) Gas-phase structure of a π-allyl-palladium complex: efficient infrared spectroscopy in a 7 T Fourier transform mass spectrometer. J Phys Chem A 111:13415–13424
156. Frison G, van der Rest G, Turecek F, Besson T, Lemaire J, Maitre P, Chamot-Rooke J (2008) Structure of electron-capture dissociation fragments from charge-tagged peptides probed by tunable infrared multiple photon dissociation. J Am Chem Soc 130(14916–14917):14916
157. Chiavarino B, Crestoni ME, Fornarini S, Lanucara F, Lemaire J, Maitre P (2007) Meisenheimer complexes positively characterized as stable intermediates in the gas phase. Angew Chem Int Ed 46:1995–1998
158. Fukui K, Takada Y, Sumiyoshi T, Imai T, Takahashi K (2006) Infrared multiphoton dissociation spectroscopic analysis of peptides and oligosaccharides by using Fourier transform ion cyclotron resonance mass spectrometry with a midinfrared free-electron laser. J Phys Chem B 110:16111–16116
159. Miller GBS, Esser TK, Knorke H, Gewinner S, Schöllkopf W, Heine N, Asmis KR, Uggerud E (2014) Spectroscopic identification of a bidentate binding motif in the anionic magnesium-CO_2 complex ([$ClMgCO_2$]-). Angew Chem Int Ed 126:14635–14638
160. Bush MF, O'Brien JT, Prell JS, Saykally RJ, Williams ER (2007) Infrared spectroscopy of cationized arginine in the gas phase: direct evidence for the transition from nonzwitterionic to zwitterionic structure. J Am Chem Soc 129:1612–1622

161. Almasian M, Grzetic J, van Maurik J, Steill JD, Berden G, Ingemann S, Buma WJ, Oomens J (2012) Non-equilibrium isomer distribution of the gas-phase Photoactive Yellow Protein chromophore. J Phys Chem Lett 3:2259–2263
162. Nosenko Y, Menges F, Riehn C, Niedner-Schatteburg G (2013) Investigation by two-color IR dissociation spectroscopy of Hoogsteen-type binding in a metalated nucleobase pair mimic. Phys Chem Chem Phys 15:8171–8178
163. Oh H-B, Lin C, Hwang HY, Zhai H, Breuker K, Zabrouskov V, Carpenter BK, McLafferty FW (2005) Infrared photodissociation spectroscopy of electrosprayed ions in a Fourier transform mass spectrometer. J Am Chem Soc 127:4076–4083
164. Fridgen TD (2009) Infrared consequence spectroscopy of gaseous protonated and metal ion cationized complexes. Mass Spectrom Rev 28:586–607
165. Cagmat EB, Szczepanski J, Pearson WL, Powell DH, Eyler JR, Polfer NC (2010) Vibrational signatures of metal-chelated monosaccharide epimers: gas-phase infrared spectroscopy of Rb^+-tagged glucuronic and iduronic acid. Phys Chem Chem Phys 12:3474–3479
166. Mukamel S, Jortner J (1976) Multiphoton molecular dissociation in intense laser fields. J Chem Phys 65(12):5204–5225
167. Black JG, Yablonovitch E, Bloembergen N (1977) Collisionless multiphoton dissociation of SF_6: a statistical thermodynamic process. Phys Rev Lett 38(20):1131–1134
168. Makarov AA, Petrova IY, Ryabov EA, Letokhov VS (1998) Statistical inhomogeneous broadening of infrared and Raman transitions in highly vibrationally excited XY_6 molecules. J Phys Chem A 102:1438–1449
169. Grant ER, Schulz PA, Sudbo AS, Shen YR, Lee YT (1978) Is multiphoton dissociation of molecules a statistical thermal process? Phys Rev Lett 40:115–118
170. Simpson TB, Black JG, Burak I, Yablonovitch E, Bloembergen N (1985) Infrared multiphoton excitation of polyatomic molecules. J Chem Phys 83:628–640
171. Parneix P, Basire M, Calvo F (2013) Accurate modeling of infrared multiple photon dissociation spectra: the dynamical role of anharmonicities. J Phys Chem A 117:3954–3959
172. Oomens J, Sartakov BG, Meijer G, Von Helden G (2006) Gas-phase infrared multiple photon dissociation spectroscopy of mass-selected molecular ions. Int J Mass Spectrom 254:1–19
173. McIllroy A, Nesbitt DJ, Kerstel ERT, Pate BH, Lehmann KK, Scoles G (1994) Sub-Doppler, infrared laser spectroscopy of the propyne $2\nu_1$ band: evidence of z-axis Coriolis dominated intramolecular state mixing in the acetylenic CH stretch overtone. J Chem Phys 100:2596–2611
174. Chirokolava A, Perry DS, Boyarkin O, Schmid M, Rizzo TR (2000) Intramolecular energy transfer in highly vibrationally excited methanol. IV. Spectroscopy and dynamics of $^{13}CH_3OH$. J Chem Phys 113:10068–10072
175. Lehmann KK, Scoles G, Pate BH (1994) Intramolecular dynamics from eigenstate-resolved infrared spectra. Annu Rev Phys Chem 45:241–274
176. Von Helden G, van Heijnsbergen D, Meijer G (2003) Resonant ionization using IR light: a new tool to study the spectroscopy and dynamics of gas-phase molecules and clusters. J Phys Chem A 107:1671–1688
177. Paech K, Jockusch RA, Williams ER (2002) Slow infrared laser dissociation in the rapid energy exchange limit. J Phys Chem A 106:9761–9766
178. Dunbar RC (1991) Kinetics of low-intensity infrared laser photodissociation. The thermal model and application of the Tolman theorem. J Chem Phys 95:2537–2548
179. Oomens J, Tielens AGGM, Sartakov BG, Von Helden G, Meijer G (2003) Laboratory infrared spectroscopy of cationic polycyclic aromatic hydrocarbon molecules. Astrophys J 91:968–985
180. Roithova J (2012) Characterization of reaction intermediates by ion spectroscopy. Chem Soc Rev 41:547–559
181. Eyler JR (2009) Infrared multiple photon dissociation spectroscopy of ions in Penning traps. Mass Spectrom Rev 28:448–467

182. Polfer NC, Oomens J (2009) Vibrational spectroscopy of bare and solvated ionic complexes of biological relevance. Mass Spectrom Rev 28:468–494
183. MacAleese L, Maitre P (2007) Infrared spectroscopy of organometallic ions in the gas phase: from model to real world complexes. Mass Spectrom Rev 26:583–605
184. Duncan MA (2000) Frontiers in the spectroscopy of mass-selected ions. Int J Mass Spectrom 200:545–569
185. Oh H-B, Breuker K, Sze SK, Ge Y, Carpenter BK, McLafferty FW (2002) Secondary and tertiary structures of gaseous protein ions characterized by electron capture dissociation mass spectrometry and photofragment spectroscopy. Proc Natl Acad Sci 99:15863–15868
186. Oomens J, Polfer N, Moore DT, van der Meer L, Marshall AG, Eyler JR, Meijer G, von Helden G (2005) Charge-state resolved mid-infrared spectroscopy of a gas-phase protein. Phys Chem Chem Phys 7:1345–1348
187. Fung YME, Besson T, Lemaire J, Maitre P, Zubarev RA (2009) Room-temperature infrared spectroscopy combined with mass spectrometry distinguishes gas-phase protein isomers. Angew Chem Int Ed 48:8340–8342
188. Odeneye MA, Stace AJ (2005) Infrared photodissociation of $(NO)_n^+ \cdot X$ cluster ions ($n \leq 5$). Phys Chem Chem Phys 7:998–1004
189. Oepts D, van der Meer AFG, van Amersfoort PW (1995) The free-electron-laser facility FELIX. Infrared Phys Technol 36:297–308
190. Glotin F, Ortega JM, Prazeres R, Rippon C (1998) Activities of the CLIO infrared facility. Nucl Instrum Methods Phys Res B 144:8–17
191. Myers LE, Bosenberg WR (1997) Periodically poled lithium niobate and quasi-phase-matched optical parametric oscillators. IEEE J Quantum Electron 33:1663–1672

Top Curr Chem (2015) 364: 43–98
DOI: 10.1007/128_2014_579

Published online: 12 March 2015

Cryogenic Methods for the Spectroscopy of Large, Biomolecular Ions

Thomas R. Rizzo and Oleg V. Boyarkin

Abstract Determining the conformation of biological molecules is key for understanding their function. The recent combination of mass spectrometry, cryogenic ion traps, and laser spectroscopy is providing new methods to interrogate individual conformations of peptides and proteins that have advantages over classical techniques of structure determination. This chapter provides an overview of these new state-of-the-art methods and illustrates several specific applications. After reviewing the fundamentals of ion production, trapping, cooling, and spectroscopic detection, we review how different combinations of these techniques have been implemented in various laboratories around the world. We then focus on applications of cryogenic ion spectroscopy from two specific laboratories to illustrate the potential of this general approach. Finally, we outline ways in which these powerful new techniques could be further improved.

Keywords spectroscopy · biomolecules · cold ion traps · mass specltrometry · lasers · conformers · structure determination

Contents

T.R. Rizzo (✉) and O.V. Boyarkin
Laboratoire de Chimie Physique Moléculaire, École Polytechnique Fédérale de Lausanne, CH-1015, Lausanne, Switzerland
e-mail: thomas.rizzo@epfl.ch

1 Introduction

1.1 Motivation

Determining both the conformation of a biological molecule and its degree of conformational flexibility is key to understanding its function in living systems. Detailed biomolecular structures have primarily been determined by X-ray diffraction and NMR, both of which have their advantages and drawbacks. While X-ray diffraction methods can determine structure at high resolution, it requires the molecule of interest to be crystallized, which is often not possible, and crystal-packing forces may alter the conformation which the molecule would adopt in solution. NMR techniques overcome some of these disadvantages in that one can examine molecules in more native environments, although this is limited to comparatively smaller molecules. Moreover, because NMR measurements are inherently slow, they can determine only the average conformation that a molecule adopts in solution. As no one method for determining the structure of biological molecules is ideal, there has been a continued search for new approaches.

Since the development of soft ionization techniques such as electrospray [1] and MALDI [2], which permit the volatilization of biological molecules of virtually any size, mass spectrometric-based methods have been playing an increasingly important role in the study of biomolecular structure. For example, in addition to the determination of the mass and primary sequence of biopolymers, mass spectrometry has demonstrated the ability to determine the stoichiometry of large molecular assemblies [3, 4]. Moreover, ion mobility spectrometry (IMS) [5–7], when used in combination with mass spectrometry, can provide further information on the structure of molecules ranging from large protein complexes [8] to small, flexible peptides [9] by determining their orientationally averaged cross sections. Of course, one of the key issues in using any gas-phase technique for the determination of biomolecular structure is the relationship between a gas-phase structure and that of the same molecule in a condensed phase environment where it carries out its function [10–14].

Gas-phase infrared spectroscopy is a more recent addition to the structure-determination toolbox for biological molecules. Using the techniques of mass spectrometry combined with cryogenic cooling [15–18], highly resolved infrared spectra of peptides with up to 17 amino acids [16, 19] have been measured. Together with theory, such spectra can precisely determine the structure of these species. The purpose of this chapter is to review the various state-of-the-art methods that have been used to combine mass spectrometry, cryogenic techniques, and infrared spectroscopy for the structural determination of biological molecules. The excellent 2008 review by Gerlich on the production and study of ultra-cold molecular ions, although focused primarily on small ions, provides the reader with an additional helpful resource [20], as does the tutorial article by Wester on radiofrequency multipole traps [21].

1.2 Necessity of Cooling for Large Molecules

High-resolution infrared spectroscopy has played a major role in the structure determination of small, gas-phase molecules. One of the limiting factors to extending such approaches to larger molecules is the need to produce them in the gas phase where the vibrational bands can be well resolved. While laser desorption techniques have considerably extended the size range of neutral molecules accessible to gas-phase spectroscopy, a major breakthrough was the development of electrospray to produce large, gas-phase ions [1]. Electrospray has demonstrated the ability to volatilize molecules ranging from atomic ions to protein complexes. With the barrier to generating gas-phase molecules of any size removed, the question remains whether one can measure highly resolved spectra of large species and extract structural information.

For small molecules, precise structural information can be determined from the measured rotational constants that one extracts from high-resolution rovibrational spectra. For larger molecules it is difficult to resolve individual rotational lines and obtain rotational constants, and even if this could be accomplished, the rotational constants of a large species may not uniquely determine its structure. However, pure vibrational spectroscopy (i.e., rotationally unresolved) can provide sufficient information to determine molecular structure when combined with theory. The specific interactions between functional groups within a molecule provide a spectroscopic fingerprint which, if of sufficiently high resolution, is uniquely related to the molecular structure. The structure is then determined by comparing measured and calculated spectra, and once a match is found (including vibrational band assignments), the corresponding calculated structure is validated. For such a procedure to work, measured vibrational spectra should be as highly resolved as possible to allow unambiguous comparison with theory, and this becomes increasing difficult as the size of the molecule, and hence the number of vibrational degrees of freedom, increases. Using vibrational spectroscopy as a structural tool for larger molecules thus requires reducing all sources of spectral congestion to the maximum degree possible.

There are three major sources of congestion in the spectrum of a large molecule that lead to overlap of lines and hence impede spectral analysis: (1) broadening from the distribution of thermally populated states (i.e., hot bands); (2) conformational heterogeneity resulting from the distribution of different stable structures; and (3) the overlap of vibrational bands with similar frequency. Sources (1) and (2) are considered inhomogeneous spectral broadening in that individual molecules in a sample have different spectra. In principle, these sources of broadening can be removed by eliminating the state or conformer distribution in one's sample. The third source of spectral congestion is considered as homogeneous broadening in that every molecule of the same type (i.e., in the same state) possesses the same vibrational bands – it is thus an inherent property of the molecule and cannot be eliminated.

Thermal inhomogeneous broadening, which arises from the thermal distribution of populated states, can be reduced to a negligible level if a molecule can be cooled to sufficiently low internal temperatures – but how low is low enough? If the ultimate goal is to obtain vibrationally (and not rotationally) resolved spectra, it suffices to eliminate all populations in excited vibrational states. For proteins, the lowest vibrational frequencies are typically on the order of 20 cm^{-1} [22], corresponding to the relative motions of large subunits with respect to one another. If one assumes a hot band transition strength similar to that from the vibrational ground state, the temperature needed to reduce the intensity of the former to 10% of the latter is ~12 K. As described in Sect. 2, this can be done with relatively standard cryogenic techniques.

While reducing or eliminating the distribution of thermally populated states by cryogenic cooling can go a long way to simplifying a vibrational spectrum, the problem of conformational heterogeneity remains. For a measured spectrum to yield a structure, it must be compared with a calculated spectrum, but the calculated spectrum corresponds to a single conformation. If a number of different stable conformers exist in the sample, the comparison with theory is difficult and the assignments tenuous. A molecule trapped in a conformation that is not the lowest in energy in a global sense gives rise to additional bands in the spectrum. Although these bands originate from higher energy states of the molecule, they are different from what we normally consider hot bands because the molecule has a different equilibrium geometry and there is a barrier on the potential energy surface separating the conformers. In principle, one could attempt to anneal the molecule by heating and re-cooling to transfer the trapped population into the lowest energy conformation – such annealing has been done both in ion mobility experiments [9, 23] and in spectroscopic experiments [24]. However, it may be that these trapped conformations are the most interesting to interrogate. A number of different types of experiments have indicated that gas-phase molecular ions produced by electrospray retain much of the structure that they had in solution when solvent is removed [10–14]. Internal hydrogen bonds can lead to barriers in the potential energy landscape such that basic structural elements are retained for some period of time [7, 10, 24–27], allowing gas-phase experiments to be performed on these kinetically trapped species that resemble their structure in solution.

Nevertheless, the presence of multiple stable conformations in a gas-phase sample complicates spectral analysis and may prevent the assignment of a structure by comparison with theory. One would thus like to be able to study individual conformations. Two basic approaches can be used to remove the conformational heterogeneity of a spectrum: physically separate the conformers before making spectroscopic measurements on a sample or use multiple-resonance techniques to assign each line in a spectrum to a particular conformer. Both types of techniques, which are independent and can be used in tandem, are discussed below.

Once one removes thermal inhomogeneous broadening and conformational heterogeneity, the remaining degree of spectral congestion is an inherent property of each molecule and cannot be removed. In this case, however, even if spectral lines overlap, a conformationally pure vibrational spectrum can be directly compared with theory to determine a structure, although the confidence in the structural determination depends on the remaining degree of spectral complexity. As discussed in Sect. 4, to go beyond this would require new types of spectroscopy.

1.3 Overview of the Chapter

In this chapter we provide an overview of the spectroscopy of cold, biomolecular ions with the goal of obtaining structural information. We first discuss the different elements that need to be combined to perform such experiments: ion production, trapping, cooling, and spectroscopic detection. For each element we briefly survey possible approaches but then quickly focus on those techniques that have found their way into practice. Having looked at the individual parts of such an experiment, we then survey how these different elements have been combined in various laboratories around the world for spectroscopic studies of cold ions and discuss the advantages of the various approaches. Following this, we discuss in more detail selected results on the spectroscopy of cold, biological ions. We finish by considering the perspectives and challenges for the future of this field.

2 Basic Experimental Approaches to Cryogenic Ion Spectroscopy

To measure the spectra of cold, biomolecular ions, one must first produce them in the gas phase and cool them to low temperatures to eliminate thermal inhomogeneous broadening. Because of their net charge, one can use electric or magnetic fields to trap the ions in space before spectroscopic interrogation. To obtain a spectrum one has to detect the absorption of light, and because the density of ions is low, it is extremely difficult to do so directly. One typically uses some type of "action spectroscopy" in which the consequences of light absorption are detected rather than the change in the transmitted light intensity itself. One can then generate a spectrum by measuring the change in the "action" as a function of the laser frequency. We summarize here the various techniques that could be used to accomplish these steps: ion production, trapping, cooling, and spectroscopic detection.

2.1 *Ion Production*

The advent of both MALDI [28, 29] and electrospray ionization (ESI) [1, 30] in the early 1990s opened the way for mass spectrometry of biological molecules, having earned their respective inventors a share of the 2002 Nobel Prize in chemistry. While both techniques have certain advantages and disadvantages, ESI seems to be the more common approach and simpler to implement. Its ability to produce multiply charged ions directly from solution allows one to study relatively large molecules at moderate mass–to-charge ratio, and there seems to be essentially no limit to the mass of the molecule that one can volatilize using this technique – even whole viruses have been produced in the gas phase by electrospray and shown to retain their activity [31].

2.2 *Ion Trapping*

2.2.1 Summary of Ion Trap Techniques

While it is not absolutely necessary to trap ions to perform spectroscopic studies, the ability to do so provides significant advantages for collection, cooling, and optical access, and virtually all studies of gas-phase biological ions have used ion traps of one sort or another. While there are many different ways to trap ions, there are a few specific requirements for an ion storage device to be suitable for cryogenic ion spectroscopy. (1) It should allow convenient loading of a significant number of ions. (2) The storage time should be long enough for cooling and the subsequent spectroscopic interrogation without a significant loss of the precursor ions. (3) In the case of photofragmentation spectroscopy (see Sect. 2.4), it should be capable of storing fragment ions of different mass-to-charge ratio and potentially with an elevated translational temperature, determined by the dissociation process. (4) It should be constructed from materials of high thermal conductivity (copper, aluminum, sapphire, etc.) to ensure efficient transfer of heat from the trap body to the cooling system. These materials must also have a low degassing rate, which is required for working at low temperatures in ultra-high vacuum. (5) Its construction must be mechanically robust to survive repetitive cooling-heating cycles. (6) It should provide clear optical access to the stored ions. Laser beams should be able to pass through the entire stored ion cloud without clipping on cold parts of a trap. This ensures maximum efficiency in ion excitation while minimizing potential heating by the laser beams. (7) If used in fluorescence experiments (see Sect. 2.4), the trap geometry has to allow the efficient collection of the light emitted by the ions. Taken together, these constraints severely limit the use of commercially available ion traps for cold ion spectroscopy and often require either a modification of such traps or a custom design.

There are many types of ion storage devices, which differ in the nature of the field(s) used to confine the ions, and the implementation of the trapping fields often has implications on the trap size and practicality of use. Perhaps the most common type of trap is the Penning trap used in mass spectrometry. By combining a strong, static magnetic field with static electric fields, Penning traps are used in ion cyclotron resonance (ICR) [32] with the purpose of analyzing a molecule's mass-to-charge ratio. While such traps have been cooled to liquid nitrogen temperatures and used for ion spectroscopy [33], cooling to 10 K has proven to be a challenge. Electrostatic storage rings use static electric fields to confine ions, and while these have been used for spectroscopy of biological molecules at ambient temperature [34], the cooling power needed to operate such instruments at low temperature is formidable. Nevertheless, efforts are currently underway to construct a cryogenic storage ring [35]. A relatively simple design for an electrostatic ion beam trap (EIBT) was developed by Zajfman, which basically consists of the charged particle equivalent of an optical resonator [36, 37]. While this design lends itself well to cryogenic experiments [38], it has not been applied to the spectroscopy of large molecular ions. Perhaps the simplest electrostatic trap is Kingdon's trap, proposed a century ago [39], but only recently implemented as an Orbitrap device [40], used for high-resolution mass analysis in commercial mass-spectrometers [41]. While the Orbitrap is small and could easily be cooled, its geometry does not lend itself to the optical access needed for spectroscopic experiments.

Perhaps the most suitable type of ion storage device for cooling and spectroscopy of cold molecular ions is the radio frequency (RF) ion trap, which uses an AC electric field at RF frequencies (typically hundreds of kilohertz to a few megahertz) in conjunction with static electric fields. Because of their importance, we describe them here in more detail.

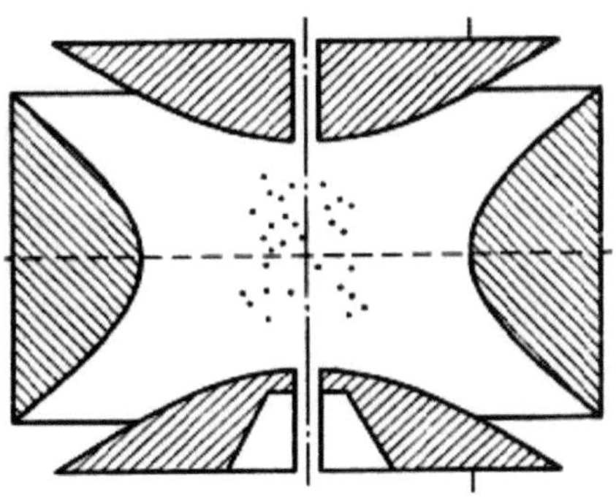

Fig. 1 Cross-section of a Paul trap. The two holes in the upper and lower end-caps serve for loading and release of stored ions

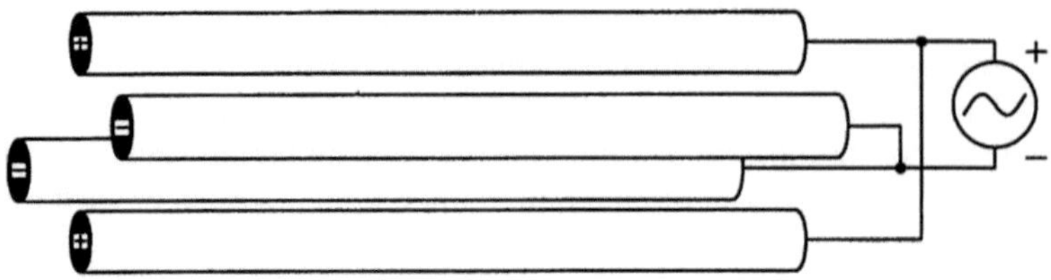

Fig. 2 Schematic of a linear quadrupole. Four rods are alternatively connected to the opposite phases of the RF driver for radial confinement of ions. The addition of two end-caps (not shown) for axial confinement allow ions to be trapped

2.2.2 RF Ion Traps

The most common RF ion trap is a Paul trap [42], a 3-D quadrupole device in which ions are confined in a small volume of typically a few tens of millimeters [2] between a hyperbolically shaped inner surface of a ring electrode and two end-cap electrodes, also of hyperbolic shape (Fig. 1). Each end-cap electrode has a central hole for loading and ejection of ions. As these traps are compact, commercially available, and allow mass-selection of stored ions, they have become an increasingly popular technically simple solution for cryogenic ion spectroscopy. Paul traps have several drawbacks for cold-ion spectroscopy, however: inefficient ion injection; an intrinsically limited ability to cool ions; low storage volume; and inconvenient optical access to the ions by laser beams.

One particular type of RF ion storage device that can overcome these limitations is the linear ion trap. In a linear ion trap of cylindrical geometry, ions are confined radially by an RF electric field created by a set of $2n$ identical linear electrodes (poles), evenly spaced on a circle with inscribed radius r_0. In linear quadrupole ion traps ($n=2$), four electrodes are alternatively connected to two sinusoidal RF waveforms of opposite sign (Fig. 2), while two end-caps are typically kept at a few volts DC relative to the electrode bias to stop ions from exiting the trap in the axial direction.

For $n>2$, a reasonably good approximation for ideal shape of the electrodes is a rod of diameter d:

$$d = \frac{2r_0}{n-1}. \tag{1}$$

The end-cap electrodes are often made as short tubes or bored discs with a hole diameter close to that of the trap inscribed circle.

Ion trajectories in linear traps can be derived using second-order differential equations for the motion of a charged particle in a harmonic electrical field. A detailed elaboration of this motion has been presented, for instance, by Gerlich [43]. As the ion motion inside the trap is relevant for ion cooling, we briefly summarize his treatment below.

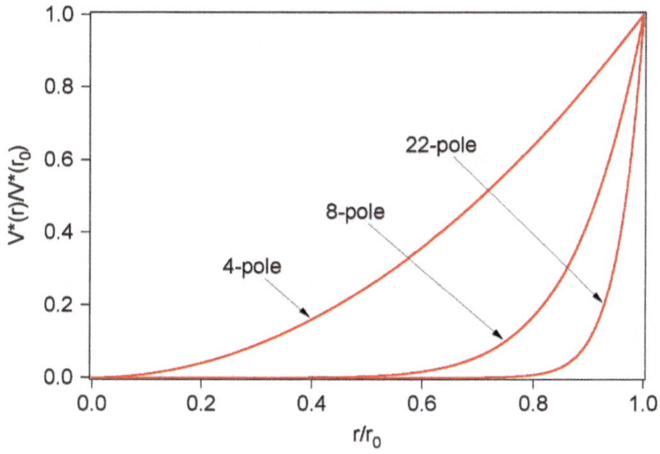

Fig. 3 Relative effective radial potentials, plotted for quadrupole, octupole and 22-pole ion traps. Reproduced with permission from [43]

The motion of an ion in an inhomogeneous RF field can be described as a superposition of slow radial motion and a fast oscillatory micro-motion at the RF frequency. In the adiabatic approximation, when the latter is much faster than the former, the two motions can be decoupled, allowing an analytical treatment of the problem. The slow radial motion is then associated with a time-independent effective electrical potential, V^*, expressed as a function of the radial position r by

$$V^*(r) = \frac{q}{4m}\left(\frac{nV}{\Omega \cdot r_0}\right)^2 \left(\frac{r}{r_0}\right)^{2n-2}, \tag{2}$$

where $2n$ is the number of electrodes, m and q are the mass and the charge of the stored ions, respectively, V and Ω are the amplitude and the frequency of RF waveforms applied to the electrodes, and r_0 is inner inscribed radius of the electrodes. As shown in Fig. 3, the shape of the effective potential depends on the number of poles, rapidly approaching a rectangular form with a flat bottom and steep walls with increasing n. As explained below, having an effective potential with a steep wall is essential for effective cooling in a linear ion trap. For trapping ions with maximum radial kinetic energy E_m, the minimum requirement is $E_m < q \cdot V^*(r_0)$. A more stringent condition, which limits ion storage volume but guarantees that the ion trajectories remain bound, is [43]

$$E_m = q \cdot V^*(r/r_0 = 0.8). \tag{3}$$

For the ion motion to be adiabatic, which ensures that an ion does not gain any additional kinetic energy from the confining RF field and its trajectories remain stable over time, the parameter η, given by the expression

$$\eta = 2n(n-1)\frac{qV}{m\Omega^2 r_0^2}\left(\frac{r}{r_0}\right)^{n-2}, \tag{4}$$

must remain small. To maintain safe trapping, Gerlich suggests [43] the following empirical criterion of adiabaticity:

$$\eta(0.8r_0) \leq 0.3. \tag{5}$$

Taken together, (2)–(5) result in the following convenient criteria:

$$\eta = 48.6 \cdot 10^9 \cdot n(n-1)\frac{z \cdot V}{M \cdot f^2 \cdot r_0^2}(0.8)^{n-2} \leq 0.3, \tag{6}$$

and

$$U_\text{m} = \frac{n \cdot \eta \cdot z \cdot V}{8(n-1)}(0.8)^n. \tag{7}$$

Here z and M are charge and mass (in Daltons) of the trapped ions, respectively, $f = \Omega/2\pi$ is the frequency of the RF field (in Hz), r_0 is the inscribed radius of the trap expressed in cm, and U_m is the maximum kinetic energy of the ions (in eV units) that can be trapped. For a particular ion trap geometry (i.e., n, r_0), Eq. (6) determines the minimum RF amplitude V_min (at $\eta = 0.3$) that can be used to trap ions with transverse kinetic energy U_m (expressed in eV). Equations (2), (3) and (5) then allow an evaluation of the minimum frequency required for trapping of ions with a particular M/z.

A detailed analysis reveals that the ion micro-motion at the applied RF frequency increases in amplitude near the turning points (close to the trap poles), while it vanishes as the ion approaches the field-free region close to the trap axis [43, 44].

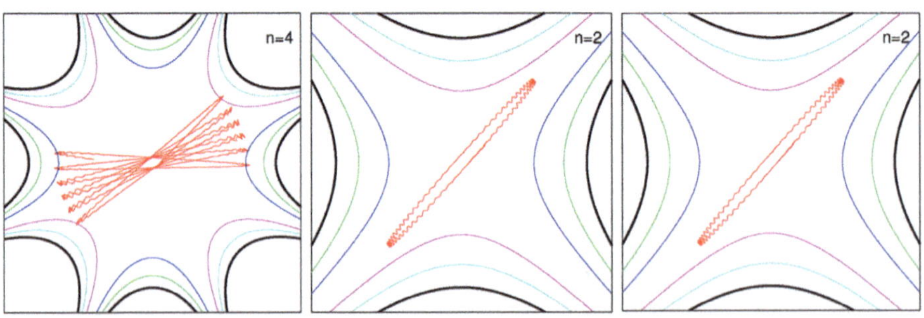

Fig. 4 Equipotential lines of the effective electric field (smooth hyperbolic *color-coded lines*) and ion trajectories (*red traces* in the middle of each figure), computed for linear ion traps of $n = 2$ (quadrupole), $n = 4$ (octupole), and $n = 11$ (22-pole) respectively from the left to the right. The region of "wiggle-free" motion centered at trap axis enlarges upon increasing the number of poles. Reproduced from [44]. Copyright (2009), with permission from Elsevier

High-order linear traps, such as a 22-pole trap, have a relatively flat radial effective potential over most of the trap volume, which minimizes the contribution of the micro-motion to ion trajectories. In contrast, ions in a quadrupole exhibit wiggling motion almost everywhere along radial ion trajectories, as illustrated in Fig. 4. While this motion makes little difference for trapping ions, it turns out to be crucial for cryogenic cooling, as discussed below.

Despite their advantages, there are certain problems associated with a use of high-order RF traps for spectroscopic studies. The most severe problem arises from inhomogeneous radial distribution of trapped ions, which exhibits a dip on the axis [45]. Numerical simulations performed for a 22-pole trap suggest that this distribution is caused by DC potentials applied to the end-cap electrodes [44], which push ions towards the trap poles, reducing the ion density on the trap axis. In addition, any imperfections and/or contamination of the poles may create local, off-axis potential wells where the slow moving cold ions tend to reside. In practice, this creates some uncertainty in the location of the stored ion cloud, complicating overlap of laser beams with a large fraction of the ions. Low-order linear RF traps, on the other hand, possess a well-defined potential minimum and concentrate ions on the trap axis, although as described below, RF heating in these devices is more of a problem. Thus, while the first vibrationally-resolved electronic spectra of cold, biomolecular ions were measured using 22-pole traps [46], in our laboratory they have recently been replaced by cold octupole ion traps [47], which strike a better compromise between the competing issues of cooling and optical access.

To load an RF trap from an external ion source, the ions must pass through one of the end-cap electrodes into the trapping region. Here, linear RF traps have a significant advantage over a 3-D RF quadrupole trap because the field on the cylindrical axis of the former is determined by the DC voltage applied to the end-caps and not by the RF field, which cancels on the center-line. Lowering the entrance end-cap voltage allows continuous injection until the point at which ions may reflect off the opposing end-cap and exit via the entrance electrode (unless kinetic energy is removed in the trap, as discussed below). In a 3-D quadrupole trap there is only a point rather than an axis where the RF field cancels, and thus injection through the end-cap is only possible for a narrow range of the RF phase where the field crosses zero [48].

Once the ions traverse the end-cap, there are two possible ways to load a linear RF trap. In one case, the kinetic energy of the injected ions doesn't change upon trapping, while in the second, collisions with buffer gas in the trap lower the kinetic energy of incoming ions. In both cases the potential of the exit end-cap must be higher than the kinetic energy of incoming ions in order to prevent them from simply passing through the trap in the axial direction. The first approach, termed gated trapping, implies raising the potential of the entrance end-cap from a value below the kinetic energy of the incoming ions to a value above it after the incoming ion bunch enters the trap, but before the ions are reflected back by the exit end-cap and return to the entrance. In this case, efficient trapping then requires that the arriving ion bunch is shorter than twice the length of the trap, which limits their maximum translational energy spread. The second approach uses translational cooling of ions in collisions with a buffer gas to lower their total energy below the potential of both end-caps, which allows ions to be continuously injected into the trap. As the same collisional trapping process can be used for cooling the internal degrees of freedom of the ion, we discus it in more detail below.

2.3 *Cooling of Molecular Ions*

2.3.1 Overview of Cooling Methods

While there are a number of methods employed to cool gas-phase ions, many of them are not suitable for large, molecular ions. For example, the most common way to cool atomic ions is laser-cooling [49–51], however the internal energy level structure of complex molecular ions does not allow the optical cycling required for this approach. One method to circumvent this problem is to use sympathetic cooling, wherein lasers are used to cool atomic ions, which subsequently cool co-trapped molecular ions by energy transfer [52–54]. While this approach has been demonstrated for complex biomolecules [55, 56], the interaction between the atomic and molecular ions in the trap is not sufficiently strong to cool the vibrational degrees of freedom of the latter, leaving them translationally cold but internally warm.

An approach that could, in principle, be used to cool the internal degrees-of-freedom of larger molecular ions is radiative cooling. Molecular ions that are confined in a low temperature ion trap eventually cool by the spontaneous emission of infrared radiation; however, the cooling rate is slow [57, 58], particularly for large, flexible biomolecules, where most of the internal energy resides in low frequency vibrations, since the radiative lifetime exhibits a λ^3 dependence. So while this approach is possible, it is not particularly practical.

A completely different approach to cooling biomolecular ions is to embed them into superfluid liquid He nanodroplets [59], which can be produced by expanding high-pressure He into vacuum through a small, refrigerated nozzle. Once in vacuum, the droplets reach and maintain a temperature of 0.38 K by evaporative cooling of helium. By passing them through an ion trap in which biomolecules are stored, the droplets can pick up the ions and thermalize them to 0.38 K, again through evaporative cooling. While there may be certain drawbacks to the interaction of the ions with the liquid helium environment, this approach has the advantage of achieving temperatures at which all internal energy is drained from the biomolecule, which is necessary for orienting molecules in an electric field, for example [60].

The most commonly used and perhaps most viable method for cooling the internal degrees of freedom of a large molecular ion is via collisions with a cold buffer gas. Here we describe the specific situation of buffer-gas cooling of ions in an RF ion trap.

2.3.2 Buffer Gas Cooling

The same collisional energy transfer process used to trap ions can be used for cryogenic cooling. This requires the buffer gas to be cooled to a low temperature, but it also requires more buffer-ion collisions to thermalize fully all degrees of freedom rather than simply the translational degrees of freedom needed for trapping. Technically, both requirements can be achieved by enclosing the trap in a thermally conductive box and cooling the entire assembly to the desired temperature. The buffer gas is either leaked or pulsed into the trap, where it can quickly thermalize with the large surface of the surrounding cold box. The box also allows the build-up of buffer gas pressure inside the trap volume, increasing the frequency of cooling collisions. The buffer molecules heated in collisions quickly re-cool upon colliding with the cold surfaces and become available again for ion cooling unless they are pumped out. Cryogenic cooling of a trap assembly is most simply done using commercial closed-cycle, two-stage He refrigerators, which offer watt-scale cooling power at 6–10 K. At these temperatures, the only buffer gas that does not condense on the cold trap surfaces is He.

The collisional cooling process outlined above has not yet considered the effect of the micro-motion of ions induced by the RF field as they begin to climb up the wall of the effective radial potential. Inelastic collisions with buffer gas can convert kinetic energy of the micro-oscillations into ion internal energy, which can counteract the collisional cooling occurring in the field-free region near the trap axis. For low-order RF multipole traps, such as a Paul trap, this can be the limiting factor in cooling the ions to low temperature. High order traps exhibit a larger field-free region, which makes them a preferable choice for cryogenic cooling. The difference in the ultimate internal temperature of the two types of traps becomes less significant for heavy ions [47, 61], however, because the collisional RF heating scales inversely with increasing ion-to-buffer gas mass ratio [43, 44, 62].

One important consideration in the design of spectroscopic experiments in buffer-gas cooled ion traps is whether the ions are interrogated in the trap itself or are first extracted to another part of the machine. If spectroscopic detection occurs outside the trap, as is the case in several of the instruments described below, one has to consider the possibility that the ions may have collisions with buffer gas molecules as they are extracted, and these collisions can warm the ions internally. In cases where buffer gas is continuously leaked into the trap, this is necessarily the case, but it also likely occurs even when the buffer gas is pulsed, since the pump-out time of cold buffer gas can be long. If maintaining the lowest possible temperature is essential to the experiment, it is better to probe the ions before extraction from the trap.

We give examples of implementation of buffer-gas cooling in RF ion traps in Sect. 2.5 and methods of temperature determination of ions in Sect. 2.6, but first we discuss spectroscopic techniques for detecting ion absorption of radiation.

2.4 Spectroscopic Detection of Ion Absorption

Once ions are stored and cooled in an appropriate ion trap, they can be interrogated by UV and/or IR spectroscopy. This is usually performed using tunable lasers, which produce beams of high photon flux, narrow bandwidth, and low divergence. The intrinsically low concentration of the stored ions and the typically short length of ion traps make direct absorption spectroscopy virtually impossible. One must thus employ sensitive, indirect spectroscopic techniques that accurately reflect the absorption transitions of gas-phase species. These so-called "action spectroscopies" rely on detecting certain physical or chemical changes in the molecule that occur in response to light absorption. Laser-induced fluorescence (LIF), in which one monitors an absorption spectrum by detecting the light spontaneously emitted by an atomic or molecular species following electronic or vibrational excitation, is a well-known example. While several groups have implemented LIF for detection and spectroscopy of large trapped ions [63–69], its application to large biomolecules remains challenging and is intrinsically limited to species with appreciably high fluorescence quantum yields. To date there have been no applications of this approach to cryogenically cooled biomolecular ions.

One general type of action spectroscopy, which can take many forms, relies on the occurrence of a chemical reaction subsequent to photon absorption and is monitored by detecting the reaction products. Different implementations of this general approach can be classified by the type of chemical change that occurs. Early applications of this technique to cryogenically cooled N_2^+ ions in a 22-pole ion trap monitored the electronic spectrum by detecting the products of the charge transfer reaction from N_2^+ to Ar; hence the name laser-induced charge transfer (LICT) [70]. The same group used the hydrogen-transfer reaction $C_2H_2^+ + H_2 \rightarrow C_2H_3^+ + H$ to measure the rovibrational spectrum of cold acetylene cation, a technique they simply called laser-induced reactions (LIR) [71].

Other manifestations of this general approach use photofragmentation reactions to monitor photon absorption, and the specific name given to the method depends upon the type of fragmentation occurring. In one such technique, called tagging spectroscopy, a neutral "messenger" molecule is condensed onto a cold ion, forming a weakly-bound ion-molecule complex [72–76]. These complexes are then isolated or selected by mass-spectrometry and interrogated by lasers. Absorption of UV or IR laser light may result in dissociation of the complex, yielding parent ions as charged fragments. A spectrum of the complex is then generated by monitoring dissociation yield as a function of the wavenumber of the dissociating laser light. Under the assumptions that the messenger molecule doesn't significantly distort the structure of the ion or absorb light itself in the wavelength region monitored, the measured spectrum may reflect that of the bare ion. The first assumption is not always evident and may require some additional studies. In some cases, however, the clusters themselves are the focus of study. For example, UV and IR spectroscopy of solvated, protonated biomolecules detect photon absorption by the detachment of H_2O, and in this case the objects of study are the water–biomolecule complexes [77–81]. In general, the use of light atoms or molecules of low polarizability, such He or H_2, reduces ion perturbations, although their low binding energies require very low temperatures to form a significant number of complexes.

An inverted version of the messenger tagging technique for detecting ion absorption uses the fact that electronic and/or vibrational excitation of ions hinders formation of weakly-bound clusters. This effect, explored years ago in relation to laser isotope separation [82], has recently been demonstrated for spectroscopy of N_2^+ ions, cooled to 10.6 K in a 22-pole trap by collisions with He and termed laser-induced inhibition of cluster growth (LIICG) [83]. An electronic spectrum is generated by monitoring the reduction of the steady-state concentration of ion–He complexes as a function of the excitation laser wavenumber.

A variant of the laser photofragmentation approach can be used for detecting spectroscopic transitions in ions doped into He nanodroplets. Light absorption by an embedded ion followed by electronic or vibrational relaxation results in the evaporation of He atoms from the host droplets, shifting their overall size distribution to lower masses. This shift can be detected by mass-spectrometric means and used for detecting UV and/or IR absorption [84]. The intrinsic non-zero background of this technique arising from the initial distribution of droplet sizes limits its sensitivity, especially for IR spectroscopy. Another related approach specific to He nanodroplets uses the fact that, under certain conditions, excitation of embedded ions may result in their ejection from the host droplet [85]. The number of ejected ions, recorded as a function of the wavenumber of an excitation laser, provides a measure of the wavelength and intensity of the transitions of the embedded ion. This zero-background technique has the potential for high sensitivity, although its applicability to large, biological molecules remains unclear.

An important variant of the photofragmentation approach for measuring a spectrum of biomolecular ions relies on the absorption of a UV photon, which, after electronic relaxation processes, leads to the breaking of a strong chemical bond (as opposed to the weak bonds discussed above for ion-molecule complexes). Monitoring the appearance of charged fragments that have m/z ratios different from that of the parent ion as a function of the photolysis UV laser wavelength generates a photodissociation spectrum. While this type of action spectroscopy accurately reflects the frequency of electronic transitions of the parent ions and has high sensitivity because of its background-free nature, the dissociation yield depends on both the absorption intensity of the parent ions as well as their dissociation probability upon electronic excitation. The latter may not always be a flat function of the UV excitation energy, which can result in a distortion of the spectral intensities. In general, all peptides and proteins absorb in the near VUV (<200 nm) at the N-CO peptide bonds. Couplings between electronic states at this high excitation energy tend to shorten the lifetimes of the optically prepared states, washing out the vibrational structure of the electronic bands [86]. The aromatic side chains of the amino acids tryptophan, tyrosine, and phenylalanine all absorb in the near-UV, and the longer lifetime of the first excited states in these residues allow the attainment of vibrational resolution in the electronic spectra of cold peptides containing one of these chromophores.

Regardless of whether one can obtain vibrational resolution in the electronic spectrum, UV photofragmentation detection of ions also allows one to obtain IR spectra. This technique, called IR-UV double resonance, requires the use of two pulsed laser sources (IR and UV) and is based on the change in UV absorption when the parent molecules are pre-excited via an infrared transition [19]. This change is particularly enhanced in the case of cryogenically cooled molecules, which all reside in the vibrational ground state prior to the excitation. Vibrational excitation in a large species such as a biomolecule is fully delocalized because of the astronomic density of vibrational states. Electronic transitions originating from the mixed vibrational states of such a pre-heated molecule are broadened. This broadening arises from differences between vibrational frequencies in the electronic ground and excited states and/or reduced Frank–Condon (FC) factors for UV transitions [19]. It can be considered as "statistical" inhomogeneous broadening, and it is fundamentally different from thermal congestion, which arises from a Boltzmann distribution of vibrational energy. There are two ways to employ this broadening for IR spectroscopy in a double resonance approach [19]. In one of them, termed "gain" spectroscopy, the wavenumber of a UV probe laser is set slightly outside a strong absorption band in the UV photofragmentation spectrum of cold ions. Whenever the preceding pulse of a tunable IR laser excites a vibrational transition, the UV absorption band broadens, overlapping the wavenumber of the UV probe laser, resulting in an increased photofragmentation yield. A vibrational absorption spectrum is recorded by monitoring this increase as a function of the IR laser wavenumber. The wavenumber of the UV laser for the optimum signal-to-noise ratio is a trade-off between minimizing the dissociation yield by the UV laser only (which requires tuning away from the UV band of the cold ions) and maximizing the yield upon IR photon absorption (tuning closer to the UV absorption band). When a UV spectrum is broad with no sharp edges (e.g., lifetime broadening), the UV probe laser can be set on the slope of the UV absorption band to detect the additional broadening caused by IR pre-heating. In such a case one should expect lower detection sensitivity because of a significant amount of dissociation from the UV laser alone.

A second way to perform IR-UV double resonance, termed "depletion" or "hole-burning" spectroscopy, requires vibrationally-resolved electronic spectra. In this case the UV probe laser is set to the maximum of a well-resolved vibronic band. When in resonance, the preceding IR laser pulse removes a fraction of population from the ground state to an upper vibrational level such that the subsequent UV pulse sees fewer molecules remaining in the ground state. Because of the broadening discussed above, the absorption cross-section of vibrationally excited molecules at the fixed wavenumber of UV probe laser is reduced, and the resulting reduction of UV-induced photofragmentation appears as dips as the IR laser is tuned in resonance with vibrational transitions. An infrared depletion spectrum is generated by monitoring the difference in the fragmentation yield between alternate experimental cycles with the IR laser "on" and "off." One important advantage of this scheme is that it employs a common ground state for the IR and UV transitions. This requirement implies that both transitions have to belong to the same conformer of the parent ion. This constraint makes IR-UV depletion spectroscopy conformer-selective, which helps eliminate sources of spectral complexity. Because spectral complexity becomes a limiting factor in the spectroscopy of large biomolecular ions, the use of such spectral simplification approaches becomes essential. It is worth noting that such double resonance schemes are not limited to the combination of IR and UV excitation or to photofragment spectroscopy. Similar approaches using IR–IR [87], UV–UV [88–91], and even IR-IR-UV schemes [92] serve as powerful means to simplify inhomogeneously broadened spectra.

A final "action spectroscopy" considered here, which also falls into the category of photofragment spectroscopy, employs IR multiple photon dissociation (IRMPD) to measure vibrational spectra of biomolecular ions [93]. Because it is a single laser technique and doesn't require that the target molecule have a UV absorption band, it is applicable to a wide range of biomolecules. However, the cleavage of a chemical bond by IR excitation requires the absorption of a large number of photons, and this has certain consequences for the resulting spectra. Multiple-photon excitation intrinsically broadens IR transitions, because the absorption frequency of a vibrational mode usually decreases upon increase of vibrational energy in a molecule because of anharmonic couplings of molecular vibrations [94], and this loss of resolution may inhibit distinguishing nearby vibrational bands, even in cold biomolecules. Also, because of its intrinsic non-linearity, spectra measured via IRMPD do not accurately reflect relative intensities of vibrational transitions. On the practical side, IRMPD spectroscopy requires a powerful laser source that is widely tunable in the range of infrared absorption by biomolecules (>2.7 μm). At present, infrared free-electron lasers (IRFEL) seem to be ideal for IRMPD spectroscopy of large species, although this requires performing experiments on instruments constructed at IRFEL facilities or transporting one's instrument to such a facility, which is not very convenient. However, the recent demonstration of IRMPD with high brightness, continuous infrared OPO holds out promise for this approach to be more widely accessible [95].

Since IR-UV double resonance and IRMPD are the major techniques currently employed for measuring infrared spectra of uncomplexed biomolecules, it is worth comparing the relative merits of the two approaches. The IR-UV double resonance approach has the advantages of being accessible with table-top lasers, works perfectly well with cold molecules, is potentially conformer-selective, and produces spectra that accurately reflect the intensities and even the absolute cross sections of infrared transitions [19]. Its disadvantages are that it requires the molecule under study to have an aromatic amino acid to serve as a UV chromophore, and even when such a chromophore exists, fast excited-state dynamics can lead to broad UV spectra and prevent conformer selectivity. Moreover, it seems that the UV spectra of closed-shell negative ions are systematically broad because of fast electron photodetachment, which competes with fragmentation. The IRMPD approach, on the other hand, has the advantage of being perfectly general, as all molecules absorb IR radiation, and it can be applied to both cations and anions. Its disadvantages are its inherent lack of conformer selectivity, uncertain spectral intensities, and lower spectral resolution. Moreover, it is not clear how easily this technique works on cold molecules, as the sharper spectra may lead to bottlenecks in the first few steps of the multiphoton pumping process.

For implementation of photofragment spectroscopy in which strong chemical bonds are broken, either in an IR-UV double resonance scheme or by IRMPD, as the size of the molecule increases one expects that the unimolecular dissociation rate will decrease, since the excitation energy is distributed among a larger number of vibrational degrees-of-freedom. This can lead to the situation in which the molecule doesn't fragment during the time in which an experiment naturally occurs (often determined by laser repetition rates). One way that this situation can be circumvented, at least for molecules in a certain size range, is to assist the dissociation process by the further excitation of an excited species via IRMPD with a CO_2 laser. This approach was first employed by Yeh et al. for vibrational spectroscopy of protonated water clusters [96] and shortly thereafter by our group for vibrational overtone spectroscopy [97, 98]. We have recently demonstrated that a similar approach can be used to enhance the dissociation yield of protonated peptides in a cold ion trap subsequent to UV excitation [99]. Maitre and coworkers have more recently demonstrated that IRMPD with a CO_2 laser can enhance the dissociation yield of peptides subsequent to IR excitation with an OPO [100, 101]. This approach, which we have called Infrared Laser Assisted Photofragment Spectroscopy (IRLAPS), will certainly play an important role as spectroscopic techniques are pushed to biological molecules of increasing size.

2.5 Some Specific Implementations of the Above Techniques

The techniques for ion trapping, cooling, detection, and spectroscopic simplification described above have been combined in various ways in different laboratories around the world. In this section we provide an overview of these approaches in a more-or-less historical fashion to give the reader an idea of how these different techniques can be employed to obtain spectroscopic information on biomolecular ions. Although the earliest applications of several of these methods were to non-biological systems, some of them were later picked up and applied in modified forms to the study of biomolecular ions, and thus we include a description of them here. We also include approaches that have the potential for application to biomolecular ions but have not yet been used for such a purpose. We do not include, however, techniques for spectroscopy of cold ions that do not have the potential to be applied to large bimolecular ions, such as laser cooling or sympathetic cooling. We also do not include supersonic molecular beam studies of neutral biological molecules, since these are treated in a separate chapter of the present volume.

In the early 1990s, Gerlich and co-workers began applying cooled, radiofrequency ion traps for laboratory studies of interstellar chemistry [102–104]. While the general design of their instrument was similar to the tandem multi-

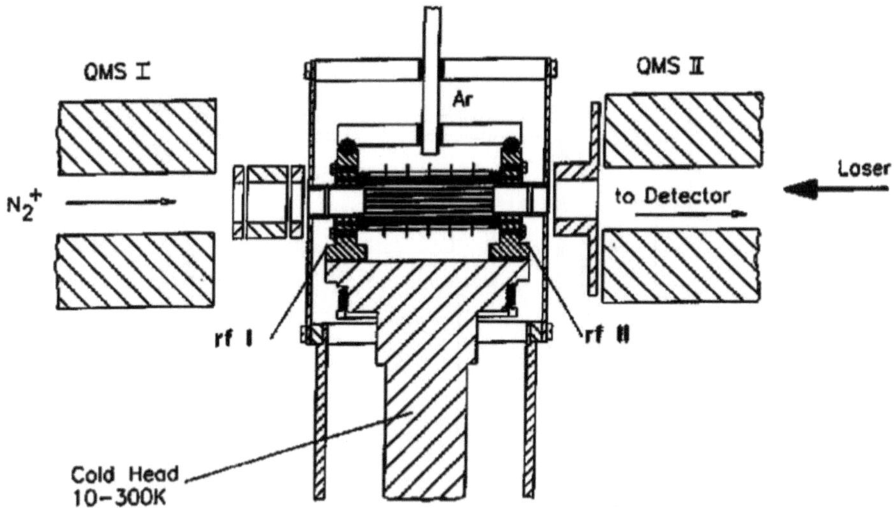

Fig. 5 Cold 22-pole ion trap machine employed by Gerlich and coworkers for spectroscopic studies of cold N_2^+ ions [70]. Quadrupoles are used for parent mass selection (QMS I) and product analysis (QMS II). The ion trap is cooled by a closed cycle helium refrigerator and helium is used a buffer gas. Copyright (1999), with permission from Elsevier

pole devices proposed years earlier [105], and the use of RF devices as ion guides [106] and traps [107] had already been demonstrated, the innovative aspect of their work was to incorporate a cryogenic 22-pole ion trap in which the ions were cooled by collisions with cold helium (Fig. 5). Although their apparatus was used primarily for measuring rate constants of ion-molecule reactions at low temperatures [108], they found that the enhancement of chemical reaction rates of laser-excited ions could be used to detect photon absorption and hence map out a spectrum (a technique that they called laser-induced reaction (LIR)). Their first spectroscopic applications of their apparatus were in measuring the Meinel band in N_2^+ [70] and the rotationally resolved antisymmetric CH stretch band in $C_2H_2^+$ [71]. While they studied only small molecular ions produced by electron-impact ionization using this approach, the general technique of using a cold, RF multi-pole ion trap in a tandem mass spectrometer served as the inspiration for applications to more complex molecules.

One such application following shortly thereafter was by Asmis et al. in which they measured the vibrational spectrum of vanadium oxide clusters [109]. Their instrument, which was very much in the spirit of that of Gerlich and coworkers, placed a cooled RF-hexadecapole trap between two mass-selecting quadrupoles (Fig. 6). Because static quadrupole benders are used on both sides of the ion trap, the machine provides particularly good spectroscopic access to the cold ions. Vibrational spectra of the buffer-gas cooled ions were monitored by IRMPD photofragment spectroscopy, detecting fragment ions passing through the analyzing quadrupole as a function of the frequency of an infrared free electron laser. They

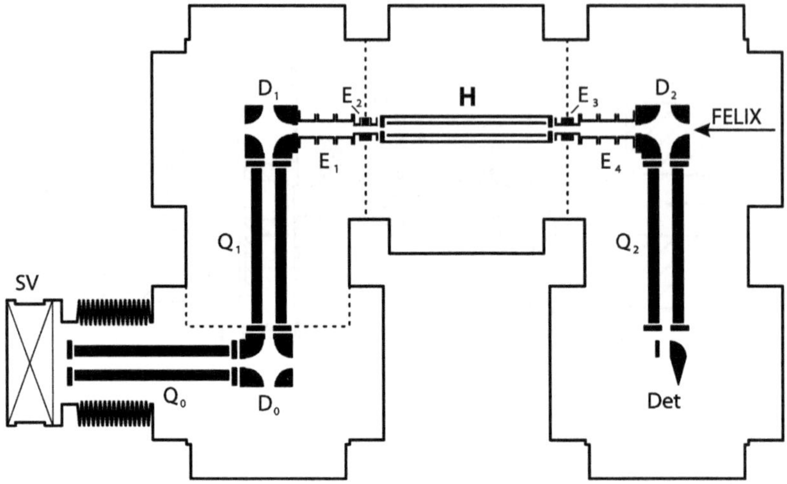

Fig. 6 Schematic of the ion trap tandem mass spectrometer employed by Asmis and coworkers [109]. The cold trap (H) is cooled by a closed cycle refrigerator and uses helium for buffer-gas cooling of the trapped ions. FELIX is the name of the infrared free electron laser

profited from both the high fluence and the wide tunability afforded by the IRFEL for measuring the IRMPD spectrum.

Both Asmis and coworkers [110] and Gerlich and coworkers [111] went on to apply their cold, multi-pole ion traps to the study of protonated water clusters. The latter produced both protonated and deprotonated water clusters by a supersonic expansion of corona-discharged H_2/H_2O mixtures and trapped them in their 22-pole ion trap cooled to liquid nitrogen temperature [111]. They focused on clusters containing between three and ten water molecules, measuring vibrational spectra via photofragment spectroscopy of the weakly-bound clusters using an infrared OPO. In this case, because the clusters are bound non-covalently, single-photon absorption is enough to remove one water molecule and lead to spectroscopic detection.

Asmis and coworkers, on the other hand, focused on the water dimer, producing it by electrospray and cooling it in their cold, RF hexadecapole ion trap, and they employed photofragment spectroscopy using the IRFEL to measure spectra [110]. The use of electrospray for producing closed-shell gas-phase ions for spectroscopic studies had first been performed by Wang and Wang [112] in their photoelectron spectroscopic study of transition metal complexes and then by Andersen and coworkers for electronic spectroscopy of fluorescent protein chromophores [34, 113, 114] and McLafferty and co-workers, who were the first to study infrared spectroscopy of gas-phase peptide ions [115]. The work of Asmis, however, was the first to combine electrospray ion production with buffer-gas cooling explicitly for spectroscopic studies, although they did not apply it to biological molecules.

The first group to put these pieces together – electrospray production of biomolecular ions, buffer-gas cooling, followed by trapping and ion spectroscopy – was that of Weinkauf [116]. They modified a commercial mass spectrometer to allow cooling of a Paul trap to liquid nitrogen temperatures, and they measured the electronic spectrum of cold, protonated tryptophan in the gas phase, detecting the absorption by monitoring photofragments [116]. While the spectrum they obtained showed some structure, the spectral features were orders of magnitude broader than expected, leading one to question the cooling efficiency in their trap. It was later discovered that the spectrum of protonated tryptophan is inherently broad because of fast excited state dynamics [46, 78]. Nevertheless, Weinkauf's experiments served as a major stimulus to study electrosprayed biomolecular ions in cold ion traps.

In a parallel effort, Wang and co-workers added a cold Paul trap to their electrospray ion source and used it for photoelectron spectroscopy of buffer-gas cooled anions [117, 118]. In this case, spectroscopic detection is achieved by collecting the photodetached electrons. While the trap could be cooled to 10 K, it is difficult to determine absolute ion temperatures because photoelectron spectra are inherently broader than IR or UV absorption spectra, preventing the resolution of low frequency hot bands. Although their focus was primarily on non-biological cluster ions of various types, elements of their machine design were later applied to the spectroscopy of biological ions [119].

In 2006, three separate groups [45, 46, 120] reported the development of cold-ion spectrometers based on the Gerlich design [108]. Maier and co-workers took the basic Gerlich design with an electron impact ion source and employed it for measuring spectra of astrophysically important molecules, using single- or two-photon photofragment spectroscopy. They clearly demonstrated cooling of both rotational and vibrational degrees of freedom in a number of small, open-shell molecular ions such as N_2O^+ [120], paradichlorobenzene [120], and 2,4-hexadiyne [121], reaching temperatures on the order of 20 K, as determined by rotational analysis of the rovibronic bands. Wester and coworkers also constructed a cold-ion spectrometer with a cryogenic 22-pole ion trap, which they used for measuring photodetachment cross-sections of OH anions [45].

In the same year, our group in Lausanne published first results from a similar instrument which was equipped with an electrospray ion source for producing closed-shell biomolecular ions, the first demonstrations of which were the measurement of the UV spectra of cold, protonated aromatic amino acids, tryptophan [46], tyrosine [46, 122], and phenylalanine [122]. Spectroscopic detection is achieved by measuring the small percentage of parent ions that fragment subsequent to UV absorption. The internal temperature of the ions was estimated to be 11–16 K from an analysis of the intensity of hot band transitions of low frequency vibrational modes. If the temperatures achieved in buffer-gas cooled ion traps are low enough and the spectra sufficiently simple, one can often resolve UV absorption spectra for different stable conformers of the molecule [122]. In this case, one can use the IR-UV double resonance techniques so profitably employed in supersonic molecular beam studies [91, 123–128] to measure conformer-specific infrared spectra, and this was applied by Stearns et al. to both individual amino acids [129] as well as peptides with up to 12 amino acid residues [130]. Subsequent improvements to the Lausanne machine (Fig. 7) included the addition of an ion funnel to

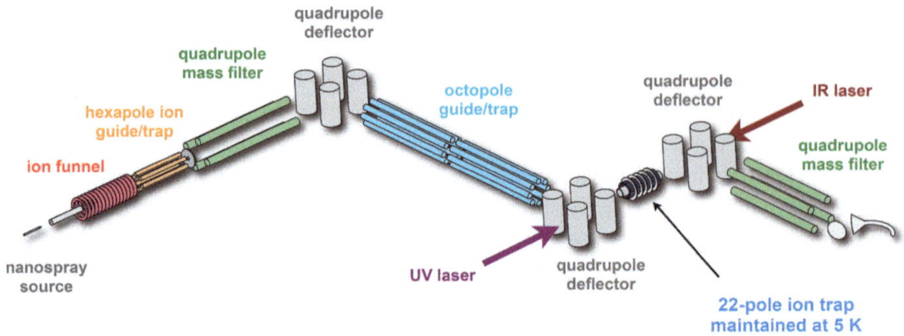

Fig. 7 The second generation Lausanne cold-ion photofragment spectrometer with a 22-pole ion trap [131]

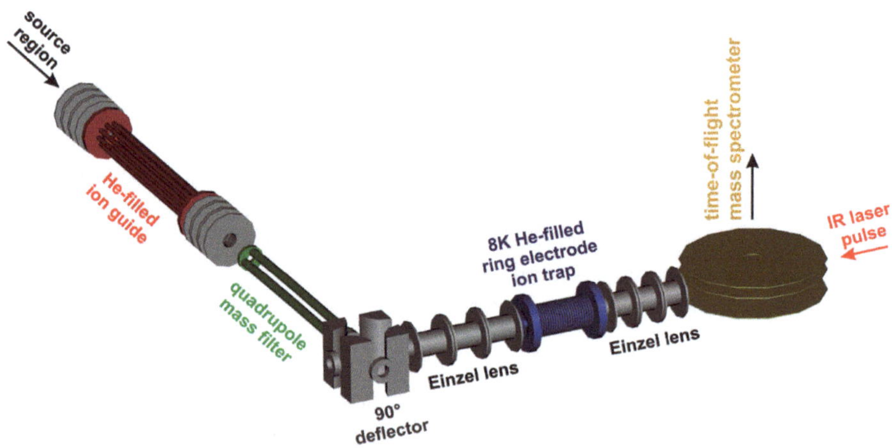

Fig. 8 Second-generation ion trap machine from the Asmis group [134]. The ion trap in this case is a ring-electrode trap, and fragment mass analysis is performed in a time-of-flight mass spectrometer. Reproduced with permission from [134]. Copyright (2009) ACS

inject the electrosprayed ions more efficiently [131]. An instrument of similar design was more recently constructed in the group of Fuke and used to measure spectra of tryptophan-containing dipeptides [132] and their clusters with solvent [133].

Several groups have innovated on the basic cold trap ion spectrometer designs mentioned above. Although they did not apply it to biomolecular ions, Asmis and coworkers modified the original design of their tandem mass spectrometer ion trap machine by replacing the RF hexadecapole with a cooled ring electrode trap (Fig. 8), originally pioneered by Gerlich [103], which allows them to extract both parent and photofragment ions from the trap and focus them into the extraction region of a time-of-flight mass spectrometer (TOFMS) for detection [134]. Substituting the TOFMS in place of the quadrupole for photofragment

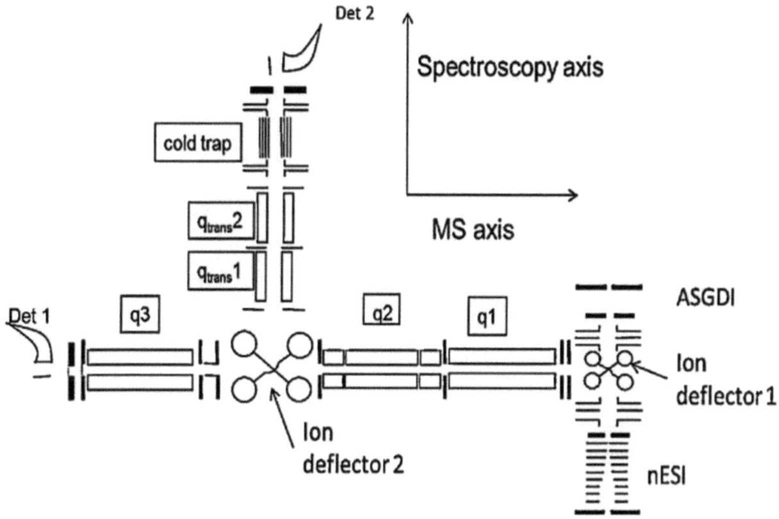

Fig. 9 Schematic of the Purdue 22-pole ion trap machine. Reproduced with permission from [18]. Copyright (2013), with permission from Elsevier

analysis allows them to monitor the entire mass spectrum each laser shot, significantly increasing their sensitivity. For detection they employed the messenger tagging technique, which was first demonstrated by Okumura et al. [72].

Along similar lines, Choi et al. combined a cold Paul trap with a TOFMS, first cooling it to 150 K [135] and later to 10 K [136]. While coupling a quadrupole ion trap with TOFMS was not new, having first been demonstrated by Lubman and coworkers in the early 1990s [137], Choi et al. were the first to do this with a cold ion trap and use it for photofragment spectroscopy [138]. Using protonated tyrosine as a benchmark [46], they estimated the internal temperature of their ions to be on the order of 50 K.

A recent collaboration between the Zwier and McLuckey groups at Purdue resulted in several innovative additions to cold ion trap spectroscopy (Fig. 9) [18]. By combining a cold, 22-pole ion trap with a triple quadrupole mass spectrometer in a T-shape design, they take full benefit of the linear quadrupoles by using them as traps, both for mass selection and mass analysis, with the latter providing some of the same advantages as a TOFMS [138]. Using protonated tyrosine as a benchmark [46], they estimated the internal temperature of their ions to be on the order of 10 K.

Johnson and coworkers have recently made a significant development in the technology of cold ion spectroscopy by combining an electrospray ion source and cryogenically cooled Paul trap with a tandem TOF mass spectrometer (Fig. 10) [139, 140]. They use the Paul trap both to cool the molecular ions and to form clusters with deuterium. They then eject a distribution of clusters from the trap and use the first stage of the tandem TOF to select one particular species before exciting it with an infrared photon and monitoring the absorption by fragmentation of the

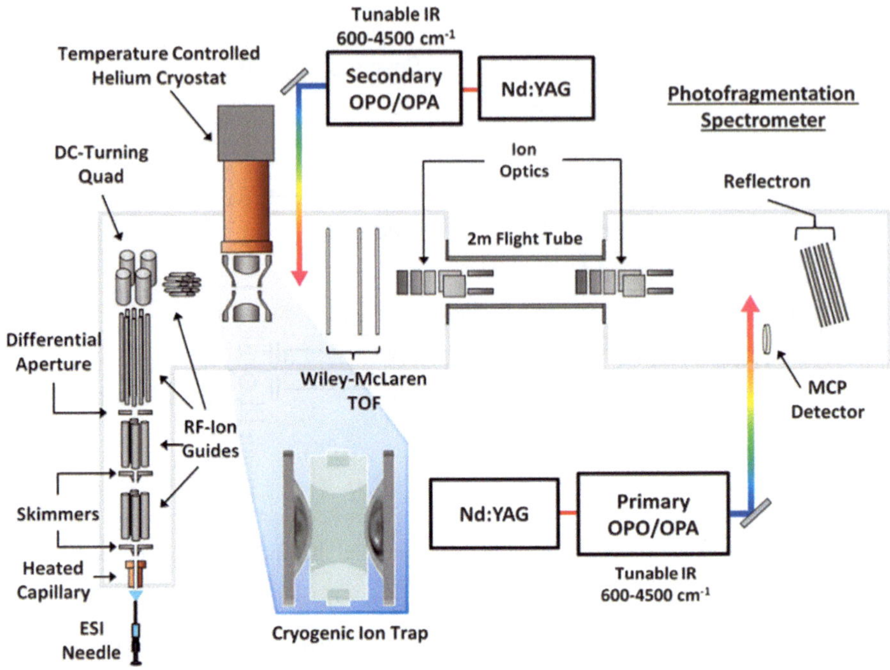

Fig. 10 The Yale cold ion spectrometer, combining a cold Paul trap with a tandem TOF mass spectrometer. Reproduced with permission from [141]. Copyright (2014) ACS

complex, which is detected in the second TOF stage. As discussed more fully in Sect. 3.1, when combined with isotope editing, this hydrogen tagging approach is proving to be an extremely powerful means to measure the spectra of cold, biomolecular ions.

As one pushes cryogenic ion spectroscopy to biological molecules of increasing size, one additional source of spectral complexity that impedes spectral analysis is the presence of multiple conformations in the ion trap. The overlap of spectra from different conformations inhibits comparison with theory, since calculations are performed on single conformations. To compare theory with experiment, one must propose a linear combination of calculated spectra of individual conformations with unknown relative populations. Given the spectral complexity of a single conformation, such a linear combination becomes far from unique. While IR-UV and IR-IR double-resonance spectroscopy in cold ion traps can help separate the spectra of different conformers, at a certain level of complexity such methods are not enough. One possibility to help deal with this problem is to separate different conformations of the ions physically before injecting them into an ion trap, and one approach to doing this is to combine ion mobility with cold ion spectroscopy. In our laboratory we have recently combined Field Asymmetric Ion Mobility Spectrometry (FAIMS) with a cold, 22-pole ion trap to perform conformer pre-selection prior to ion trapping and UV photofragment spectroscopy [24, 142]. Since conformer overlap is a major

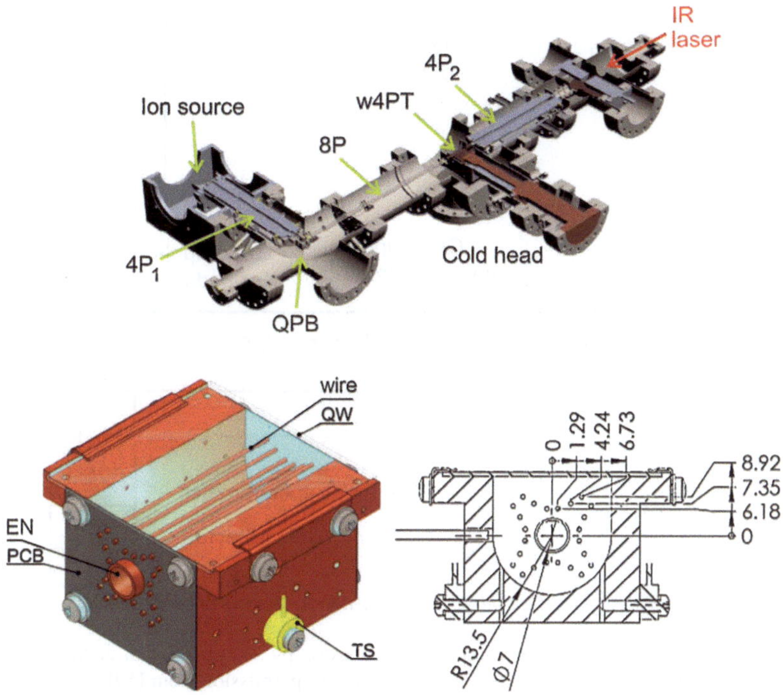

Fig. 11 Wire quadrupole ion trap constructed in the Roithova group in Prague. Reproduced with permission from [61]. Copyright (2013), Elsevier

impediment to extending spectroscopic techniques to larger molecules, this subject is discussed further in Sects. 3.2 and 4.

Returning to where much of the impetus for buffer-gas cooling of ions began, Dieter Gerlich, in collaboration with the Roithova group in Prague, has recently developed a new wire-based quadrupole ion trap, which they inserted into a commercial tandem mass spectrometer (Fig. 11) and cooled to 3.8 K [61]. The use of wires placed in such a fashion as to approximate a quadrupole field results in an open design that in principle lends itself both to detection by fluorescence emission as well as by tagging and photofragment spectroscopy, the latter using the capabilities of the tandem mass spectrometer. The first application of this new machine has been to measuring the infrared spectrum of cold, $C_7H_6^{2+}$, detecting the absorption of the IR radiation by helium tagging and predissociation spectroscopy [61]. While not applied to biological molecules, if used with an electrospray source, such a machine would be well suited for such studies.

In a completely different approach to all the above-mentioned studies, von Helden and coworkers have combined ion trap technology with superfluid helium nanodroplets to measure spectra of cold biomolecular ions [59]. As illustrated schematically in Fig. 12, after producing gas-phase biomolecules via electrospray, they mass select them, bend them 90° with a static quadrupole deflector, and then trap them in a room temperature ion trap. A pulsed, helium droplet source produces

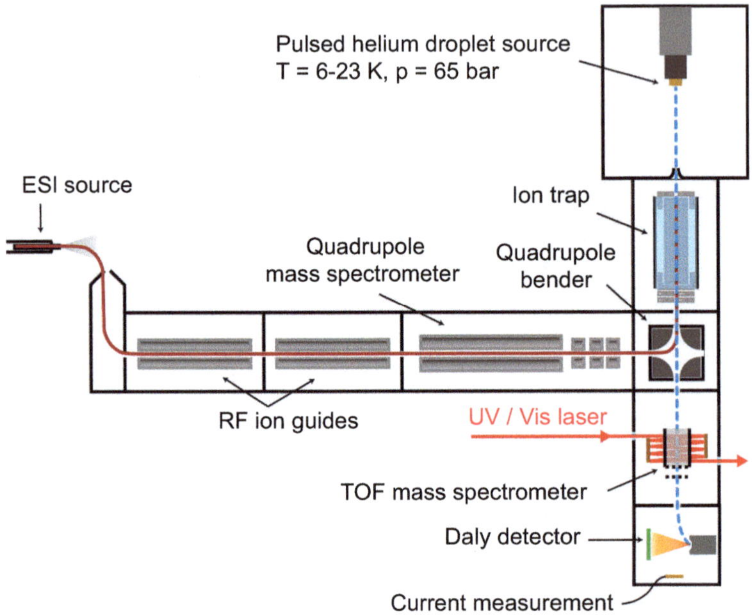

Fig. 12 Helium droplet machine for cold-ion spectroscopy, constructed in at the Fritz Haber Institute by von Helden and coworkers. Reproduced with permission from [59]

droplets which traverse the trap and pick up warm ions, cooling them as they carry them out of the trap and into a region of the machine in which spectroscopic detection is performed. Spectra are measured by detecting the change in droplet size (in a TOF MS) caused by helium boil-off as the IR or UV excitation is deposited into the droplet. The great advantage of this approach is the low temperature (0.37 K) achievable in the superfluid helium droplets, although one must consider the potential influence of the helium environment on the molecule.

Before we discuss selected results from a few of these different approaches in more detail, it might be helpful to consider first the issue of ion temperature – how it is measured and what the limiting factors are in achieving the lowest temperature.

2.6 *Determination of the Ion Vibrational Temperature*

From the point of view of structure determination via vibrational spectroscopy, the goal of cryogenic cooling is to achieve vibrational temperatures sufficiently low such that hot bands from low frequency vibrational modes are not populated and thus do not complicate the spectra. As discussed in Sect. 1.2, for vibrational modes with frequencies on the order of 20 cm^{-1}, a vibrational temperature of 12 K would reduce the population of molecules in this state to 10% of the ground state. But how does one measure the internal temperature of a large molecule?

A rough estimate of the ion temperature can be obtained by the observation of weakly-bound molecular clusters. If the binding energy between an ion and an adduct is known, then the equilibrium population of the cluster relative to the unclustered molecule provides an estimate of the temperature. For the lowest temperature this requires weakly-bound complexes, and the most sensitive measure would be to use complexes between ions and helium. Binding energies of helium to polyatomic ions can be calculated or estimated from measurements made on simpler ions. For example, when helium binds to an amino acid that is charged on the N-terminus, it is almost certain it binds to the ammonium group, and in this case an estimate of the binding energy can be obtained from that of NH^{4+}• He [143]. Sometimes the observation of ion-helium clusters in a trap is simply used as a qualitative indication that the ions are cold, without providing a precise temperature determination. For example, Jasik et al. [61] observed that the hydrocarbon dication $C_7H_6^{2+}$ can be clustered with as many as five helium atoms in their wire-based quadrupole ion trap cooled to 4 K.

Spectroscopic measurements are able to provide a much more accurate determination of ion temperature. If we consider molecules cooled by a cold buffer gas, one can assume that, given a sufficient number of collisions, the translational, rotational, and vibrational degrees of freedom are in thermal equilibrium. In this case, a temperature measurement corresponding to any one of these degrees of freedom is indicative of the vibrational temperature. Measurement of the Doppler profile of an isolated rovibronic transition can be used to determine the translational temperature [83, 144]. Chakrabarty et al. [83] employed both the LIICG and LICT techniques (see Sect. 2.4) in a 22-pole ion trap cooled to 5 K to measure the Doppler profile of cold N_2^+ ions, and from this they calculate a rotational temperature of 10.6 K. This approach is very difficult for larger molecular ions, however, since the small rotational constants would require both extremely low temperatures and extremely high spectral resolution to observe single rovibrational transitions.

An alternative spectroscopic approach would be to determine the rotational temperature from the intensities in a rotationally resolved vibrational or electronic spectrum. In this case, even if one cannot resolve individual rovibrational transitions, one can still estimate the temperature by simulating the rotational contour of an individual vibronic band [120]. For this, one needs to know the rotational constants of the molecule and the direction of the transition moment; however, even rough estimates of these quantities can lead to a reasonable temperature estimate. In measuring either Doppler widths or rotational band contours, the linewidth one obtains may contain a contribution from the finite lifetime of the molecule, determined by its intramolecular vibrational energy redistribution and/or dissociation rate if some type of photofragment spectroscopy is used, and this can make the temperature appear to be higher than it really is.

Another approach is to determine the vibrational temperature by measuring the intensities of low-frequency hot-bands. Following our initial work [46], a few groups have used the hot-band intensities in the UV photofragment spectrum of cold, protonated tyrosine to estimate the degree of internal cooling of their respective cold traps [18, 138], making this molecule a benchmark for comparing

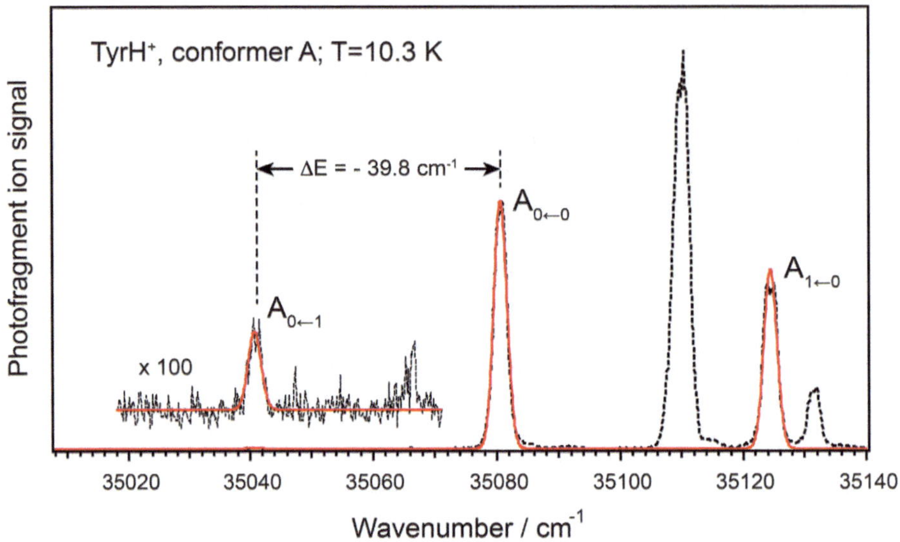

Fig. 13 Hot band transition in conformer A of protonated tyrosine used to determine the vibrational temperature. This spectrum was measured in a cold octupole ion trap, although almost identical results were obtained with our 22-pole ion trap

biomolecular ion temperature. This approach is illustrated schematically in Fig. 13 using the electronic spectrum of a particular conformer (A) of protonated tyrosine. One can determine the energy of a low frequency vibrational mode by observing the shift of the hot-band transition ($\mathrm{A}_{0\leftarrow 1}$) from the band origin ($A_{0\leftarrow 0}$), which, in this case, is 39.8 cm^{-1}. To be able to determine the temperature, one needs to know the Franck–Condon factor of the hot-band transition relative to the band origin. As a general rule of thumb, the Franck–Condon factor for such a transition should be similar to that of the vibrational band ($A_{1\leftarrow 0}$), which originates from the same state as the band origin.

In this case the temperature is given by:

$$T = \frac{-\Delta E}{k \times \ln\left(\frac{I_{A_{0\leftarrow 1}}}{I_{A_{1\leftarrow 0}}}\right)} = \frac{-\ 39.8\ \mathrm{cm}^{-1}}{\left(0.693\,\mathrm{cm}^{-1}\mathrm{K}^{-1}\right) \times \ln(0.0038)} = 10.3\ \mathrm{K}.$$

Using this procedure, Redwine et al. obtained an upper limit for the temperature of 10–16 K for their 22-pole ion trap, consistent with what we measured in our 22-pole trap [46] as well as in a newly constructed cold octupole trap. In contrast, the 3-D quadrupole ion trap employed by Choi et al. attained temperatures of 45–54 K for protonated tyrosine [138], which is likely to reflect RF heating of the ions.

Heating by the RF trapping field, which is more problematic for lower-order multipole traps, can result in the situation in which the translational temperature of the ions is different from that of the cold buffer gas. As Gerlich pointed out [20], if there is a sufficient number of collisions with the buffer gas, then it is the "collisional temperature" that determines the internal temperature of the ions and not their translational temperature in the laboratory frame. Under these conditions, the ion internal temperature is determined by the mass-weighted collision temperature,

$$T = \frac{(m_1 T_2 + m_2 T_1)}{(m_1 + m_2)},$$

where subscripts 1 and 2 refer to the buffer gas and the ion. This means that one can cool the internal degrees of freedom of a large ion with a light buffer gas, even if the ions are translationally warm. For example, with helium as buffer gas cooled to 5 K, if a biomolecular ion of mass 500 has a translational temperature of 150 K, the internal temperature of the ions should be able to reach a temperature of 6.2 K. Our recent experiments in a cold octupole ion trap confirm this [47].

3 Examples of Cryogenic Spectroscopy of Biomolecular Ions

While cryogenic ion spectroscopy has been used extensively on small molecular ions, its application to biomolecular ions is still in its infancy. Several groups now have operating cold ion trap machines which they are applying to the spectroscopy of biomolecules, and the next few years should witness steep growth in the field. In this section we highlight some recent results from two different laboratories, the group of Mark Johnson at Yale and our own group at EPFL in Lausanne, which should give the reader a good overview of the capabilities of cryogenic spectroscopy of biomolecular ions.

3.1 Results from the Yale cold-ion spectrometer

3.1.1 H_2 Tagging and Vibrational Predissociation Spectra of Singly Charged Ions

As discussed in Sect. 2.4, in the absence of an electronic chromophore, vibrational predissociation of a weakly attached messenger or "tag" species provides a universal method for obtaining linear action spectra of cold molecules. Because single photon photodissociation requires the dissociation energy of the tag to be less than the photon energy, best results are obtained for tags such as rare gases and molecular hydrogen. Wang and Wang first demonstrated cooling [118] and H_2 tagging [145] of ESI ions injected into a cryogenically cooled Paul trap, but their first reports concentrated on doubly charged ions because of the apparent difficulty generating the more weakly bound complexes of H_2 with singly charged species [145]. By simply introducing a pulsed buffer gas and delayed extraction strategy,

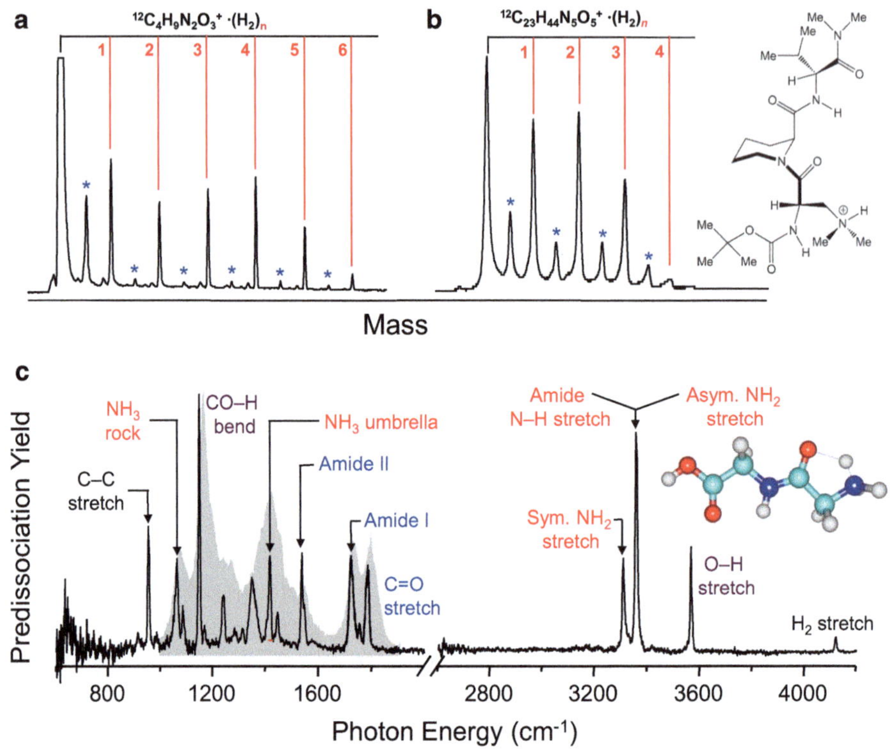

Fig. 14 *Top panel*: Time-of-flight mass spectra illustrating formation of H_2 adducts to ESI generated ions in a 10 K ion trap (**a**) the dipeptide ($GlyGlyH^+$) and (**b**) the tripeptide indicated in the *inset*. *Bottom panel*: Vibrational predissociation spectrum of H_2-tagged $GlyGlyH^+$ (*solid black line*). The gray overlay in trace (**c**) is the previously reported IRMPD spectrum [146] of the bare ion at 300 K. *Stars* in top panel indicate presence of naturally occurring ^{13}C isotopologues

the Yale group was able to generalize the method for application to essentially any charged target [140]. The performance of the source is illustrated for protonated peptides in Fig. 14, where the cluster distribution is optimized for production of the first few adducts. The fact that these weakly-bound clusters form is clear evidence that the ions must be cold. Note that the tripeptide clusters displayed in Fig. 14b are typical in that most of the ion distribution can be shifted to the tagged regime.

To carry out vibrational spectroscopy, the Yale team interfaced the Paul trap tagging source to a time-of-flight based double- (or triple-) focusing photo-fragmentation mass spectrometer (Fig. 10), which yields spectra by predissociation of the weakly-bound H_2 mass "tags". Comparison of the spectrum of H_2-tagged H^+GlyGly in Fig. 14c with the IRMPD spectrum of the same molecule [146] establishes that much sharper features are resolved when working with a single-photon action approach applied to cold ions. The dependence of the number of photo-evaporated tags on the photon energy establishes that the binding energies of the H_2 molecules are in the range 350–500 cm^{-1}. The H_2 stretching band is weak but evident in the spectra (see the highest energy peak in Fig. 14c) with shifts from the bare molecule of about 50 cm^{-1}. This is consistent with a vibration-induced charge-transfer mechanism for the activation of the IR forbidden fundamental in bare H_2. The stretching fundamentals of the NH groups directly contacting the H_2 ligands display shifts on the order of 10 cm^{-1}, while the transitions arising from more remote domains of the molecular ion do not exhibit a discernible shift. Through careful control of the trap temperature, several other molecular tags have been employed, such as N_2 and CO_2, which condense at 25 and 100 K, respectively. This capability is especially useful in the application of ion spectroscopy to study intermediates in the activation of small molecules by organometallic catalysts [147].

3.1.2 Band Assignments Using Isotope-Edited Spectroscopy

After a series of demonstration experiments on protonated dipeptide derivatives [148], the Yale group carried out an intensive spectroscopic study of the synthetic tripeptide displayed in Fig. 14b, which is a bio-inspired stereoselective bromination catalyst [149]. The vibrational predissociation spectrum of the H_2-tagged ion is presented in Fig. 15a, highlighting the region of the amide I and II stretches in addition to the NH stretching fundamentals. Interestingly, the amide I region displays two closely spaced doublets, strongly suggesting that these are derived from the four C=O groups in the molecule. Of course, there is always the possibility that such multiplets could also be caused by the formation of different conformers in the ion source and/or tag-induced shifts. Even in the case where these split features are attributed to a single conformer, it is not evident a priori the degree to which the oscillators are coupled such that transitions reflect collective delocalized motions rather than the frequency of a single oscillator. All these issues have been addressed experimentally, however, to achieve a unique assignment scheme and identify the folded structure and the intramolecular contacts that support it.

The issue of multiple conformer contributions was resolved using an IR-IR hole-burning scheme in which one laser probes the population in the species resonant at the probe laser energy while another powerful IR laser intercepts the same ion packet upstream and is scanned throughout the entire spectrum [17]. Because resonant excitation by both lasers is destructive, any time the scanned pump laser excites one of the transitions associated with the isomer that is monitored by the probe, the probe predissociation signal is reduced. Consequently, the entire spectrum of one conformer is recorded as a series of "dips" in the fixed frequency probe signal. Because both lasers yield the same fragment ions, however, it is essential to introduce an intermediate stage of mass selection between the pump and probe, and hence the method is commonly referred to as an IR^2MS^3 approach. Although three conformers were observed when this method was applied to the dipeptide H^+SarSar [17], the tripeptide catalyst appeared in a single isomeric form, with the only

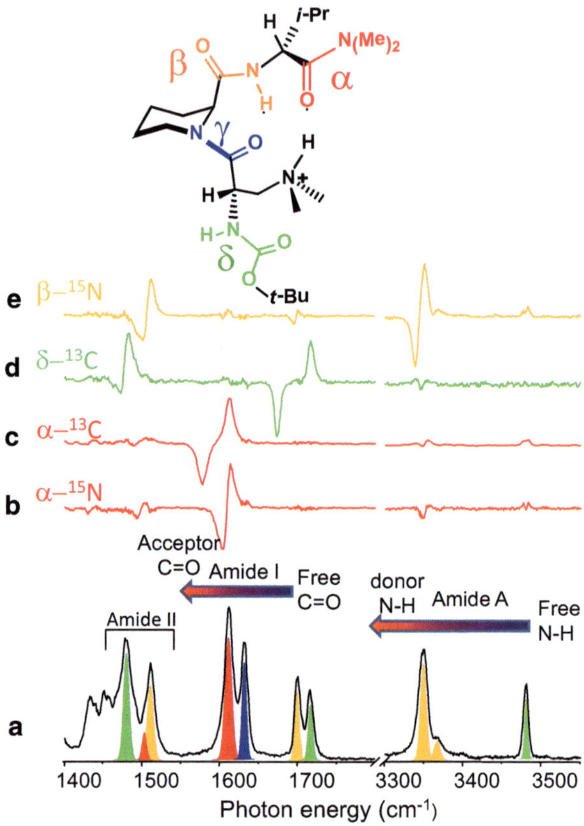

Fig. 15 (**a**) IR D_2-predissociation spectrum of the tripeptide indicated at the *top* with all ^{12}C and ^{14}N. The spectra with heavy isotopes at the indicated positions were subtracted from trace (**a**) to obtain the color-coded difference spectra (**b-e**) in the four upper traces. The color coding of the traces and peaks corresponds to the functional groups as indicated in the *top schematic structure*. Reproduced from [149]

evidence for a minor additional species being the asymmetric NH stretching doublet at 3,200 cm^{-1} [149].

The band assignments for the single conformer were then carried out using a variation of the site-specific isotopic labeling method described by Stearns et al. [150]. Heavy isotopes of carbon and nitrogen were introduced at the centers as indicated by the color-coded structure at the top of Fig. 15. Spectra of each isotopologue were then obtained and subtracted from the one corresponding to the compound with all light, naturally abundant isotopes to reveal features that either change intensity or are shifted when the mass changes at a particular site. Four such difference spectra are presented in Fig. 15, where positive peaks indicate transitions that disappear upon heavy isotope incorporation and the negative peaks record the locations of new features not present in the light isotopologue. This selective identification of bands whose position or intensity depends on the displacement of a particular atom is commonly used in the assignment of protein spectra and is typically referred to as “isotope edited” spectroscopy. One of the main conclusions from the behavior of the difference spectra obtained from the tripeptide catalyst is that the amide I bands respond individually to the mass changes along the backbone, indicating that they are largely uncoupled. For example, trace c (red) establishes that only one of the four bands responds, and the resulting shift is quantitatively that expected from the reduced mass of an isolated C=O diatomic oscillator (35 vs 36 cm^{-1}, respectively). Interestingly, one can also change the mass of the N atom connected to the carbon of the same C=O, which results in a much smaller shift of the same transition, thus revealing the varying degrees to which surrounding atoms contribute to the normal mode responsible for the transition. Note that assignment of the NH bands is particularly convenient when the mass change occurs at the heavy N atom rather than the more usual H/D substitution, since the shifts are so small that the derivative lineshapes immediately identify the shifted features (yellow trace e in Fig. 15).

The fact that the amide I bands in the tripeptide catalyst are largely decoupled is very useful in the identification of the intramolecular H-bonding configuration, since red-shifted C=O and NH stretching transitions are associated with the linkages. As such, the locations of the local bond frequencies revealed by isotope-edited spectroscopy directly encode the local environments of the bonds. Although this alone can yield a clear picture of the contacts driving the fold in simple cases where the options are limited by proximity of plausible complementary H-bonding targets, the fact that one can essentially "color code" the observed bands with specific bonds provides a severe constraint on the theoretical prediction of the spectral pattern by forcing candidate structures to recover the isotope shifts at many key sites in the molecules. A particularly exciting future direction for this method is to apply it to more symmetrical H-bonding arrangements, where one expects the amide I region to reflect strong excitonic coupling between the C=O oscillators. In this case, the coupling of isotope editing with cold, conformer-specific vibrational spectroscopy promises to provide a direct means to establish both the coupling matrix elements as well as the unperturbed locations of local oscillators.

3.2 *Results from the EPFL Cold Ion Spectrometer*

While the Yale group has focused on the technique of H_2 tagging spectroscopy, our work at EPFL has up to now employed UV photofragmentation and IR-UV double resonance for detecting spectroscopic transitions in cold biomolecular ions. Early experiments demonstrated the degree of spectral simplification that occurs upon cooling and focused on the spectroscopy of individual amino acids [46, 122], dipeptides [129], and the effects of clustering water molecules onto simple species [78]. We then turned to slightly more complex peptides, trying to determine spectroscopic signatures of secondary structures. For this we examined peptides of the form $AcFA_nKH^+$, which, on the basis of ion mobility studies, are expected to form helical structures in the gas-phase [16, 130]. Using IR-UV double resonance together with nitrogen-15 labeling of specific peptides, we were able to assign spectra of four conformers of $AcFA_kKH^+$ present in our ion trap and demonstrate that they do indeed form helices in the gas phase.

This early work was reviewed in 2009, and we refer the reader to that review [16], as well as the original publications cited above. Our work since that time has been focused on pushing spectroscopic techniques to larger, more complex biological molecules. This involves developing a tight connection between experiment and theory, developing techniques to deal with increasing spectral complexity, and understanding the correlation of the structures of large molecules produced by electrospray and those in solution. We address some of these issues in the sections below.

3.2.1 Gramicidin S: A Case Study in Spectroscopic Structure Determination

As an example of spectroscopic structure determination of a moderately complex biomolecular ion, we describe here recent results from our laboratory on gramicidin S (GS), a ten amino acid cyclic peptide (cyclo-VOLFPVOLFP, where "O" designates ornithine and Phe is the D- rather than the L-enantiomer). Gramicidin S is a natural antibiotic that serves as a template for the design of new drugs [151]. Under physiological conditions, GS is doubly protonated, which is the pharmacologically active form. NMR and X-ray analyses have revealed an anti-parallel β-sheet structure in solution [152] and in crystals [153], respectively. This highly stable β-sheet makes it a benchmark for studying such structures in proteins. Although the fully desolvated peptide in the gas phase may adopt different three-dimensional conformations than in solution, the structure of the isolated molecule provides a stringent test of theoretical approaches to structure calculations.

For a large molecule such as gramicidin, the measured spectra must provide a sufficient number of constraints to serve as a rigorous test of theory. This implies obtaining at least vibrational resolution in the IR and/or UV spectra, which for molecules of this size can only be achieved if they are cooled to low temperatures. De Vries and co-workers [154] measured an infrared spectrum of neutral GS in the amide *A* region, using jet-cooling of laser-desorbed neutrals. The cold gas-phase molecules were then interrogated by double-resonance IR-UV spectroscopy combined with the resonance-enhanced two-photon ionization (R2PI). Because of the limited number of collisions in their supersonic expansion, they were able to achieve only partial vibrational resolution in the IR spectrum, which was not sufficient for unambiguous comparison with structural calculations. Von Helden and co-workers [155] measured IR spectra of doubly protonated GS and its deuterated analog in the amide I and II regions, employing IRMPD of room-temperature ions by a free-electron laser. While the spectra were used to validate structures calculated for a few low-energy conformers, the low spectral resolution in these experiments limited the number of the derived spectroscopic constraints with which to test theoretical predictions. This emphasizes the importance of high vibrational resolution as a prerequisite for structural determination of large species.

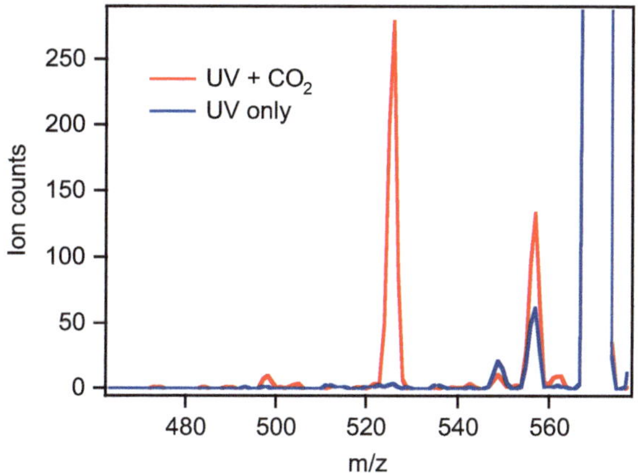

Fig. 16 Photofragmentation mass spectra of $[GS + 2H]^{2+}$ ($m/z = 571.36$ Da) measured with the UV laser only (*blue trace*) and with subsequent CO_2 laser excitation (*red trace*). The UV laser wavenumber is fixed at 37,723.3 cm^{-1} (the band origin of conformer *A*, see Fig. 19). Two of the most abundant fragments in the UV photofragmentation mass-spectrum correspond to loss of CO and CO-NH neutral fragments. The two most abundant fragments in the UV + CO_2 mass-spectrum correspond to the loss of a CO group ($m/z = 557.3$ Da) and to cleavage of the C_α–C_β bond of a phenylalanine residue ($m/z = 525.8$ Da)

Our experiments combine cryogenic cooling in a 22-pole ion trap and high-resolution table-top lasers for measuring vibrationally-resolved IR and UV spectra of GS [15]. The protonated gas-phase peptides are generated from solution using a nanospray ion source, mass-selected, and stored in a 22-pole ion trap kept at 6 K, where they are cooled by collisions with helium. The trapped ions are then interrogated by UV photofragmentation or by photofragment-detected IR-UV double-resonance. In addition to this, some experiments use a pulsed TEA CO_2 laser, which fires 200 ns after the UV laser, to increase the dissociation yield after UV excitation, as described at the end of Sect. 2.4.

Figure 16 presents photofragmentation mass-spectra of doubly protonated GS, measured with UV excitation only and with UV excitation followed by CO_2 laser excitation. Two – the most abundant fragments in the "UV-only" mass-spectrum – correspond to the loss of CO and of CONH. These fragmentation channels result from prompt dissociation of $[GS + 2H]^{2+}$ on an electronically excited potential energy surface. In addition to the loss of CO, IRMPD of the UV pre-excited ions greatly enhances the dissociation channel that corresponds to the loss of the Phe side-chain. This phenomenon, observed in the case of many Phe- and Tyr-containing peptides, is believed to originate from intersystem crossing of the UV-excited peptides to a long-lived triplet state [156, 157]. The long triplet lifetime makes the dissociation of the UV-excited ions by a subsequent CO_2 laser pulse highly efficient, even several milliseconds after UV excitation [156, 157].

Figure 17 shows electronic spectra of the doubly-protonated GS measured by monitoring fragments with $m/z = 557$ as a function of the UV laser wavenumber. The spectra were obtained at the lowest trap temperature of 6 K (Fig. 17a) and under

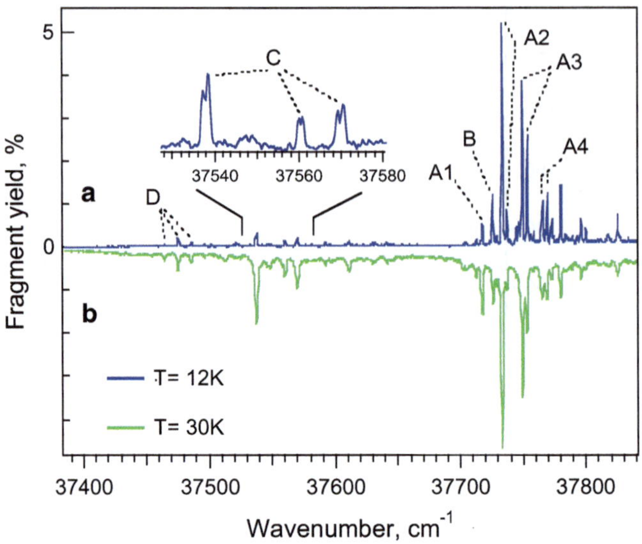

Fig. 17 Electronic spectra of doubly-protonated GS cooled to (**a**) ~12 K, and (**b**) ~30 K. The *insert* shows details of the three peaks labeled C. Reproduced with permission from [158]. Copyright (2010) ACS

slightly warmer conditions (Fig. 17b). The transitions in these spectra have been assigned to three different conformers, designated A, B, and C, and the corresponding transitions in Fig. 17 are labeled by the same letters and enumerated. Intensities of different peaks change differently upon the increase of temperature. This observation helps in their assignment to hot bands or to different conformers. For instance, the relative intensity of only the first peak of the progression (A1) increases with increasing temperature, suggesting that A1 is a hot band of this progression and the next peak, A2, is its band origin. The hot band allows us to determine vibrational temperature of the ions, which is ~12 K.

The intensities of the peaks labeled C and D grow relative to those of conformer A with increasing temperature, indicating that they correspond to two higher-energy conformers, and, as described below, this is confirmed by conformer-specific infrared spectra. The transitions A2–A4 are each split by 3.9 cm^{-1}, while the peaks labeled C are split by ~1 cm^{-1}. Both splittings originate from exciton coupling between the two Phe chromophores. A detailed analysis suggests that the difference between the splittings in A and C arise from a larger spacing between chromophores in conformer C and/or a difference in their relative orientations. The position of the band origin of conformer A in the electronic spectrum is significantly shifted to a higher energy with respect to that of neutral, gas phase Phe (37,535 cm^{-1}) [159] and of several protonated helices where the charge is remote from Phe chromophore [16]. This large blue-shift is caused by the close proximity of the phenyl groups to the charges, which are located on the side-chains of the ornithine residues. Experiments in which solvation of the charges with crown ether

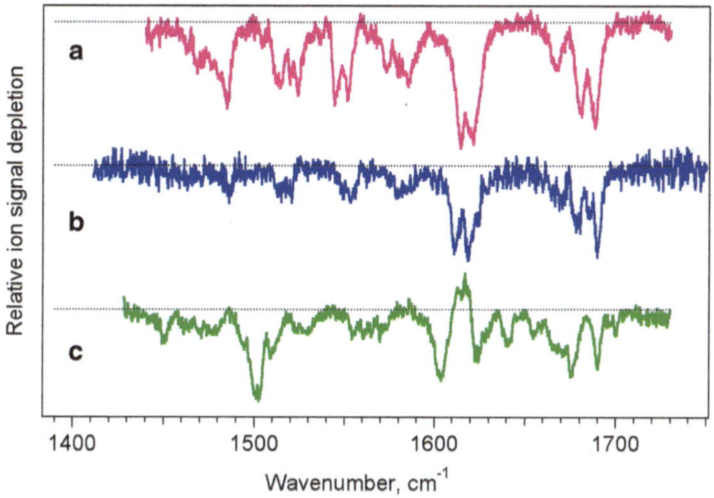

Fig. 18 IR-UV depletion spectra of doubly protonated GS: (**a**) conformer A; (**b**) conformer B; (**c**) conformer C; measured by fixing UV laser wavenumber on their respective peaks A1–A4, B and C (the largest peak only) in Fig. 18

molecules (18-crown-6) removes this shift confirm our assessment that it arises from a cation-π interaction. Moreover, the appearance of a single, split electronic band for conformer A indicates that both phenyl groups interact with a charged ornithine side-chain in an identical manner, implying a symmetrical structure. This structural constraint, confirmed by IR spectroscopy of conformer A, was later used in pre-selection of calculated structures prior to high-level optimizations.

Figure 18 shows IR-UV depletion spectra of $[GS+2H]^{2+}$ measured by fixing the UV laser wavenumber to transitions assigned to conformers A, B, and C in the spectrum of Fig. 17a. Each IR spectrum reflects the specific vibrational frequencies associated with the different conformers giving rise to the probed UV transitions. Several IR bands in the spectrum of conformer A appear to be doubly degenerate. For example, the five peaks lying between 1,600 and 1,700 cm^{-1} belong to the ten bands that have primarily C=O stretch character. This double degeneracy implies that pairs of oscillators are situated in equivalent positions in the molecule, and since GS is cyclic, this conformer must have C_2 symmetry [158].

Several IR transitions belonging to conformer A have been assigned using isotopic substitution. For example, isotopic shifts caused by substitution of ^{14}N by ^{15}N in Val and Leu permit distinguishing their NH-stretches from those in Phe and Orn (Fig. 19a, b) [15]. Moreover, NH-stretch transitions of NH_3^+ have been identified by complexing $[GS+2H]^{2+}$ with two crown ether molecules (18-crown-6), which form particularly strong hydrogen bonds with the ammonium groups. As shown in Fig. 19c, d, binding of the crown ether shifts two peaks near 3,240 cm^{-1} by 120–150 cm^{-1}, suggesting their assignment to NH_3^+ groups. Finally, deuteration of the exchangeable protons shifts frequencies of NH-stretch and NH-bend vibrations to the red (compare Fig. 20a, b). This allows an unambiguous assignment of

Fig. 19 Infrared spectra of conformer A of $[GS+2H]^{2+}$: (**a**, **c**) the natural isotopologue, (**b**) isotopologues with ^{15}N substituted for the Leu and Val residues, and (**d**) $[GS+2H]^{2+}$/[18]crown-6 complex. Reprinted from [15]. Copyright (2011), with permission from John Wiley and Sons

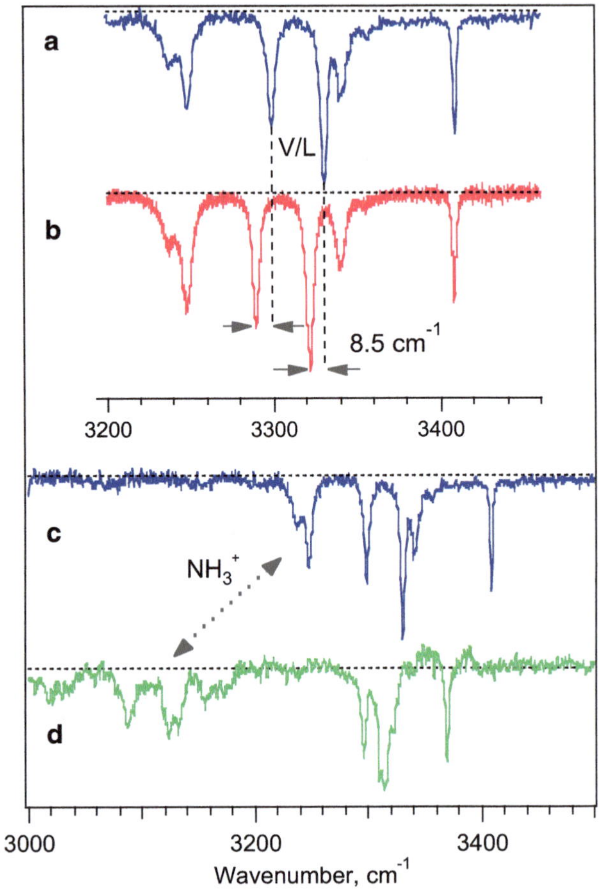

the remaining peaks to CH- and CO-stretch vibrations. All these assignments, together with the symmetry inferred from the degenerate bands, provide constraints for validating calculated structures. Figure 20 summarizes the IR spectroscopy of $[GS+2H]^{2+}$ conformer A and compares the experimental data with the calculated spectra for the lowest energy structure of $[GS+2H]^{2+}$ and its deuterated analog (see [15] for details of the calculations). Because deuteration induces a significant change in IR spectra but no change of peptide structure, a comparison of the measured and the calculated spectra for the deuterated ions constitutes an additional test for calculations. The calculated and the measured spectra demonstrate a remarkably good match for both isotopologues of doubly protonated GS, validating the calculated structure of the most stable conformer of this species.

Figure 21 compares the calculated structure of the isolated molecule, validated by cold ion spectroscopy, with that of the condensed-phase structure, solved by X-ray diffraction. The structures are similar, each containing a characteristic anti-parallel beta-sheet motif. The near-symmetrical (C2) structure of the isolated,

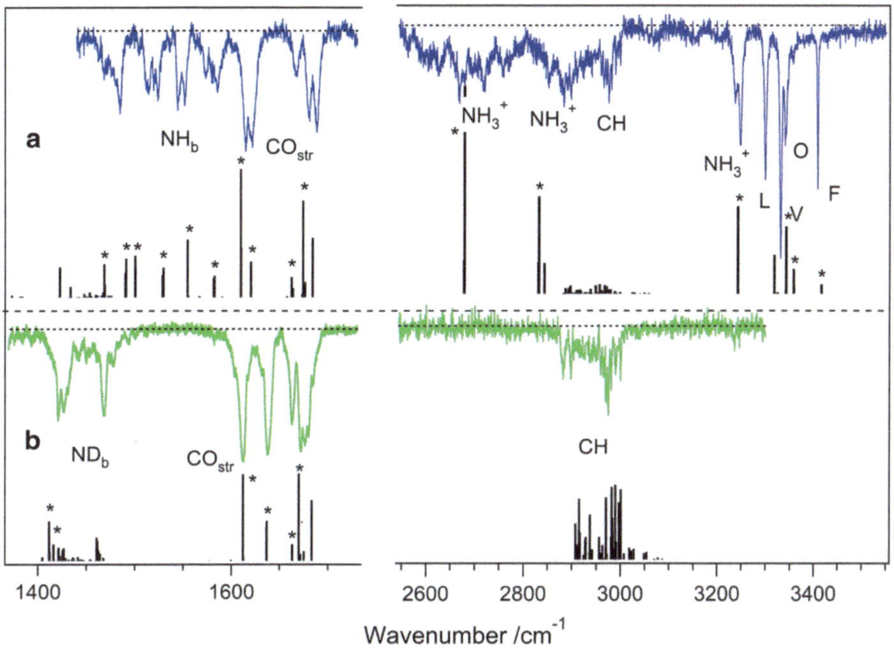

Fig. 20 Infrared spectra of conformer A of (**a**) $[GS+2H]^{2+}$ and (**b**) its deuterated analogue, together with the corresponding calculated vibrational spectra for the most stable calculated structure of these species. Reprinted from [15]. Copyright (2011), with permission from John Wiley and Sons

gas-phase peptide appears more compact. It exhibits a parallel alignment of the two Phe rings, each of which is in close proximity to an ammonium group of an Orn side chain. This difference largely results from solvation of the charged Orn side chains in the crystal, which prevents their participation in cation-π hydrogen bonds with the Phe rings. In the gas-phase structure, the ammonium groups also make hydrogen bonds with the carbonyl oxygens of the Phe and Orn residues, which anchor them to the peptide backbone.

While the structures of isolated biomolecules provide benchmark standards for structural calculations, biomolecules naturally reside in water solutions, and their functionality in vivo is largely determined by the structures they adopt upon solvation. Interactions of biomolecules with the surrounding bulk water are difficult to study on an atomic level in liquids because of the inhomogeneity of these interactions and their dynamical nature. Instead, one can do this by studying peptide-water clusters of controlled stoichiometry, prepared in the gas phase. Application of this approach to gramicidin S, for which the structures have been solved both in the gas phase and in solution, offers the unique opportunity to investigate how the former evolves to the latter upon gradual solvation.

While electrospray ionization has been designed primarily for mass-spectrometric measurements, which require full desolvation of the analyte ions,

a b

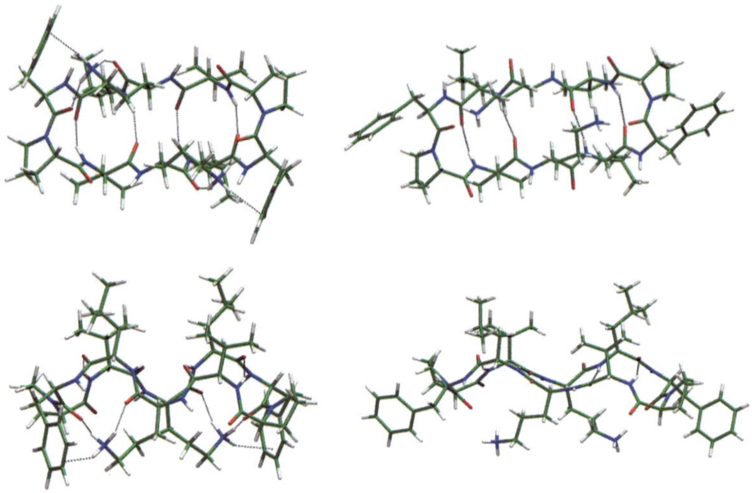

Fig. 21 Three-dimensional views of the $[GS+2H]^{2+}$ structures: (**a**) determined by cold-ion spectroscopy for the lowest-energy conformer of the gas-phase species, and (**b**) obtained by X-ray diffraction for crystallized species (reconstructed from the data of [8]). Reprinted from [15]. Copyright (2011), with permission from John Wiley and Sons

by sampling the electrospray plume under the most gentle conditions, one can generate a significant number of water-gramicidin S clusters for spectroscopic investigation. Clusters with a specific number of water molecules can be selected by a quadrupole mass-filter and then isolated and cooled in a 22-pole ion trap. Figure 22 shows sample electronic spectra of bare $[GS+2H]^{2+}$ and its complexes with up to 50 water molecules. Detection of the spectroscopic transitions is achieved by photofragment spectroscopy, since subsequent to absorption of a UV photon the clusters boil off up to 15 water molecules.

The first important result of this data is the observation of vibrationally resolved bands in the spectra of complexes with up to 30 water molecules. Even without calculations, the measured UV spectra provide some qualitative understanding of how the structure of the peptide changes upon solvation. Binding of the first two water molecules to $[GS+2H]^{2+}$ leads to a ~100 cm^{-1} red-shift of the UV band origin of the clusters towards that of the isolated phenylalanine chromophore, while further hydration shows little additional influence. In the bare peptide, the ~200 cm^{-1} blue shift of the UV band origin relative to that in neutral Phe results from a strong NH_3^+–π interaction. The shift halfway back upon solvation by two water molecules implies their binding to protonated NH_3^+ groups, which weakens the cation–π interaction and leads to decoupling of the two aromatic rings from the peptide backbone. This observation suggests that addition of the first two water molecules removes the most significant difference between the gas phase and the native structures of $[GS+2H]^{2+}$ (Fig. 21), and that the subsequent solvation by

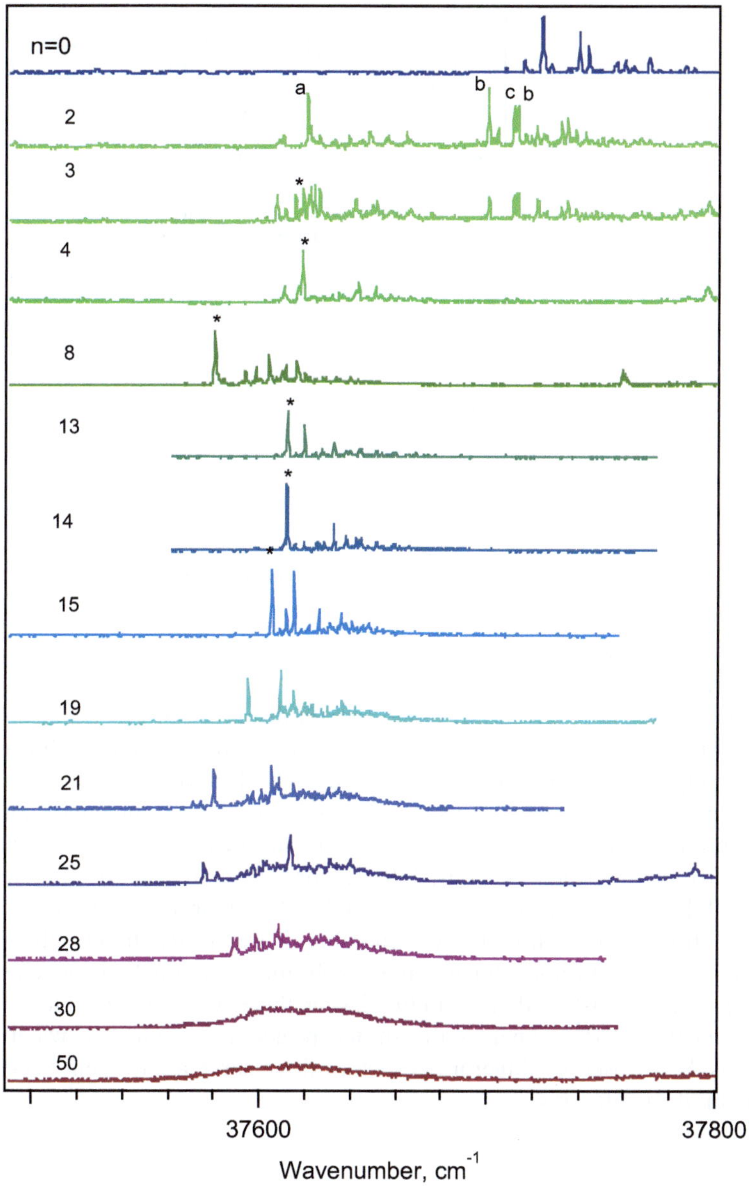

Fig. 22 Sample electronic spectra of cold $[GS + 2H]^{2+}$ and its complexes with n water molecules, $(2 \leq n \leq 50)$. Letters *a, b, c* and *asterisks* label the transitions employed for measuring conformer-selective IR spectra

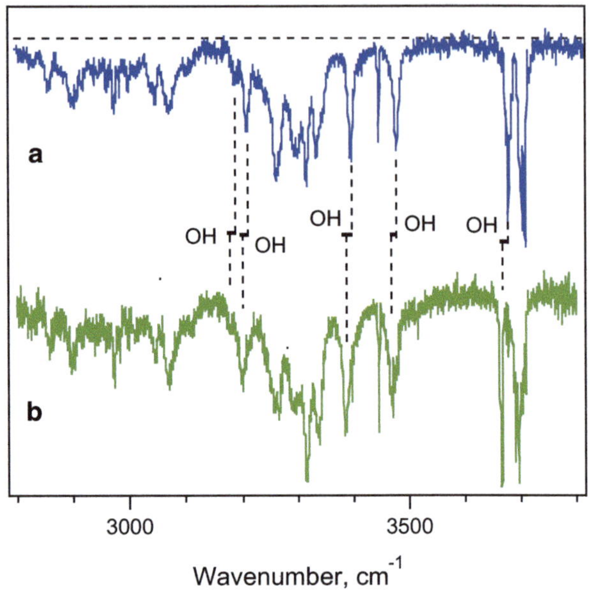

Fig. 23 IR-UV depletion spectra of the (**a**) $[GS + 2H + 8H_2\,[^{16}O]^{2+}$ and (**b**) $[GS + 2H + 8H_2\,[^{18}O]^{2+}$ isotopologue complexes

additional water molecules is less important for reshaping the peptide towards its final, native structure [79].

The vibrationally resolved electronic spectra displayed in Fig. 22 also provide the opportunity to measure conformer-specific vibrational spectra of $[GS{+}2H]^{2+}\bullet(H_2O)_n$ complexes via IR-UV double resonance [79]. Moreover, when combined with isotopic labeling of water, this method facilitates the assignment of transitions in the vibrational spectrum of the solvated biomolecules to either the parent ion or to the surrounding water network [79]. Figure 23 compares IR-UV depletion spectra of $[GS{+}2H]^{2+}$ complexed with eight $H_2{}^{16}O$ or $H_2{}^{18}O$ molecules. The isotopic substitution redshifts the wavenumber of water OH stretch vibrations by ~9.5 cm^{-1} but should not influence noticeably the vibrational frequencies of the peptide itself. The unshifted peaks in Fig. 23 can therefore be assigned to vibrations of solvated $[GS{+}2H]^{2+}$, whereas the shifted peaks are assigned to water stretch vibrations. Spectroscopic fingerprints such as these serve as stringent benchmarks for structural calculations.

3.2.2 Bradykinin – Dealing with Conformational Complexity

The preceding section on gramicidin S demonstrates how the combination of buffer-gas cooling and IR-UV double-resonance spectroscopy can simplify the spectrum of a decapeptide and allow its precise structural determination in the gas phase as well as elucidate the key structural changes upon stepwise solvation. As one pushes these techniques to larger biomolecules, the increasing number of stable conformations they can adopt increase the degree of spectral complexity,

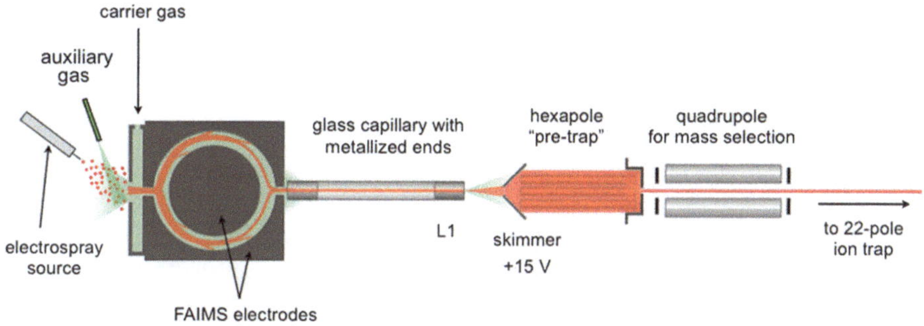

Fig. 24 Schematic of FAIMS device coupled to the front end of our ion trap photofragment spectrometer. Reprinted from [24]. Copyright (2012), with permission from Springer

making it more difficult to obtain conformer-specific infrared spectra, which is the key measurement needed to validate theoretically generated structures. Pushing spectroscopic structure determination to larger peptides thus requires dealing with the increase in conformational heterogeneity. One way this could be accomplished would be to use the mobility of a biomolecular ion in a finite pressure gas to separate conformers physically before injecting them into a cold ion trap. Indeed, ion mobility spectrometry (IMS) has demonstrated the ability to separate different conformational states of rather complex molecules [9, 160–164], and its coupling with cold ion traps could be exploited to extend the range of application of cryogenic ion spectroscopy.

Ion mobility spectrometry has been implemented in a number of different forms, and for a proof-of-principle demonstration of its coupling with cryogenic ion spectroscopy we chose to use field-asymmetric ion mobility spectrometry (FAIMS) [165–167]. FAIMS uses the difference in ion mobility in a gas at atmospheric pressure at high and low electric fields to deflect different conformational states of a molecule perpendicular to their drift direction. Figure 24 shows a schematic of a commercial FAIMS source (Thermo Fisher) coupled to the inlet of our cold-ion spectrometer [24]. The electrosprayed ions are entrained in nitrogen and drawn into the gap between an inner cylindrical electrode and an outer electrode. An asymmetric high-voltage waveform is applied between the two electrodes, causing conformers of different "differential mobility" to travel through the channel with a different radius of curvature in their trajectory. A DC potential applied across the electrode gap is used to compensate this deflection and allow a specific conformer to traverse the channel successfully and be drawn into the capillary inlet of our spectrometer. A FAIMS "spectrum" consists of plotting the ion transmission as a function of the DC compensation voltage.

We chose to investigate the doubly charged peptide bradykinin ($[bk+2H]^{2+}$) as an example of a molecule with a high degree of spectral complexity arising from multiple conformations. Although it has only nine amino acids, part of the molecule is believed to be intrinsically disordered, which should lead to a distribution of gas-phase conformers. Indeed both previous work using FAIMS [168, 169] as well as

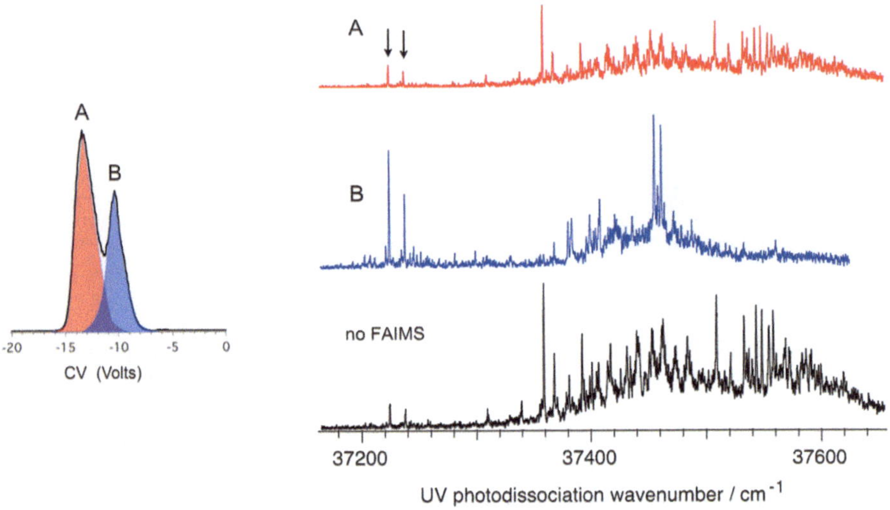

Fig. 25 *Left panel*: FAIMS spectrum of $[bk + 2H]^{2+}$, showing the total ion current as a function of the compensation voltage. The two peaks correspond to two different conformational families, A and B. *Right panel*: Electronic spectrum of $[bk + 2H]^{2+}$ with no FAIMS separation (*bottom*) and that obtained transmitting conformational family A (*top*) or B (*middle*)

recent spectroscopic studies [24, 142, 170] attest to the presence of at least five stable gas-phase conformers of $[bk+2H]^{2+}$. The key question that we sought to answer is whether conformers can be separated by their ion mobility and injected into a cold ion trap without a significant degree of interconversion.

The left panel of Fig. 25 shows a FAIMS spectrum where the transmitted ion current of $[bk+2H]^{2+}$ is plotted as a function of the compensation voltage [142]. While the resolution of our FAIMS device is not high enough to distinguish all the conformations present, it is sufficient to resolve two families of conformations transmitted at different values of the compensation voltage, which we label A and B. The top (red) and middle (blue) electronic spectra in the right hand panel of Fig. 25 are those in which only conformer family A or B is transmitted through the FAIMS device and admitted into our spectrometer, while the bottom spectrum shows the electronic spectrum of $[bk+2H]^{2+}$ without any conformer pre-selection [24]. One thing that is immediately clear is that the spectra obtained selecting either family A or B are different, and that the overall spectrum can be constructed from a weighted sum of the two. There are, however, some small peaks in the spectrum of conformer B which appear in the spectrum of conformer A (marked with arrows), indicating that there is a slight degree of interconversion among conformers during the injection process. Care must be taken to ensure that the electrosprayed ions are subjected to the gentlest conditions possible, and this is largely determined by the FAIMS separation parameters and the injection energy into our instrument. The latter can be controlled by the difference of the potential of the end of the glass capillary (L1) and the skimmer (see Fig. 24). Because the pressure in this region is

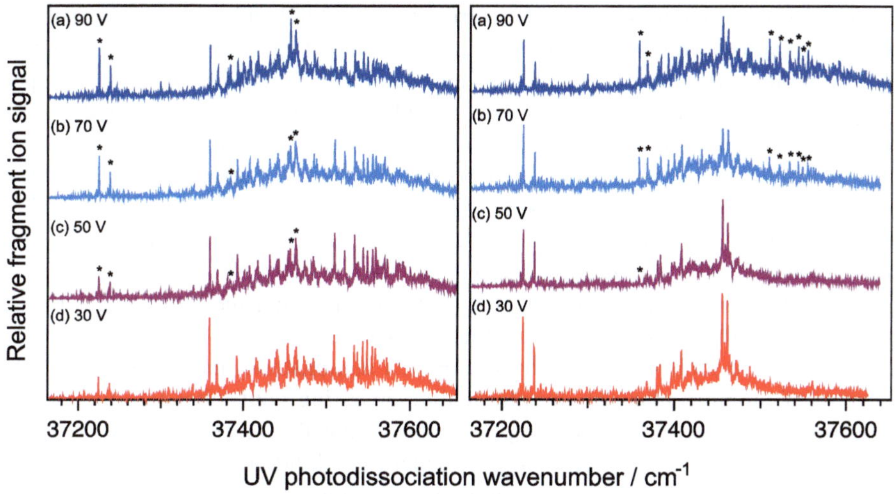

Fig. 26 Electronic photofragment excitation spectra of $[bk+2H]^{2+}$ recorded as a function of the injection voltage into the hexapole ion trap. One conformational family is initially introduced into the mass spectrometer: *left*–conformational family A; *right*–conformational family B. The *starred peaks* are those corresponding to the non-selected family, which grows in upon collisional excitation. Reprinted from [24]. Copyright (2012), with permission from Springer

on the order of 1 mbar, there are energizing collisions upon injection, and these must be made as gentle as possible. Despite the limited resolution of our FAIMS device and a small amount of conformer interconversion between conformer families, these data successfully demonstrate that it is possible to separate gas-phase conformational families, cool them in an ion trap, and measure their spectra without completely interconverting them.

In addition to demonstrating our ability to separate conformers of biomolecular ions for spectroscopic study, we can learn something about the nascent conformational distribution of the electrosprayed sample by energizing the ions upon injection. In Fig. 26d we show the electronic spectra obtained by selecting either conformer family A (left) or B (right) under gentle injection conditions. As we increase injection energy of the ions by varying the potential on L1 (i.e., spectra c–a), we warm them up before they are cooled down in our cold ion trap, and one can observe the emergence of new peaks corresponding to the conformer family that was not initially selected (indicated by asterisks). At an injection potential of 90 V, the spectra of the two conformational families appear almost identical, and they do not change significantly upon further increase of the collision energy. This implies that the injection process serves to anneal the gas-phase conformer distribution before the ions are cooled, resulting in a "quasi-equilibrium" distribution. What is most important to note is that the nascent conformer distribution produced by electrospray without any conformer selection (bottom spectrum in Fig. 25) is different from the annealed distribution. This means that the conformations initially produced in the electrospray process are "kinetically trapped" [9, 23].

The issue of kinetic trapping of electrosprayed biomolecules is an active subject of investigation with important consequences, and it may be that these trapped conformations are the most interesting to investigate. A number of different types of experiments have indicated that gas-phase molecular ions produced by electrospray retain much of the structure that they had in solution when solvent is removed [10–14]. Internal hydrogen bonds can lead to barriers in the potential energy landscape such that basic structural elements are retained for some period of time [7, 10, 24–27], allowing gas-phase experiments to be performed on species that resemble their structure in solution. Whether or not they retain their solution conformation or simply a conformation they adopt during the latter stages of the electrospray process is currently a subject of debate.

While these proof-of-principle experiments demonstrate that one can simplify spectra of biomolecular ions by physically separating conformer families before injecting them into a cold ion trap, it is clear that the FAIMS device we have employed doesn't have the resolution to separate individual conformers of doubly charged bradykinin, no less larger, more complicated biomolecules. Other FAIMS devices have achieved higher resolution by using longer ion residence times [169], higher fields [171, 172], or different separation gas [172, 173]. However, in most of these cases, the degree of heating of the ions caused by the separation fields increases, and this can cause conformer isomerization.

Drift tube ion mobility spectrometry (DT-IMS) has the advantages of achieving higher resolution using significantly lower fields, and thus is perhaps better suited for coupling with cold ion spectroscopy [9, 160–164]. For example, Clemmer and coworkers use IMS-IMS techniques to isolate different conformations of the small protein ubiquitin and demonstrate that they don't interconvert under normal conditions [27, 174–176]. However, upon activation these conformations can be made to unfold, revealing a range of intermediate states which themselves could be isolated and analyzed in subsequent IMS stages [176]. Experiments currently underway in our laboratory, in collaboration with the Clemmer group, couple a multi-stage IMS drift tube to a cold ion trap to extend the range of cryogenic ion spectroscopy experiments to more complex molecules.

In closing this section, it is perhaps worth noting that FAIMS and drift-tube IMS select molecular conformations on the basis of different, although related, physical properties. The former uses the difference in ion mobility at high field and low field to separate different conformers, while DT-IMS simply uses the difference in orientationally averaged cross sections. Indeed two-dimensional experiments combining FAIMS and DT-IMS by Shvartsburg et al. demonstrated that the combination of the two are able to identify more stable conformations than either technique alone [177]. It may be that using FAIMS at not too high field strengths in combination with DT-IMS would be the best way to preselect the conformational states of biomolecular ions before injecting into a cold ion trap for spectroscopic studies.

4 Challenges and Perspectives for Application to Larger Molecules

In concluding this chapter, it is worth considering the limits of the techniques described here in terms of the size of molecules to which they are applicable for obtaining structural information. It should be clear from the examples given that cryogenic cooling is essential for one to be able to determine the structures of moderately sized peptides with any degree of confidence. This comes from the need to measure vibrational spectra of individual conformers, since this is what provides the most critical validation of calculated structures. Without cooling, double-resonance techniques for spectral simplification simply won't work, because thermal congestion prohibits conformer selection for all but the smallest molecules. Moreover, for molecules above a certain size, conformer-sorting methods such as ion mobility may be necessary in addition to cooling to have any hope of obtaining conformer-resolved infrared spectra. Even with all the combined power of cryogenic cooling, spectroscopic simplification techniques, and conformer separation techniques, obtaining structural information for peptides of larger than ~20 amino acid residues still faces some challenges. Some of these challenges are on the experimental side, related to the methods used for spectroscopic detection; others are related to the ability of experiment to generate sufficient spectroscopic constraints to validate calculated structures, while still others concern the ability of theory to determine structures that can be validated by experiment.

Certain types of action spectroscopy described in Sect. 2.4 may encounter difficulties as one applies them to significantly larger molecules. Consider the use of UV photofragmentation as a means of spectroscopic detection as described above in the cases of gramicidin S and bradykinin. The dissociation of a large molecule in the electronic ground state during a limited time delay between UV excitation (with subsequent fast internal conversion) and fragment detection requires a significant excess of vibrational energy above the lowest dissociation threshold – a concept called "kinetic shift" in the mass spectrometry community [178–180]. For example, a small protein such as cytochrome C contains $\sim 2 \times 10^5$ cm^{-1} of vibrational energy at room temperature but remains chemically stable for quite a long time. A much higher internal energy is thus required for dissociating this protein on the timescale of a spectroscopy experiment, which is typically on the order of milliseconds. For gramicidin S this problem was overcome by detecting specific fragments arising from prompt dissociation on an electronically excited surface subsequent to UV excitation. While this seems to occur when using aromatic amino acids as chromophores [86, 99, 129, 150, 181, 182], it may not be general for all biomolecules.

One way to avoid this problem would be to use a lower energy photodissociation process for spectroscopic detection such as the tagging technique discussed in Sects. 2.4 and 3.1.1. In this case, one would have to use IR-IR double resonance instead of IR-UV double resonance for conformer selection. While this is entirely feasible for a certain size range of biological molecules and has distinct advantages

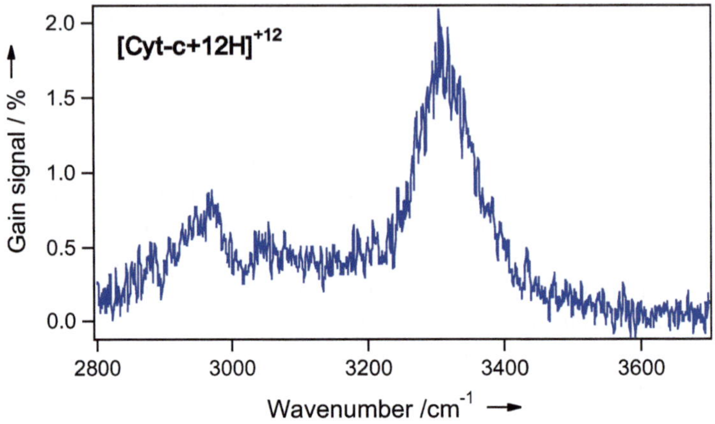

Fig. 27 IR-UV gain spectrum of $[Cyt\text{-}C + H]^{12+}$ protein cooled in a 6 K 22-pole ion trap. The UV laser is fixed at 288.75 nm, while detecting ion signal at $m/z = 1{,}022$ as a function of IR wavenumber. The plotted ion signal is normalized to the parent signal at $m/z = 1{,}033$. Reprinted from [19]. Copyright (2013), with permission from John Wiley and Sons

in cases where the molecule of interest does not have a convenient UV chromophore, it is ultimately limited by the inherent degree of spectral congestion in the vibrational spectrum.

Ultimately, if one restricts oneself to determining biomolecular structure from vibrational spectra, one is limited by the large number of overlapping, infrared-allowed transitions as the molecule of interest increases in size. This can be seen in the case of multiply-charged cytochrome C [19]. Even though its electronic spectrum remains unresolved after cooling to a low temperature, the lowest energy UV band can be used to measure an IR gain spectrum, since it broadens upon IR excitation [19], although this is not conformer-specific. Figure 27 shows such a gain spectrum for $[Cyt\text{-}C + 12H]^{12+}$ cooled in our 22-pole ion trap. The spectrum contains three bands of ca. 100 cm^{-1} width, which arise from absorption by CH-, NH-, and OH-stretch vibrations. The width of these bands results from the fact that there are more than a hundred NH oscillators alone in this protein, all of which absorb within a range of a few hundred cm^{-1}. This high density of vibrational transitions should be multiplied by the number of conformers likely to be present under the experimental conditions. Nevertheless, even a partially resolved spectrum such as this may still provide certain tests of theory if conformational heterogeneity could be reduced using ion mobility techniques.

The spectrum in Fig. 27 underlines the difficulty of using infrared spectroscopy alone for determining biomolecular structure. While each vibrational band in an infrared spectrum provides a frequency and intensity for validating calculated structures, and in some cases certain frequencies are indicative of strong interactions between specific functional groups, in general the connection with theory is not directly geometrical in nature. The vibrational spectrum provides a test of the final calculated structure but gives no guidance during the conformational search performed to find that structure. As the size of the molecule increases, the sampling of conformation space becomes increasingly difficult, and thus an infrared spectrum alone, no matter how highly resolved, does not help guide the theory to the correct neighborhood of conformational space. To extend cryogenic ion spectroscopy techniques successfully to larger molecules, one needs information to use as constraints for guiding the conformational search.

One source of such constraints could be the orientationally averaged cross section obtained from ion mobility. At the end of Sect. 3.2.2 we discussed the use of drift tube ion mobility as a conformational filter for cryogenic ion spectroscopy but didn't mention the information content obtained from IMS. The cross section determined from the drift time in such an experiment could, in principle, be used as a filter in the conformational search procedure by comparing the cross sections of candidate structures to the measured value. While this would require a fast calculation of the cross section, for use as a conformational filter this could be done with reduced accuracy.

Another type of constraint that has been little exploited in gas-phase infrared spectroscopy is related to the direction of the vibrational transition moments. In addition to being assigned a frequency and intensity, each vibrational band can be characterized by the direction of its transition moment relative to some fixed axis in the molecule. If a molecule could be oriented in space, one could in principle use polarized lasers to obtain information on the transition moment direction. If this quantity has a predicable relationship to a bond axis (for example using a bond-dipole approximation for isolated light-atom stretch vibrations), such information could be used as a geometrical constraint in the conformational search process. Indeed, Miller and coworkers demonstrated that one could perform "brute force" orientation of DNA bases in superfluid liquid helium droplets, and this allowed them to measure vibrational transition moment angles [60, 183, 184]. Large biomolecular ions, even though they have a net charge, can have extremely high dipole moments, and if one could cool them to sufficiently low temperatures and transiently orient them, one might be able to extract similar information. Having transition moment directions in addition to vibrational band frequencies and intensities could then be used to constrain the conformational search process and allow extension of cryogenic ion spectroscopy to significantly larger molecules.

Another challenge to the conformational search problem is related to the phenomenon of kinetic trapping [25]. If, when solvent is removed, large molecules tend to be kinetically trapped in conformations that are not the lowest energy in the gas phase but related to their conformation in solution, then the search process may have no way to find them if it considers only the isolated molecule. In this case, one may have to start the search process in a solvent environment and try to simulate the kinetic trapping process. Both IMS experiments [7, 10, 14, 26, 27] and spectroscopic experiments [24] have identified kinetically trapped conformations, but more quantitative information, such as their relative energetics, may be needed to guide conformational searching.

Conclusions

In this chapter we have attempted to give the reader an overview of the various techniques that can be used to measure spectra of cold, gas-phase biomolecular ions. The goal has been to provide an appreciation of the various methods for cooling and spectroscopic detection as well as an idea of how they have been combined in creative ways in different laboratories around the world to measure spectra of cold ions. It should be noted that there is nothing that restricts these techniques to biomolecular ions – in fact, applications to molecules such as cation–crown ether complexes [185–187], reaction intermediates [188], and organometallic catalysts [147] have proven to be particularly powerful.

From the specific examples cited, it should be clear that these techniques can provide detailed information on the structure of moderately sized peptides (or molecules of equivalent size). There are challenges with them to peptides with more than ~20 amino acids, for example, from both the experimental and theoretical points of view. As this field is still relatively new, one can expect creative solutions to many of these obstacles to emerge from the laboratories currently applying these techniques.

Acknowledgements We gratefully acknowledge and thank the École Polytechnique Fédérale de Lausanne and the Fond National Suisse (through grants 120065, 130579 and 140344) for their generous support of this work. The examples from our laboratory discussed in this chapter are the result of the work of many talented group members: Natalia Nagornova, Georgios Papadopoulos, and many others who developed the techniques on which their work was based.

References

1. Fenn JB, Mann M, Meng CK, Wong SF, Whitehouse CM (1989) Science 246:64
2. Tanaka K, Waki H, Ido Y, Akita S, Yoshida Y, Yoshida T, Matsuo T (1988) Rapid Commun Mass Spectrom 2:151
3. Laganowsky A, Reading E, Hopper JTS, Robinson CV (2013) Nat Protoc 8:639
4. Uetrecht C, Heck AJR (2011) Angew Chem Int Ed 50:8248
5. Clemmer DE, Jarrold MF (1997) J Mass Spectrom 32:577
6. Vonhelden G, Wyttenbach T, Bowers MT (1995) Science 267:1483
7. Shelimov KB, Clemmer DE, Hudgins RR, Jarrold MF (1997) J Am Chem Soc 119:2240
8. Hall Z, Politis A, Robinson CV (2012) Structure 20:1596
9. Bohrer BC, Mererbloom SI, Koeniger SL, Hilderbrand AE, Clemmer DE (2008) Annu Rev Anal Chem 1:293
10. Pierson NA, Chen LX, Valentine SJ, Russell DH, Clemmer DE (2011) J Am Chem Soc 133:13810
11. Florance HV, Stopford AP, Kalapothakis JM, McCullough BJ, Bretherick A, Barran PE (2011) Analyst 136:3446
12. Ruotolo BT, Giles K, Campuzano I, Sandercock AM, Bateman RH, Robinson CV (2005) Science 310:1658

13. Breuker K, McLafferty FW (2008) Proc Natl Acad Sci U S A 105:18145
14. Silveira JA, Servage KA, Gamage CM, Russell DH (2013) J Phys Chem A 117:953
15. Nagornova NS, Guglielmi M, Doemer M, Tavernelli I, Rothlisberger U, Rizzo TR, Boyarkin OV (2011) Angew Chem Int Ed 50:5383
16. Rizzo TR, Stearns JA, Boyarkin OV (2009) Int Rev Phys Chem 28:481
17. Leavitt CM, Wolk AB, Fournier JA, Kamrath MZ, Garand E, Van Stipdonk MJ, Johnson MA (2012) J Phys Chem Lett 3:1099
18. Redwine JG, Davis ZA, Burke NL, Oglesbee RA, McLuckey SA, Zwier TS (2013) Int J Mass Spectrom 348:9
19. Nagornova NS, Rizzo TR, Boyarkin OV (2013) Angew Chem Int Ed 52:6002
20. Gerlich D (2008) In: Smith IWM (ed) Low temperatures and cold molecules. World Scientific, Singapore, p 295
21. Wester R (2009) J Phys B Atom Mol Optical Phys 42
22. Painter PC, Mosher LE, Rhoads C (1982) Biopolymers 21:1469
23. Pierson NA, Valentine SJ, Clemmer DE (2010) J Chem Phys B 114:7777
24. Papadopoulos G, Svendsen A, Boyarkin OV, Rizzo TR (2012) J Am Soc Mass Spectrom 23: 1173
25. van der Spoel D, Marklund EG, Larsson DSD, Caleman C (2011) Macromol Biosci 11:50
26. Hudgins RR, Woenckhaus J, Jarrold MF (1997) Int J Mass Spectrom 165:497
27. Koeniger SL, Clemmer DE (2007) J Am Soc Mass Spectrom 18:322
28. Karas M, Hillenkamp F (1988) Anal Chem 60:2299
29. Karas M, Bachmann D, Bahr U, Hillenkamp F (1987) Int J Mass Spectrom 78:53
30. Fenn JB, Mann M, Meng CK, Wong SF, Whitehouse CM (1990) Mass Spectrom Rev 9:37
31. Fuerstenau SD, Benner WH, Thomas JJ, Brugidou C, Bothner B, Siuzdak G (2001) Angew Chem Int Ed 40:542
32. Comisarow MB, Marshall AG (1974) Chem Phys Lett 25:282
33. Wong RL, Paech K, Williams ER (2004) Int J Mass Spectrom 232:59
34. Nielsen SB, Lapierre A, Andersen JU, Pedersen UV, Tomita S, Andersen LH (2001) Phys Rev Lett 8722:228102
35. von Hahn R, Berg F, Blaum K, Lopez-Urrutia JRC, Fellenberger F, Froese M, Grieser M, Krantz C, Kuhnel KU, Lange M, Menk S, Laux F, Orlov DA, Repnow R, Schroter CD, Shornikov A, Sieber T, Ullrich J, Wolf A, Rappaport M, Zajfman D (2011) Nucl Instruments Meth Phys Res Sect B Beam Interact Mater Atoms 269:2871
36. Dahan M, Fishman R, Heber O, Rappaport M, Altstein N, Zajfman D, van der Zande WJ (1998) Rev Sci Instrum 69:76
37. Zajfman D, Heber O, VejbyChristensen L, BenItzhak I, Rappaport M, Fishman R, Dahan M (1997) Phys Rev A 55:R1577
38. Lange M, Froese M, Menk S, Varju J, Bastert R, Blaum K, Lopez-Urrutia JRC, Fellenberger F, Grieser M, von Hahn R, Heber O, Kuhnel KU, Laux F, Orlov DA, Rappaport ML, Repnow R, Schroter CD, Schwalm D, Shornikov A, Sieber T, Toker Y, Ullrich J, Wolf A, Zajfman D (2010) Rev Sci Instrum 81
39. Kingdon KH (1923) Phys Rev 21:408
40. Makarov AA (1999) US patent US5,886,346 A
41. Michalski A, Damoc E, Lange O, Denisov E, Nolting D, Muller M, Viner R, Schwartz J, Remes P, Belford M, Dunyach JJ, Cox J, Horning S, Mann M, Makarov A (2012) Mol Cell Proteomics 11
42. Paul W, Steinwedel H (1953) Zeitschrift Fur Naturforschung Section a-a. J Phys Sci 8:448
43. Gerlich D (1992) In: Baer C-YNaM (ed) State-selected and state-to-state ion-molecule reaction dynamics, part 1. Experiment. Wiley & Sons, Inc. (New York, Chichester, Brisbane, Toronto, Singapore), p 1
44. Asvany O, Schlemmer S (2009) Int J Mass Spectrom 279:147
45. Trippel S, Mikosch J, Berhane R, Otto R, Weidemuller M, Wester R (2006) Phys Rev Lett 97
46. Boyarkin OV, Mercier SR, Kamariotis A, Rizzo TR (2006) J Am Chem Soc 128:2816

47. Boyarkin OV, Kopysov V (2014) Rev Sci Instrum 85
48. Quarmby ST, Yost RA (1999) Int J Mass Spectrom 191:81
49. Itano WM, Wineland DJ (1982) Phys Rev A 25:35
50. Eschner JR, Morigi G, Schmidt-Kaler F, Blatt R (2003) J Optical Soc Am B 20:1003
51. Hänsch TW, Schawlow AL (1975) Opt Commun 13:68
52. Bowe P, Hornekaer L, Brodersen C, Drewsen M, Hangst JS, Schiffer JP (1999) Phys Rev Lett 82:2071
53. Vogelius IS, Madsen LB, Drewsen M (2002) Phys Rev Lett 89
54. Ostendorf A, Zhang CB, Wilson MA, Offenberg D, Roth B, Schiller S (2006) Phys Rev Lett 9724:3005
55. Offenberg D, Zhang CB, Wellers C, Roth B, Schiller S (2008) Phys Rev A 78:061401
56. Wellers C, Borodin A, Vasilyev S, Offenberg D, Schiller S (2011) Phys Chem Chem Phys 13:18799
57. Dunbar RC (1989) J Chem Phys 90:7369
58. Ho YP, Dunbar RC (1996) Int J Mass Spectrom 154:133
59. Filsinger F, Ahn DS, Meijer G, von Helden G (2012) Phys Chem Chem Phys 14:13370
60. Choi MY, Miller RE (2007) J Phys Chem A 111:2475
61. Jašík J, Žabka J, Roithová J, Gerlich D (2013) Int J Mass Spectrom 354–355:204
62. Gerlich D (2008) In: Smith IWM (ed) Low temperatures and cold molecules. World Scientific Publishing, Singapore, p 121
63. Khoury JT, Rodriguez-Cruz SE, Parks JH (2002) J Am Soc Mass Spectrom 13:696
64. Cage B, Friedrich J, Little RB, Wang YS, McFarland MA, Hendrickson CL, Dalal N, Marshall AG (2004) Chem Phys Lett 394:188
65. Shi XG, Parks JH (2010) J Am Soc Mass Spectrom 21:707
66. Sagulenko P, Frankevich V, Steinhoff R, Zenobi R (2013) Int J Mass Spectrom 338:11
67. Yao H, Jockusch RA (2013) J Phys Chem A 117:1351
68. Talbot FO, Rullo A, Yao HH, Jockusch RA (2010) J Am Chem Soc 132:16156
69. Forbes MW, Jockusch RA (2009) J Am Chem Soc 131:17038
70. Schlemmer S, Kuhn T, Lescop E, Gerlich D (1999) Int J Mass Spectrom 187:589
71. Schlemmer S, Lescop E, von Richthofen J, Gerlich D, Smith MA (2002) J Chem Phys 117: 2068
72. Okumura M, Yeh LI, Myers JD, Lee YT (1986) J Chem Phys 85:2328
73. Lisy JM (1997) Int Rev Phys Chem 16:267
74. Johnson MA (2003) In: Armentrout PB, Gross ML, Caprioli R (eds) The encyclopedia of mass spectrometry: theory and ion chemistry. Elsevier, Oxford
75. Duncan MA (2003) Int Rev Phys Chem 22:407
76. Bieske EJ, Dopfer O (2000) Chem Rev 100:3963
77. Kamariotis A, Boyarkin OV, Mercier SR, Beck RD, Bush MF, Williams ER, Rizzo TR (2006) J Am Chem Soc 128:905
78. Mercier SR, Boyarkin OV, Kamariotis A, Guglielmi M, Tavernelli I, Cascella M, Rothlisberger U, Rizzo TR (2006) J Am Chem Soc 128:16938
79. Nagornova NS, Rizzo TR, Boyarkin OV (2012) Science 336:320
80. Saigusa H, Urashima S, Asami H (2009) J Phys Chem A 113:3455
81. Ebata T, Hashimoto T, Ito T, Inokuchi Y, Altunsu F, Brutschy B, Tarakeshwar P (2006) Phys Chem Chem Phys 8:4783
82. Zellweger JM, Philippoz JM, Melinon P, Monot R, Vandenbergh H (1984) Phys Rev Lett 52: 522
83. Chakrabarty S, Holz M, Campbell EK, Banerjee A, Gerlich D, Maier JP (2013) J Chem Phys Lett 4:4051
84. Goyal S, Schutt DL, Scoles G (1992) Phys Rev Lett 69:933
85. Smolarek S, Brauer NB, Buma WJ, Drabbels M (2010) J Am Chem Soc 132:14086
86. Antoine R, Dugourd P (2011) Phys Chem Chem Phys 13:16494

87. Gerardi HK, Breen KJ, Guasco TL, Weddle GH, Gardenier GH, Laaser JE, Johnson MA (2010) J Phys Chem A 114:1592
88. Hunig I, Seefeld KA, Kleinermanns K (2003) Chem Phys Lett 369:173
89. Florio GM, Zwier TS (2003) J Phys Chem A 107:974
90. Zwier TS (2001) J Phys Chem A 105:8827
91. Carney JR, Zwier TS (2000) J Phys Chem A 104:8677
92. Shubert VA, Zwier TS (2007) J Phys Chem A 111:13283
93. Bagratashvili VN (1985) Multiple photon infrared laser photophysics and photochemistry. Harwood Academic Publishers, New York
94. Letokhov VS (1989) Laser spectroscopy of highly vibrationally excited molecules. Adam Hilger, Boston, p 384
95. Gulyuz K, Stedwell CN, Wang D, Polfer NC (2011) Rev Sci Instrum 82:54101
96. Yeh LI, Okumura M, Myers JD, Price JM, Lee YT (1989) J Chem Phys 91:7319
97. Boyarkin OV, Settle RDF, Rizzo TR (1995) Berichte Der Bunsen-Gesellschaft-Phys Chem Chem Phys 99:504
98. Settle RDF, Rizzo TR (1992) J Chem Phys 97:2823
99. Guidi M, Lorenz UJ, Papadopoulos G, Boyarkin OV, Rizzo TR (2009) J Phys Chem A 113:797
100. Durand S, Rossa M, Hernandez O, Paizs B, Maître P (2013) J Phys Chem A 117:2508
101. Sinha RK, Erlekam U, Bythell BJ, Paizs B, Maitre P (2011) J Am Soc Mass Spectrom 22: 1645
102. Gerlich D, Horning S (1992) Chem Rev 92:1509
103. Gerlich D (1992) Adv Chem Phys 82:1
104. Paul W, Lucke B, Schlemmer S, Gerlich D (1995) Int J Mass Spectrom 149:373
105. Vestal ML, Futrell JH (1974) Chem Phys Lett 28:559
106. Teloy E, Gerlich D (1974) Chem Phys 4:417
107. Bustamente SW, Okumura M, Gerlich D, Kwok HS, Carlson LR, Lee YT (1987) J Chem Phys 86:508
108. Gerlich D (1993) J Chem Soc Faraday Trans 89:2199
109. Asmis KR, Brummer M, Kaposta C, Santambrogio G, von Helden G, Meijer G, Rademann K, Woste L (2002) Phys Chem Chem Phys 4:1101
110. Asmis KR, Pivonka NL, Santambrogio G, Brummer M, Kaposta C, Neumark DM, Woste L (2003) Science 299:1375
111. Wang YS, Tsai CH, Lee YT, Chang HC, Jiang JC, Asvany O, Schlemmer S, Gerlich D (2003) J Phys Chem A 107:4217
112. Wang XB, Wang LS (2000) J Am Chem Soc 122:2339
113. Andersen LH, Lapierre A, Nielsen SB, Nielsen IB, Pedersen SU, Pedersen UV, Tomita S (2002) Eur Phys J D 20:597
114. Andersen JU, Hvelplund P, Nielsen SB, Tomita S, Wahlgreen H, Moller SP, Pedersen UV, Forster JS, Jorgensen TJD (2002) Rev Sci Instrum 73:1284
115. Oh H, Breuker K, Sze SK, Ge Y, Carpenter BK, McLafferty FW (2002) Proc Natl Acad Sci U S A 99:15863
116. Nolting D, Marian C, Weinkauf R (2004) Phys Chem Chem Phys 6:2633
117. Wang XB, Wang LS (2008) Rev Sci Instrum 79:73108
118. Wang XB, Woo HK, Wang LS (2005) J Chem Phys 123:51106
119. Alata I, Bert J, Broquier M, Dedonder C, Feraud G, Gregoire G, Soorkia S, Marceca E, Jouvet C (2013) J Phys Chem A 117:4420
120. Dzhonson A, Gerlich D, Bieske EJ, Maier JP (2006) J Mol Struct 795:93
121. Dhzonson A, Maier JP (2006) Int J Mass Spectrom 255:139
122. Stearns JA, Mercier S, Seaiby C, Guidi M, Boyarkin OV, Rizzo TR (2007) J Am Chem Soc 129:11814
123. Snoek LC, Kroemer RT, Hockridge MR, Simons JP (2001) Phys Chem Chem Phys 3:1819

124. Chin W, Mons M, Dognon JP, Piuzzi F, Tardivel B, Dimicoli I (2004) Phys Chem Chem Phys 6:2700
125. Fricke H, Gerlach A, Unterberg C, Rzepecki P, Schrader T, Gerhards M (2004) Phys Chem Chem Phys 6:4636
126. Hunig I, Kleinermanns K (2004) Phys Chem Chem Phys 6:2650
127. Bakker JM, Plutzer C, Hunig I, Haber T, Compagnon I, von Helden G, Meijer G, Kleinermanns K (2005) ChemPhysChem 6:120
128. Reha D, Valdes H, Vondrasek J, Hobza P, Abu-Riziq A, Crews B, de Vries MS (2005) Chem Eur J 11:6803
129. Stearns JA, Guidi M, Boyarkin OV, Rizzo TR (2007) J Chem Phys 127:154322
130. Stearns JA, Boyarkin OV, Rizzo TR (2007) J Am Chem Soc 129:13820
131. Svendsen A, Lorenz UJ, Boyarkin OV, Rizzo TR (2010) Rev Sci Instrum 81:073107
132. Fujihara A, Matsumoto H, Shibata Y, Ishikawa H, Fuke K (2008) J Phys Chem A 112:1457
133. Fujihara A, Noguchi N, Yamada Y, Ishikawa H, Fuke K (2009) J Phys Chem A 113:8169
134. Goebbert DJ, Wende T, Bergmann R, Meijer G, Asmis KR (2009) J Phys Chem A 113:5874
135. Choi CM, Lee JH, Choi YH, Kim HJ, Kim NJ, Heo J (2010) J Phys Chem A 114:11167
136. Choi CM, Choi DH, Heo J, Kim NJ, Kim SK (2012) Angew Chem Int Ed 51:7297
137. Michael SM, Chien M, Lubman DM (1992) Rev Sci Instrum 63:4277
138. Choi CM, Choi DH, Kim NJ, Heo J (2012) Int J Mass Spectrom 314:18
139. Leavitt CM, Wolk AB, Kamrath MZ, Garand E, Van Stipdonk MJ, Johnson MA (2011) J Am Soc Mass Spectrom 22:1941
140. Kamrath MZ, Relph RA, Guasco TL, Leavitt CM, Johnson MA (2011) Int J Mass Spectrom 300:91
141. Wolk AB, Leavitt CM, Garand E, Johnson MA (2013) Acc Chem Res 47:202
142. Papadopoulos G, Svendsen A, Boyarkin OV, Rizzo TR (2011) Faraday Discuss Chem Soc 150:243
143. Lakin NM, Olkhov RV, Dopfer O (2001) Faraday Discuss Chem Soc 118:455
144. Glosik J, Hlavenka P, Plasil R, Windisch F, Gerlich D, Wolf A, Kreckel H (2006) Philos Trans R Soc A Math Phys Eng Sci 364:2931
145. Wang XB, Xing XP, Wang LS (2008) J Phys Chem A 112:13271
146. Wu R, McMahon TB (2009) J Chem Phys B 113:8767
147. Ingram AJ, Wolk AB, Flender C, Zhang JL, Johnson CJ, Hintermair U, Crabtree RH, Johnson MA, Zare RN (2014) Inorg Chem 53:423
148. Kamrath MZ, Garand E, Jordan PA, Leavitt CM, Wolk AB, Van Stipdonk MJ, Miller SJ, Johnson MA (2011) J Am Chem Soc 133:6440
149. Garand E, Kamrath MZ, Jordan PA, Wolk AB, Leavitt CM, McCoy AB, Miller SJ, Johnson MA (2012) Science 335:694
150. Stearns JA, Seaiby C, Boyarkin OV, Rizzo TR (2009) Phys Chem Chem Phys 11:125
151. Tamaki M, Sasaki I, Nakao Y, Shindo M, Kimura M, Uchida Y (2009) J Antibiot
152. Kawai M, Yamamura H, Tanaka R, Umemoto H, Ohmizo C, Higuchi S, Katsu T (2005) J Pept Res 65:98
153. Hull SE, Karlsson R, Main P, Woolfson MM, Dodson EJ (1978) Nature 275:206
154. Abo-Riziq A, Crews BO, Callahan MP, Grace L, de Vries MS (2006) Angew Chem Int Ed 45:5166
155. Kupser P, Pagel K, Oomens J, Polfer N, Koksch B, Meijer G, von Helden G (2010) J Am Chem Soc 132:2085
156. Guidi M (2008) Ph.D. thesis, Ecole Polytechnique Fédérale de Lausanne
157. Zabuga A, Kamrath MZ, Boyarkin OV, Rizzo TR (2014) J Chem Phys 141:154309
158. Nagornova NS, Rizzo TR, Boyarkin OV (2010) J Am Chem Soc 132:4040
159. Snoek LC, Robertson EG, Kroemer RT, Simons JP (2000) Chem Phys Lett 321:49
160. Jarrold MF (2000) Annu Rev Phys Chem 51:179
161. Wyttenbach T, Bowers MT (2003) Top Curr Chem 225:207
162. McLean JA, Ruotolo BT, Gillig KJ, Russell DH (2005) Int J Mass Spectrom 240:301
163. Tang K, Shvartsburg AA, Lee HN, Prior DC, Buschbach MA, Li FM, Tolmachev AV, Anderson GA, Smith RD (2005) Anal Chem 77:3330

164. Kanu AB, Dwivedi P, Tam M, Matz L, Hill HH (2008) J Mass Spectrom 43:1
165. Purves RW, Guevremont R, Day S, Pipich CW, Matyjaszczyk MS (1998) Rev Sci Instrum 69:4094
166. Guevremont R, Purves RW (1999) J Am Soc Mass Spectrom 10:492
167. Shvartsburg AA, Tang K, Smith RD (2005) J Am Soc Mass Spectrom 16:2
168. Purves RW, Barnett DA, Ells B, Guevremont R (2001) Rapid Commun Mass Spectrom 15: 1453
169. Shvartsburg AA, Li FM, Tang KQ, Smith RD (2006) Anal Chem 78:3706
170. Papadopoulos G (2012) PhD Thesis, Ecole Polytechnique Fédérale de Lausanne
171. Shvartsburg AA, Tang K, Smith RD, Holden M, Rush M, Thompson A, Toutoungi D (2009) Anal Chem 81:8048
172. Shvartsburg AA, Prior DC, Tang KQ, Smith RD (2010) Anal Chem 82:7649
173. Shvartsburg AA, Tang K, Smith RD (2010) Anal Chem 82:32
174. Koeniger SL, Merenbloom SI, Sevugarajan S, Clemmer DE (2006) J Am Chem Soc 128: 11713
175. Koeniger SL, Merenbloom SI, Clemmer DE (2006) J Phys Chem B 110:7017
176. Koeniger SL, Merenbloom SI, Valentine SJ, Jarrold MF, Udseth HR, Smith RD, Clemmer DE (2006) Anal Chem 78:4161
177. Shvartsburg AA, Li FM, Tang KQ, Smith RD (2006) Anal Chem 78:3304
178. Chupka WA (1959) J Chem Phys 30:191
179. Klots CE (1973) J Chem Phys 58:5364
180. Gordon SM, Reid NW (1975) Int J Mass Spectrom 18:379
181. Gregoire G, Kang H, Dedonder-Lardeux C, Jouvet C, Desfrancois C, Onidas D, Lepere V, Fayeton JA (2006) Phys Chem Chem Phys 8:122
182. Perot M, Lucas B, Barat M, Fayeton JA, Jouvet C (2010) J Phys Chem A 114:3147
183. Choi MY, Dong F, Han SW, Miller RE (2008) J Phys Chem A 112:7185
184. Choi MY, Dong F, Miller RE (2005) Philos Trans R Soc Lond Ser A Math Phys Eng Sci 363:393
185. Inokuchi Y, Kusaka R, Ebata T, Boyarkin OV, Rizzo TR (2013) ChemPhysChem 14:649
186. Inokuchi Y, Boyarkin OV, Kusaka R, Haino T, Ebata T, Rizzo TR (2012) J Phys Chem A 116:4057
187. Inokuchi Y, Boyarkin OV, Kusaka R, Haino T, Ebata T, Rizzo TR (2011) J Am Chem Soc 133:12256
188. Fedorov A, Couzijn EPA, Nagornova NS, Boyarkin OV, Rizzo TR, Chen P (2010) J Am Chem Soc 132:13789

Top Curr Chem (2015) 364: 99–152
DOI: 10.1007/128_2014_620

Published online: 14 April 2015

Theoretical Methods for Vibrational Spectroscopy and Collision Induced Dissociation in the Gas Phase

Marie-Pierre Gaigeot and Riccardo Spezia

Abstract In this chapter we review recent advances in theoretical methods to understand and rationalize anharmonic vibrational spectroscopy (IR-MPD and IR-PD) and collision induced dissociations (CID) in the gas phase. We focused our attention on the application of molecular dynamics-based methods. DFT-based molecular dynamics was shown to be able to reproduce InfraRed Multi-Photon Dissociation (IR-MPD) and InfraRed Pre-Dissociation (IR-PD) action spectroscopy experiments, and help assign the vibrational bands, taking into account finite temperature, conformational dynamics, and various anharmonicities. Crucial examples of dynamical vibrational spectroscopy are given on the protonated Ala_nH^+ series (related to IR-MPD in the 800–4,000 cm^{-1} domain), ionic clusters (related to IR-PD in the 3,000–4,000 cm^{-1} region), and neutral peptides (related to IR-MPD in the far-IR). We give examples from simple (e.g., cationized urea) to more complex (e.g., peptides and carbohydrates) molecular systems where molecular dynamics was particularly suited to understanding CID experiments.

Keywords Anharmonic spectra · Chemical dynamics · DFT-based dynamics · Gas phase fragmentation · IR-MPD · IR-PD · RRKM · Vibrational spectroscopy

Contents

M.-P. Gaigeot (✉) and R. Spezia (✉)
LAMBE UMR 8587, Université d'Evry-Val-d'Essonne, Paris, France
e-mail: mgaigeot@univ-evry.fr; riccardo.spezia@univ-evry.fr

1 Introduction

Vibrational spectroscopy is one of the most important and crucial experimental tools broadly applied in chemistry, physics, and biology. Vibrational spectroscopy is unique in the sense that it provides a relationship between the three-dimensional organization of molecular structures and their vibrational fingerprints. This book provides a snapshot of the tremendous developments that have been made in the field of gas phase spectroscopy for the structural characterization of (bio)-molecules and assemblies of molecules within the past decade. A preceding snapshot was given in the book by Schermann [1]. Vibrational action spectroscopy has in particular experienced an amazingly successful era, with many new experimental developments and applications in many different areas in chemistry, analytical chemistry, physics, and biology. Previous reviews include [2–6], and updated reviews can be found in the present book.

Vibrational action spectroscopy can be roughly divided into experiments performed at low (and extremely low) temperatures and experiments performed at finite temperature. A further division concerns neutral and charged molecules (and clusters) being investigated, in relation to the initial stage of production of the molecules. One more division amongst the experimental set-ups is the number of photons used for laser interrogation and subsequent molecular fragmentation. IR-MPD (InfraRed Multi-Photon Dissociation) experiments use several photons before the fragmentation of the investigated molecule is achieved, whereas IR-PD (InfraRed Photo-Dissociation or InfraRed Pre-Dissociation) experiments use one photon for the departure of the tagged entity from the investigated molecular assembly. IR-MPD experiments have been more traditionally applied in the domain of assemblies of biomolecules, possibly interacting with metal ions and/or a few solvent molecules, whereas IR-PD experiments have been more traditionally applied in the domain of clusters and small biomolecules. See again [1–6] and the present chapters for more details and illustrations.

As in many other experimental areas, vibrational action spectroscopy strongly relies on theoretical calculations to provide a clear and definitive picture of the structures of the molecular assemblies. The synergy between experiments and theoretical calculations is visible in the publications: almost 100% of the papers in the gas phase IR-PD/IR-MPD communities are now published with associated theoretical calculations to help analyze the experimental fingerprints.

In this chapter, our goal is to present theoretical methods applied to gas phase vibrational spectroscopy. This is reviewed in Sect. 3 where we present harmonic and anharmonic spectra calculations, with special emphasis on dynamical approaches to anharmonic spectroscopy. In particular, we present the many reasons and advantages of dynamical anharmonic theoretical spectroscopy over harmonic/ anharmonic non-dynamical methods in Sect. 3.1. Illustrations taken from our work on dynamical theoretical spectroscopy are then presented in Sects. 4–6 in relation to action spectroscopy experiments. All examples presented here are conducted either in relation to finite temperature IR-MPD and IR-PD experiments, or to cold IR-MPD experiments. Beyond the conformational dynamics provided by the finite temperature trajectories, the chosen examples illustrate how the dynamical spectra manage to capture vibrational anharmonicities of different origins, of different strengths, in various domains of the vibrations from 100 to 4,000 cm^{-1}, and on various molecular systems. Other comprehensive reviews on theoretical anharmonic spectroscopy can be found in [7–9].

The examples we have taken from our work have been chosen to illustrate a few issues, where anharmonicities are the main ingredients and where dynamical theoretical spectroscopy plays a crucial role in elucidating the fingerprints. The first examples presented in Sect. 4 on protonated alanine peptides Ala_nH^+ of increasing size and structural complexity illustrate the importance of temperature and entropic effects in structural equilibria, and how conformational dynamics at finite temperature is the key ingredient for unraveling the structures-vibrational fingerprints relationships for floppy peptide molecules. Also illustrated in these examples is the importance of vibrational anharmonicities in $N-H^+\cdots O=C$ hydrogen bonds, and how their precise characterization in the dynamical spectra is a definitive asset in unraveling the peptidic structures. These theoretical investigations have been carried in relation with finite temperature IR-MPD action spectroscopy experiments, where temperatures varied from ~300 K to about 500 K, depending on the actual set-ups.

Our second example in Sect. 5 illustrates how theoretical dynamical spectroscopy is successful at capturing strong anharmonic hydrogen bonds of the type $M^-\cdots$solvent and the associated subtly enhanced solvent–solvent H-bonds of varying strengths in ionic clusters (where M^- is an ion and the solvent can be water or methanol). Strong anharmonicities of ~700–900 cm^{-1} have been correctly captured by the dynamical spectra, in relation to IR-PD experiments conducted at ~100 K temperature.

The third example presented in Sect. 6 illustrates our preliminary theoretical dynamical spectra in the far-IR domain below 1,000 cm^{-1}, where vibrational anharmonicities arising from mode-couplings and delocalized motions have been captured with great accuracy on a model phenylalanine neutral di-peptide, in relation to cold IR-MPD experiments.

Gas phase vibrational spectroscopy methods are often coupled to activation methods which allow molecular dissociation. After, e.g., an activated reaction,

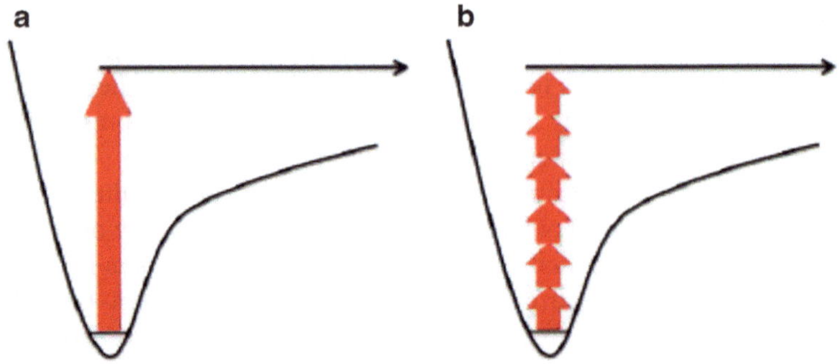

Fig. 1 Different activation schemes leading to dissociation: (**a**) single reactive collision; (**b**) multiple collisions activation

vibrational spectroscopy can be used to characterize fragments obtained and to help in defining reaction mechanisms and products. A popular way to activate ions in the gas phase is collision induced dissociation (CID). The experimental side is reviewed in one of the chapters (see Patrick and Polfer [10]), whereas we review here theoretical methods to be applied for CID modeling.

From a theoretical point of view, we can distinguish two large categories of activation processes and subsequent dissociation: (1) dissociation obtained in a single collision limit; (2) multiple collision activation. In the first activation mechanism, the energy is put into the ion in a short time; because a unique shock can dissociate the ion. This is also what happens in surface induced dissociation (SID) experiments, where the ion collides with an inert surface and can easily dissociate after the shock. When using collision with an inert gas, as in CID, the single collision regime is a limit regime not totally reached in practice, but which can be used as a good approximation when the gas pressure is very low. In this regime the molecule gets the energy and then two mechanisms can be responsible for subsequent reactivity: (1) a shattering mechanism, where dissociation occurs over a shorter time and the bond where the energy is initially stored is broken before the internal vibrational relaxation (IVR) takes place; (2) an IVR (or almost IVR) mechanism, where the energy has time to flow through the modes being redistributed and thus the dissociation channel is activated and reaction occurs. This second regime, or, better, the full IVR regime, is, on the other hand, characteristic of multiple collisions which "softly" heat the system towards the reaction channel. These two activation modes are schematically shown in Fig. 1.

Collision-induced dissociation is a particular case of unimolecular reaction, such that in the statistical limit the unimolecular RRKM theory can be applied [11]. This theory underlines that the micro canonical distribution is preserved along the reaction path, and thus the IVR regime is supposedly achieved within the characteristic reaction time. As pointed out by Schag and Levin some years ago, the unimolecular dissociation of large molecules reflects bottlenecks to intramolecular vibrational energy distribution and the available phase space is too large to allow dissociation from a uniform distribution of the excitation [12]. As we have discussed in the literature, and discuss more extensively in Sect. 7, reactivity induced by dissociation is a dynamical process which can or cannot be described by statistical methods.

Thus, we have divided our discussion in Sect. 7 into three parts: (1) on the identification of the potential energy surface (PES) on which the dynamics occur (Sect. 7.1); (2) on the statistical methods employed and on the framework of their applicability (Sect. 7.2); (3) on direct dynamics applied to understand CID (Sect. 7.3).

In particular, we show how direct dynamics can be useful for a more complete understanding of the fragmentation processes. This is essentially related to four aspects: (1) exploring the PES "on-the-fly" is particularly useful for complex and large molecules whose full description (all possible conformers and reaction channels) cannot be achieved manually; (2) revealing the transferred energy after collision and how it is partitioned between rotational and vibrational energy – this information can then be plugged into RRKM equations (once the PES is known) and thus provide the lifetime distributions of reactant, products and intermediates; (3) obtaining information on energy flow through modes after collision and before dissociation; (4) achieving direct access to dissociation, thus identifying the reaction mechanisms and whether or not reactivity is statistical.

2 Brief Survey of Some Relevant Theoretical Methods for Vibrational Spectroscopy and Collision-Induced Dissociation in the Gas Phase

As we see in the next sections, the Potential Energy Surface (PES) of a gas phase molecule or cluster is the crucial entity for unraveling molecular structures. A PES is a complex multidimensional surface composed of $3N-6$ dimensions, where N is the number of atoms in the system; see a schematic illustration of a PES in Fig. 2. Of special interest in a PES are the local minima (valleys) which represent energetically stable structures, and the saddle points or transition states. The transition states connect two local minima. Minima and transition states are topologically characterized, respectively, by all harmonic vibrational frequencies being real and by one imaginary frequency. Harmonic frequencies are calculated by diagonalization of the Hessian matrix of the system, i.e., a matrix composed of second derivatives of the potential energy with respect to the 3N coordinates of the system. Six of these frequencies are zero, corresponding to global translational and rotational motions of the system in space.

Exploration of the PES to localize the minima and the associated structures is now a routine job performed by all well-known quantum chemistry packages (Gaussian, ADF, Gamess, NWChem, ORCA, ...); see for instance [13] for detailed discussions on the methods and algorithms for conformational search on the PES. Conformational search is also called geometry optimization. This is a rather "easy" task for molecular systems with a reasonable number of internal degrees of

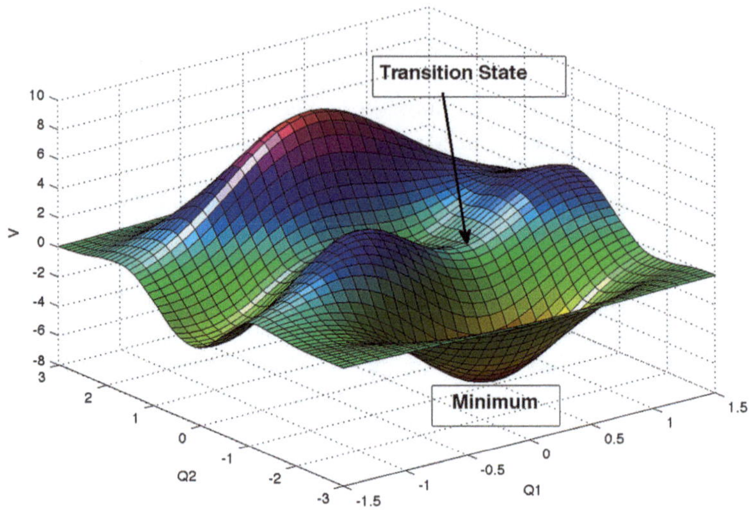

Fig. 2 Schematical potential energy surface landscape showing a typical minimum and transition state connecting two local minima

freedom, but such a conformational search is complex and time-consuming as soon as the size (and number of internal degrees of freedom) of the system increases. The number of conformations to be searched to sample the PES properly indeed increases dramatically as soon as the molecular system becomes larger. Only the lowest energy structures are of interest anyway, but it might be time-consuming and difficult to find these specific structures on the complex PES.

Nowadays, a preliminary conformational search is generally performed with a low-level representation of the interactions, typically employing classical force fields or semi-empirical-based quantum methods, followed by clustering of structures in terms of structural families, and providing a first ranking of molecular structures in terms of energy order, from lowest to highest energies. It should be noted that this conformational search can be performed with standard algorithms locating minima on a PES as described in the book by Cramer [13], or by Monte-Carlo (MC) exploration of the PES, thus allowing relatively large motions in the conformational sampling, or by Molecular Dynamics (MD) simulations usually coupled with elaborate Parallel Tempering methods using replicas of the same system at different temperatures and allowing MC jumps between replicas, thus enhancing the PES sampling. MC and MD methods, together with elaborate enhancement of PES sampling, are described in the book by Smit and Frenkel [14]. Any other theoretical method for enhancing the PES sampling can be applied, for instance metadynamics [15].

For structural families within a certain energy range, obtained with this first low-level screening, subsequent geometry optimizations are performed at a higher level of theory, i.e., using electronic ab initio methods. The DFT (Density Functional Theory) electronic representation is nowadays probably the most widely used in these calculations, as it provides an electronic framework capable of dealing with molecular systems composed of (several) hundreds of atoms. The energy ordering obtained with electronic representations is usually very different from the ordering obtained with lower levels of representations, and very often minima obtained at the low level PES representations no longer exist on the quantum PES, whereas new local conformations appear with the quantum representation. The low level PES exploration is therefore considered more as a first screening of the PES, providing guidelines for performing higher level quantum calculations on selected and supposedly relevant portions of the PES. For the lowest energy structures obtained with the quantum representation, harmonic frequencies are calculated through the diagonalization of a Hessian matrix, and are used to discriminate minima, transition states, or any conformation with more than one imaginary frequency (which represents no specific topological point on the PES).

To go beyond the knowledge of the structures at the minima and transition states, and learn more about the evolution with time of the structures and/or the exploration of the PES "on the fly," one has to employ Molecular Dynamics (MD) methods. See again the book by Smit and Frenkel [14] for further details, the book by Marx and Hutter [16] on MD based on the electronic DFT framework, and the book by Cramer [13] for semi-empirical and QM-MM MD methods.

The applications of quantum-based MD which we present throughout this chapter in the context of infrared vibrational spectroscopy and collision-induced dissociation rely on Born–Oppenheimer MD simulations, briefly outlined below. Within Born–Oppenheimer MD, the static (time-independent) Schrödinger equation is solved at each step of the dynamics, i.e., for each configuration of the nuclei:

$$\hat{H}_e(\mathbf{r};\mathbf{q})\Psi(\mathbf{r};\mathbf{q}) = E_e(\mathbf{q})\Psi(\mathbf{r};\mathbf{q}) \tag{1}$$

where $\mathbf{r}$ are the electron coordinates, $\hat{H}_e(\mathbf{r};\mathbf{q})$ the electronic Hamiltonian operator which depends parametrically on the nuclei positions, and $E_e(\mathbf{q})$ the ground state energy at a given atomic/nuclei configuration. E_e can be calculated at the semi-empirical, DFT, or higher ab initio levels (typically MP2). Once this is known, forces acting on the nuclei can be calculated and the nuclear degrees of freedom are propagated with classical molecular dynamics according to

$$-\frac{\partial E_e}{\partial q_i} = m_i \frac{d^2 q_i}{dt^2} \tag{2}$$

where positions q_i (and momenta p_i) of each of the nuclei of mass m_i evolve on the potential energy surface E_e obtained from (1).

We also encourage readers to go to [1] where one chapter surveys in slightly more details the methods mentioned above.

3 Theoretical Methods for Vibrational IR Spectroscopy

3.1 *Many Good Reasons to Prefer Dynamical Anharmonic Spectroscopy*

The Fermi Golden Rule provides the definition for the calculation of an infrared (IR) spectrum $I(\omega)$ as a function of the reciprocal wave-number ω [17]:

$$I(\omega) = 3\sum_{i}\sum_{f}\rho_i|\langle f|\mathbf{E.M}|i\rangle|^2\delta(\omega_{fi}-\omega) \tag{3}$$

where **E** is the applied external field vector, **M** is the dipole vector of the absorbing molecular system, $|i\rangle$ and $|f\rangle$ are, respectively, the initial and final vibrational states of the system (eigenstates of the system excluding the radiation), ρ_i is the density of molecules in the initial vibrational state $|i\rangle$, and ω_{fi} is the reciprocal wave-number of the transition between the initial and final vibrational states.

In the double harmonic approximation (mechanical and electrical), the Fermi Golden Rule reduces to (4) [18]:

$$\kappa = \sum_{\alpha=x,y,z}\sum_{k=1}^{3N-6}\left|\left(\frac{\partial\mu_\alpha}{\partial q_k}\right)_{eq}\right|^2\delta_{v_k,v_k\pm1}. \tag{4}$$

where κ is the absolute intensity of the active infrared transitions. In the above expression, the sums run over the three directions of space and over the 3N−6 normal modes q_k of the system and $\left(\frac{\partial\mu_\alpha}{\partial q_k}\right)_{eq}$ are the transition dipole derivatives in all three directions of space. Normal modes and transition dipole moments are defined at the equilibrium geometry (eq.) of the system. See for instance [18–20] for full derivations. The normal modes q_k give the frequency values at which the $0\rightarrow1$ ($v_k\rightarrow v_{k+1}$) vibrational transitions take place and the associated atomic movements, whereas dipole derivatives give the infrared intensities associated with each mode. These two quantities are routinely obtained with geometry optimization and harmonic frequency calculations (diagonalization of the Hessian matrix), typically performed by quantum chemistry packages (in case of ab initio electronic representations) such as the well-known Gaussian, ADF, Gamess, NWChem, ORCA (…) packages. DFT (Density Functional Theory) electronic representation is nowadays probably the most used in these calculations, as it provides an electronic framework capable of dealing with molecular systems composed of (several) hundreds of atoms.

The conventional way to proceed for the evaluation of (4) is to perform a conformational search on the potential energy surface (PES) of the molecular system to assess stable structures (see Sect. 2 in this chapter for more details). For the lowest energy structures, harmonic frequencies are calculated, and are used to eliminate transition states or any conformation with more than one imaginary frequency, and thus keep only the structures associated with minima on the PES for further analyses. For these minima, harmonic frequencies of the normal modes are scaled according to the electronic representation and basis sets used in the calculations [21–23], possibly also depending on the frequency range [24], and a final vibrational spectrum is obtained, either IR or Raman depending on rules applied ((4) is for IR). The main purpose of scaling the calculated frequencies is to compensate for the double harmonic approximations (potential energy surface and dipole moment; see (4)) and compensate for the approximations in the level of the ab initio representation. The calculated absorption spectra are finally convoluted with a Gaussian or Lorentzian band profile, generally adjusted for experimental conditions. Once these steps have been achieved, comparisons between the calculated IR vibrational spectra of the identified lower energy conformers and the experimental spectrum can be made. A match is searched, i.e., a one-to-one match that one structure gives the experimental signatures. To that end, the calculated band-positions and band-intensities are compared with the experiment.

The paradigm here is that the lowest energy conformation should be the one providing the (best) match with the experimental signatures. Experimentally, the formation of the lowest energy conformation is indeed believed to be driven by kinetic and entropic effects. Although this might prove right in many experiments performed in the gas phase, especially when they are based on supersonic expansions for the production of the molecules and clusters, this is not a systematic statement. Exceptions in supersonic expansions have been reported (see for instance [25, 26]), and clear deviations have been observed by the group of J. Lisy in IR-PD argon-tagging experiments where high energy conformers can be kinetically trapped [27, 28].

The main limitations of the theoretical harmonic approach (referred to in the remainder of the text as "static harmonic") to vibrational spectroscopy are:

- The search for the minima of lower energy on the potential energy surface, which can typically be tricky for floppy molecules. This is also a difficult search to perform for large and complex molecular assemblies, and for molecules surrounded by a few solvent molecules. See Sect. 2. Last but not least, in order to fulfill the paradigm mentioned above, one has to make sure that the lowest energy conformer has indeed been found, which is by no means trivial.
- The double harmonic approximations entering (4).
- The lack of temperature. By construction, the search for the minima on the PES provides frozen structures. This is especially a very crude approximation for floppy molecules which can undergo conformational dynamics at finite temperature, as is likely to occur in the IR-MPD and IR-PD experiments of interest to us.

We return to these points in more detail later on.

In Statistical Mechanics, the Fermi Golden Rule (3) can be reformulated using Linear Response Theory [17, 29], and can thereby be rewritten as the Fourier

Transform of the time correlation function of the fluctuating dipole moment vector of the absorbing molecular system:

$$I(\omega) = \frac{2\pi\beta\omega^2}{3cV} \int_{-\infty}^{\infty} \mathrm{d}t \langle \delta\mathbf{M}(t) \cdot \delta\mathbf{M}(0) \rangle \ \exp(i\omega t) \tag{5}$$

where $\beta = 1/kT$, T is the temperature, c is the speed of light in vacuum, and V is the volume. The angular brackets represent a statistical average of the correlation function, where $\delta\mathbf{M}(t) = \mathbf{M}(t) - <\mathbf{M}>$ with $<\mathbf{M}>$ the time average of $\mathbf{M}$(t). The calculation in (5) is done in the absence of an applied external field. For the prefactor in (5), we have taken into account an empirical quantum correction factor (multiplying the classical line shape) of the form $\beta\hbar\omega/(1 - \exp(-\beta\hbar\omega))$, which was shown by us and others to give accurate results on calculated IR intensities [30–32]. For more detailed discussions on quantum corrections, see for instance [33–36].

The main advantages of the molecular dynamics (MD) approach in (5) for the calculation of infrared spectra (also called "dynamical spectra" in the remainder of the text) can be listed as follows:

- There are no approximations made in (5) apart from the hypothesis of linear response theory, i.e., a small perturbation from the applied external electric field on the absorbing molecular system. Such a condition is always fulfilled in vibrational spectroscopy of interest here. There are no harmonic approximations made, be they on the potential energy surface or on the dipole moment, contrary to the static calculations entering into (4). These approximations are not needed in (5). As a consequence, vibrational anharmonicities are naturally taken into account in (5). One thus needs to know the time evolution of the dipole moment of the system to calculate an anharmonic IR spectrum. This is naturally done with molecular dynamics simulations. In fact, the finite temperature dynamics takes place on all accessible parts of the potential energy surface, be they harmonic or anharmonic. The quality of the potential energy surface is entirely contained in the "ab-initio" force field used in the dynamics, calculated at the DFT/BLYP (+dispersion when needed) level in works presented here (see details in Sect. 3.2). The good to excellent agreements of the absolute (and relative) positions of the different active bands obtained in our theoretical works (see for instance dynamical spectra in the gas phase [37–41], in the liquid phase [31, 42–45], and at solid–liquid and liquid–air interfaces [46–48]) is a demonstration (though a posteriori) that this level of theory is correct.
- Even more crucial in this discussion, the calculation of IR spectra with MD is related *only* to the time-dependent dipole moment of the molecular system, requiring neither any harmonic expansion of the transition dipole moment nor the knowledge of normal modes, in contrast to harmonic calculations. Therefore, if the dipole moments and their fluctuations are accurately calculated along the trajectory, the resulting IR spectrum should be reliable. The vibrations therefore do not rely directly on the curvature of the potential energy surface at the minima

on the PES (i.e., normal modes and derivatives using these normal modes) but rather on the time evolution of the electric dipole moment of the molecular system, which is governed by the conformational dynamics at the finite temperature of the simulation. As a consequence, dynamical anharmonic spectra from (5) and harmonic spectra from (4) rely on strictly different properties, and presumably require different levels of accuracy for the evaluation of these properties. See a more detailed discussion on these points later.

- Equation (5) gives the whole infrared spectrum of a molecular system in a *single calculation*, i.e., the band-positions, the band-intensities, and the band-shapes, through the Fourier transform of a time correlation function. There are no approximations applied, particularly the shape and broadening of the vibrational bands result from the underlying dynamics and mode-couplings of the system at a given temperature.
- Dynamics simulations are performed at finite temperature. At a given temperature, and depending on the energy barriers on the potential energy surface (PES), conformational dynamics between different isomeric forms of the absorbing gas phase molecular entity can be sampled by MD. All conformations populated over time are thus taken into account in the final calculation of the infrared spectrum from (5). Beyond the conformational dynamics between conformers and isomers of the molecule of interest, any population dynamics (around the minima structures), typically H-bond dynamics, local geometry rearrangements, also give rise to a natural broadening of the calculated IR active bands, which is essential for the comparison to the experimental spectra. These points have been demonstrated in some of our work, typically in our seminal case-study of the gas phase floppy protonated peptide Ala_2H^+ [40, 49], as reviewed in Sect. 4. Very good agreement of the ~300 K dynamical infrared spectra of this peptide with the IR-MPD experimental spectra, could be achieved thanks to the simulated room-temperature dynamics of the gas phase peptide, thus enabling us to capture and to take into account the continual conformational dynamics between the two major conformers of the molecule over time. This could not be achieved with standard static calculations [49, 50]. One pivotal issue we have emphasized in this work is that we found the most populated conformations of Ala_2H^+ not to be the geometries at the bottom wells on the PES, but rather all conformations explored over time in going from one isomer to the other. A Boltzmann-weighted IR spectrum based on the harmonic spectra of the two minima structures on the PES was therefore of no use for matching the experimental signatures.

Here we pause for a more general discussion on vibrational anharmonicities. As already pointed out in the introduction, static quantum chemistry calculations can be performed beyond the vibrational harmonic approximation, though they are more rarely applied in the communities of IR-MPD and IR-PD action spectroscopy. There are exceptions however; see works by the group of A.B. McCoy in collaboration with IR-PD experiments from the group of M.A. Johnson at Yale University

[51, 52], or works from R.B. Gerber's group in collaboration with the IR-MPD experiments of van Helden and Oomens on the IR-MPD FELIX set-up [53].

Static vibrational anharmonic calculations have been reviewed very recently by Gerber [8], a pioneer in the field. See also [54, 55] for slightly older reviews in the domain, showing how fast the field is still evolving. The vibrational self-consistent field (VSCF) method is probably the most extensively used method, providing accuracy and moderate computational cost. Developed in the late 1970s [56–59], its renewed interest lies in recent algorithmic developments of VSCF and its variant form VSCF-PT2 (based on Perturbation Theory), also called CC-VSCF in the literature (Correlation Corrected), thus making the method amenable to larger biomolecular systems at affordable computational cost. See for instance [60, 61] for the latest developments concerning linear scaling of the VSCF-PT2 method with an increase in the number of degrees of freedom and pairwise normal interactions simplifications.

Advantages and disadvantages of these theoretical methods over the dynamical method in accounting for vibrational anharmonicities can be roughly summarized as follows.

The anharmonic methods cited above are still based on geometry optimizations (0 K structures), therefore not including temperature, whereas MD does include temperature in the final spectrum. This is obviously an important attractive argument in favor of MD when the objective of the computations is the interpretation of finite temperature spectroscopy as in IR-MPD and IR-PD experiments. As already pointed out above in the case of the Ala_2H^+ peptide, a Boltzmann-weighted IR spectrum based on the anharmonic spectra of the two minima structures on the PES did not help match the experimental signatures. Finite temperature dynamics was mandatory to achieve a match and provide understanding of the experimental signatures [40]. See Sect. 4 for more details.

Size of the molecules is another issue. Resolution of the full-dimensional vibrational Schrödinger equation for systems larger than a few atoms is too complex to be solved directly and some approximations are needed, where VSCF-PT2 is based on the replacement of the explicit correlation in the N-body system by a series of single-particle problems coupled through an effective potential which is dependent on all the other degrees of freedom. Even in this case, the size of the molecular systems which can be investigated remains limited, on the order of typically 20–30 atoms [55]. A further simplified VSCF-PT2 algorithm has been developed by Gerber [61], based on a criterion related to geometric properties of the normal modes so as to keep a priori only the relevant mode–mode couplings. Tri-peptides have hence been investigated [61] and bigger peptides should also be accessible with this method. Size of the peptides is less critical in DFT-based MD, especially (see description of DFT-based MD in Sect. 3.2) using the CP2K package [62–64] where a dual electronic representation in Gaussian and Plane Wave (GPW) basis sets is employed, and clever and fast algorithms have been implemented to solve the electronic Schrödinger equation at each time-step of the dynamics. Hence, vibrational spectroscopy of peptides containing up to 10–20 residues is currently investigated in our group and others with DFT-MD.

One crucial point in a comparative study of VSCF-PT2 and DFT-MD for anharmonic spectroscopy, is spectroscopic accuracy, or in other words the quality of the underlying PES. This is crucial for VSCF calculations, as they require a description of the PES curvature(s) at the minima. On the one hand, VSCF-PT2 anharmonic frequencies can be performed with high quality PES, i.e., at least B3LYP and even at the MP2 level of theory (though much more costly), and very accurate spectroscopic data have been obtained [55]. When increasing the size of the systems, i.e., towards peptides of reasonable size (for the IR-MPD/IR-PD experimental communities), these authors nonetheless often resort to much simpler and less expensive PES, such as the semi-empirical PM3 method [61]. On the other hand, DFT-MD is based on the DFT electronic representation and the final quality of the PES depends on the functional used. All investigations performed in our group used the BLYP functional, augmented by dispersion whenever it is needed for hydrogen bonded systems, and one might argue that this level is not sufficient (see [55] for comments on this issue). As reviewed in the present chapter, DFT-MD/BLYP (+dispersion) has been successful and accurate at capturing the main vibrational anharmonicities relevant to the spectroscopy of gas phase protonated peptide chains [39–41], highly anharmonic ionic clusters [37, 65], and vibrational couplings arising from solute-solvent interactions [31, 44] but also solid–liquid and liquid–air interfaces [46, 47]. In other words, this electronic representation has been found to be accurate for various molecular systems, in various environments, and in various domains of vibrational frequencies from 100 to 4,000 cm^{-1}. Note that B3LYP-MD trajectories are computationally more costly than non-hybrid functionals-based MD, typically a factor of ~40 per time-step (using the CP2K package).

We are thus tempted to argue that we have been able to achieve spectroscopic accuracy with the DFT/BLYP representation in most applications to date, although some failures have also been noted [66].

Here again we emphasize that the DFT-MD spectra are *not* calculated from normal modes, and thus are *not* calculated from the anharmonic curvatures at the minima on the PES, but they *are* calculated from the time evolution of dipole moments. The quality of the PES is obviously of importance for DFT-MD trajectories, in terms of energy differences between conformers and transition states, in terms of forces acting on the atoms, properties thus governing the conformational dynamics, and consequently the relevance of the conformations dipole moments over time. These are the driving forces of DFT-MD-based vibrational spectroscopy rather than the intrinsic quality of the curvatures at the minima on the PES. In other words, the requirements on accuracy to be reached in the time evolution of dipole moments entering (5) for dynamical spectroscopy are different from what is required in (4) from the PES curvatures at the minima (static harmonic and anharmonic spectra calculations alike). As a consequence, any debate on the BLYP functional used in our DFT-MD trajectories for dynamical spectroscopy in comparison to the B3LYP functional widely used in static harmonic/anharmonic spectra calculations might not be so relevant. Furthermore, as illustrated in Sect. 6, where we employ DFT-MD for spectra in the anharmonic 100–1,000 cm^{-1} domain, mode couplings are amazingly accurately obtained by DFT-MD at the BLYP level, once again proving that the time evolution of the dipole moment of the peptide, even on a "low level BLYP PES representation," is extremely well provided by this electronic representation.

Gas phase dynamical anharmonic vibrational spectroscopy using DFT-based MD in relation to action spectroscopy has been initiated in our group, with work combining IR-MPD experiment and DFT-MD spectroscopy on the protonated Ala_2H^+ peptide [40]. This is reviewed in Sect. 4. We believe this investigation and the subsequent associated investigations on Ala_3H^+ [39] and Ala_7H^+ [41] have triggered interest not only in the IR-MPD/IR-PD experimental communities but also in the theoretical community. Hence, the group of V. Blum in Berlin later followed our strategies on $Ala_{15}H^+$ [67]. In relation to IR-MPD experiments, the group of R.B. Gerber has also applied DFT-MD to the anharmonic spectroscopy of protonated sugars [68, 69]. In a different area of experiments, Spiegelman et al. applied these strategies to gas phase polycyclic aromatic hydrocarbons (PAH) and their hydrates [70]. In our group, IR-MPD and anharmonic DFT-MD spectroscopy have been combined on the flexible peptides $Ala_{(n=2,3,7)}H^+$ [39–41], on the protonated and deprotonated phosphorylated amino acid serine [38, 66], and more recently we have been interested in highly anharmonic ionic clusters [37, 65] in collaboration with IR-PD argon-tagging experiments. Some remarkable results on these clusters are reviewed in Sect. 5. Also of pivotal interest to us is the combination of IR-MPD and DFT-MD spectroscopy in the far-IR region, where anharmonicities arising from delocalized modes and from mode couplings are crucial, thus pushing forward the quest of the DFT-MD theoretical methodology for anharmonic spectroscopy. Such investigations are highlighted in Sect. 6. Quantum effects of the (hydrogen) nuclei might be of importance for molecules in which O–H and N–H vibrational signatures are the main features used for assigning structures. The group of D. Marx is pioneering the application of Path Integral DFT-based MD to that end; see for instance the publication on the vibrational spectroscopy of CH_5^+ in that context [71]. Rather than performing these accurate but rather expensive simulations, we have chosen to incorporate quantum effects in the initial conditions of classical nuclear dynamics, especially through the insertion of zero point vibrational energy (ZPE) for initial positions and velocities of the dynamics [37, 72].

As stated, (5) gives the whole infrared spectrum of a molecular system with a *single calculation*, i.e., band-positions, band-intensities, and band-shapes. Some more comments are required.

The positions of the vibrational bands directly reflect the quality of the representation of the intramolecular interactions. As already pointed out, our work has shown that the DFT/BLYP (+dispersion) representation works extremely well, not only for band-spacings but also for the absolute values of vibrational bands. Intermolecular hydrogen bonding leads to vibrational shifts, the extent of which reflects the strengths of the interactions. Our work has shown how dynamical spectra can be accurate in reproducing bandshifts produced by hydrogen bonding of various strengths, as illustrated in the next sections. One very recent illustration concerns the very strong anharmonic bandshift of the N–H stretch motion in $Cl^{-}\cdots NMA(H_2O)_n$ clusters, where the dynamical spectra were able to account for the large 900 cm^{-1} red-shift upon hydrogen bonding.

Band-intensities and their matching to the experiments are commonly seen as a probe of the quality of the electrostatic interactions of the theoretical model. On one hand, (5) is strictly valid for one-photon linear IR absorption spectroscopy: this signal is thus identical to the signal measured in linear IR absorption spectroscopy in the liquid phase. Theoretical and experimental signals are thus directly comparable in the liquid condensed phase, in terms of band-positions, band-intensities, and band-shapes, and work on liquid water and solutes immersed in liquid water has indeed shown how excellent this agreement can be [31, 43, 44, 73]. On the other hand, temperature and energy distribution (equipartition) among the vibrational degrees of freedom play a non-trivial role in the accuracy of band-intensities derived from MD simulations. This can become crucial in gas phase spectroscopy calculations. Equipartition of the energy among the 3N−6 vibrational degrees of freedom is indeed not an easy task to achieve. We have discussed this point for low-temperature 20 K (quasi-)harmonic dynamics of gas phase amino acids and peptides [38, 74]. If modes are difficult to couple, equipartition is almost impossible to achieve, especially during the rather short timescales affordable in DFT-MD. The thermalization process is improved by surrounding solvent, as this increases the possibilities of energy transfers, not only through vibrational couplings but also through intermolecular interactions. Some discrepancies in band-intensities between dynamical spectra and experiments can therefore possibly be traced back to the equipartition of energy. Tavan et al. [75] have for instance devised a thermal correction to dynamical intensities based on the ratio of the average temperature of the dynamics and the actual temperature of the individual vibrational modes which might be able to correct deficiencies arising in the equipartition of energy in short timescale dynamics.

The situation in the gas phase is even more complicated by the actual signals recorded in the action spectroscopy experiments. IR-MPD and IR-PD experiments are multi-photon IR absorption processes leading to fragmentation of the molecule or cluster. The recorded signal is the fragmentation yield as a function of the IR excitation wavelength. It is thus an indirect measurement of IR absorption, in contrast to the linear IR absorption. So, although they reflect the same underlying vibrational properties, stationary IR absorption ((4) and (5)) and IR-MPD/IR-PD experiments are by no means equivalent. Calculations and experiments are therefore not directly comparable for band-intensities, giving rise to another source of possible discrepancies. The direct simulation of IR-MPD fragmentation spectra, with a clear theoretical expression of signal intensity in terms of dynamical quantities, remains an open question. See [76, 77] for kinetic models attempting to provide such a theoretical framework. It should be noted that it is commonly believed that the relative intensities of the active bands in IR-MPD are governed by the absorption of the few first photons, which would then be closer to calculations. The agreement between our dynamical spectra and IR-MPD/IR-PD experiments nonetheless appears very satisfactory from the point of view of spectral intensities.

DFT electronic representation routinely allows the treatment of molecular systems containing hundreds of heavy atoms, and the related DFT-MD trajectories can be reasonably accumulated over a few tens of picoseconds for systems composed of typically 100–500 atoms. Going beyond these scales of time and length would require simpler electronic representations, such as semi-empirical, or classical force fields.

In order for classical force fields to be reliable for vibrational spectroscopy, they have to be based on anharmonic analytical expressions for the intramolecular stretch, bend, and torsional motions, as well as stretch-bend (and all possible combinations between the modes) cross-terms for the intramolecular interactions. Anharmonic expressions and mode-couplings are indeed mandatory to be included in the force field expression to be able to achieve reliable vibrational band-positions. Such refinements are not included in the well-established CHARMM, AMBER, GROMOS, or OPLS force fields, routinely employed in MD simulations of biomolecules and organic molecules. Furthermore, as shown in [78, 79], any force field for spectroscopy should include at least fluctuating atomic charges in the electrostatic model, and also possibly fluctuating dipole moments (and maybe beyond), in order to recover reliable IR band-intensities. Again, such refinements are not included in the classical force fields routinely used in the literature. Any advance in the area of classical force fields for vibrational spectroscopy thus requires new force field developments, which is a huge amount of work. Such developments can be found in [80, 81] for instance, related to specific molecules and/or specific vibrational modes.

Semi-empirical dynamics seem more appealing for increasing timescales and lengths, because they are still based on an electronic representation, although simplified. The group of R.B. Gerber has in particular shown that PM3 representations can be accurate enough for vibrational spectroscopy of simple molecules [82] (static calculations). DFTB (DFT Tight Binding) also shows promising results for vibrational spectroscopy; see for instance [83–87].

DFT-MD simulations can also be applied to IR, Raman, and nonlinear Sum Frequency Generation vibrational spectroscopies in the liquid phase and at interfaces. This is beyond the scope of the present review, but we refer the reader to [7, 9, 31, 47, 73] for more information on the time-dependent signals to be calculated, and results.

3.2 *DFT-Based MD*

Our methodology consists of DFT-based molecular dynamics simulations, performed within the Born–Oppenheimer (BOMD) framework [62–64] (mainly using the CP2K package) or the Car-Parrinello [62, 88] framework (using the CPMD package). All applications presented in Sects. 4–6 employ BOMD, so that we only present here simulation details related to BOMD performed with the CP2K package. The methods and algorithms employed in the CP2K package are described in detail in [64]. A detailed presentation of BOMD and CPMD methods can be found in [89, 90].

In our dynamics, the nuclei are treated classically and the electrons quantum mechanically within the DFT formalism. Dynamics consist of solving Newton's equations of motion at finite temperature, with the forces acting on the nuclei deriving from the Kohn–Sham energy. In BOMD the Schrödinger equation for the electronic configuration of the system is solved at each time step of the dynamics (i.e., at each new configuration of the nuclei); see Sect. 2. Mixed plane waves and Gaussian basis sets are used in CP2K. Only the valence electrons are taken into account and pseudo-potentials of the Goedecker–Tetter–Hutter (GTH) form are used [91–93]. We use the Becke, Lee, Yang, and Parr (BLYP) gradient-corrected functional [94, 95] for the exchange and correlation terms. Dispersion interactions have been included with the Grimme D2 and D3 corrections [96] in certain applications. Calculations are restricted to the Γ point of the Brillouin zone. We employ plane-wave basis sets with a kinetic energy cut-off usually around 340 Ry and Gaussian basis sets of double-ζ (DZVP) to triple-ζ (TZVP) qualities from the CP2K library. The kinetic energy cut-off and basis set sizes are systematically optimized on each investigated system by careful energy convergence tests.

Our dynamics are strictly microcanonical (NVE ensemble), once thermalization has been achieved (through NVE and velocity rescaling periods of time). Our gas phase simulations use the decoupling technique of Martyna and Tuckerman [97] to eliminate the effect of the periodic images of the charge density, relevant to charged gas phase systems.

The knowledge of the evolution with time of the molecular dipole moments is mandatory for the calculation of IR spectra with MD simulations. In the modern theory of polarization, the dipole moment of the (periodic) box cell is calculated with the Berry phase representation, as implemented in the CPMD and CP2K packages [98]. Briefly, in the limit where the Γ point approximation applies, the electronic contribution to the cell dipole moment $\mathbf{M}_\alpha^{el}$ (where $\alpha = x, y, z$) is given by [99]

$$\mathbf{M}_\alpha^{el} = \frac{e}{|\mathbf{G}_\alpha|} \Im \ln z_N \tag{6}$$

where $\Im \ln z_N$ is the imaginary part of the logarithm of the dimensionless complex number $z_N = \left\langle \Psi \middle| e^{-i\mathbf{G}_\alpha \cdot \hat{\mathbf{R}}} \middle| \Psi \right\rangle$, $\mathbf{G}_\alpha$ is a reciprocal lattice vector of the simple cubic supercell of length L ($\mathbf{G}_1 = 2\pi/L(1,0,0)$, $\mathbf{G}_2 = 2\pi/L(0,1,0)$, $\mathbf{G}_3 = 2\pi/L(0,0,1)$), and $\hat{\mathbf{R}} = \sum_{i=1}^{N} \hat{\mathbf{r}}_i$ denotes the collective position operator of the N electrons (or in other words the center of the electronic charge distribution). Ψ is the ground-state wave function. The quantity $\Im \ln z_N$ is the Berry phase, which in terms of a set of occupied Kohn–Sham orbitals $\psi_k(\mathbf{r})$ is computed as $\Im \ln z_N = 2\Im \ln det\mathbf{S}$ with elements of the matrix $\mathbf{S}$ given by $S_{kl} = \langle \psi_k | e^{-i\mathbf{G}_\alpha \cdot \hat{\mathbf{r}}} | \psi_l \rangle$ [99].

IR spectra are given as products $\alpha(\omega)n(\omega)$ expressed in cm^{-1} (decadic linear absorption coefficient) as a function of wavenumber, ω, in cm^{-1}. The spectra are smoothed with a window filtering applied in the time domain, i.e., each term of the correlation function $C(t)$ is multiplied by a Gaussian function $exp(-0.5\sigma(t/t_{max})^2)$, where t_{max} is the length of the simulation, and σ is usually chosen around a value of 10 for gas phase simulations. This convolution only has the purpose to remove the numerical noise arising from the finite length of the Fourier transform of (5). These calculations are performed with our home-made code.

The average times easily affordable with DFT-based MD are of the order of a few tens of picoseconds for systems composed of a few hundred atoms, running on massively parallel machines. This is the timescale on which stretching, bending, and torsional motions can be properly sampled. For rather rigid molecules in the gas phase or molecules, the geometry of which is somewhat 'constrained' by the surrounding aqueous phase environment, one trajectory accumulated over this timescale is enough to get a reliable theoretical infrared spectrum [31, 38, 44]. When floppy peptides undergo conformational dynamics over the tens of picoseconds period of time, which is especially the case at room temperature in the gas phase, several trajectories have to be accumulated in order to take into account the conformational diversity. The final IR spectra presented in our work are averaged over all trajectories, so the conformational diversity and heterogeneity of the dynamics at finite temperature can be accounted for, in a natural way. Each of the dynamics performed gives slightly different IR features, reflecting the actual conformational sampling and properties of the isomeric/conformational family, and the statistical average over all trajectories is the only meaningful and relevant quantity, which can thus be compared to the spectrum recorded in the experiments. Such statistics is also relevant for H-bond dynamics.

The length of trajectories is also related to the vibrational domain to be sampled. One has to keep in mind that the length of time has to be commensurate with the investigated vibrational motions. Hence, trajectories around 5 ps are just enough to sample stretching motions in the high frequency domain of 3,000–4,000 cm^{-1}, provided that several trajectories starting from different initial conformations (structure and/or velocities) are accumulated and averaged for the final IR dynamical spectrum. In the mid-IR domain, trajectories of at least 10 ps each are needed to sample the slower stretching and bending motions of the 1,000–2,000 cm^{-1} domain. In the far-IR below 1,000 cm^{-1}, much longer trajectories are needed to sample properly the much slower motions typical of that domain, i.e., torsional motions and possibly opening/closure of structures (typically for peptide chains). These delocalized and highly coupled motions require trajectories of 20–50 ps on average to be safely sampled, again with a final average over a few trajectories. We illustrate dynamical spectroscopy for all these three domains in the next sections of that review.

All investigations presented in Sects. 4–6 have employed Born–Oppenheimer dynamics (BOMD). We apply no scaling factor of any kind to the vibrations extracted from the dynamics. The sampling of vibrational anharmonicities, i.e., potential energy surface, dipole anharmonicities, mode couplings, anharmonic modes, being included in our simulations, *by construction*, application of a scaling factor to the band-positions is therefore not required. As reviewed afterwards, excellent agreements between dynamical spectra and IR-MPD and IR-PD experiments have been achieved. Any remaining discrepancies between dynamical and experiment spectra should mainly be because of the choice of the DFT/BLYP (+dispersion) functional, as DFT-based dynamics are only as good as the functional itself allows. Conformational sampling might be another source of discrepancy: sampling is limited by the length of the trajectories we can afford and the number of initial isomers we are including in the investigation. It is also clear that vibrational anharmonic effects probed in molecular dynamics are intrinsically linked to the temperature of the simulation, as pointed out in [37, 74, 100]. Closely related is the influence of quantum effects such as Zero Point Vibrational Energy (ZPE) in the initial conditions of the dynamics, playing a role in the dynamics temperature and in the final dynamical vibrational frequency values. This has been discussed in [72] in the context of semi-classical dynamics for anharmonic spectroscopy and in [37] in the context of highly anharmonic red-shifts in ionic clusters from DFT-MD simulations.

3.3 Assignment of Modes with Vibrational Density of States VDOS

An accurate calculation of anharmonic infrared spectra is one goal to achieve, the assignment of the active bands into individual atomic displacements or vibrational modes is another. This issue is essential to the understanding of the underlying molecular structural and dynamical properties. In molecular dynamics simulations, interpretation of the infrared active bands into individual atomic displacements is traditionally and easily done using the vibrational density of states (VDOS) formalism. The VDOS is obtained through the Fourier transform of the atomic velocity auto-correlation function:

$$VDOS(\omega) = \sum_{i=1,N} \int_{-\infty}^{\infty} \langle \mathbf{v_i}(t) \cdot \mathbf{v_i}(0) \rangle \; \exp(i\omega t) \; dt \tag{7}$$

where i runs over all atoms of the investigated system and $\mathbf{v_i}(t)$ is the velocity vector of atom i at time t. As in (5), the angular brackets in (7) represent a statistical average of the correlation function. The VDOS spectrum provides all vibrational modes of the molecular system. However, only some of these modes are either infrared or Raman active, so VDOS spectra can by no means directly substitute IR or Raman spectra.

The VDOS can be further decomposed according to atom types, or to groups of atoms, or to chemical groups of interest, to obtain a detailed assignment of the vibrational bands in terms of individual atomic motions. This is done by restricting the sum over i in (7) to the atoms of interest only. Such individual signatures are easy to interpret in terms of movements for localized vibrational modes involving only a few atomic groups, typically in the 3,000–4,000 cm^{-1} domain, but the interpretation of the VDOS becomes more complicated for delocalized modes or for highly coupled modes. In the far-IR domain below 1,000 cm^{-1}, where such motions are typical, we resort to the Fourier transforms of intramolecular coordinates (IC) time correlation functions instead of the VDOS: $\int_{-\infty}^{\infty} \langle IC(t) \cdot IC(0) \rangle \exp(i\omega t)\, dt$. Of course, such an approach requires some a priori knowledge of the relevant IC(s), as one cannot easily analyze all possible combinations.

We and others have developed methods to extract vibrational modes from the dynamics, especially bypassing the limitations from VDOS assignments [19, 101–103]. Such methods usually provide "effective normal modes," similar to the well-known normal modes obtained by a Hessian diagonalization in static harmonic calculations, but maintaining a certain degree of mode couplings and temperature from the dynamics in the final modes. Describing these methods is beyond the scope of this chapter; all applications discussed in Sects. 4–6 use the simplest VDOS and ICs for assigning the vibrational modes.

4 IR-MPD Spectroscopy and Conformational Dynamics of Floppy Ala_nH^+ Protonated Peptides

The vibrational spectroscopy of gas phase protonated peptides Ala_nH^+ is the first example that illustrates the crucial need to take into account the conformational dynamics of molecules into the final vibrational spectrum calculation. See [39–41, 49] for a complete description of our theoretical investigations on the spectroscopy of Ala_2H^+, Ala_3H^+, and Ala_7H^+.

Our room temperature trajectories on Ala_2H^+ showed a highly floppy molecule with continuous isomerization dynamics between transA1 and transA2 conformers (see Fig. 3, right). This is followed by the time evolution of the dihedral angle $\Phi = C_2 - N - C_3 - C_4$ (top of Fig. 3): this angle distinguishes the two lower energy conformers of Ala_2H^+, transA1 with $\Phi = 198.1°$ and transA2 with $\Phi = 284.2°$ (from geometry optimizations at the BLYP/6-31G* level [40]). The continuous evolution of Φ seen on the plot roughly between 180° and 300° during the dynamics reveals the continuous exploration of the two wells of transA1 and transA2 at room temperature.

The free-energy curve along the dihedral coordinate Φ extracted from the dynamics (calculated as $G(\Phi) = -k_B T \ lnP(\Phi)$, where $P(\Phi)$ is the probability of occurrence of a certain angle Φ in the course of the dynamics, bottom of Fig. 3) shows a rather flat well for angles between 190° and 300°. There is thus no evident barrier between transA1 and transA2, only an overall free energy difference of

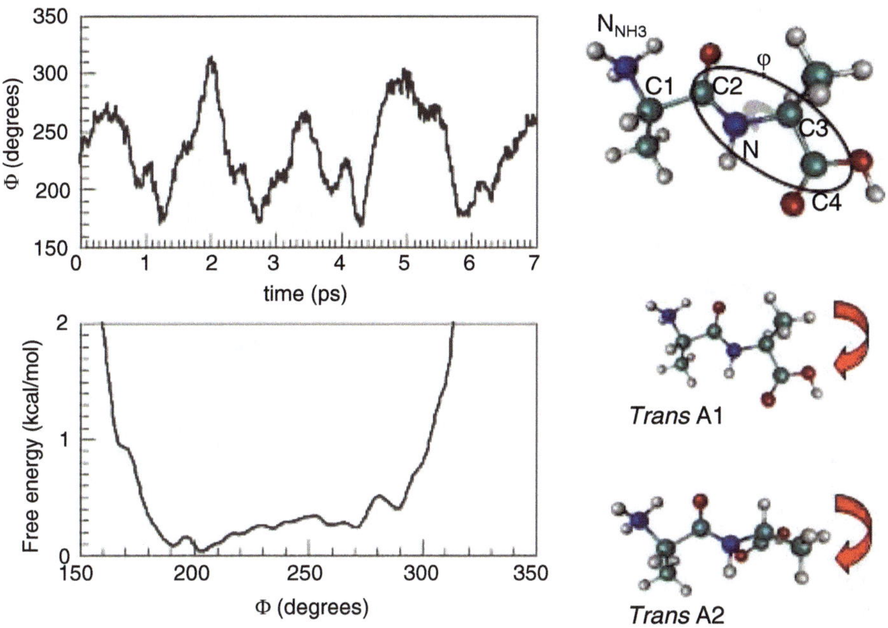

Fig. 3 Ala_2H^+ gas phase dynamics from [40]. *Top*: time evolution of the dihedral angle $\Phi = C_2 - N - C_3 - C_4$ for a typical trajectory of the trans isomer at 300 K. *Bottom*: free energy profile along the Φ dihedral angle averaged over all trans trajectories. *Right*: atom labeling for Ala_2H^+ and schematic representations of transA1/transA2 conformers

~0.3 kcal/mol between the two structures. This value is quite a distance away from the ~2 kcal/mol energy difference between the two optimized conformers, and from the ~2.5 kcal/mol [50] energy difference between transA1 and the transA1-transA2 transition state, computed at 0 K. This emphasizes the importance of entropic contributions in structural equilibria, naturally taken into account in the dynamics.

Overall, we found that the conformations where the N-terminal of the peptide is protonated $\left(NH_3^+\right)$ are predominantly populated, with a strong hydrogen bond between NH_3^+ and the neighboring carbonyl group. Moreover, there is enough energy for the NH_3^+ group to rotate and exchange the hydrogen atom that can be hydrogen bonded to the neighboring carbonyl, at room temperature. This is done a few times over the tens of picoseconds trajectories.

The dynamical anharmonic infrared spectra of gas phase Ala_2H^+ are presented in Figs. 4 and 5 (using (5)) and compared to the IR-MPD experiments from [104] in the 2,800–4,000 cm^{-1} domain (Fig. 4) and from [50] in the 1,000–2,000 cm^{-1} domain (Fig. 5). See [39, 40] for a more detailed description of results and assignments.

In the 3,000–4,000 cm^{-1} domain we can see that our dynamical spectrum reproduces very well the IR-MPD spectrum. The band located at 3,560 cm^{-1} corresponds to the O–H stretch of the C-terminal COOH group of the peptide, the

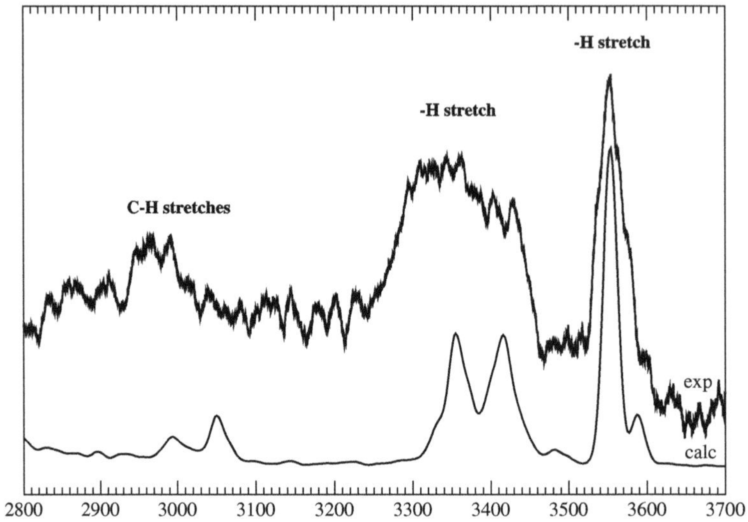

Fig. 4 Infrared spectrum of gas phase Ala_2H^+ in the trans form in the 2,800–4,000 cm^{-1} vibrational domain. Taken from [39]. IR-MPD Experiment (*top*), theoretical calculation from DFT-MD (*bottom*)

calculation matching the experimental band in terms of position and band shape. The shoulder located at ~3,590 cm^{-1} nicely reflects the feature also present in the experiment. The OH group is free of hydrogen bonding throughout the dynamics. The broad band located between 3,300 and 3,500 cm^{-1} (including the small feature at ~3,490 cm^{-1}) is produced by the N–H stretches from the amide N–H group and from the free N–H groups of the NH_3^+ N-terminal. The symmetric and antisymmetric stretches of the free N–H of the NH_3^+ can be seen at 3,415 and 3,370 cm^{-1}, respectively. The backbone amide N–H stretch also appears on the higher frequency part of the band, around 3,415 cm^{-1}. Although the calculated band nicely reflects the two parts which can be observed in the experimental band, it is nonetheless not broad enough in comparison to the experiment. The width of IR bands from molecular dynamics arises from a combination of temperature, conformational dynamics of the molecule, and anharmonicities (mode couplings, dipole anharmonicity, PES anharmonicity). The difference observed here is likely to be because the temperature of the simulations is not high enough.

Interestingly, the IR signature of the NH_3^+ hydrogen atom involved in the N–$H^+\cdots$O=C H-bond with the neighboring C=O carbonyl throughout the trajectories is spread over the broad 2,500–2,850 cm^{-1} region of the spectrum, strongly downshifted from the symmetric and antisymmetric modes of NH_3^+. Strong anharmonicity of the N–$H^+\cdots$O=C hydrogen bond are responsible for the displacement of this band to low frequencies, whereas mode-couplings anharmonicities and the dynamics of the hydrogen bond at finite temperature are responsible for its ~350 cm^{-1} spread. Indeed, the N–$H^+\cdots$O=C H-bond distance fluctuates by 0.3–

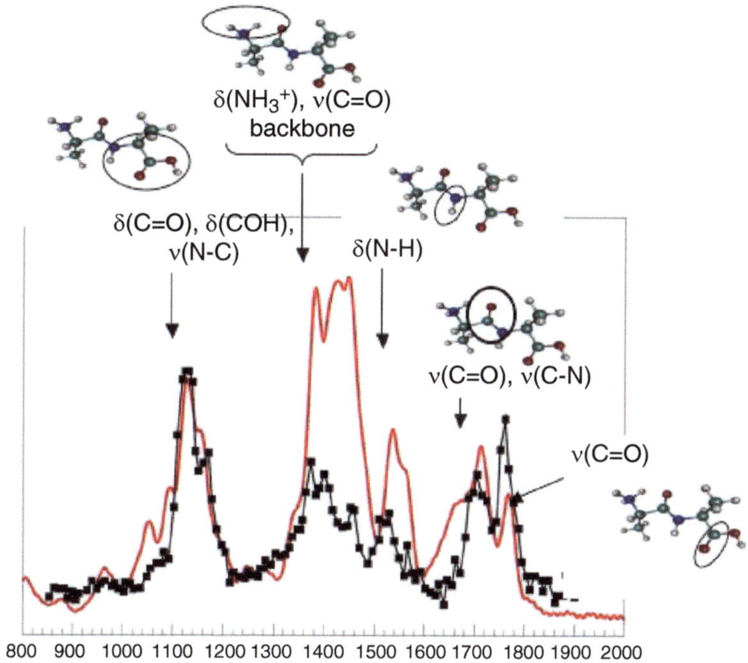

Fig. 5 Infrared spectrum of gas phase Ala_2H^+ in the trans form in the 800–2,000 cm^{-1} vibrational domain. Taken from [40]. *Squares*: IR-MPD experiment taken from [50]. *Solid curve*: dynamical spectrum from DFT-based molecular dynamics at ~300 K. The band assignment deduced from the vibrational density of state (VDOS) analysis is illustrated with schemes on *top of the bands*

0.4 Å around its mean value and there is enough energy for the NH_3^+ group to rotate and exchange the hydrogen atom, which can be hydrogen bonded to the neighboring carbonyl. These complex H-bond dynamics therefore lead to vibrational signatures which are spread over a large frequency domain. At the time of our publications [39, 40] there was no experimental signal recorded in this domain of 2,500–2,850 cm^{-1} to validate our theoretical spectral data. This has been recorded by the group of M.A. Johnson in 2013 [52] with tagging IR-PD experiments, confirming our dynamical spectral assignments.

The double band around 3,000–3,100 cm^{-1} is produced by the combined C_α–H and C–H stretch modes of the methyl groups of the peptide. These bands are up-shifted by ~40 cm^{-1} from experiment, and the band spacing (~50 cm^{-1}) is slightly bigger than the experimental one (~30 cm^{-1}). It is nonetheless remarkable that our calculation predicts an intensity of the band so close to experiment. The harmonic calculation [105] does not give such intensity to this mode.

In the 1,000–2,000 cm^{-1} mid-IR (see Fig. 5) the dynamical spectrum again matches the IR-MPD experiment extremely well, again without any scaling factor applied. Both experiment and calculation display three separate absorption domains, which are reproduced with very good accuracy by the dynamical spectrum in terms of both relative positions of the bands and bandwidths. The relative positions of the different bands and fine details such as the two sub-bands of the 1,100–1,200 cm^{-1} domain or the two distinct IR active bands of the 1,300–1,600 cm^{-1} domain are also very well reproduced. The two peaks located at 1,130 cm^{-1} and 1,150 cm^{-1} are produced by stretching and bending of the N- and C-terminals of the peptide. The band between 1,340 and 1,500 cm^{-1} arises from vibrations on the N-terminal side of the peptide, namely $\delta(N-H)$ bending motions of the NH_3^+ group coupled with skeletal $\nu(C-C)$ stretchings. The narrow band extending between 1,500 and 1,600 cm^{-1} is composed of the $\delta(N-H)$ amide II motion coupled with $\nu(N-C_2)$ stretching of the backbone. The 1,620–1,800 cm^{-1} domain is composed of three separate peaks in the dynamical spectrum: the 1,670 cm^{-1} peak comes from the amide I of the $C_2{=}O_2$ carbonyl group, the 1,720 cm^{-1} is related to a superposition of the two amide I motions, and the peak at 1,770 cm^{-1} involves $C_4{=}O_4$. The amide I domain is clearly composed of three bands of high intensity in our calculation, whereas the experiment offers only two such bands, although a third sub-peak of far lower intensity may be distinguished in the 1,660 cm^{-1} tail of this domain. Discrepancies in intensities between experiment and simulated dynamical spectrum can be traced back to the difference in the signals experimentally recorded (fragmentation yield) and calculated (linear IR absorption). See discussion in Sect. 3.1.

An important aspect from Fig. 3 that we have not discussed yet but now becomes pivotal in the discussion of the dynamical IR spectra concerns the sampling of the two lower energy conformers of Ala_2H^+ over time. This sampling can be seen in Fig. 3 by the number of times dihedral angle values typical of transA1 and transA2 conformers (respectively $\Phi \sim 198°$ and ~280°) have been obtained over the length of the trajectory. One can see that the majority of conformations sampled over time have angle values different from these two targets: in other words, the dynamics sample conformations outside the minima on the PES, which is also clear from the free energy profile in Fig. 3 with the rather flat transA1-transA2 well. The direct conclusion is that a Boltzmann weighted spectrum of the harmonic spectra of transA1 and transA2 is not useful for the interpretation of the IR-MPD experiment. These two conformers are indeed clearly not the ones of interest once temperature is included, but rather all conformers explored during the continuous transA1-transA2 conformational dynamics participate in the final vibrational features. The weight of each conformation explored during the dynamics is naturally taken into account in the vibrational spectrum.

We continue the exploration of the vibrational signatures of protonated alanine peptides by investigating Ala_3H^+. Three main structural families have been identified by Vaden et al. [104] and McMahon et al. [106] which can be summarized in the "NH_2 family" where the proton is located on the N-terminal carbonyl group of the tripeptide, "elongated NH_3^+" (denoted as the 'NH_3^+' family) where the proton is located on the N-terminal NH_3^+ and the peptide chain is extended (similar to transA1/transA2 structures for Ala_2H^+ above), and "folded NH_3^+" (denoted as 'folded' family) where the peptide chain is folded through an $NH_3^+ \ldots O=C$

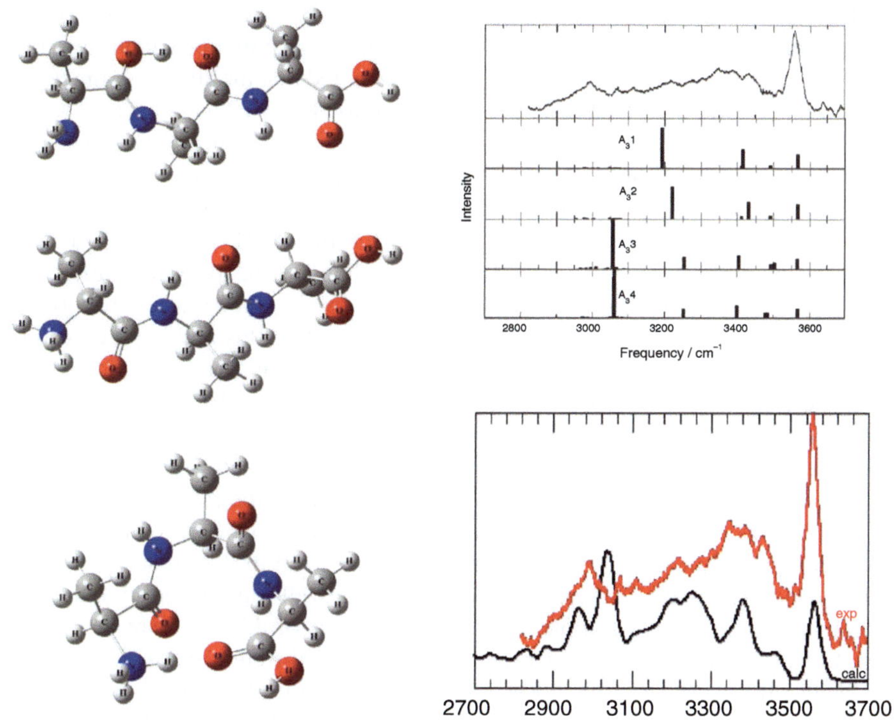

Fig. 6 Infrared spectrum of gas phase Ala_3H^+ for the 2,500–4,000 cm^{-1}. Taken from [39]. *Left*: conformational families of Ala_3H^+ taken into account for the dynamics. *Right*: (*top*) harmonic calculations from [104] and (*bottom*) comparison of IR-MPD experiment (exp) and dynamical spectrum from DFT-MD simulations (calc) [39]

H-bond. See Fig. 6 for illustrations. Energy barriers have not been characterized, but are presumably high enough with respect to 380–530 K temperatures of relevance for the IR-MPD experiments performed by the group of J.P. Simons in Oxford [107], so that conformational dynamics are not likely to occur during the timescale of our trajectories. Overall, we indeed do not observe spontaneous conformational interconversion/isomerization during the dynamics around room temperatures, but this becomes permitted around 500 K.

Considering the initial photochemical scheme used for the production of the ions in the experiment [107], trajectories have been accumulated at temperatures in the range 380–530 K, and have all been used in the statistical average for the dynamical IR spectrum (each trajectory has the same weight in the final average). The main dynamical behaviors seen in the trajectories correspond to H-bond dynamics and local geometrical reorganizations, including rotations of NH_3^+, CH_3, and COOH groups, which influence the vibrational features of the peptide as detailed below. The final spectrum of the protonated alanine tripeptide obtained from our DFT-MD trajectories is presented in Fig. 6 (bottom right). It agrees remarkably well with the IR-MPD experimental spectrum (see discussion on band-intensities in Sect. 3.1).

The band located at 3,560 cm^{-1} is produced by the O–H stretch of the C-terminus, identical to that in Ala_2H^+ and Ala_3H^+, corresponding to conformations with a similar free C-Terminal O–H. It should be noted that the slight asymmetry of this band is correctly reproduced by the calculations. The 3,100–3,500 cm^{-1} region can be separated into two parts. The 3,300–3,500 cm^{-1} frequency region is produced by the stretches of the N–H groups of the tripeptide that are not involved in hydrogen bonds along the dynamics, i.e., the symmetric and antisymmetric stretch of the N-terminal amine NH_2, the stretches of the free N–H of the NH_3^+, and the stretches of the C-terminal amide N–H, depending on the peptide family. The 3,100–3,300 cm^{-1} domain is uniquely produced by the stretching motion of the N-terminal amide N–H group. Our trajectories show that this amide group can be weakly hydrogen bonded to the amine N-terminal in the NH_2 tripeptide family (through a distorted H-bond), and also weakly hydrogen bonded to the C-terminal carbonyl of the peptide in the NH_3^+ family (again distorted H-bond). These H-bonds are, however, strong enough to induce the shift towards lower energy of the N-terminal N–H stretch in comparison to the positions of the free N–Hs.

The conformational diversity of Ala_3H^+ and the conformational dynamics of the N–H groups within these structures thus nicely show up in the broad N–H vibrational band. The overall shape of the N–H band is well reproduced by our dynamical spectrum, and the complex vibrational patterns of this band are therefore a result of the local dynamics of the N–H groups in the different Ala_3H^+ families.

Identically to Ala_2H^+, the 2,800–3,100 cm^{-1} domain of Ala_3H^+ is assigned to the C–H stretches arising from the methyls and the C_α–H groups of the tripeptide. Also, similar to the Ala_2H^+ peptide, the vibrational signatures of $N\text{–}H^+\cdots O{=}C$ in the NH_3^+ families are strongly red-shifted to a lower frequency compared to the other N–H stretches, and appear over the extended 2,000–2,800 cm^{-1} domain. Such a large 800 cm^{-1} spread is again entirely caused by the vibrational anharmonicities, mode-couplings, and dynamics of the $N\text{–}H^+\cdots O{=}C$ H-bond at finite temperature. Remarkably, the stretching of the protonated skeleton $C\text{–}O\text{–}H^+$ in the NH_2 family is superimposed on the $N\text{–}H^+$ stretches of the NH_3^+ families. It is consequently impossible to distinguish both families using the stretching patterns of the H-bonded $N\text{–}H^+$ or $C\text{–}O\text{–}H^+$ groups in the high frequency domain alone.

It is interesting to compare the scaled harmonic spectra of the three Ala_3H^+ peptide families with the dynamical anharmonic spectrum discussed above. This is done in Fig. 6, where the scaled harmonic spectra are taken from [104] for the four lowest optimized geometries (top right of Fig. 6). As can be immediately observed, the N–H broad band of the experimental spectrum is systematically associated with only two main intense harmonic bands, greatly separated by 200 and 150 cm^{-1} for the NH_2 and NH_3^+ families, respectively; hints of a third band located close to the ~3,490 cm^{-1} experimental band can also be seen, with a very low intensity. The interpretation given here from MD simulations that the broad and complex N–H vibrational band comes from the intrinsic local dynamics of the N–H groups in the different conformers/isomers of Ala_3H^+, including the breaking and forming of these H-bonds, can only be achieved when performing molecular dynamics simulations. Scaled harmonic spectra, possibly including a Boltzmann weighted average, are unable to give such a broad dynamical band. Last but not least, C–H harmonic modes predicted around 3,000 cm^{-1} have no intensity in the harmonic spectra, whereas the anharmonic spectrum extracted from MD correctly predicts the intensity in this region.

Comparing the spectra of Ala_2H^+ and Ala_3H^+ peptides, the first instructive result is that both peptides display a C-terminal O–H group free of any hydrogen bonding. This band is maintained with the size of the Ala_nH^+ peptide [104]. The N–H band is broadened in going from Ala_2H^+ to Ala_3H^+, which reflects more diverse conformations and conformational dynamics of the hydrogen bonds of the N–H groups of these peptides. For both peptides, the amides not involved in any H-bond (the single N–H of Ala_2H^+, the C-terminal N–H of Ala_3H^+, and the NH_2 group of Ala_3H^+) and the N–H groups of NH_3^+ also free of H-bonds, are collectively responsible for the 3,500–3,300 cm^{-1} vibrational band. The longer chain Ala_3H^+ has one amide at the N-terminal which can now be involved in several kinds of hydrogen bond patterns depending on the conformations. It is remarkable that this amide group is responsible for the extension of the N–H band towards the lower 3,100 cm^{-1} range. Such an extended band is experimentally observed [104] for increasing size of Ala_nH^+, even gaining more intensity for Ala_7H^+. The C–H vibrational band is conserved from Ala_2H^+ to Ala_3H^+ and throughout larger sizes, greatly extending between 3,100 and 2,900 cm^{-1}.

Going to the larger Ala_7H^+ peptide, the initial structures for the trajectories were taken from [104], and the three lower energy conformers were used as starting conformations for the dynamics (they lie within less than 20 kJ/mol of energy [104]). Ala_7H^+ are globular folded structures in which the N-terminal NH_3^+ is the central element for the folding. Considering the photochemical scheme used for the production of Ala_7H^+ [104], it is very likely that these conformations would equally participate in the IR-MPD spectrum. The three lower energy optimized structures contain charge-solvating $NH^+ \rightarrow O{=}C$ hydrogen bonds involving one (A_71), two (A_72), and three (A_73) NH groups. Seven trajectories have been accumulated, two initiated from A_73 optimized structure, two from A_72, and three from A_71, each for 3–5 ps.

The average conformation at room temperature differs from the frozen 0 K well-defined structures. Although the 0 K temperature drives the formation of simultaneous multiple hydrogen bonds between the NH_3^+ charged group and its immediate surroundings, dynamics at finite temperature does not favor such multiple H-bonds. Out of the seven BOMD trajectories, two provide a single $NH^+\cdots O{=}C$ H-bond formed on average as the most relevant statistical event, five provide two such H-bonds on average, and only one trajectory provides a three $NH\cdots O{=}C$ H-bonds situation over transient periods of time.

The two $NH\cdots O{=}C$ H-bonded situation is thus statistically the most probable at the temperature of 350 K. These H-bonds can form and break easily. It is not surprising that maintaining three $NH\cdots O{=}C$ H-bonds at finite temperature is rather difficult, as such a configuration provides a huge constraint over the whole

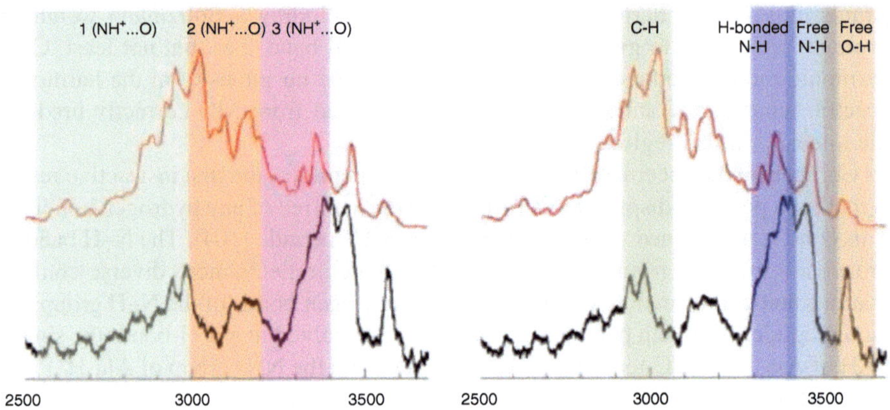

Fig. 7 Experimental IR-MPD spectrum from [104] (*bottom, black line*) and dynamical spectrum (*top, red line*) obtained as an average over the seven trajectories of Ala_7H^+ protonated peptide. Taken from [41]. The *left figure* has color-codings and labels explaining the assignments of the three vibrational domains in terms of average number of $NH^+\cdots O=C$ H-bonds formed in Ala_7H^+ peptide. The *right figure* displays the assignments in terms of stretchings of the O–H, N–H, N–H^+, and C–H (C_α–H and CH_3) motions

geometry of the peptide: the peptide does not comply easily with such a constraint, once temperature and entropic effects are taken into account. The (CO)OH C-terminal hydroxyl of Ala_7H^+ is also on average not hydrogen bonded to its surroundings. As expected, over the rather limited periods of time of the dynamics (5 ps), only the dynamical behavior of forming and breaking H-bonds has been observed in the dynamics and no other remarkable structural reorganization of the chain could be seen (this would require much longer simulations and possibly higher temperatures, as already seen for Ala_3H^+).

Figure 7 presents the IR-MPD experimental data extracted from [104] and the IR dynamical spectrum extracted from the seven trajectories of Ala_7H^+ peptide. Bands have been colored according to the nature of the stretching assignments. It is clear that all experimental spectral features are obtained in the dynamical spectrum. When comparing IR-MPD and calculated spectra, one has to be cautious discussing IR intensities, as they are not comparable between experiments and simulations; see Sect. 3.1. The band shapes and positions agree well. In particular, the new experimental feature appearing at 3,100–3,230 cm^{-1} for Ala_7H^+, not seen for smaller peptide chains [39], is present in the dynamical spectrum.

We do not discuss the vibrational assignments in detail; they can be found in [41] and they follow the trends found for the shorter peptides. One band discussed here, though, is the band observed in the 3,030–3,300 cm^{-1} which was absent for the lower peptide chain lengths. It is remarkably reproduced by the present dynamical spectrum, though slightly less broad. This band is entirely produced by N–H^+ stretches arising from conformations where the NH_3^+ N-terminal is involved in *two* hydrogen bonds on average. The region between 2,500 and 3,000 cm^{-1} is more complex for Ala_7H^+, as the highly anharmonic vibrations of the *singly* H-bonded

NH_3^+ conformations and H-bonded hydroxyl groups provide vibrational signatures in that domain on top of the C–H stretchings.

One can go one step further in the discussion of the N–H$^+$···O=C vibrational signatures in the Ala$_n$H$^+$ series. Ala$_2$H$^+$, Ala$_3$H$^+$, and some Ala$_7$H$^+$ conformers have one such H-bond on average at room temperature: its vibrational signature appears in the 2,000–3,000 cm^{-1} domain. This N–H$^+$···O=C H-bond is highly non-harmonic, and the vibrational domain reflects the strength and degree of anharmonicity of such H-bond. The strongest H-bonds are formed in the Ala$_3$H$^+$ peptide, where vibrational signatures down to 2,000 cm^{-1} have been characterized, whereas the two other peptides have signatures above 2,500 cm^{-1}. For Ala$_7$H$^+$, we have seen that the N–H$^+$···O=C strength decreases with increase in the number of H-bonds formed between NH_3^+ and the C=O groups. Hence, 1-H-bond conformers provide signatures in the 2,500–3,000 cm^{-1}, whereas 2-H-bonds conformers provide signatures in the 3,000–3,300 cm^{-1}, and 3-H-bonds conformers in the 3,300–3,400 cm^{-1}. In this latter case, the N–H$^+$···O=C vibrational signatures overlap with those arising from H-bonded N–H (neutral) amide groups.

It is also clear from the above analyses of the dynamics of Ala$_n$H$^+$ peptides that the knowledge of finite temperature properties is mandatory to characterize vibrational signatures of floppy molecules. Entropic effects are naturally taken into account in the dynamics, with consequences on conformational equilibria. Anharmonic vibrational effects, i.e., mode couplings, anharmonicities from H-bonds, and anharmonic motions on the PES, are naturally taken into account in the dynamics, without a priori knowledge. Furthermore, at finite temperatures, conformers with simultaneous N–H···O=C H-bonds might not be statistically relevant, as shown here for Ala$_7$H$^+$. Although the 0 K frozen picture provided conformers in which the NH_3^+ group was involved in three simultaneous H-bonds as the most stable structure, the finite temperature shows that only 2- and 1-H-bonds structures are statistically relevant. Once this conformational property is taken into account, the IR dynamical vibrational spectrum is in good agreement with the experiment.

5 IR-PD and Highly Anharmonic Ionic Clusters

The above examples of Ala$_n$H$^+$ peptides demonstrate that dynamical spectra are capable of capturing the dynamical behavior of the N–H$^+$···O=C H-bonds and their vibrational anharmonicities with remarkable accuracy, thus providing definitive assignments of the protonated alanine peptide structures produced in the gas phase. We now turn to the vibrational spectroscopy of even more challenging molecular systems, ionic clusters, displaying large vibrational anharmonicities. Details of our combined IR-PD experimental and theoretical dynamical investigations can be found in [37, 65], where the structures of Cl$^-$-(Methanol)$_{1,2}$ and Cl$^-$NMA$(H_2O)_{n=0\text{–}2}$ clusters (NMA = *N*-methyl-acetamide) have been unraveled

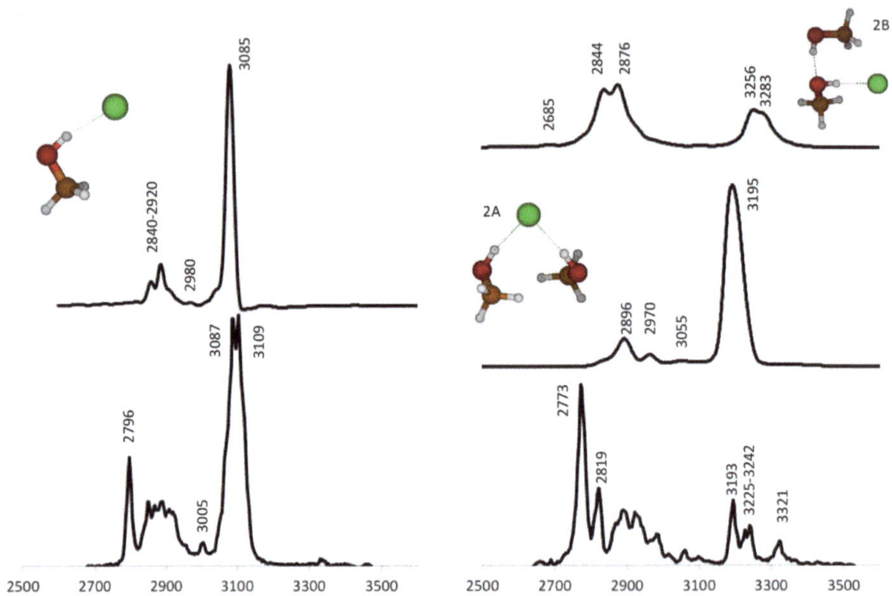

Fig. 8 Dynamical IR spectrum (100 K) of Cl^-–CH_3OH (*top left*) and Cl^-–$(CH_3OH)_2$ (*top right*) compared to the associated IR-PD experimental spectra of Cl^-–$(CH_3OH)Ar$ and Cl^-–$(CH_3OH)_2Ar$ (*bottom of each figure*). The *inserts* are illustrations of the structures of Cl^-–CH_3OH and Cl^-–$(CH_3OH)_2$. Taken from [65]

combining InfraRed Predissociation (IR-PD) experiments from the group of J. Lisy (University of Illinois at Urbana-Champaign) and DFT-based molecular dynamics simulations (DFT-MD). We highlight below some results on Cl^--(Methanol)$_{n=1-2}$ systems.

The experimental IR-PD and the dynamical spectrum of Cl^--CH_3OH complex are reported in Fig. 8 (left). The dynamical spectrum shows a prominent peak at 3,085 cm^{-1} closely matching the experimental 3,087–3,109 cm^{-1} band. The experiment displays a doublet, believed to result either from multiple argon binding sites or from the simultaneous presence of the staggered and eclipsed conformations of Cl^--$CH_3OH(Ar)$ in the experiment [108]. At the 100 K temperature of the dynamics, we have shown that there is a continuous conformational dynamics (rotation) of the methyl group without geometrical distinction between staggered and non-staggered orientations of this group over time. There are thus no distinct vibrational signatures of these specific orientations to be expected in the vibrational signatures of the cluster at finite temperature. The multiple binding sites of argon in the cluster (not taken into account in the dynamics) therefore remain the most probable source of the doublet.

The O–H stretch calculated from the dynamical IR spectrum is red-shifted from 3,682 cm^{-1} in uncomplexed methanol (our calculation with the same BOMD representation to be compared to 3,681 cm^{-1} in gas phase experiment [109]) to 3,085 cm^{-1} when the chloride anion binds to the methanol OH. This is in agreement with experiment: the dynamical IR spectrum thus provides a proper account of the large 600 cm^{-1} down shift of the O–H stretching once O–H is bound to Cl^-. No scaling factor has been applied in the dynamical spectrum. The methyl stretching motions provide the 2,840–2,920 cm^{-1} broad band in the dynamical spectrum, also in agreement with the IRPD features. Remarkably, one can distinguish a feature of low intensity in the dynamical spectrum located at 2,980 cm^{-1}, in agreement with the IRPD feature at 3,005 cm^{-1}. This was interpreted as a combination hot band in [110].

The IR-PD spectrum and dynamical spectra of the two isomers of Cl^--(Methanol)$_2$ are presented in Fig. 8 (right). The BOMD IR spectrum of isomer 2A shows two main bands in the 2,500–3,500 cm^{-1} domain: a very intense band located at 3,195 cm^{-1} produced by the O–H stretching motions of the two equivalent O–H groups, and two bands between 2,800 and 3,000 cm^{-1} produced by the methyl C–H motions. The peak of very small amplitude between 3,000 and 3,100 cm^{-1} is also produced by O–H. For isomer 2B, the double peak (3,256 and 3,283 cm^{-1}) arises from the O–H stretching motion of the hydrogen-bonded O–H group of the second methanol molecule. The broad 2,750–3,000 cm^{-1} band has overlapping contributions from the CH_3 stretching motions (2,850–3,000 cm^{-1}) and from the O–H stretching of the methanol hydrogen bonded to the Cl^- anion. This last contribution is at 2,844 cm^{-1} and is broadly active in the 2,750–2,900 cm^{-1} domain.

Clearly, the addition of a second methanol to Cl^--(Methanol)$_1$ in the second shell of Cl^- in isomer 2B, strengthens the strong OH···Cl^- hydrogen bond and thus induces a strong red-shift of the O–H stretch from 3,085 cm^{-1} (Cl^--(Methanol)$_1$) to 2,750–2,900 cm^{-1} (isomer 2B of Cl^--(Methanol)$_2$). For isomer 2A, the addition of a second methanol to Cl^--(Methanol)$_1$, with the two methanol molecules located in the first hydration shell of the anion, forming two ionic hydrogen bonds with Cl^-, then results in a weakening of the H-bond and therefore provides a shift of the O–H stretch to higher frequency, from 3,085 cm^{-1} (Cl^--(Methanol)$_1$) to 3,195 cm^{-1} (isomer 2A of Cl^--(Methanol)$_2$).

The O–H bands above 3,000 cm^{-1} in isomers 2A (a single intense peak) and 2B (double peak) match the positions and band-shapes of the bands observed experimentally between 3,150 and 3,300 cm^{-1}. The calculated O–H stretch for isomer 2A might be too intense in comparison to the experiment, but the presence of one band (3,193 cm^{-1} in the IR-PD) and a doublet (3,225 and 3,242 cm^{-1} in the IR-PD) are well reproduced by the dynamical spectra arising from the superposition of the spectra of both isomers. Intensity depends on the ratio of population of the two isomers; this ratio has been taken as 1:1 here. Similarly, the O–H stretch band coming from the O–H hydrogen bonded to Cl^- in isomer 2B, localized at 2,844 cm^{-1}, is very close to the 2,819 cm^{-1} band in the IR-PD.

Our dynamical spectra thus clearly suggest that the two isomers 2A and 2B of Cl^--(Methanol)$_2$ are simultaneously present in the experiment. Our dynamics and the RRKM calculations presented in [65] show that they do not spontaneously interconvert over the short 10 ps timescale of the ab initio dynamics, but that they can possibly interconvert over the hundred picosecond and nanosecond timescales at the 100 K temperature relevant to the experiment.

6 Cold IR-MPD and Vibrational Anharmonicities Below 1,000 cm^{-1}

The previous examples illustrated the strength of DFT-based molecular dynamics for vibrational spectroscopy, in particular for getting vibrational anharmonicities with good accuracy in the 1,000–4,000 cm^{-1} domain. In these examples, anharmonicities resulted from the formation of medium to very strong hydrogen bonds, from their associated local dynamics at finite temperature, and from mode-couplings in this high frequency domain. We now illustrate how dynamical spectroscopy can provide agreement with experiments in the domain below 1,000 cm^{-1}. That domain is expected to be strongly anharmonic, so dynamics-based theoretical spectroscopy should be a relevant method to apply.

This is demonstrated here with DFT-MD simulations to extract the vibrational spectrum of the Ac-Phe-Ala-NH_2 peptide (capped peptide composed of phenyl and alanine residues, labeled FA in the remainder of the text; see Fig. 9) in the far-IR domain, in relation to IR-UV ion-dip experiments [111]. With dynamical spectroscopy, one has to keep in mind that the length of the trajectories has to be commensurate with the domain investigated. Hence, to sample the vibrational modes and their couplings in the 100–1,000 cm^{-1} region, we have accumulated trajectories of 30 ps and make an average for the final theoretical spectrum over two trajectories (i.e., 60 ps sampling in total).

Figure 9 presents the dynamical spectrum calculated for structure FA1 along with the experimental spectrum in the 100–800 cm^{-1} region. The far-IR vibrations are extremely well reproduced by the DFT-MD simulation. Especially the vibrations below 500 cm^{-1} are in excellent agreement, with maximum deviations below 10 cm^{-1}, almost within the spectral experimental resolution, and the agreement in

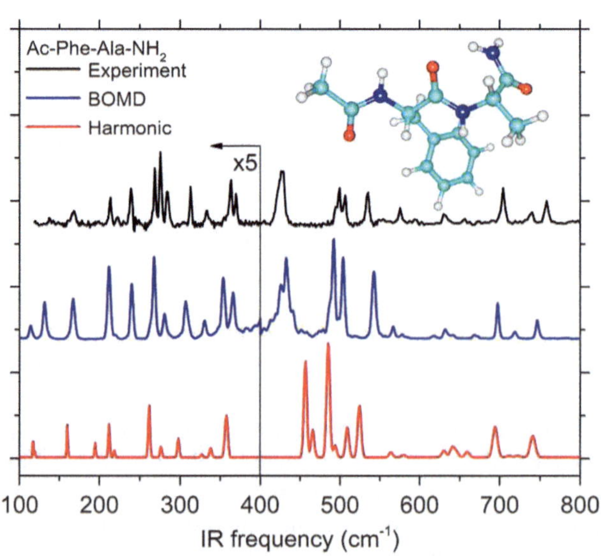

Fig. 9 IR-MPD experiment (*black*), static harmonic IR spectrum (*red*), and dynamical IR spectrum (*blue*) of Ac-Phe-Ala-NH_2 peptide in the far-IR domain. For clarity, amplitudes are multiplied by 5 below 400 cm^{-1} in all spectra. Currently submitted to Angew Chemie Int [111]

the 500–1,000 cm^{-1} domain is within 30 cm^{-1}. The dynamical anharmonic spectrum undoubtedly proves that the FA1 structure is responsible for the experimental vibrational features. See for comparison the harmonic spectrum of this structure, calculated with the same BLYP representation, where it can be seen that neither the positions nor the number of active bands match the experiment. In contrast, the dynamical anharmonic far-IR spectrum is in excellent agreement with the experiment, for the number of active bands, band-positions and shapes. Absolute band-intensities seem underestimated, but relative intensities are well reproduced by the simulations.

DFT-MD-based IR spectroscopy can be obtained with high accuracy in the far-IR vibrational domain, providing a definitive assignment of the experimental vibrational features. No scaling factors entered our final theoretical spectrum. Such an achievement is out of reach of harmonic spectra in that vibrational domain. DFT-MD simulations combined with experimental far-IR spectra are thus a promising accurate tool for the structural characterization of peptides structures in the far-IR vibrational region. Particularly for larger peptides, where the diagnostic amide vibrations in the mid-IR spectrum (amide I, II and III) may suffer from spectral congestion, the set of delocalized vibrations in the far-IR spectrum may form an interesting alternative for unraveling peptides structures from vibrational spectroscopy. For the small peptide shown here, DFT-MD has been able to discriminate between similar secondary structures using far-IR anharmonic signatures only (not shown; see [111]), which cannot be differentiated in the mid-IR spectrum.

7 Collision Induced Dissociation

In collision induced dissociation (CID) the ions selected by choosing a given m/z are sent with a given translational energy to the collision room where they collide with an inert gas (Ar, Ne, N_2, Xe ...). Thus, translational energy is converted in internal rovibrational energy, and then the ion can subsequently fragment. There are different techniques for activating ions which differ for the energy range employed, for the instrument, and for the activation mechanism, as reviewed by Sleno and Volmer [112], Laskin and Futrell [113], or Mayer and Poon [114]. In this domain again, theoretical calculations can help one to understand the CID processes and in particular provide an atomistic comprehension of the mechanisms and pathways leading to the final fragments. We thus review in Sect. 7 the most popular theoretical tools applied in the domain, i.e., RRKM statistical theory and direct molecular dynamics, and simultaneously review how and when these methods can be used and what knowledge can be gained. We illustrate with some examples taken from the literature and from our works.

As we have discussed in the introduction of this chapter, dissociation of ions induced by collision (with an inert gas, CID, or with a surface, SID) can be theoretically described by following the PES (Potential Energy Surface) from reactants to products. On this PES the reactivity can be statistical, thus being described by RRKM theory, or not-statistical. Here we thus review how the knowledge of PES is used to rationalize CID (Sect. 7.1), then the RRKM theory and its application to CID reactivity (Sect. 7.2), and finally how chemical dynamics can provide a framework for an interpretation of experiments taking into account reactivity on both short and long timescales (Sect. 7.3).

7.1 Potential Energy Surface

Following Born–Oppenheimer approximation, the reactivity of an activated system can be described as dynamically evolving on the potential energy surface, i.e., the multidimensional surface (a 3N−6 surface in internal coordinates, where N is the number of atoms of the dissociating molecule) obtained by solving the time-independent Schroedinger equation for the molecular system at fixed nuclei positions. The well-known methods of quantum chemistry are used [13] and they give access to energies and vibrational frequencies of reactants, products, intermediates, and transition states connecting them. A schematical PES is shown in Fig. 2, representing a model system. The reactants and (eventually) intermediates are geometries characterized by 3N−6 real frequencies (i.e., minima in the PES) and are obtained by means of well-know geometry optimization procedures. The transition states are points on the PES that connect two minima and, from a topological point of view, are obtained as saddle points on the PES characterized by one imaginary frequency (these are called “tight” transition states; we see later on how another type of TS can be defined).

As previously mentioned, the energy is obtained by solving the electronic structure problem. This gives the internal electronic energy to which the zero-point energy should be added to describe the actual potential felt by the molecule during the reaction. The zero point energy correction is obtained from vibrational normal modes, adding the energy of each harmonic oscillator associated to normal modes:

$$E^{ZPE} = \sum_{i}^{3N-6} \frac{1}{2}\hbar\omega_i \tag{8}$$

where ω_i is the frequency of the i normal mode and $\hbar = h/2\pi$ with h the Plank constant. Solving the PES problem thus needs an efficient algorithm to identify minima and saddle points (a problem common to the conformational sampling) and also the reaction channel(s). Furthermore, to have a full picture of the multidimensional surface, several energy points should be calculated, and thus the electronic structure problem should be solved in a relatively fast way. Thus the PES problem can be divided into two (connected) problems: (1) an efficient sampling of different conformers and reaction channels; (2) efficient way of calculating electronic energy of the molecular system. For small molecules both problems are solved by considering all possible isomerizations and reaction channels. For large systems, enhanced sampling techniques should be considered, whereas generally the conformational sampling is done by molecular dynamics at high temperature (with a simple Hamiltonian) and fragmentation pathways are obtained by moving the bonds and angles manually.

The reaction pathways identification (quite a tricky problem for large systems) is often guided by experiments: only products compatible with m/z observed in CID experiments are studied. This was the case for Ca^{2+}/formamide [115], Ca^{2+}/urea [116], or uracil interacting with Ca^{2+}, Pb^{2+}, Cu^{2+}, or Na^{+} [117] dissociation studies of Yanez, Salpin, and co-workers.

Traditionally, computational studies of reaction paths are performed by constructing a PES as extended as possible, i.e., locating minima and transition states. With this approach it was possible to understand the fragmentation mechanisms of metal-biomolecular systems, as reported in a series of studies by Ohanessian and co-workers on glycine-Cu^{+} [118], Zn^{2+} binding amino acids [119] or for a series of singly and doubly charged cations binding glycine [120].

Peptide fragmentation mechanism is non-trivial [113, 121, 122] and some studies have used PES to understand the different complex reaction channels. Several authors have thus studied PES of protonated glycine [123], glycylglycine [124], diglycine [125], small peptide models to understand the competition between b_2^+ and a_2^+ formation [126–128], glycylglycylglycine [129], or to investigate the model of mobile proton suggested to be at the basis of peptide gas phase reactivity [130, 131]. These calculations are often coupled with vibrational spectroscopy in the gas phase, which is used to validate the structure of the reaction products [132, 133].

When calculating the electronic energy, the level of theory chosen depends mainly on the system size. Although for small systems high correlated methods (such as CCSD(T) or CASPT2) can be used, when the system is larger DFT becomes compulsory. MP2 is also an option, but it needs an extended basis set to provide converged results. Furthermore, because PES deals with dissociation, basis set superposition error should be considered, and this is generally done by employing the Boys and Bernardi method [134]. In this field, in particular for systems containing metals, often the DFT method is first compared with high level calculations on small systems of the same kind (e.g., organic molecules in interaction with a metal) to identify the best functional (and basis set) which can be then applied to larger systems. This was done, for example, by the Yanez group when studying Sr^{2+}–formamide, pointing out that the G96LYP functional reproduces CCSD(T) calculations [135] and then uses this functional when studying the PES [136]. An important aspect is that to calculate the ZPE it is necessary to obtain harmonic frequencies. To this end the Hessian matrix has to be constructed and then diagonalized, and this is often possible only at a low level of theory. Thus, for example, in the case of La^{3+}, Gd^{3+}, and Sm^{3+} interacting with NH_3, the DFT and MP2 electronic energies (without inclusion of ZPE) were first validated against CCSD(T) and CASPT2 calculations and then used to describe reactivity

[137]. Furthermore, when dealing with large systems, it is possible to obtain the electronic energy at a level of theory and then add the ZPE obtained at a lower one.

Before discussing RRKM statistical theory and how it is applied to study CID reactivity, it should be noted that transition states located by topological analysis of the PES (i.e., saddle points) are a particular kind of transition state. In fact, a more general definition of transition state comes from the variational transition state theory (V-TST) which is related to reaction kinetics [138, 139]. Saddle points are always transition states but they are special cases, called tight transition states. In the case of dissociation pathways, when the two parts of the molecule are broken apart there is no saddle point in the reaction coordinate but a loose transition state exists. This is defined better in the next section. Finally, it should be mentioned that a relatively different formulation of transition state is given by Vanden-Eijnden, Ciccotti, and co-workers to study rare event sampling [140]. In this framework, a transition state is defined as an iso-committor surface (in both position and momenta) where the probability to go to the products is equal to those to come back to the reactants. Although it has never been applied to CID reactivity, it would be particularly intriguing, especially in the most general formulation which considers both position and momenta of the system.

7.2 RRKM Theory

RRKM theory is the well-known and consolidated statistical theory for unimolecular dissociation. It was developed in the late 1920s by Rice and Ramsperger [141, 142] and Kassel [143], who treated a system as an assembly of s identical harmonic oscillators. One oscillator is truncated at the activation energy E_0. The theory disregards any quantum effect and the approximation of having all identical is too crude, such that the derived equation for micro canonical rate constant, $k(E)$,

$$k(E) = \nu \left(\frac{E - E_0}{E} \right)^{s-1} \tag{9}$$

is incapable of providing the correct rate. The theory is based on a concept still present in the actual formulation: energy flows statistically among all the oscillators and the chance of finding the system with a particular arrangement of its internal energy is equivalent to any other. Later, Marcus and Rice [144] and Rosenstock, Wallenstein, Wahrhaftig, and Eyring [145] developed the actual version of the theory which takes into account the vibrational (and rotational) degrees of freedom in detail, leading to the well-known RRKM/QET (for quasi-equilibrium theory) expression. In its simple formulation (i.e., neglecting rotational or tunneling effects), reads simply

$$k(E) = \frac{\sigma N^{\ddagger}(E - E_0)}{h\rho(E)} \tag{10}$$

where $\rho(E)$ is the reactant vibrational density of states, $N^{\ddagger}(E - E_0)$ is the sum of states in the transition state, and σ is the reaction symmetry factor. This expression is very simple and powerful: from geometry and vibrational frequencies of reactant and transition state it is possible to obtain reaction rate constants as a function of the energy injected into the system. The calculation of sum and density of states can be done using the direct count method with the implementation proposed by Beyer and Swineheart [146]. Earlier, Whitten and Rabinovich proposed a method based on classical sum of states formulation with empirical parameterization to save computing time [147, 148]. Nowadays it is rarely used because the computational power available makes both calculation of frequencies and direct counting accessible. Thus, the question, from theoretical point of view, is to identify minima and transition states; this comes back to correct exploration of the PES (see Sect. 7.1), considering minima and transition states for all possible isomerizations and reaction channels, and having access to the energy given to the system in the collision process. An important aspect to highlight is that the resulting rate constant depends only on the total energy E and the total angular momentum in the more general formulation that takes into account rotational energy. In fact, it is assumed that the rate constant does not depend upon where the energy is initially located and that a microcanonical ensemble is maintained as the molecule dissociates. This is equivalent to the assumption that IVR is rapid compared to the lifetime with respect to dissociation. That is, vibrations are assumed to be strongly coupled by higher order terms in the expansion of the potential energy function. This means that after a collision the energy is likely to be in proportion to its equilibrium probability and that the states at that energy are all equally probable. To what extent this "ideal" is attained and what are the possible other situations was described in detail by Bunker and Hase [149]. Despite the intrinsic limitation of an "ideal" theory, RRKM is able finally to take into account many examples of unimolecular dissociation produced by collisional activation, as reviewed, for example by Baer and Mayer [150]. The simple RRKM formulation of (10) can be extended, taking into account rotational energy [151], tunneling effects [152, 153], and non-adiabatic transitions between different spin states [154].

RRKM theory is also at the basis of localization of "loose" transition states in the PES. Another assumption of the theory is that a "critical configuration" exists (commonly called transition state or activated complex) which separates internal states of the reactant from those of the products. In classical dynamics this is what is represented by a dividing surface separating reactant and product phase spaces. Furthermore, RRKM theory makes use of the transition state theory assumption: once the system has passed this barrier it never comes back. Here we do not want to discuss the limits of this assumption (this was done extensively for the liquid phase [155] but less in the gas phase; for large molecules we can have a situation similar to systems in a dynamical solvent, where the non-reacting sub-system plays the role of the solvent in liquid phase), but how RRKM theory is useful for TS searching. We have to remember that a reaction in statistical theory is represented by a flux in phase space and the TS corresponds to a dividing surface at which this flux is a minimum. In the previous section we described the TS localization on the PES, a point with a particular topological feature: one imaginary and 3N−7 real frequencies. This is called "tight" transition state and the localization of this dividing surface is dominated by the energy contribution to determining the TS sum of states. However, in dissociation phenomena, we often have a PES from reactant to product without any saddle point. In this case, the transition state is defined and thus located by the "variational transition state theory" (V-TST) [138, 139]. The TS is a minimum in the sum of states along the reaction coordinate:

$$\frac{dN^{\ddagger}(E,J,R)}{dR} = 0 \tag{11}$$

Thus all points of the PES can be obtained (in principle) and unimolecular dissociation rate constants expressed as a function of internal energy. Nowadays, the vibrational frequencies can be calculated at a high level of theory, and the sum and density of states obtained by direct count without any particular problem for small or medium sized systems, and thus RRKM kinetics were employed to examine several unimolecular dissociations [150]. Even complex kinetic schemes were solved to obtain the rate constant for product formation through different reaction channels [156, 157].

7.3 Chemical Dynamics

Static (PES) and kinetic (RRKM) information can be complemented by chemical dynamics simulations, which are able to fill some aspects of gas phase reactivity not considered by the previous approaches. In particular, chemical dynamics can be used to model explicitly the collision between the ion and the target atom and thus it is possible to obtain the energy transferred in the collision and (eventually) the reactions. The molecular system, represented as an ensemble of atoms each bearing a mass m_i, evolves on the Born–Oppenheimer potential energy surface through Newton's equation of motions:

$$-\frac{\partial V}{\partial q_i} = m_i \frac{d^2 q_i}{dt^2} \tag{12}$$

This means that the molecule's nuclei positions, q_i, and momenta, p_i, evolve on the potential energy obtained by solving the time-independent Schroedinger equation at each configuration.

This general chemical dynamics (or direct dynamics) approach can be applied to study different molecular problems (see, e.g., the recent review by Hase, Song, and

Gordon [158]), and the application to problems related to CID was pioneered mainly by Hase, Bunker, and co-workers. In particular, the random collision of an atom with a molecule was discussed (and an operative algorithm is given) for atom–molecule reaction on pre-determined potential energy surfaces [159] and the initial conditions of the molecule were discussed in the framework of state-selected unimolecular dissociation [160, 161]. The same procedure, also valid for non-reactive collisions, and the general chemical dynamics code VENUS, developed by Hase and co-workers [162, 163], can be used for different reactivity situations. In particular for CID, one has to consider initial conditions of the ion and the rare gas atom (or molecule). Initial internal energy of the ion can be considered in different ways: (1) sampling normal mode vibrational energies from a Boltzmann distribution at a given temperature (T_{vib}); (2) sampling a microcanonical ensemble by orthant or normal mode sampling [164, 165]; (3) sampling fixed vibrational normal modes; and (4) adding energy to a particular mode (local mode sampling [166]) on a previously determined total energy. These initial conditions are generally classically sampled (or semi-classically because one considers normal mode zero point energies) and recently the effect of quantum initial conditions on unimolecular reactions was also considered [167].

Rotational energy and angular momentum for the polyatomic molecule is selected by assuming separability of vibrational and rotational motion. Thus initial rotational conditions are obtained either by assuming a thermal partitioning of RT/2 about each internal rotational axis or by assuming that the molecule is a symmetric top ($I_x < I_y = I_z$) and then the total angular momentum and its x component are sampled from the probability distributions:

$$P(J_x) = \exp\left[-J_x^2/2I_xk_bT\right] \quad 0 \leq J_x \leq \infty \tag{13}$$

$$P(J) = J\exp\left[-J^2/2I_zk_bT\right] \quad J_x \leq J \leq \infty \tag{14}$$

as described by Bunker and Goring-Simpson [168].

Then the ion-projectile relative energy is set and random orientations in Euler angles between the (rigid body) ion and the projectile (often an atom, in the case of N_2 the center of mass is considered and vibrational and rotational sampling is done also to set its internal initial conditions) are sampled. Then the possible impact parameters are considered. They can either be set to a unique value or an ensemble of possible values can be sampled in a defined range generally corresponding to the molecular size. Finally, the collision is carried out at a given energy defined in the center-of-mass of the system composed by the ion and the projectile, E_{CM}, which is in relation to the laboratory framework energy, E_{LAB} (directly set in the instrument):

$$E_{CM} = \frac{m_2}{m_1 + m_2} E_{LAB} \tag{15}$$

where m_1 is the mass of ion and m_2 of the projectile. A detailed review of collision dynamics in CID is reported by Douglas [169].

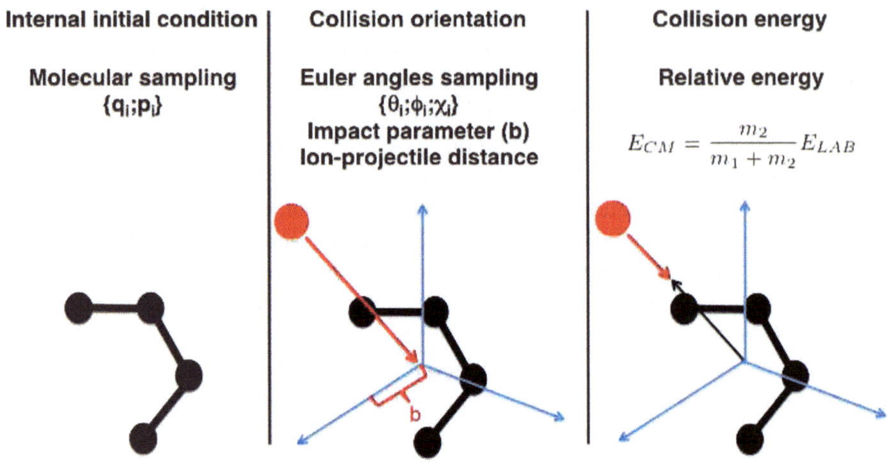

Fig. 10 Schematic representation of how the CID molecular dynamics is generally set-up (see Hase et al. [159] for details)

This procedure implemented in the chemical dynamics code VENUS [162, 163] is suited to mimic the collision (in the single collision limit hypothesis). Simulation parameters (temperature, energy, etc.. . .) can be tuned as needed. A schematic picture chemical dynamics set-up for simulating collisions is shown in Fig. 10.

Initially, direct dynamics simulations were performed using analytical potential energy functions for both intra- and intermolecular interactions. Reactivity (generally of a cluster) was initially taken into account by analytical Morse functions. Energy transfer was studied by using non-reactive molecular mechanics force fields, examining its dependence on the collision energy, the nature of the projectile, or the molecular shape of the ion. Energy transfer was studied, for example, in the case of trans-stilbene by Bolton and Nordholm [170]. Lim and Garret studied the collision between different gases and vibrationally excited azulene, finding that using simple ion-projectile analytical function results are in good agreement with experiments in particular for heavy gases (Ar, Kr, Xe) and that the energy transfer is dominated by the repulsive part of the potential [171]. In this way, energy transfer in Al clusters was studied as a function of cluster size and shape, pointing out that, when in the spherical structure, the translational energy is mainly transferred into internal vibrational energy; on the other hand, in planar clusters the rotational excitation is relevant [172]. A clear characterization of rotational vs vibrational activation is important when the RRKM reactivity is considered and in particular if the reaction channels have different rotational barriers (rotational energy is important, particularly when reactant and TS do not have the same moment of inertia). Furthermore, the energy transfer in octahedral Al_6 clusters was studied with different projectiles [173] (Ne, Ar, Xe) finding similar results, and the energy transfer could be qualitatively interpreted by means of the simple refined impulsive model developed by Mahan [174]. More recently, Hase and co-workers applied chemical dynamics to peptide CID, using AMBER force field [175] for intramolecular potential energy and an analytical function for the ion–projectile (Ar) interaction obtained from high level quantum chemistry calculations on building blocks [176], being purely repulsive:

$$V_{Ar-i} = A_{Ar-i}\exp(-b_{Ar-i}r_{Ar-i}) + \frac{C_{Ar-i}}{r_{Ar-i}^9}. \tag{16}$$

In this way, they were able to study energy transfer in peptide CID as a function of peptide shape [177].

By employing an analytical potential describing Cr^+–CO interaction, Martinez-Nunez et al. [178] have studied the CID of a $Cr(CO)_6^+$ cluster colliding with Xe, where they found that reaction dynamics is highly affected by "impulsive character." This corresponds to the so-called shattering reaction mechanism, where the bond is broken largely before energy transfer between vibrational modes. It should be noted that the shattering mechanism can be important in driving the reactivity, as was, for example, noticed experimentally and theoretically by studying the dissociation of CH_3SH^+ and $CH_3SCH_3^+$ [179, 180]. Shattering is a reaction mechanism that shirks the basic assumptions of statistical theory and thus any prediction based on this theoretical framework. Dynamics can thus be fundamental to point out whether this kind of direct dissociation mechanism plays a role in the appearance of reaction products which cannot be explained from merely statistical considerations.

Later, Hase and co-workers developed couplings between VENUS and quantum chemistry codes such as Gaussian, GAMESS-US, MOPAC, and NW-Chem, such that the ion reactivity could be considered directly (i.e., without any prior knowledge of the PES). A QM description can be used to treat either the whole system (ion and projectile) or only the intramolecular ion potential, whereas the ion-projectile intermolecular interaction is treated via an analytical potential, similar to that of (16). A full QM description, at B3LYP level, was employed in studying fragmentation of H_2CO^+ colliding with Ne [181], finding good agreement with experiments for the collisional cross section, σ_{CID}. This quantity is obtained from simulations, simply by evaluating the reactive probability as a function of impact parameter, $P(b)$, and then integrating over the whole b range:

$$\sigma_{CID} = 2\pi\int_0^{b_{max}} P(b)bdb \tag{17}$$

This study was able to show the main processes that can be activated by collision: (1) conversion of collisional energy into internal energy of scattered H_2CO^+; (2) sequential activation of a bond and reactivity observed in the simulation time length (200 fs); (3) direct reaction after the collision (here H atom knock out). It is also interesting to note how the energy transfer is affected by collision orientation and, to some extent, the initial vibrational preparation of the ion. Full QM studies can be applied to relatively small systems, with the advantage of being able to describe in detail the effects of projectile nature in reactivity without any

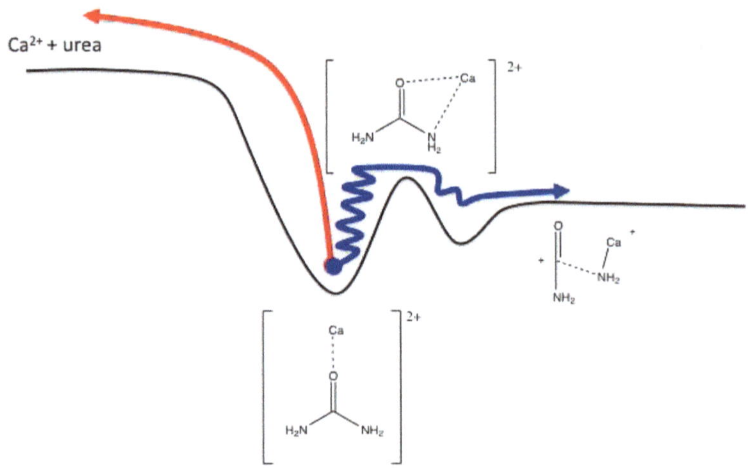

Fig. 11 Schematic representation of reactivity obtained in CID of $[CaUrea]^{2+}$ by Spezia et al. [185], where neutral loss (Ca^{2+} + urea) pathway is obtained by a "direct" mechanism whereas the Coulomb explosion $\left(caNH_2^+ + NH_2CO^+\right)$ pathway is obtained by a structural reorganization

parameterization and taking into account the possibility of charge transfer to the ion, as pointed out by Anderson and co-workers studying the reactivity of NO_2^+ with rare gases [182].

By using a QM + MM approach it was possible to treat bigger systems, with a reasonable statistical sampling. With this approach our group studied the reactivity of protonated urea (the QM part was done at the MP2 level of theory) showing how the shattering mechanism is responsible for opening the reaction channel, leading to the high energy product observed experimentally [183]. We have further studied the role of the projectile in energy transfer: a diatomic projectile (N_2) transfers less energy with respect to a monoatomic gas (Ar) [184]. Interestingly, it was found that the rotational activation decreases as the initial rotational quantum number of the projectile increases. This was qualitatively justified by the constraint of conserving total angular momentum, but more studies on rotational activation are needed. The same approach was extended to study reactivity of doubly-charged cations, $CaUrea^{2+}$ [185]. Here, dynamics showed how the neutral loss pathway is obtained in direct activation whereas the Coulomb explosion, occurring after molecular reorganization, is obtained at longer timescales (schematically shown in Fig. 11). In the same study, energy transfer obtained from dynamics was used in statistical RRKM approaches.

To improve statistics, in both length of time and system size, the QM part can be treated by employing semi-empirical Hamiltonians. In this way, it was possible to study peptide reactivity, as done by Hase and co-workers for reactivity of N-protonated glycine [186] in both CID and SID, suggesting that "nonstatistical fragmentation dynamics may be important in the collisional dissociation of protonated amino acids and peptides." In this study the CID impact parameter was set to zero, and the shattering was observed mainly in SID, showing that, also in this case, it is initiated by particular collisional orientations. AM1 results were in agreement with both Amber- (for energy transfer) and (for reactivity) MP2-based simulations [187], thus paving the way for using semi-empirical Hamiltonians in the puzzling study of gas phase reactivity of peptides. Protonated octaglycine SID was studied by Park et al. [188], showing a dependence on the relevance of shattering on the collisional orientation between the ion and the surface. More recently, in our group we have studied the CID reactivity of *N*-formylalanylamide (a model dipeptide), pointing out that the semi-empirical Hamiltonian is able to provide a picture of reactivity in agreement with experiments, such that it becomes possible to provide a semi-quantitative theoretical MS-MS spectra [189]. Furthermore, these results strengthen the "mobile" proton model [190, 191] which triggers the peptides gas phase reactivity. Currently, we are studying different related systems to understand better the molecular details of such a rich reactivity.

The use of a semi-empirical Hamiltonian also makes possible the study of negative polyanions. By coupling QM + MM chemical dynamics simulations with ESI-MS/MS experiments, we have recently investigated the fragmentation of a sulfated saccharide (the galactose-6-sulfate), determining for the first time reaction mechanisms of a sugar by a combined computational and experimental approach [192]. Furthermore, Bednarski et al. studied dissociation of poly[(*R*,*S*)-3-hydroxybutanoic acid] anions, in which molecular dynamics was able to rationalize the discrepancy between experiments and RRKM analysis [193]. The reactivity of Li^+–Uracil was recently studied with a specifically adjusted semi-empirical AM1 Hamiltonian, as reported by Martínez-Nunez and co-workers [194]. It should be noted that in this case, as well as for the protonated urea reactivity [184], once the transition state was identified and the energy transfer obtained, it was possible to use dynamics to study how a system evolves from the transition states towards different reaction channels. In particular, post-transition state dynamics show that, starting from a transition state, other products can also be obtained and not just that identified by the transition state.

A QM (or QM + MM) approach in studying CID reactivity makes possible, in principle, the study of the whole process: collision, energy transfer, energy flow through modes (if fast enough), and eventually fragmentation. Thus, we require a tool to provide us with a complete answer to the question of what is the reaction of an ion in the CID process: unfortunately there are several issues making this goal difficult to achieve. The main limitation is in the time available in simulations and in a lack of sufficient statistical sampling. Even when employing faster methods (such as a semi-empirical Hamiltonian), we can observe processes in the pico-seconds timescale, whereas reactivity (especially for lager systems) may occur on much longer timescales. In other words, direct dynamics can be used to obtain, on the one hand, fast direct reactivity (from hundreds of femtoseconds to tens of picoseconds) and on the other the energy transfer which can be used in statistical kinetics theory (such as RRKM) to evaluate reactivity. The combination of fast and slow reaction timescales is what physically provides the final product distribution obtained in tandem mass spectrometry experiments, and chemical dynamics can be used to rationalize the appearance of different fragments. For example, as stated in the previously mentioned studies from the groups of Hase, Anderson, and ourselves, the "shattering" reaction mechanisms are often strongly related to the orientation of the collision. Thus the shape of the ion can be determinant for the appearance (and the relative population) of some reaction pathways.

8 Conclusions and Some Future Directions

This chapter was intended to give some flavors of molecular dynamics-based methods for the calculation of gas phase vibrational spectra and collision induced dissociations of gas phase molecules and complexes. The examples were taken from our own work. We hope we have convincingly shown the essential ingredients entering these simulations, their usefulness in relation to IR-MPD and IR-PD action spectroscopic experiments, their usefulness in relation to CID experiments, and how essential these methods are to produce definitive assignments and microscopic interpretation of experimental features.

We now conclude this chapter by a discussion of what we think are the challenges to be tackled by theoreticians in the years to come in the two domains of gas phase spectroscopy and collision-induced dissociation modeling, but also some challenges that, we theoreticians, would like to suggest to the experimentalists to strengthen our knowledge of structural and dynamical information of gas phase molecular assemblies.

One systematic issue in molecular dynamics (MD) simulations is the timescales and lengths. Our anharmonic spectra were calculated with DFT-based MD, where typical timescales and lengths are a few $10'$ of picoseconds and molecular systems composed of ~500 atoms. Investigations of anharmonic spectra of peptides in the far-IR domain require trajectories of at least 30–50 ps for relevant sampling of the mode-couplings. As shown in Sect. 6, dynamical spectra are of excellent accuracy and thus can help unravel peptidic conformations using a vibrational domain where the vibrational signatures are expected to be less congested than in the 1,000–2,000 cm^{-1} and 3,000–4,000 cm^{-1} domains. Going beyond these timescales, which might evidently be necessary for more complex peptides than those presented in Sect. 6, and beyond the current sizes in order to investigate peptides composed of several tens of residues, semi-empirical-based MD should be the method to use. Its advantage is that it is still based on electronic representations, though simplified, and does not rely as heavily as classical force fields on large scale parameterization of the analytical force field expressions. Semi-empirical MD has been applied to our trajectories for collision induced dissociation, with great success. This thus seems a promising avenue for dynamical spectroscopy which we have already engaged.

Quantum effects of the nuclei have not been taken into account in the classical nuclear dynamics presented here. We currently follow strategies based on the introduction of Zero Point Energy (ZPE) within the modes at the initial time of the dynamics to include quantum effects in the classical nuclear trajectories. We have implemented such strategies [37, 72] and applied them to the dynamical spectroscopy of ionic clusters [37] in relation to IR-PD experiments. Theoretical issues such as Zero Point Leak [195, 196] can be encountered, however, and we are currently working in that domain. Closely related are Fermi resonances, also investigated by our group.

The experimental methods employed for the production of the molecular assemblies in the gas phase can be an issue for the subsequent spectroscopic characterization. In some of the IR-MPD experiments, ions are produced with the typical techniques of mass spectrometry, typically ESI (Electro Spray Ionization) and MALDI (Matrix-Assisted Laser Desorption/Ionization). In others, supersonic expansions are used to form peptides at rather low temperature. In some of the IR-PD experiments, the ionic clusters are formed by a combination of supersonic expansion and subsequent collisional processes. In these experiments in particular, it has been shown that higher energy conformers can be kinetically trapped. Dynamics simulations and associated RRKM calculations can be used to clarify the processes and mechanisms leading to the formation of higher energy conformers, and prevent isomerization towards lower energy conformers. Such simulations are currently underway in our group.

Certainly of importance to theoreticians (and surely to experimentalists) is the direct modeling of the action spectroscopy intensity signals. It should be remembered that static and dynamical spectra calculations are related to the linear one-photon absorption signal, whereas the experiments record the consequent action of this absorption in terms of molecular fragmentation. Theory and experiments are by no means equivalent, and, as discussed in the introduction of this chapter, deviation between experiment and theory is expected on band-intensities. Theoretical work has to be done for a direct modeling of the signal. Kinetic-based models and kinetic modeling are certainly the approaches worth following.

What challenges can we suggest to IR-MPD/IR-PD experimentalists? We strongly encourage developments of set-ups in the area of temperature-dependent spectroscopy. This would open up the path to the direct experimental sampling of the different conformers of molecules and clusters with an increase of internal energy (of course this is highly dependent on energy barriers) through the direct temperature-evolution of the IR signatures. This would also be one way to probe vibrational anharmonicities gradually, and the appearance of different kinds of anharmonicities. Temperature is the natural parameter of molecular dynamics simulations, so the interplay with MD simulations is undoubtedly pivotal for the interpretation of temperature-dependent IR features. This is already under way in our group in synergy with experiments of ions trapping from the Lisy group, where temperature effects can be investigated by switching between different fragmentation channels. Another crucial challenge is multi-dimensional gas phase IR spectroscopy, following the spirit of the now rather well-established 2D-IR spectroscopy conducted in the liquid phase and at solid-liquid and liquid-air interfaces. These two-dimensional spectroscopies provide direct information on mode-couplings, and they simultaneously reduce the possible congestion of spectral information seen in linear (1D) spectra of complex and large peptides. There again, MD-based spectroscopy is an essential theoretical tool for the interpretation of the experiments, as already seen in several publications of 2D-IR spectroscopy in the liquid phase. Closely related are vibrational excitations and subsequent time-dependent IR spectroscopy, the development of which we also strongly encourage, following the spirit of the latest experimental developments of time-dependent linear IR in liquids, or time-dependent 2D-IR in liquids and at interfaces. There again, MD time-dependent simulations are essential as theoretical support. Two-dimensional and time-dependent spectroscopic experiments not only reveal time-dependent properties but also reveal couplings, both properties being the essential ingredients of MD-based theoretical spectroscopy.

In CID modeling we believe chemical dynamics simulations can be the basis for defining a complete theoretical procedure to obtain the tandem mass spectrometry (MS/MS) spectrum of a given molecule from nothing. In fact, although for other experiments (e.g., InfraRed, NMR, Raman, UV-Vis) theoretical and computational methods can provide the final spectrum of a given molecular species, this is not currently the case for MS/MS. Thus, if we write down a new or experimentally unknown species, it is almost impossible to infer a theoretical MS/MS spectrum prior to experiments. We believe this is the challenge for theoretical modeling in collision-induced dissociation for years to come. Combining different dynamics approaches for both short and long timescales (as already presented here), it should be possible to devise a theoretical procedure to obtain the MS/MS spectrum of any given molecular species from its chemical and reactive properties. To accomplish this, chemical dynamics and RRKM statistical theory of unimolecular dissociation can be combined, providing the means for the theoretical framework of the different reaction pathways leading to gas phase fragmentation. To this end, developments in the quantum description, in particular for complex biomolecules, are needed, together with a theoretical framework for ro-vibrational partitioning of energy transfer. Connected with the energy transfer problem, particular care should be devoted to model multiple collisions. This further bridges the gap between different timescales involved in the fragmentation processes. These are general directions in which we are currently working.

Acknowledgement MPG would like to dedicate this chapter to the memory of Prof Jean-Pierre Schermann, mentor and friend, and one of the pioneers of gas phase spectroscopy. The theoretical work from our group presented here could not have been done without PhD and Post-Doc collaborators, past and present, among whom we would like to acknowledge Dr Codruta Marinica, Dr Michael Martinez, Dr Alvaro Cimas, Morgane Pfeiffer, Sacha Abadie, Dr Marie-Laure Bonnet, Dr Vincent Brites, Dr. Daniel Ortiz, Dr. Ane Eizaguirre, Ana Martin Somer, and Estefania Rossich. All theoretical spectroscopy investigations have been devised in collaboration with experimentalists from the IR-MPD and IR-PD communities, among whom we are extremely grateful to Prof James M. Lisy, Dr Jordan Beck, Prof Jean-Pierre Schermann, Dr Gilles Grégoire, Dr Gilles Ohanessian, Dr Philippe Maître, Prof John Paul Simons, Dr Lavina Snoek, Dr Tim Vaden, Dr Isabelle Compagnon, Prof Anouk Rijs, Prof Jos Oomens, and Sander Jaeqx. In CID modeling, collaborations with Prof. William L. Hase, Prof. Kihyung Song, Dr. Jean-Yves Salpin, Prof. Jeanine Tortajada, Prof. Manuel Yanez, and Prof Jos Oomens have been of utmost importance to our work. Computational resources from GENCI-France (Grand Equipement National de

Calcul Intensif) are acknowledged. Fundings from ANR-PROBIO, ANR-NSF SPIONCLUS, 'ATIGE' Genopole-France, Labex Charm3at, are acknowledged.

References

1. Schermann JP (2007) Spectroscopy and modeling of biomolecular building blocks. Elsevier, Amsterdam, The Netherlands
2. Polfer NC (2011) Chem Soc Rev 40:2211
3. Polfer NC, Oomens J (2009) Mass Spectrom Rev 28:468
4. Rizzo TR, Stearns JA, Boyarkin OV (2009) Int Rev Phys Chem 28:481
5. Simons JP (2009) Mol Phys 107:2435
6. Stedwell CN, Galindo JF, Roitberg AE, Polfer NC (2013) Annu Rev Anal Chem 6:267
7. Gaigeot MP (2010) Theoretical spectroscopy of floppy peptides at room temperature. A DFTMD perspective: gas and aqueous phase. Phys Chem Chem Phys 12:3336
8. Roy TK, Gerber RB (2013) Phys Chem Chem Phys 15:9468
9. Thomas M, Brehm M, Fligg R, Vohringer P, Kirchner B (2013) Phys Chem Chem Phys 15:6608
10. Patrick AL, Polfer NC (2014) Top Curr Chem. doi:10.1007/128_2014_576
11. Baer T, Hase WL (1996) Unimolecular reaction dynamics. Theory and experiments. Oxford University Press, New York
12. Schlag EW, Levine RD (1989) On the unimolecular dissociation of large molecules. Chem Phys Lett 163(6):523–530
13. Cramer CJ (2004) Essentials of computational chemistry. Wiley, New York
14. Frenkel D, Smit B (2001) Understanding molecular simulation. Academic, London
15. Iannuzzi M, Laio A, Parrinello M (2003) Phys Rev Lett 90:238302
16. Marx D, Hutter J (2009) Ab Initio molecular dynamics. Cambridge University Press, Cambridge
17. McQuarrie DA (1976) Statistical mechanics. Harper-Collins, New York
18. Wilson EB, Decius J, Cross P (1955) Molecular vibrations. McGraw-Hill, New York
19. Gaigeot MP, Martinez M, Vuilleumier R (2007) Mol Phys 105:2857
20. Zerbi G (1982) Introduction to the theory of vibrational frequencies and vibrational intensities. In: Person WB, Zerbi G (eds) Vibrational intensities in infrared and Raman spectroscopy. Elsevier, Amsterdam/Oxford/New York
21. Irikura KK, Johnson RD, Kacker RN (2005) J Phys Chem A 109:8430
22. Johnson RD, Irikura KK, Kacker RN, Kessel R (2010) J Chem Theory Comput 6:2822
23. Scott AP, Radom L (1996) J Phys Chem 100:16502
24. Bouteiller Y, Gillet JC, Gregoire G, Schermann JP (2008) J Phys Chem A 112:11656
25. Sohn WY, Kim M, Kim SS, Park YD, Kang H (2011) Phys Chem Chem Phys 13:7037
26. Zehnacker A, Suhm MA (2008) Angew Chem Int Ed 47:6970
27. Rodriguez JD, Lisy JM (2009) Int J Mass Spectrom 283:135
28. Rodriguez O, Lisy JM (2011) J Phys Chem Lett 2:1444
29. Kubo R, Toda M, Hashitsume N (1991) Statistical physics, vol II, 2nd edn. Springer, Berlin
30. Ahlborn H, Space B, Moore PB (2000) J Chem Phys 112:8083
31. Gaigeot MP, Sprik M (2003) Ab initio molecular dynamics computation of the infrared spectrum of aqueous uracil. J Phys Chem B 107:10344
32. Iftimie R, Tuckerman M (2005) Decomposing total IR spectra of aqueous systems into solute and solvent contributions: a computational approach using maximally localized Wannier orbitals. J Chem Phys 122:214508
33. Borysow J, Moraldi M, Frommhold L (1985) Mol Phys 56:913
34. Kim H, Rossky PJ (2006) J Chem Phys 125:074107

35. Lawrence CP, Skinner JL (2005) Proc Natl Acad Sci 102:6720
36. Ramirez R, Lopez-Ciudad T, Kumar P, Marx D (2004) J Chem Phys 121:3973
37. Beck JP, Gaigeot MP, Lisy JM (2013) Phys Chem Chem Phys 15:16736
38. Cimas A, Maitre P, Ohanessian G, Gaigeot MP (2009) J Chem Theory Comput 5:2388
39. Cimas A, Vaden TD, de Boer TSJA, Snoek LC, Gaigeot MP (2009) J Chem Theory Comput 5:1068
40. Marinica C, Grgoire G, Desfranois C, Schermann JP, Borgis D, Gaigeot MP (2006) J Phys Chem A 110:8802
41. Sediki A, Snoek LC, Gaigeot MP (2011) Int J Mass Spectrom 308:281
42. Bovi D, Mezzetti A, Vuilleumier R, Gaigeot MP, Chazallon B, Spezia R, Guidoni L (2011) Phys Chem Chem Phys 13:20954
43. Gaigeot MP (2010) Phys Chem Chem Phys 12:10198
44. Gaigeot MP, Vuilleumier R, Sprik M, Borgis D (2005) J Chem Theory Comput 1:772
45. Handgraaf JW, Meijer EJ, Gaigeot MP (2004) J Chem Phys 121(20):10111–10119
46. Gaigeot MP, Sprik M, Sulpizi M (2012) J Phys Condens Matter 24:124106
47. Sulpizi L, Salanne M, Sprik M, Gaigeot M (2013) J Phys Chem Lett 4:83
48. Sulpizi M, Gaigeot M, Sprik M (2012) The silica/water interface: how the silanols determine the surface acidity and modulate the water properties. J Chem Theory Comput 8:1037
49. Grégoire G, Gaigeot MP, Marinica DC, Lemaire J, Schermann JP, Desfrançois C (2007) Phys Chem Chem Phys 9:3082
50. Lucas B, Gregoire G, Lemaire J, Maitre P, Ortega JM, Rupenyan A, Reimann B, Schermann JP, Desfrancois C (2004) Phys Chem Chem Phys 6:2659
51. Garand E, Kamrath MZ, Jordan PA, Wolk AB, Leavitt CM, McCoy AB, Miller SJ, Johnson MA (2012) Science 335:694
52. Leavitt CM, DeBlase AF, Johnson CJ, van Stipdonk M, McCoy AB, Johnson MA (2013) J Phys Chem Lett 4:3450
53. von Helden G, Compagnon I, Blom MN, Frankowski M, Erlekam U, Oomens J, Brauer B, Gerber RB, Meijer G (2008) Phys Chem Chem Phys 10:1248
54. Benoit DM (2009) Front Biosci 14:4229
55. Chaban GM, Gerber RB (2008) Theor Chem Acc 120:273
56. Bowman JM (1978) J Chem Phys 68:608
57. Carney GD, Sprandel LL, Kern CW (1978) Adv Chem Phys 37:305
58. Cohen M, Greita S, McEarchran RD (1970) Chem Phys Lett 60:445
59. Gerber RB, Ratner MA (1979) Chem Phys Lett 68:195
60. Pele L, Brauer B, Gerber RB (2007) Theor Chem Acc 117:69
61. Pele L, Gerber RB (2008) J Chem Phys 128:165105
62. CPMD (2013) http://www.cpmd.org
63. The CP2K Developers Group (2013) http://www.cp2k.org/
64. VandeVondele J, Krack M, Mohamed F, Parrinello M, Chassaing T, Hutter J (2005) Comput Phys Commun 167:103
65. Beck JP, Gaigeot MP, Lisy JM (2014) Spectrochim Acta A Mol Biomol Spectrosc 119:12
66. Cimas A, Gaigeot MP (2010) Phys Chem Chem Phys 12:3501
67. Rossi M, Blum V, Kupser P, von Helden G, Bierau F, Pagel K, Meijer G, Scheffler M (2010) J Phys Chem Lett 1:3465
68. Brauer B, Pincu M, Buch V, Bar I, Simons JP, Gerber RB (2011) J Phys Chem A 115:5859
69. Xie HB, Jin L, Rudic S, Simons JP, Gerber RB (2012) J Phys Chem B 116:4851
70. Simon A, Spiegelman F (2013) J Chem Phys 138:194309
71. Ivanov SD, Witt A, Marx D (2013) Phys Chem Chem Phys 15:10270
72. Oanh NTV, Falvo C, Calvo F, Lauvergnat D, Basire M, Gaigeot MP, Parneix P (2012) Phys Chem Chem Phys 14:2381
73. Sieffert N, Buhl M, Gaigeot MP, Morrison CA (2013) Liquid methanol from DFT and DFT/MM molecular dynamics simulations. J Chem Theor Comput 9:106
74. Gaigeot MP (2008) J Phys Chem A 112:13507

75. Schmitz M, Tavan P (2004) J Chem Phys 121:12247
76. Calvo F, Parneix P (2012) Eur Phys J D 66:165
77. Parneix P, Basire M, Calvo F (2013) J Phys Chem A 117:3954
78. Bellencontre F (2008) Contributions au developpement d'un champ de forces classique pour la spectroscopie vibrationnelle de biomolcules. Ph.D. thesis, University of Evry val d'Essonne, France
79. Palmo K, Krimm S (1998) J Comput Chem 19:754
80. Schropp B, Wichmann C, Tavan P (2010) J Phys Chem B 114:6740
81. Semrouni D, Balaj OP, Calvo F, Correia CF, Clavaguera C, Ohanessian G (2010) J Am Soc Mass Spectrom 21:728
82. Adesokan AA, Gerber RB (2009) J Phys Chem A 113:1905
83. Abdali S, Niehaus T, Jalkanen KJ, Cao X, Nafie LA, Frauenheim T, Suhai S, Bohr H (2003) Phys Chem Chem Phys 5:1295
84. Elstner M, Frauenheim T, Kaxiras E, Suhai S (2000) Phys Status Solidi B 217:357
85. Gaus M, Goez A, Elstner M (2013) J Chem Theory Comput 9:338
86. Simon A, Rapacioli M, Mascetti J, Spiegelman F (2012) Phys Chem Chem Phys 14:6771
87. Van-Oanh NT, Parneix P, Bréchignac P (2002) J Phys Chem A 106:10144
88. Car R, Parrinello M (1985) Unified approach for molecular dynamics and density-functional theory. Phys Rev Lett 55:2471
89. Marx D, Tuckerman ME, Parrinello M (2000) Solvated excess protons in water: quantum effects on the hydration structure. J Phys Condens Matter A 12:153–159
90. Vuilleumier R (2006) Density functional theory based ab initio molecular dynamics using the Car-Parrinello approach. In: Ferrario M, Binder K, Ciccotti G (eds) Computer simulations in condensed matter systems: from materials to chemical biology, vol 1. Springer, Berlin
91. Goedecker S, Teter M, Hutter J (1996) Phys Rev B 54:1703
92. Hartwigsen C, Goedecker S, Hutter J (1998) Phys Rev B 58:3641
93. Krack M (2005) Theor Chem Acc 114:145
94. Becke A (1988) Density-functional exchange-energy approximation with correct asymptotic behavior. Phys Rev A 38:3098
95. Lee C, Yang W, Parr RG (1988) Development of the Colle-Salvetti correlation-energy formula into a functional of the electron density. Phys Rev B 37:785
96. Grimme S, Antony J, Ehrlich S, Krieg H (2010) A consistent and accurate ab initio parametrization of density functional dispersion correction (DFT-D) for the 94 elements H-Pu. J Chem Phys 132:154104
97. Martyna GJ, Tuckerman ME (1999) J Chem Phys 110:2810
98. Bernasconi M, Silvestrelli PL, Pamnello M (1998) Phys Rev Lett 81:1235
99. Resta R (1998) Phys Rev Lett 80:1800
100. Carbonniere P, Dargelos A, Ciofini I, Adamo C, Pouchan C (2009) Phys Chem Chem Phys 11:4375
101. Martinez M, Gaigeot MP, Borgis D, Vuilleumier R (2006) J Chem Phys 125:144106
102. Mathias G, Ivanov SD, Witt A, Baer MD, Marx D (2012) J Chem Theory Comput 8:224
103. Nonella M, Mathias G, Tavan P (2003) J Phys Chem A 107:8638
104. Vaden TD, de Boer TSJA, Simons JP, Snoek LC, Suhai S, Paizs B (2008) J Phys Chem A 112:4608
105. Vaden TD, de Boer TSJA, MacLeod NA, Marzluff EM, Simons JP, Snoek LC (2007) Phys Chem Chem Phys 9:2549
106. Wu R, McMahon TB (2007) J Am Chem Soc 129:11312
107. MacLeod N, Simons J (2007) Mol Phys 105:689
108. Rodriguez JD, Kim D, Tarakeshwar P, Lisy JM (2010) J Phys Chem A 114:1514
109. Falk M, Whalley E (1961) J Chem Phys 34:1554
110. Beck JP, Lisy JM (2010) J Phys Chem A 114:10011
111. Jaeqx S, Oomens J, Cimas A, Gaigeot M, Rijs AM (2014) Angew Chem Int Ed 53:3663

112. Sleno L, Volmer DA (2004) Ion activation methods for tandem mass spectrometry. J Mass Spectrom 39:1091–1112
113. Laskin J, Futrell JH (2003) Collisional activation of peptide ions in FT-ICR mass spectrometry. Mass Spectrom Rev 22:158–181
114. Mayer PM, Poon C (2009) The mechanisms of collisional activation of ions in mass spectrometry. Mass Spectrom Rev 28:608–639
115. Eizaguirre A, Mo O, Yanez M, Salpin JY, Tortajada J (2012) Modelling peptidemetal dication interactions: formamide–Ca^{2+} reactions in the gas phase. Org Biomol Chem 10: 7552–7561
116. Corral I, Mo O, Yanez M, Salpin JY, Tortajada J, Radom L (2004) Gas-phase reactions between urea and Ca^{2+}: the importance of Coulomb explosions. J Phys Chem A 108: 10080–10088
117. Trujillo C, Lamsabhi A, Mo O, Yanez M, Salpin JY (2011) Unimolecular reactivity upon collision of uracil–Ca^{2+} complexes in the gas phase: comparison with uracil–M^+ (M = H, alkali metals) and uracil–M^{2+} (M = Cu, Pb) systems. Int J Mass Spectrom 306:27–36
118. Hoppilliard Y, Ohanessian G, Bourcier S (2004) Fragmentation mechanisms of glycine-Cu in the gas phase. An experimental and theoretical study. J Phys Chem A 108:9687
119. Hoppilliard Y, Rogalewicz F, Ohanessian G (2001) Structures and fragmentations of zinc (II) complexes of amino acids in the gas phase. II. Decompositions of glycine-Zn(II) complexes. Int J Mass Spectrom 204:267–280
120. Hoyau S, Pelicier JP, Rogalewicz F, Hoppilliard Y, Ohanessian G (2001) Complexation of glycine by atomic metal cations in the gas phase. Eur J Mass Spectrom 7:303–311
121. Harrison AG (2009) To b or not to b: the ongoing saga of peptide b ions. Mass Spectrom Rev 28:640–654
122. Paizs B, Suhai S (2005) Fragmentation pathways of protonated peptides. Mass Spectrom Rev 24:508–548
123. Rogalewicz F, Hoppilliard Y (2000) Low energy fragmentation of protonated glycine. An ab initio theoretical study. Int J Mass Spectrom 199:235–252
124. Paizs B, Suhai S (2001) Theoretical study of the main fragmentation pathways for protonated glycylglycine. Rapid Commun Mass Spectrom 15:651–663
125. Armentrout PB, Heaton AL (2012) Thermodynamics and mechanisms of protonated diglycine decomposition: a computational study. J Am Soc Mass Spectrom 23:621–631
126. Bythell BJ, Somogyi A, Paizs B (2009) What is the structure of b_2 ions generated from doubly protonated tryptic peptides? J Am Soc Mass Spectrom 20:618–624
127. Paizs B, Lendvay G, Vekey K, Suhai S (1999) Formation of b_2^+ ions from protonated peptides: an ab initio study. Rapid Commun Mass Spectrom 13:525–533
128. Paizs B, Szlavik Z, Lendvay G, Vekey K, Suhai S (2000) Formation of a_2^+ ions of protonated peptides. An ab initio study. Rapid Commun Mass Spectrom 14:746–755
129. El Aribi H, Rodriquez CF, Almeida DRP, Ling Y, Mak WWN, Hopkinson AC, Siu KWM (2003) Elucidation of fragmentation mechanisms of protonated peptide ions and their products: a case study on glycylglycylglycine using density functional theory and threshold collision-induced dissociation. J Am Chem Soc 125:9229–9236
130. Paizs B, Csonka IP, Lendvay G, Suhai S (2001) Proton mobility in protonated glycylglycine and formylglycylglycinamide: a combined quantum chemical and RRKM study. Rapid Commun Mass Spectrom 15:637–650
131. Polfer NC, Oomens J, Suhai S, Paizs B (2007) Infrared spectroscopy and theoretical studies on gas-phase protonated leu-enkephalin and its fragments: direct experimental evidence for the mobile proton. J Am Chem Soc 129:5887–5897
132. Bythell BJ, Matre P, Paizs B (2010) Cyclization and rearrangement reactions of an fragment ions of protonated peptides. J Am Chem Soc 132:14766–14779
133. Polfer NC, Oomens J, Suhai S, Paizs B (2005) Spectroscopic and theoretical evidence for oxazolone ring formation in collision-induced dissociation of peptides. J Am Chem Soc 127:17154–17155

134. Boys SF, Bernardi F (1970) The calculation of small molecular interactions by the differences of separate total energies. some procedures with reduced errors. Mol Phys 19:553–566
135. Eizaguirre A, Yanez M, Tortajada J, Salpin JY (2009) Sr^{2+} -neutral molecules interactions: an assessment of theoretical procedures. Chem Phys Lett 464:240–244
136. Eizaguirre A, Mo O, Yanez M, Salpin JY (2011) Modeling the interactions between peptide functions and Sr^{2+}: formamide–Sr^{2+} reactions in the gas phase. Phys Chem Chem Phys 13: 18409–18417
137. Quemet A, Vitorge P, Cimas A, Liu S, Salpin J-Y, Marsden C, Tortajada J, Gagliardi L, Spezia R, Gaigeot M-P, Brennetot R (2013) Reactivity of lanthanoid mono-cations with ammonia: a combined inductively coupled plasma mass spectrometry and computational investigation. Int J Mass Spectrom 334:27–37
138. Hase WL (1983) Variational unimolecular rate theory. Acc Chem Res 16:258–264
139. Truhlar DG, Garret BC (1980) Variational transition-state theory. Acc Chem Res 13:440–448
140. Maragliano L, Fischer A, Vanden-Eijnden E, Ciccotti G (2006) String method in collective variables: minimum free energy paths and isocommittor surfaces. J Chem Phys 125:024106
141. Rice OK, Ramsperger HC (1927) Theories of unimolecular gas reactions at low pressures. J Am Chem Soc 49:1617
142. Rice OK, Ramsperger HC (1928) Theories of unimolecular gas reactions at low pressures. II. J Am Chem Soc 50:617
143. Kassel LS (1928) Studies in homogeneous gas reactions. I. J Phys Chem 32:225
144. Marcus RA, Rice OK (1951) The kinetics of the recombination of methyl radicals and iodine atoms. J Phys Chem 55:894
145. Rosenstock HM, Wallenstein MB, Wahrhaftig AL, Eyring H (1952) Absolute rate theory for isolated systems and the mass spectra of polyatomic molecules. Proc Natl Acad Sci U S A 38:667
146. Beyer T, Swinehart DR (1973) Number of multiply-restricted partitions. ACM Commun 16:379
147. Whitten GZ, Rabinovitch BS (1963) Accurate and facile approximation for vibrational energy level sums. J Chem Phys 38:2466
148. Whitten GZ, Rabinovitch BS (1964) Approximation for rotation-vibration energy level sums. J Chem Phys 41:1883
149. Bunker DL, Hase WL (1973) On non-RRKM unimolecular kinetics: molecules in general and CH_3NC in particular. J Chem Phys 59:4621–4632
150. Baer T, Mayer PM (1997) Statistical Rice-Ramsperger-Kassel-Marcus quasiequilibrium theory calculations in mass spectrometry. J Am Soc Mass Spectrom 8:103–115
151. Zhu L, Hase WL (1990) Comparison of models for calculating the RRKM unimolecular rate constant k(E, J). Chem Phys Lett 175:117–124
152. Miller WH (1979) Tunneling corrections to unimolecular rate constants, with application to formaldehyde. J Am Chem Soc 101:6810–6814
153. Miller WH (1987) Tunneling and state specificity in unimolecular reactions. Chem Rev 87: 19–27
154. Harvey JN, Aschi M (1999) Spin-forbidden dehydrogenation of methoxy cation: a statistical view. Phys Chem Chem Phys 1:5555–5563
155. Pu J, Gao J, Truhlar DG (2006) Multidimensional tunneling, recrossing, and the transmission coefficient for enzymatic reactions. Chem Rev 106:3140–3169
156. Aschi M, Grandinetti F (1999) Unimolecular decay of the thiomethoxy cation, CH_3S^+: a computational study of the detailed mechanistic aspects. J Chem Phys 111:6759–6768
157. Cimas A, Gamez JA, Mo O, Yanez M, Salpin JY (2008) Computational study on the kinetics of the reaction between Ca^{2+} and urea. Chem Phys Lett 456:156–161
158. Hase WL, Song K, Gordon MS (2003) Direct dynamics simulations. Comput Sci Eng 5: 36–44
159. Hase WL, Ludlow DM, Wolf RJ, Schlick T (1981) Translational and vibrational energy dependence of the cross section for $H + C_2H_4 \rightarrow C_2H_5*$. J Phys Chem 85:958–968

160. Chapman S, Bunker DL (1975) An exploratory study of reactant vibrational effects in CH_3 + H_2 and its isotopic variants. J Chem Phys 62:2890–2899
161. Sloane CS, Hase WL (1977) On the dynamics of state selected unimolecular reactions: chloroacetylene dissociation and predissociation. J Chem Phys 66:1523–1533
162. Hase WL, Duchovic RJ, Hu X, Komornicki A, Lim KF, Lu DH, Peslherbe GH, Swamy KN, VandeLinde SR, Varandas A, Wang H, Wolf RJ (1996) VENUS96: a general chemical dynamics computer program. Quantum Chem Program Exchange 16:671
163. Hu X, Hase WL, Pirraglia T (1991) Vectorization of the general Monte Carlo classical trajectory program VENUS. J Comput Chem 12:1014–1024
164. Hase WL, Buckowski DG (1980) Monte Carlo sampling of a microcanonical ensemble of classical harmonic oscillators. Chem Phys Lett 74:284
165. Wolf R, Hase WL (1980) Trajectory studies of model H-C-C → H + C=C dissociation. I. Random vibrational excitation. J Chem Phys 72:316
166. Lu DH, Hase WL (1988) Classical trajectory calculation of the benzene overtone spectra. J Phys Chem 92:3217
167. Sun L, Hase WL (2010) Comparisons of classical and Wigner sampling of transition state energy levels for quasiclassical trajectory chemical dynamics simulations. J Chem Phys 133: 044313
168. Bunker DL, Goring-Simpson E (1973) Alkali-methyl iodide reactions. Faraday Discuss Chem Soc 55:93
169. Douglas DJ (1998) Applications of collision dynamics in quadrupole mass spectrometry. J Am Soc Mass Spectrom 9:101–113
170. Bolton K, Nordholm S (1996) Classical trajectory study of collision induced intramolecular energy transfer in trans-stilbene. Chem Phys 206:103–128
171. Lim KF, Gilbert RG (1990) Trajectory simulations of collisional energy transfer of highly vibrationally excited azulene. J Phys Chem 94:77–84
172. de Sainte Claire P, Peslherbe GH, Hase WL (1995) Energy transfer dynamics in the collision-induced dissociation of Al_6 and Al_{13} clusters. J Phys Chem 99:81478161
173. de Sainte Claire P, Hase WL (1996) Thresholds for the collision-induced dissociation of clusters by rare gas impact. J Phys Chem 100:8190–8196
174. Mahan BH (1970) Refined impulse approximation for the collisional excitation of the classical anharmonic oscillator. J Chem Phys 52:5221
175. Cornell WD, Cieplak P, Bayly C, Gould I, Merz KM Jr, Ferguson D, Spellmeyer D, Fox T, Caldwell J, Kollman P (1995) A second generation force field for the simulation of proteins, nucleic acids, and organic molecules. J Am Chem Soc 117:5179
176. Meroueh O, Hase WL (1999) Collisional activation of small peptides. J Phys Chem A 103:3981–3990
177. Meroueh O, Hase WL (2000) Energy transfer pathways in the collisional activation of peptides. Int J Mass Spectrom 201:233–244
178. Martinez-Nunez E, Fernandez-Ramos A, Vazquez SA, Marques JMC, Xue M, Hase WL (2005) Quasiclassical dynamics simulation of the collision-induced dissociation of $Cr(CO)_6^+$ with Xe. J Chem Phys 123:154311
179. Chen YJ, Fenn PT, Lau KC, Ng CY, Law CK, Li WK (2002) Study of the dissociation of $CH_3SCH_3^+$ by collisional activation: evidence of non statistical behavior. J Phys Chem A 106:9729–9736
180. Fenn PT, Chen YJ, Stimson S, Ng CY (1997) Dissociation of CH_3SH^+ by collisional activation: evidence of non statistical behavior. J Phys Chem A 101:6513–6522
181. Liu J, Song K, Hase WL, Anderson SL (2003) Direct dynamics study of energy transfer and collision-induced dissociation: effects of impact energy, geometry, and reactant vibrational mode in H_2CO^+–Ne collisions. J Chem Phys 119:3040–3050
182. Liu J, Uselman BW, Boyle JM, Anderson SL (2006) The effects of collision energy, vibrational mode, and vibrational angular momentum on energy transfer and dissociation in NO_2^+-rare gas collisions: an experimental and trajectory study. J Chem Phys 125:133115

183. Spezia R, Salpin JY, Gaigeot MP, Hase W, Song K (2009) Protonated urea collision-induced dissociation. Comparison of experiments and chemical dynamics simulations. J Phys Chem A 113:13853–13862
184. Jeanvoine Y, Gaigeot M-P, Hase WL, Song K, Spezia R (2011) Collision induced dissociation of protonated urea with a diatomic projectile: effects on energy transfer and reactivity via chemical dynamics simulations. Int J Mass Spectrom 308:289–298
185. Spezia R, Cimas A, Gaigeot MP, Salpin JY, Song K, Hase WL (2012) Collision induced dissociation of doubly-charged ions: Coulomb explosion vs neutral loss in $[Ca(urea)]^{2+}$ gas phase unimolecular reactivity via chemical dynamics simulations. Phys Chem Chem Phys 14:11724–11736
186. Meroueh O, Wang Y, Hase WL (2002) Direct dynamics simulations of collision- and surface-induced dissociation of N-protonated glycine. Shattering fragmentation. J Phys Chem A 106:9983–9992
187. Park K, Song K, Hase WL (2007) An ab initio direct dynamics simulation of protonated glycine surface-induced dissociation. Int J Mass Spectrom 265:326–336
188. Park K, Deb B, Song K, Hase WL (2009) Importance of shattering fragmentation in the surface-induced dissociation of protonated octaglycine. J Am Soc Mass Spectrom 20: 939–948
189. Ortiz D, Martin-Gago P, Riera A, Song K, Salpin JY, Spezia R (2013) Gas-phase collision induced dissociation mechanisms of peptides. Theoretical and experimental study of *N*-formylalanylamide fragmentation. Int J Mass Spectrom 335:33–44
190. Dongre A, Jones J, Somogyi A, Wysocki V (1996) Influence of peptide composition, gas-phase basicity, and chemical modification on fragmentation efficiency: evidence for the mobile proton model. J Am Chem Soc 118:8365–8374
191. Wysocki V, Tsaprailis G, Smith L, Breci L (2000) Mobile and localized protons: a framework for understanding peptide dissociation. J Mass Spectrom 35:1399–1406
192. Ortiz D, Salpin JY, Song K, Spezia R (2014) Galactose 6-sulfate collision induced dissociation using QM+MM chemical dynamics simulations and ESI-MS/MS experiments. Int J Mass Spectrom 358:25–35
193. Bednarski H, Sohlberg K, Domanski M, Weszka J, Adamus G, Kowalczuk M, Cozan V (2011) A combined theoretical and experimental study of mechanisms of fragmentation active for PHB oligomers in negative-ion mode multistage mass spectrometry. Int J Mass Spectrom 304:15–24
194. Rodriguez-Fernandez R, Vazquez SA, Martinez-Nunez E (2013) Collision-induced dissociation mechanisms of $[Li(uracil)]^{+}$. Phys Chem Chem Phys 15:7628–7637
195. Czako G, Kaledin AL, Bowman JM (2010) J Chem Phys 132:164103
196. Habershon S, Manolopoulos DE (2009) J Chem Phys 131:244518

Top Curr Chem (2015) 364: 153–182
DOI: 10.1007/128_2014_576

Published online: 5 November 2014

Peptide Fragmentation Products in Mass Spectrometry Probed by Infrared Spectroscopy

Amanda L. Patrick and Nicolas C. Polfer

Abstract Vibrational spectroscopy offers detailed insights, by virtue of diagnostic infrared bands, into the chemical structures and moieties which are formed during peptide fragmentation inside mass spectrometers. Over the past few years, IRMPD spectroscopy has led to a greatly improved understanding of the chemistry that takes place during collision-induced dissociation (CID) of protonated peptides. For instance, the rearrangement chemistry of **b**- and **a**-type ions, which is relevant in sequence scrambling pathways, has been directly confirmed with the technique. In this chapter, we provide a brief background on peptide fragmentation chemistry, and give an overview of areas where vibrational spectroscopy has been successfully implemented, such as CID of protonated and de-protonated peptides. We also discuss the potential of the technique for elucidating lesser-studied radical dissociation processes, such as electron capture dissociation (ECD), electron transfer dissociation (ETD), and laser photodissociation.

Keywords IRMPD spectroscopy · Peptide sequencing · Ion structures · CID · Fragmentation mechanisms

Contents

A.L. Patrick and N.C. Polfer (✉)
Department of Chemistry, University of Florida, Buckman Drive, P.O. Box 117200, Gainesville, FL 32611, USA
e-mail: apatrick@chem.ufl.edu; polfer@chem.ufl.edu

1 Peptide Sequencing

The amino acid sequence of a peptide cannot be determined by mass measurement alone, but rather is obtained by dissociating the molecule inside a mass spectrometer and weighing the resulting fragments. In mass spectrometry jargon, these sequencing methods are referred to as *tandem* mass spectrometry, often written as MS/MS for mass isolation followed by fragmentation, or MS^n for multiple stages of mass isolation and fragmentation. These techniques have become the gold standard in identifying peptides and proteins in the developing field of proteomics, because of their sensitivity and ease of automation, amongst other advantages [1, 2].

Forming fragment ions requires ion activation. This can be caused by the ion colliding with a background gas, absorbing photons, or interacting with electrons. The most useful bond cleavages in terms of sequencing are those which occur on the peptide backbone. The nomenclature for peptide bond cleavages is depicted in Fig. 1 [3]. By convention, the N-terminus (in this case a free amine) is shown on the left and the C-terminus (in this case a free acid) is on the right. The types of product ions observed depend on the activation method and whether the peptide is positively (e.g., protonated) or negatively (e.g., deprotonated) charged. The numerical indices indicate the number of residues incorporated in the fragment, counting from either the N- or the C-terminus. For complementary fragments, such as $\mathbf{b_2}$ and $\mathbf{y_1}$,

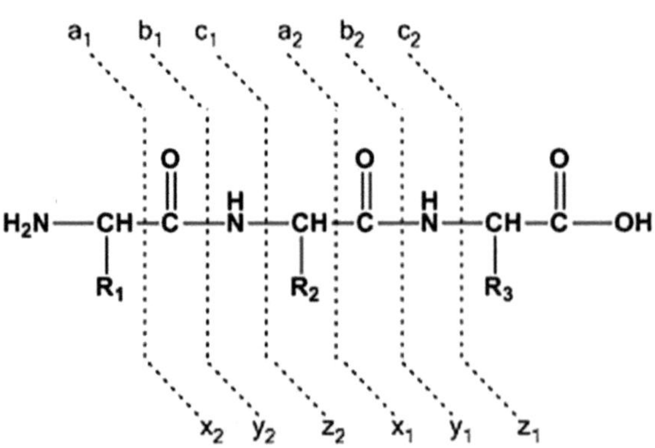

Fig. 1 Nomenclature of *a*, *b*, *c*, *x*, *y*, *z* fragments, showing backbone cleavages for a model peptide

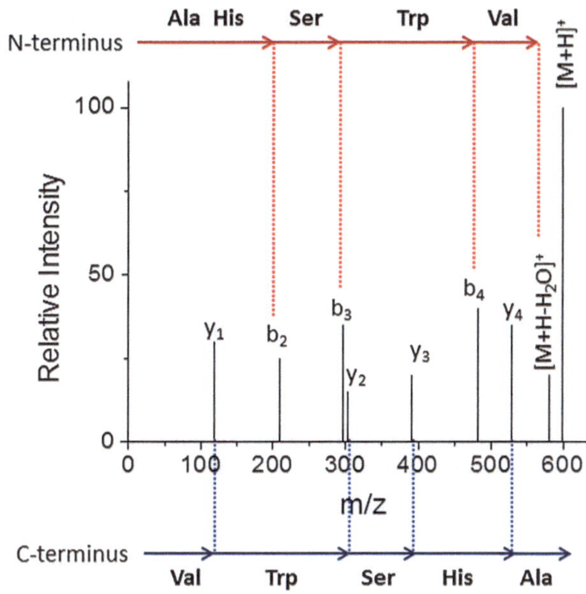

Fig. 2 Sequence determination from a hypothetical CID mass spectrum based on **b** and **y** ion series

the appearance of either or both in a mass spectrum is dependent on the competition for the charge carrier (e.g., proton).

The most established ion activation method is collision-induced dissociation (CID), where the peptide ions are heated by collisions with a background gas, often in an ion trap. The premise of peptide sequencing in CID is shown for an example peptide in Fig. 2. The peptide bonds are often the most labile to cleavage, resulting in **b** and/or **y** ion series. The mass difference of consecutive fragments in these series can be used to read off the sequence, either based on the **b** ion series from the N-terminus or the **y** ion series from the C-terminus. Figure 2 gives a cartoonish representation of a sequencing experiment and does not account for the many caveats associated with this approach. In real experiments, many backbone cleavages are often missed, limiting the sequence information which can be obtained. In addition, the mass spectra do not directly reveal which fragments represent **b** vs **y** ions. In fact, a number of other product ions are also commonly observed in CID mass spectra, such as **a** ions (which are thought to be consecutive fragments from **b** ions, via the loss of CO) and water (e.g., **b**-H_2O) or ammonia (e.g., **a**-NH_3) loss ions.

Despite the large number of potential product ions included in sequencing analysis, on average only 20% of product ions can be identified in typical MS/MS spectra from proteomics studies [4]. This suggests that other product ions are produced which are not as well documented. Various experimental techniques have been employed to obtain structural, and thus mechanistic, insights on the fragmentation of peptide ions, including ion mobility, gas-phase H/D exchange, isotopic labeling, and IRMPD spectroscopy. Of these, IRMPD spectroscopy is particularly powerful, as chemical moieties can be directly confirmed by virtue of diagnostic vibrational modes. The experimental data can also be compared to computed IR spectra of putative candidate structures and/or experimental IR spectra of synthetically made reference compounds to allow an unambiguous assignment.

In the following sections we describe structural and mechanistic insights for peptide sequencing chemistry in the light of measurements from IRMPD spectroscopy. We begin with the fragmentation chemistry of protonated peptides yielding **b** ions, as to date these have been investigated most thoroughly by IRMPD spectroscopy.

2 CID of Protonated Peptides

The chemistry of protonated peptides is strongly affected by the mobilities of protons in these peptides. The "mobile proton" model rationalizes the fact that peptides which have the same number of protons as basic residues (e.g., arginine) require higher internal energies to induce fragmentation [5, 6]. This suggests that protons are tightly sequestered on these basic sites, and that they need to be made mobile to participate in the fragmentation chemistry on the backbone. In other words, the mobile proton model is based on the premise that protons play a crucial role in facilitating and directing bond cleavages.

2.1 b Ions

2.1.1 Small b Ions

The structural identification of **b** ions has been a long-standing pursuit in mass spectrometry. Scheme 1 depicts four mechanisms which could, in principle, rationalize peptide bond cleavage, illustrated for the example of a $\mathbf{b_2}$ ion. In mechanism 1, direct bond cleavage results in an *acylium* structure. All other proposed mechanisms involve a nucleophilic attack on the backbone carbonyl C, while the excess proton is localized on the adjacent amide NH. In mechanism 2, a carbonyl oxygen induces nucleophilic attack, leading to a five-membered ring *oxazolone* structure. In mechanism 3, the N-terminus N acts as the nucleophile, yielding a six-membered ring *diketopiperazine* structure. Finally, in mechanism 4, a nucleophile on a side chain (either residue 1 or 2) participates in a nucleophilic attack, leading to a cyclized fragment. All of these putative fragment structures are chemically distinct, and would hence be expected to exhibit differences in their vibrational spectra. The location of the proton is left unassigned, as there may be multiple sites of high proton affinity, such as the N-terminus, the oxazolone N, or some basic groups on the side chain. IRMPD spectroscopic results are presented to prove/disprove each mechanism in turn.

Chemically, acylium ions are known to be labile to CO loss, and are therefore rarely observed. The vicinity of aromatic moieties can stabilize the acylium structure, because of resonance stabilization with the aromatic ring. A prime example for

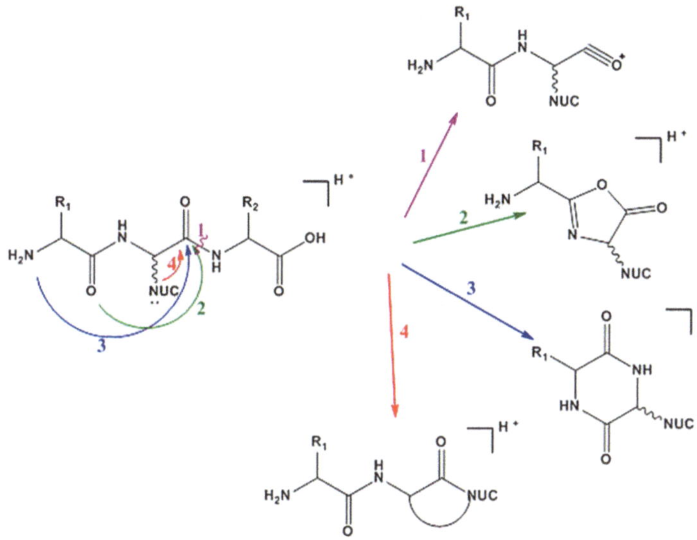

Scheme 1 Proposed mechanisms for formation of **b** ions in protonated peptides

this is the p-aminobenzoyl cation fragment generated from para-aminobenzoic acid (PABA). The acylium CO stretch, which has partial triple-bond character, is expected to appear at a high frequency. Figure 3 shows that the CO bond stretch frequency of the p-aminobenzoyl cation is ~2,170 cm^{-1}, in the same range as the free CO stretch of carbon monoxide [7]. The IRMPD spectrum of a **b** ion, in this case the $\mathbf{b_4}$ ion generated from TyrGlyGlyPheLeu, shows no such band anywhere in the 2,000–2,500 cm^{-1} range, clearly disproving the acylium ion hypothesis for **b** ions. In fact, previous CID experiments had shown that acylium structures spontaneously lose CO to produce a corresponding **a** ion [8, 9]. The structural stabilization of **b** ions requires some form of cyclization on its C-terminal side, for which each of the other three proposed mechanisms presents a potential alternative.

The IRMPD spectrum of the simplest $\mathbf{b_2}$ ion, generated from the tripeptide GlyGlyGly, is shown in Fig. 4 [10]. The experimental spectrum in the mid-IR range (in black) is compared to computed IR spectra for various putative structures (in red), namely the diketopiperazine structure (protonated on a carbonyl O) and two oxazolone structures, one protonated on the oxazolone ring N and the other on the N-terminus. As there are no side-chain groups, no mechanism 4-type structures need to be considered. The final geometry optimizations and frequency calculations in Fig. 4 were carried out using density functional theory at the B3LYP/6-31G+(*d,p*) level using a scaling factor of 0.98. It should be noted that the procedure for searching the conformational spaces of larger molecules often requires molecular dynamics simulations to generate a large number of candidate structures, which are then further optimized with quantum-chemical methods for accurate thermochemical information, as well as harmonic frequency calculations. A more thorough discussion of the theoretical methods employed in interpreting gas-phase IR spectroscopy results is

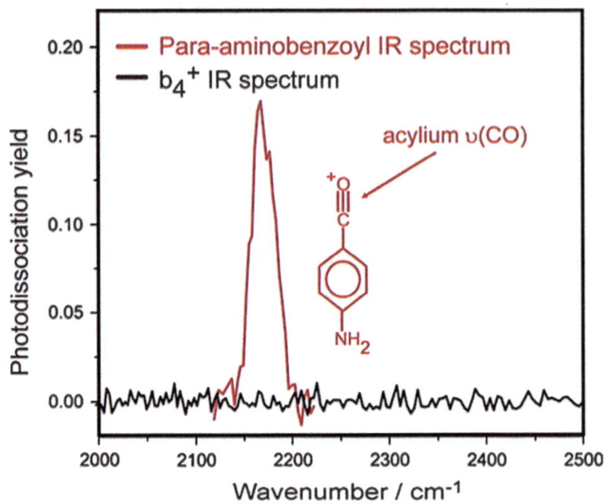

Fig. 3 IRMPD spectra of p-aminobenzoyl cation and b_4 ion from TyrGlyGlyPheLeu in acylium CO stretch region. Reprinted with permission from Polfer et al. [7]. Copyright (2007) American Chemical Society

given in Chap. 3 in this book. The prominent band at ~1,960 cm^{-1} is assigned to the characteristic oxazolone ring CO stretch, and is hence a strong identifier for oxazolone structures. This band is generally observed at a much higher frequency than regular amide I bands (i.e., backbone CO stretches). Conversely, the absence of diketopiperazine structures is confirmed by the lack of CO stretch bands in the 1,700–1,800 cm^{-1} range. It is interesting to note that the diketopiperazine structure is in fact the lowest-energy configuration, implying that the relative stabilities of the product ions are often poor indicators for which pathway is favored. The key kinetic barrier which is thought to impede an attack from the N-terminus is the *trans–cis* isomerization of the peptide bond between residues 1 and 2. In other simple $\mathbf{b_2}$ fragments (e.g., AlaAla and AlaGly), the oxazolone pathway is dominant, despite the fact that the diketopiperazine structures are energetically favored [11, 12].

The presence of basic residues and/or proline can change this picture. Figure 5 shows the IRMPD spectrum of the $\mathbf{b_2}$ fragment from the tripeptide HisProAla [13]. The experimental IRMPD spectrum closely matches the computed spectra of the diketopiperazine structures, while exhibiting no oxazolone band(s). Moreover, a direct comparison to the IRMPD spectrum of the synthetic analog, protonated cyclo(HisPro), yields a convincing match. In other IRMPD experiments on $\mathbf{b_2}$ ions with basic residues, such as arginine [14] and histidine [15], the diketopiperazine product was also found to be dominant. To rationalize these findings, it was proposed that the basic residue may lower the barrier of the *trans–cis* isomerization of the peptide bond, thus making the diketopiperazine pathway more competitive.

In addition to the carbonyl O or the N-terminal N, groups on the side chain may also act as nucleophiles, giving rise to different cyclized products. For instance, the amide groups of asparagine and glutamine side chains have O and N atoms which could act as nucleophiles. Figure 6 compares the IRMPD spectrum of the $\mathbf{b_2}$ ion

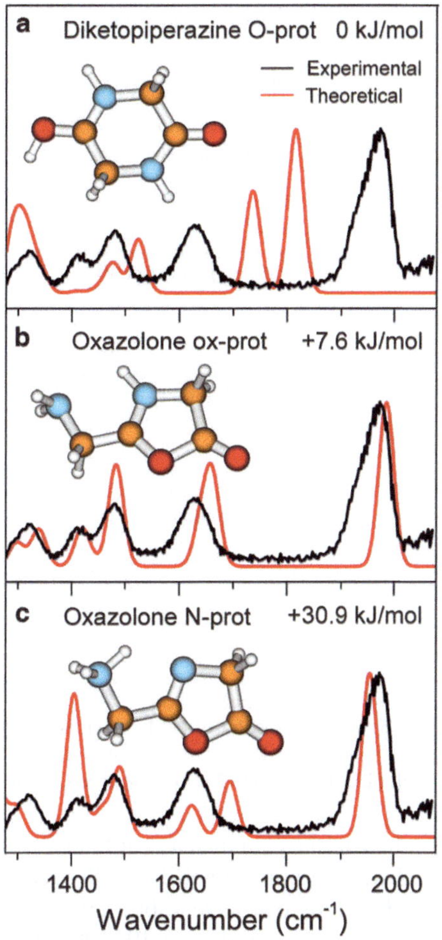

Fig. 4 IRMPD spectrum of $\mathbf{b_2}$ from GlyGlyGly compared to computed IR spectra of putative candidates: (**a**) O-protonated diketopiperazine, (**b**) oxazolone protonated oxazolone, and (**c**) N protonated oxazolone. Reprinted with permission from Chen et al. [10]. Copyright (2009) American Chemical Society

from PheGlnAla to various putative product ion structures [16]. The nucleophile is highlighted by color coding to discern more easily the mechanism that leads to the product ion structure. The most compelling match is observed for the cyclic *imide* structure (Fig. 6b), involving the amide N as the nucleophile. None of the other putative structures – the diketopiperazine, the oxazolone, nor the cyclic *isoimide* (i.e., involving nucleophilic attack from the side-chain amide O) – seem to be present, suggesting that in this case the side-chain N pathway is dominant. The relative thermochemistries of the product ions are again a poor predictor of which pathway is observed experimentally. The diketopiperazine product is the most stable, followed by the cyclic imide, but the oxazolone structures are only marginally higher in energy. Only for the cyclic isoimide product does the thermochemistry possibly come into play, given that it is significantly less stable, and it can hence be excluded on energetic grounds.

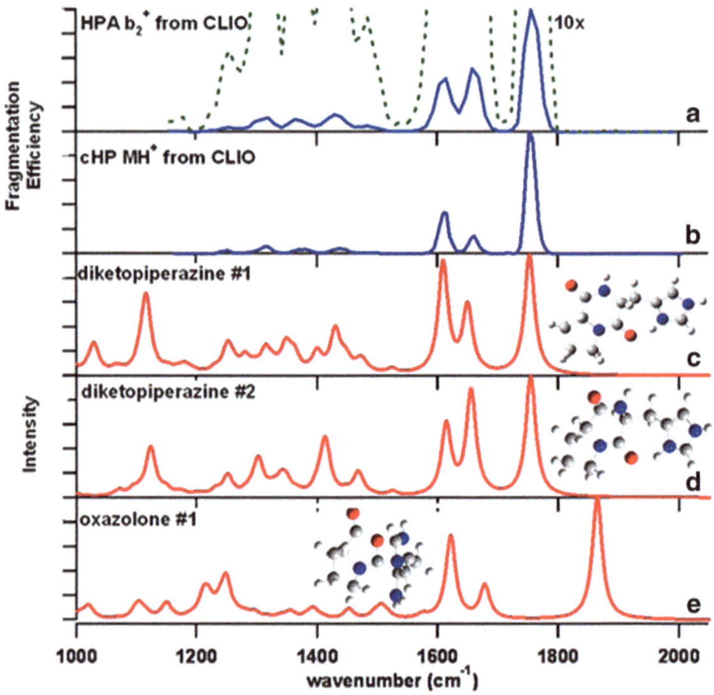

Fig. 5 IRMPD spectra of (**a**) $\mathbf{b_2}$ from HisProAla and (**b**) protonated cyclo(HisPro) compared to (**c**–**e**) computed IR spectra of putative structures. Reprinted with permission from Gucinski et al. [13]. Copyright (2013) American Chemical Society

In summary, in the case of **b** ions from protonated peptides, multiple cyclization pathways are in competition with each other. The structures of the precursor and the product ions can have subtle effects as to which pathway predominates. Table 1 summarizes a number of IRMPD spectroscopy studies on $\mathbf{b_2}$ ions [11–17]. As a general trend, residues with simple residues prefer oxazolone product ions, while the presence of basic residues appears to favor the diketopiperazine pathway. The presence of side-chain nucleophiles opens up other competing pathways. More systematic studies are required to better establish the general trends in this chemistry. Moreover, the effect of experimental parameters on the dissociation chemistry has so far been largely neglected. For instance, the extent and the timescale of ion activation are known to play an important role in controlling the kinetics of competing pathways. Most of the IRMPD spectroscopy studies to date have involved slow heating of ions in traps by multiple collisions, as opposed to fast ion activation by single collisions. The results summarized herein are, therefore, most pertinent for ion trap measurements, but it is not yet clear how relevant these observations are for different instrumental designs for ion activation, such as those involving collision cells.

The experimental results from IRMPD spectroscopy are ideally suited for comparisons with theory, particularly quantum-chemical calculations. The interpretation of experimental IRMPD spectra often relies on a comparison to calculated

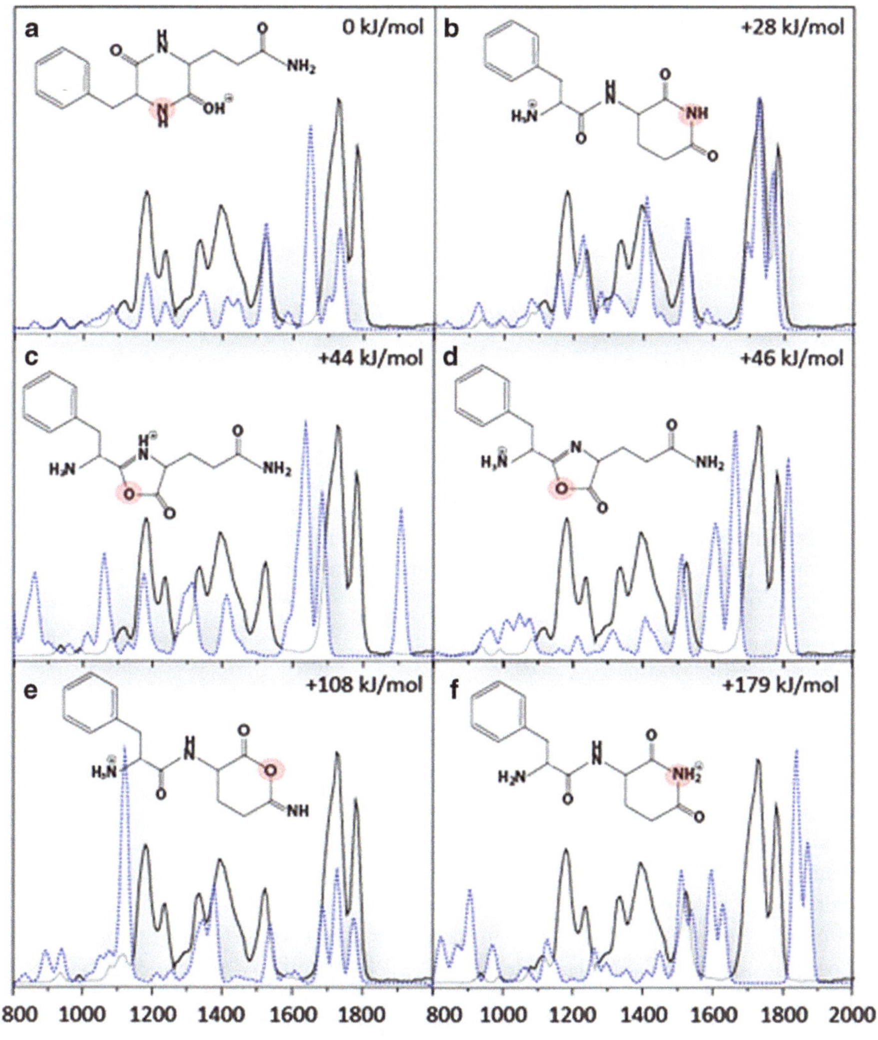

Fig. 6 The experimental IRMPD spectrum (*solid black lines*) of the $\mathbf{b_2}$ from PheGlnAla compared to the computed IR spectra (*dashed blue lines*) of various possible structures, specifically (**a**) carbonyl oxygen-protonated diketopiperazine, (**b**) N-terminus-protonated glutarimide, (**c**) oxazolone nitrogen-protonated oxazolone, (**d**) N-terminus-protonated oxazolone, (**e**) N-terminus-protonated 6-imino-δ-valerolactone, and (**f**) glutarimide nitrogen-protonated glutarimide. Adapted from [16]. Copyright 2013, with kind permission from Springer Science and Business Media

spectra of putative candidates. These quantum-chemical approaches also allow a detailed analysis of the potential energy surfaces of competing reactions [18]. While IRMPD spectroscopy does not give direct insights into the transition states of the various processes, the branching ratios of product ions can be determined, which can serve as a constraint for studying competitive processes.

Table 1 Summary of IRMPD results for $\mathbf{b_2}$ ions with various sequence motifs. (Compilation of studies from [11–17])

Sequence	Oxazolone	Diketopiperazine	Side-chain nucleophile
AlaGly	✓	✗	✗
AlaAla	✓	✗	✗
GlyArg	✗	✓	✗
ArgGly	✓	✓	✗
HisAla	✓	✓	✗
LysAla	✗	✓	✗
GlyPro	✓	✗	✗
AlaPro	✓	✓	✗
IlePro	✓	✓	✗
ValPro	✓	✓	✗
HisPro	✗	✓	✗
PheGln	✗	✗	✓

Complementary experimental approaches, such as threshold collision-induced dissociation (TCID), yield more direct information on the transition states in peptide fragmentation chemistry [19].

2.1.2 Mid-Sized to Larger b Ions

Current theoretical approaches are amenable to understanding the fragmentation chemistry of small peptides and their corresponding product ions at a very fundamental level. Yet, the dissociation chemistry of larger peptides is often more relevant for determining the repercussions of these fundamental studies for sequencing in proteomics.

We may thus ask how the different pathways above are affected by the length of the peptide sequence. Figure 7 depicts IRMPD spectra of a series of **b** ions of different sizes, based on the repeating peptide motif GlyAlaTyr. The smallest **b** fragment in the series, $\mathbf{b_4}$, displays a distinctive oxazolone CO stretch band at ~1,915 cm^{-1}. This band is, however, not observed in any of the larger **b** ions, nor is it observed for $\mathbf{b_6}$, $\mathbf{b_9}$, or $\mathbf{b_{12}}$. Note that a wider range from 1,760 to 1,960 cm^{-1} is shown. The diagnostic position of an oxazolone CO stretch band in larger **b** ions depends on the site of proton attachment in the peptide, and the extent of hydrogen bonding to the CO moiety. In previous measurements of the $\mathbf{b_4}$ fragment from TyrGlyGlyPheLeu, three distinct oxazolone CO stretches were observed, at 1,780, 1,890, and 1,930 cm^{-1} [20]. The 1,780-cm^{-1} band was assigned to an oxazolone conformation with the proton located at the N-terminus, whereas the 1,930-cm^{-1} band was matched to an oxazolone with the proton located on the oxazolone ring N, thus the reason for including a wider diagnostic range for oxazolone structures. Fortunately, this region is still far removed from other background modes in peptides, such as the amide I band, and can hence be considered distinctive.

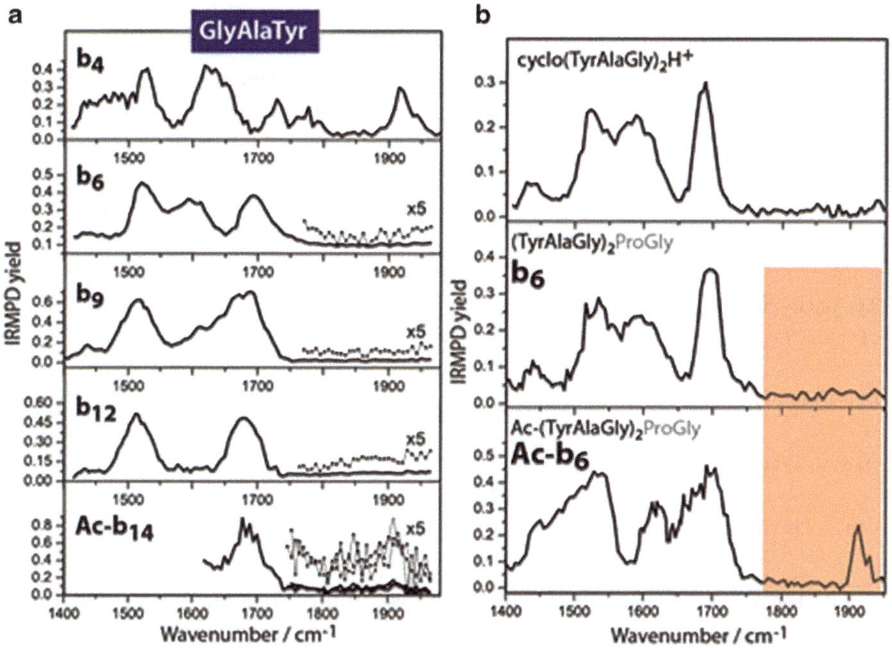

Fig. 7 (**a**) IRMPD spectra of series of **b** ions with sequence motif GlyAlaTyr compared to *N*-acetylated $\mathbf{b_{14}}$. (**b**) Comparison of IRMPD spectra of $\mathbf{b_6}$ to protonated cyclo(TyrAlaGly-TyrAlaGly) and acetylated $\mathbf{b_6}$. Diagnostic oxazolone band region is highlighted by (**a**) 5x magnification and (**b**) color coding. Reprinted with permission from Tirado and Polfer [21]. Copyright© 2012 Wiley-VCH Verlag GmbH & Co. KGaA, Weinheim

The sudden switch, going from prevalent oxazolone structures for $\mathbf{b_4}$ to non-oxazolone structures for $\mathbf{b_6}$, $\mathbf{b_9}$, or $\mathbf{b_{12}}$ for the motif GlyAlaTyr, indicates a size effect in the dissociation chemistry. In order to verify the chemical identity of these non-oxazolone structures, the IR spectra of $\mathbf{b_6}$ ions were compared to IR spectra of synthetically made cyclic peptides. This comparison is shown for the example of $\mathbf{b_6}$ with the sequence motif (TyrAlaGly) in Fig. 7b, demonstrating that, in fact, *macrocyclic* **b** ion structures are generated [21]. It is possible to prevent head-to-tail macrocyclization by chemically altering the N-terminus by acetylation. The corresponding IR spectrum of an acetylated $\mathbf{b_6}$ ion is shown in Fig. 7b (lower panel), exhibiting a distinct oxazolone band at ~1,900 cm^{-1}. A similar control acetylation experiment is shown in the lowest panel in Fig. 7a for a $\mathbf{b_{14}}$ fragment. A small, yet discernible band indicates that oxazolone structures are formed in this control experiment. This suggests that the lack of an oxazolone band for larger **b** ions, namely $\mathbf{b_9}$ and $\mathbf{b_{12}}$, is caused by an absence of oxazolone structures, not an inability to photodissociate these larger molecules.

Table 2 summarizes a number of IRMPD spectroscopy results from series of **b** ions [10, 21–25], confirming the general preference for forming macrocycles for mid-sized to larger **b** ions.

Table 2 Summary of IRMPD spectroscopy studies of larger **b** ions with various sequence motifs oxazolone vs macrocycle structures. (Compilation of studies from [10, 21–25])

Sequence motif	x (for b_x)	Oxazolone	Macrocycle
$(Gly)_n$	2, 3	✔	✗
	4–7	✔	✔
	8	✗	✔
AlaAlaAlaAlaAla	4	✔	✗
TyrGlyGlyPheLeu	2, 3	✔	✗
	4	✔	✔
TyrAlaGlyPheLeu	5	✗	✔
$(TyrAlaGly)_n$	4	✔	✗
	6	✗	✔
$(AlaTyrGly)_n$	4	✔	✗
	6	✗	✔
$(GlyAlaTyr)_n$	4	✔	✗
	6, 9, 12	✗	✔
GlnTrpPheGlyLeuMet	6	✗	✔

The formation of macrocycles raises the possibility for sequence scrambling in CID [26, 27]. As shown in Fig. 8, linear oxazolones isomerize to macrocycles, which can then open up at different positions, resulting in a permutation of the sequence [26]. A number of IRMPD spectroscopic studies have shown that macrocycles (or a mixture of macrocycles and oxazolones) are formed for mid-sized to larger **b** ions. The trend thus seems to favor macrocycles for larger **b** ions. At least statistically, larger **b** ions are also more likely to open up at a different residue, giving rise to permutation of the sequence. This raises the question whether sequence scrambling/permutation is related to the sequence length of the precursor peptide which is fragmented.

For sequence scrambling to be registered in CID mass spectra, fragments with permuted sequences need to be sequentially dissociated to smaller fragments. Figure 9 shows the possible result of sequence scrambling on CID mass spectra. A distinction is made between original sequence ions (in blue), as opposed to permuted sequence ions (in red), which originate from sequence scrambling. In order to test how prevalent sequence scrambling is, the dissociation chemistry of a large number of peptides needs to be investigated. Proteomics studies typically involve thousands of peptides, and hence constitute a statistically meaningful data set which can be mined for trends. Figure 9 shows plots of the percentages of original vs permuted sequence ions as a function of precursor ion length [4]. This data was obtained by extracting the information on original and permuted sequence ions out of each MS/MS mass spectrum for thousands of tryptic digest peptides. Original sequence ions account for just fewer than 20% of all ions in CID mass spectra, meaning that roughly 80% of ions are not accounted for. The ratio of original sequence ion identification does not seem to vary significantly with peptide length. Permuted sequence ions seem to make up an increasing share of the product ions, and approach/exceed original sequence ions for longer sequences.

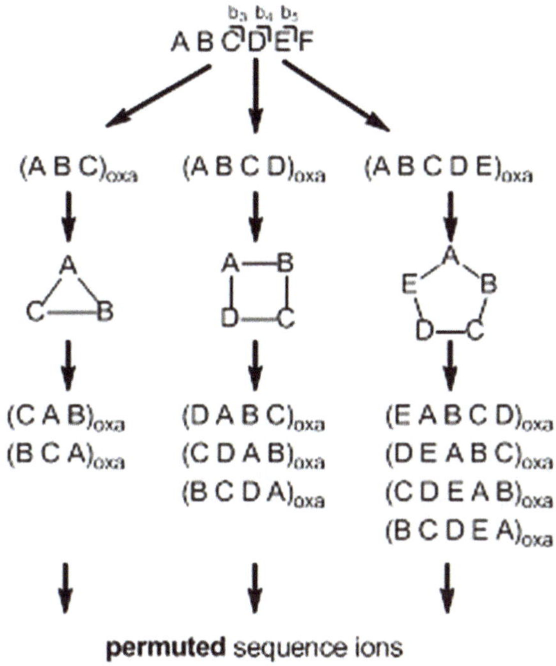

Fig. 8 Formation of permuted sequence ions from **b** ions as a function of sequence length. Adapted with permission from Yu et al. [4]. Copyright (2011) American Chemical Society

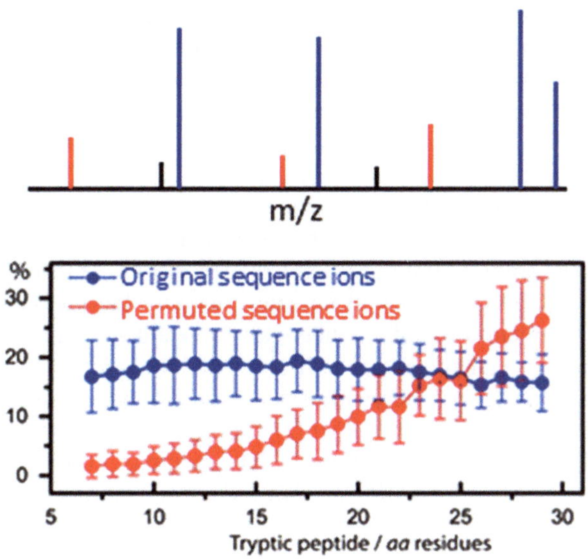

Fig. 9 *Top*: Assignment of original (*blue*) and permuted (*red*) sequence ions, as well as unassigned fragments (*black*), in each MS/MS mass spectrum. *Bottom*: Mean percentages of original and permuted sequence ions in MS/MS mass spectra as a function of peptide length. Adapted with permission from Yu et al. [4]. Copyright (2011) American Chemical Society

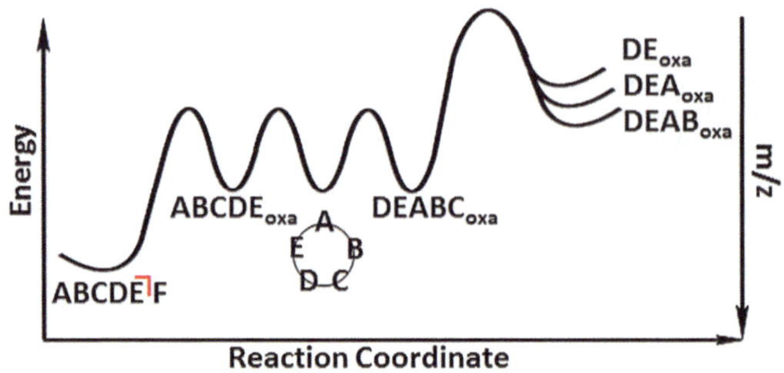

Fig. 10 Potential energy surface of sequence scrambling of **b** ions via macrocycle formation

While there appears to be a correlation between longer peptide length and a higher incidence of scrambling in CID mass spectra, permuted sequence ions only account for a minor percentage of all fragmentation products, at least for the most abundant peptides with less than 16 residues. The IRMPD spectroscopy studies on medium-sized **b** (i.e., 4–7) to larger **b** ions (>7) seem to indicate ubiquitous macrocyclization, and it is thus noteworthy that the rate of sequence scrambling is not more pronounced. In fact, it has been found that, in most proteomic studies, sequence-scrambled ions can safely be ignored as a source of erroneous sequence assignments [28, 29]. Figure 10 attempts to rationalize why the prevalence of scrambling in CID mass spectra may be lower. The figure shows the hypothetical potential energy surface of a peptide fragmenting to a **b** ion which is subject to macrocyclization and sequence scrambling. As the peptide overcomes the dissociation barrier to form an oxazolone **b** ion, this structure can probably also overcome the barriers to macrocyclization and potentially to ring opening to a permuted sequence oxazolone; all of these barriers are expected to be within a similar energy range, and hence accessible. On the other hand, the consecutive fragmentations to smaller-mass, permuted sequence ions involve much more substantial barriers. As a result, many of the fragments get trapped as macrocycles or oxazolones, and only a few progress onto permuted sequence ions. This behavior is especially important for CID conditions in ion traps, as only the precursor ion is resonantly excited. Fragment ions, which inherently have a different m/z, are no longer activated, but are instead cooled by collisions with background (e.g., helium) gas. It is thus not surprising that the yield of permuted sequence ions is low.

In summary, the IRMPD experiments have confirmed that macrocyclization of **b** ions is highly prevalent in CID conditions. Fortunately, this does not seem to pose a considerable problem for sequencing in MS^2 experiments, as the masses of oxazolones, whether of original or permuted sequence, and macrocycles are all identical. Nonetheless, the prevalent macrocyclization is a concern for higher order MS^n experiments. In an MS^3 experiment, the **b** product ion formed during MS^2 is excited again, and would hence be able to follow the pathway(s) to permuted sequence ions. For peptides which yield only limited sequence information during MS^2, one would have to be very cautious when using MS^3 in terms of identifying the sequence.

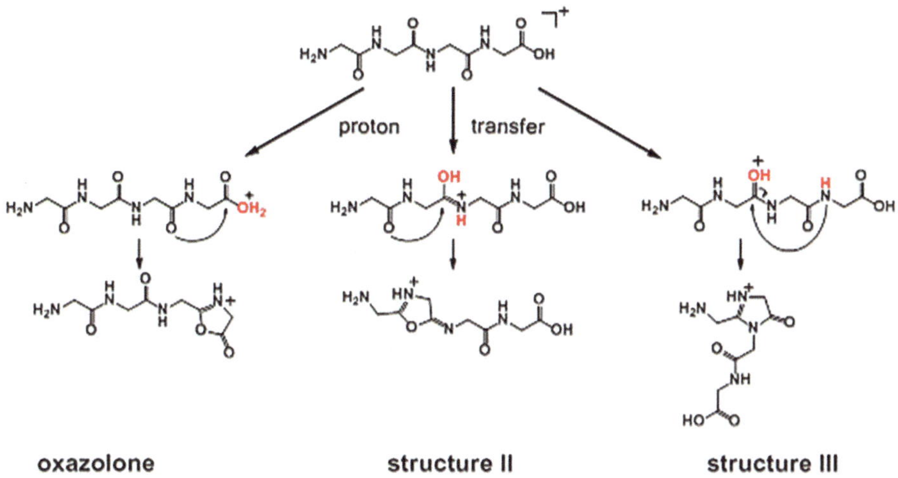

Scheme 2 Proposed mechanisms for water loss from protonated peptide. Adapted with permission from Verkerk et al. [31]. Copyright (2011) American Chemical Society

2.1.3 $[M+H-H_2O]^+$ Ions

Water loss from a protonated peptide ion of *n* residues forms a product ion which is isomeric with the $\mathbf{b}_n$ fragment ion. This may arise from the oxazolone or diketopiperazine fragmentation pathways, such as the case of ArgGly and GlyArg, leading to the expected **b** ion structures; [14] however, it can also be caused by other mechanisms. Three potential pathways for (non-residue specific) water loss are summarized in Scheme 2. The first pathway (on the left) gives the familiar oxazolone structure through loss of water from the C-terminus. The second pathway (middle) is a "Retro-Ritter" type reaction which is achieved through water loss at the second peptide bond, giving "structure II" [30]. The final possible pathway (on the right) also proceeds via internal water loss, this time from the first peptide bond, and results in an N_1 protonated 3,5-dihydro-4*H*-imdazol-4-one, denoted here as "structure III" [31].

While tandem mass spectrometry [30], including isotope labeling experiments [31], has shed some light on possible reaction pathways, especially ruling out the oxazolone pathway as dominant, it is insufficient for pinpointing whether structures II or III are dominant. Again, IRMPD spectroscopy can provide direct structural evidence to help determine the most likely fragmentation pathway. The water loss product from tetraglycine was probed by IRMPD spectroscopy (see Fig. 11a). The comparison between the experimental IRMPD spectrum of the water loss ion and the computed IR spectra for oxazolone structures with different protonation sites clearly demonstrated the absence of oxazolones, but it was not clear whether structure II or structure III was formed [30].

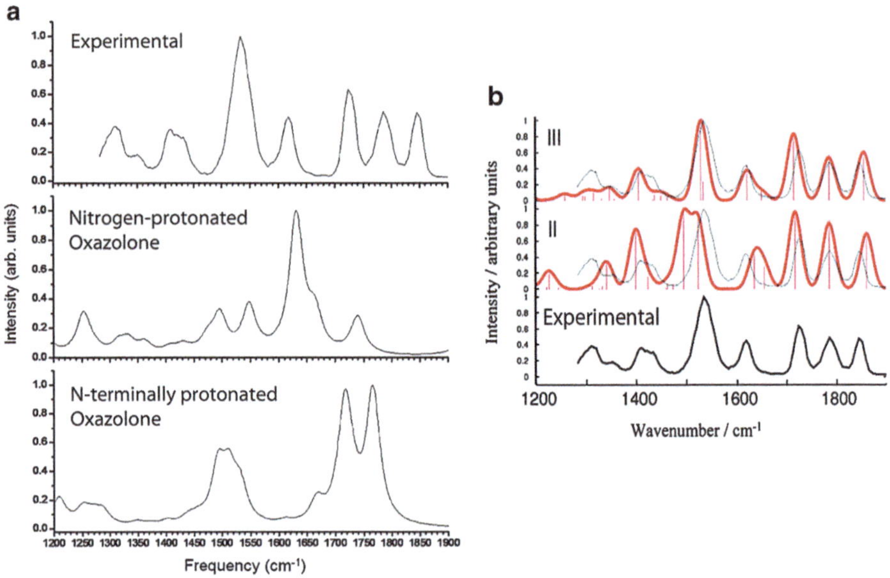

Fig. 11 IRMPD spectrum of water loss product from GlyGlyGlyGly compared to computed IR spectra for (**a**) oxazolone structures, and (**b**) structures II and III (Scheme 2). (**a**) Adapted with permission from Bythell et al. [30]. Copyright (2010) American Chemical Society. (**b**) Adapted with permission from Verkerk et al. 31. Copyright (2011) American Chemical Society

A follow-up study allowed for distinction between structure II and structure III (amongst others) as possible candidates for the water-loss structure [31]. Figure 11b shows the comparison of theoretical spectra for these two structures with the experimental spectrum for the ion. While many IR bands are similar between the two theoretical spectra, the band at ~1,530 cm^{-1} in the experimental spectrum most closely matches structure III's computed spectrum. This band is assigned to the C–N stretch in the *imidazolidinone* ring and the N–H wag of the amide group. It should be noted that, in addition to the closest band in the computed spectrum of structure II being shifted, it also exhibits a lower-energy unresolved band which is completely unobserved in the experimental spectrum.

This evidence clearly shows that, at least in the simple case of GlyGlyGlyGly, water-loss products are not predominantly composed of an oxazolone structure caused by the loss of water from the C-terminus, but instead are composed of a structure formed by loss of water from across peptide bonds of the peptide's backbone, in this case probably an imidazolidinone. While this specific fragmentation pathway should not obscure MS/MS sequencing experiments, it is another potential source of confusion for further (e.g., MS^3) stages of tandem mass spectrometry. Furthermore, this example underscores the fact that in most instances there are multiple potential fragmentation pathways in competition.

Scheme 3 Proposed mechanism for **a** ion formation from **b** ion, followed by rearrangement to N-term imine structure

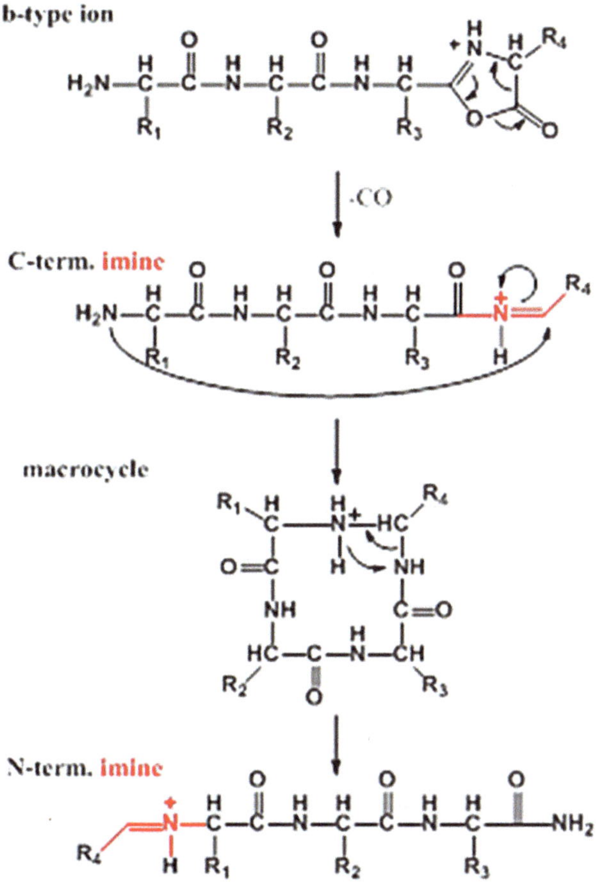

2.2 *a Ions*

The generation of **a** ions during CID is thought to proceed via the loss of CO from **b** ions. Scheme 3 summarizes the current understanding of the predominant rearrangement chemistry that takes place for **a** ions [32], largely based on experimental results from IRMPD spectroscopy and ion mobility mass spectrometry, as well as computations at the DFT level. The loss of CO from an oxazolone structure leads to a linear structure with an imine at the C-terminus, named here *C-term imine*, for the sake of convenience. A head-to-tail attack from the N-terminus results in a macrocycle. A proton migration to an amide NH makes the adjacent amide bond labile for cleavage, and can thus result in a linear structure with the imine located at the N-terminus, and an amide $CONH_2$ at the C-terminus, named here *N-term imine*. In this latter structure, the sequence order has been changed from the original sequence, and thus sequential dissociation product ions of this **a** ion would rationalize elimination of internal residues or sequence scrambling.

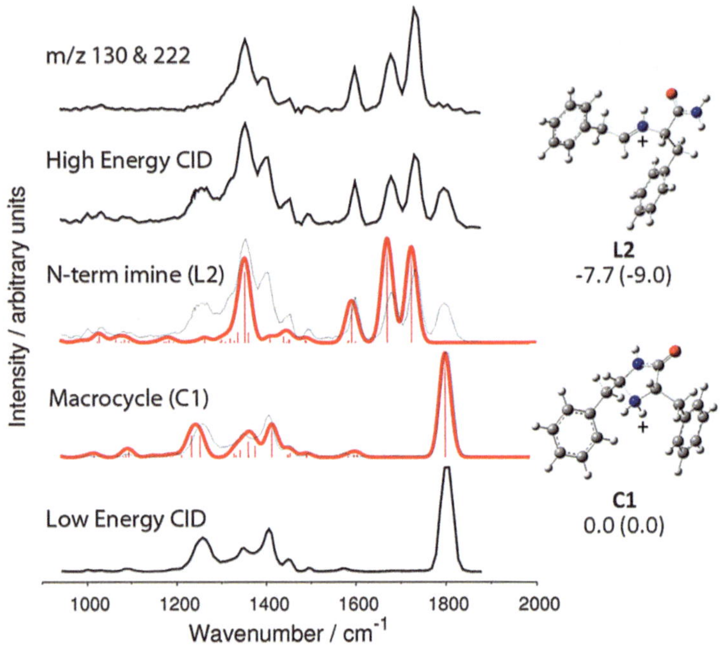

Fig. 12 IRMPD spectra of $\mathbf{a_2}$ from PhePhePhe under low- and higher-energy CID conditions (*black*) compared to computed IR spectra for macrocycle and N-term imine (*red*). Adapted from Verkerk et al. [33]. Copyright 2012, with permission from Elsevier

The fragmentation chemistry of **a** ions appears to be more complex than for **b** ions. In fact, a number of other competitive fragmentation pathways have been proposed [32], which are not shown in Scheme 3, but which may be important for other **a**-type ions which have not yet been studied. In analogy to **b** ions, the sequence length is probably going to play a role in the fragmentation chemistry of **a** ions.

IRMPD spectra of a small **a** ion, the $\mathbf{a_2}$ ion from PhePhePhe, are depicted in Fig. 12 [33]. The product ion was generated by nozzle-skimmer dissociation, using either lower- or higher-energy CID conditions, by changing the voltage drop between the cone and the hexapole entrance. While the IRMPD spectrum under lower-energy conditions closely mirrors the computed spectrum for the macrocycle, the IRMPD spectrum under higher-energy conditions exhibits a broader envelope, involving a number of additional bands. The extra bands strongly indicate the presence of multiple structures. Comparisons to theory show that N-term imine could account for all of the additional bands. In further corroboration of this interpretation, if only selective photofragment ions are monitored (here *m/z* 130 and 222), as shown in the top IRMPD spectrum, the pattern strongly resembles the computed spectrum for N-term imine. In summary, the experimental results suggest that the macrocycle is exclusively formed in softer fragmentation conditions, whereas a mixture of macrocycle and N-term imine structures are generated under more energetic conditions. This is consistent with the proposed reaction mechanism, where the N-term imine structure lies further along on the potential energy reaction coordinate, thus requiring more energetic conditions. It should be noted that the initially formed C-term imine structure is not confirmed in these experiments, suggesting its instability to N-terminal attack.

Figure 13 shows a comparison of IRMPD spectra of the $\mathbf{a_4}$ ion from TyrGlyGlyPheLeu, recorded using different experimental set-ups and free electron lasers: the spectrum was first recorded using the free electron laser FELIX [7], and later recorded using the free electron laser CLIO [34]. While both spectra have some similarities, the FELIX IRMPD spectrum clearly has many more features than the corresponding CLIO IRMPD spectrum. These differences are unlikely to be caused by the different laser systems, as FELIX and CLIO are in fact fairly similar in output power and fluence. Rather, a notable difference between both experiments is the fact that both fragmentations were carried out on different types of mass spectrometers, involving differences in their ion activation and ion cooling dynamics. In the FELIX experiments, the precursor ions were activated by sustained off-resonance irradiation collision-induced dissociation (SORI-CID) inside the Penning trap of a Fourier transform mass spectrometer. The CLIO experiments, on the other hand, were carried out in a 3D quadrupole ion trap. One key difference in both set-ups is the background pressure, which was in the 10^{-7} mbar range for the FELIX measurements, as opposed to 10^{-3} mbar for the CLIO set-up. The more congested FELIX spectrum indicates the presence of an (or multiple) additional structure(s) that was/were not seen in the CLIO spectra, and thus complicates the interpretation. The observation of one main product in the CLIO experiments, which also happens to be the terminal product of a multi-barrier reaction process, suggests slightly more energetic conditions in those experiments, effectively providing ions with sufficient energy for the reaction to go to completion.

The same $\mathbf{a_4}$ ion was also structurally interrogated by ion mobility mass spectrometry (middle panel, Fig. 13) [35]. Ion mobility measures the collision cross sections of ions, and at least three distinct structures could be separated in those measurements. A comparison of the measured collision cross sections to cross sections computed (bottom panel, Fig. 13) for different chemical structures suggests that the most compact configuration is assigned to macrocycle (I), while the most abundant configuration, II, matches the N-term imine conformers. Finally, the C-term imine represents the most extended conformation (III). The relative contributions of these different structures is also related to their relative thermodynamic stabilities, represented on the vertical scale in this figure: the N-term imine conformers (estimated at 61%) are much more stable than either the macrocycle (31%) or the C-term imine (8%).

The combined information from IR spectroscopy and ion mobility allows a detailed interpretation of the product ions which are formed. In the CLIO spectra, the predominant *N-term* imine structure is confirmed. In contrast, the FELIX spectrum shows evidence for all three structures, but the spectral congestion makes it difficult to establish their relative abundances. The ion mobility results confirm that all three structures can be generated in CID conditions, and allows the

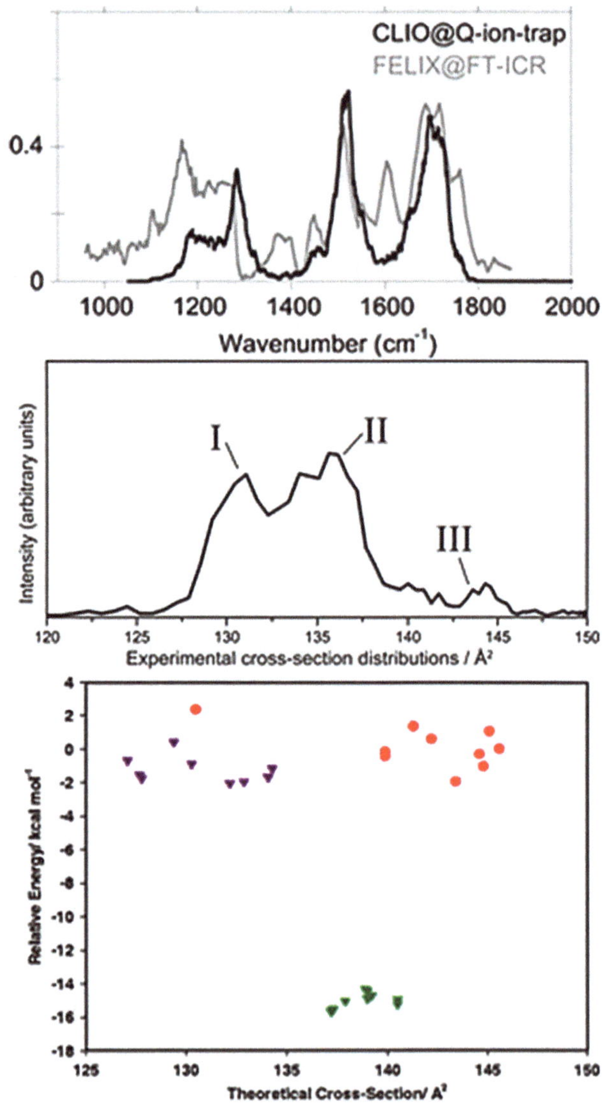

Fig. 13 TyrGlyGlyPheLeu $\mathbf{a_4}$ ion. *Top*: IRMPD spectra. *Middle*: Experimental collision cross section distributions from ion mobility measurements. *Bottom*: Computed collision cross sections for low-energy conformers of macrocycle (*purple*), N-term imine (*green*), and C-term imine (*red*). *Top*: Reprinted from [34]. Copyright 2012, with kind permission from Springer Science and Business Media. *Middle*: Adapted with permission from Polfer et al. [35]. Copyright (2008) American Chemical Society. *Bottom*: Adapted from [34]. Copyright 2012, with kind permission from Springer Science and Business Media

relative abundances to be approximated. Nonetheless, the ion mobility results by themselves yield little insight into the detailed chemical structures of the ions, for which the IRMPD spectroscopy results are much more helpful. In many peptide dissociation processes, a number of competing mechanisms and rearrangements can take place, potentially giving rise to a mixture of isomeric structures. In these cases, IRMPD spectra often become too congested to allow a simple interpretation, and alternate gas-phase structural tools are required. Ion mobility is well suited for this purpose, but gas-phase H/D exchange has also been used successfully in this regard. [10]

3 CID of Deprotonated Peptides

To date, most peptide sequencing has been carried out in the positive ionization mode. Therefore, the majority of fragmentation mechanism and fragment ion structure studies have focused on protonated systems. However, most modern instruments are capable of operating in both positive and negative modes. Furthermore, it is likely that the combined use of both modes would allow for greater sequence coverage and a more unbiased ionization for analytes, which preferentially ionize in one mode over the other. Thus, there is increasing interest in the fragmentation patterns and product ion structures of deprotonated peptides. Here, summaries from the currently available studies are presented and paths forward are laid out.

The first deprotonated peptide system to be studied by IRMPD spectroscopy was AlaAlaAla and its $\mathbf{a}_3$ fragment anion [36]. The precursor ion took a carboxylate form. The $\mathbf{a}_3$ ion contained amide I and II bands which matched all calculated spectra, except that of the *enolate*, and an amide stretching mode at 1,560 cm^{-1} which only matched theoretical spectra of the *amidate* and enolate forms. Therefore, these results suggest the amidate form is the observed structure. Further comparison in the 1,200–1,400 cm^{-1} region also produced good matches between theory and experiment. This is in agreement with calculations that show the amidate is lower in energy, likely because of its increased resonance stabilization. It should be noted that in this case only a single peptide (and a single fragment) was studied and it was with the simplest alkyl side-chain groups. Therefore, no generalizations for other sequence motifs (or chain lengths) can be drawn from this specific study.

More recently, **c**-type fragments have been studied in the negative mode [37]. In this case, several peptide systems were studied to determine the relative influence of amino acid chain length and side-chain makeup. Because **c**-type ions are known to form structures corresponding to linear peptide fragments with a C-terminus amide moiety, the important question for IRMPD to answer in this study is where the site of deprotonation resides for various systems. The possible structures are given in Fig. 14.

The simplest structure studied was the $\mathbf{c}_1$ ion of peptides that do not contain ionizable side-chains (e.g., Ala and Gly). The IRMPD spectrum for the $\mathbf{c}_1$ of one such peptide, AlaPhe, is shown in Fig. 15, along with comparisons to calculated spectra for various deprotonation sites. For this simple system, the spectrum calculated for the structure deprotonated at the C-terminus amide moiety (Fig. 15a) clearly gives the best match. This corresponds to the presence of structure I from Fig. 14. It should be noted that these simple systems do not contain an amide peptide bond, since they contain only a single peptide residue, and hence this is not a possible deprotonation site.

The longer $\mathbf{c}_2$ ion was shown to form a similar linear structure, but the deprotonation site was on the internal amide of the peptide bond. In cases where ionizable side-chains (e.g., tyrosine and tryptophan) were present, both $\mathbf{c}_1$ and $\mathbf{c}_2$ ions were preferentially deprotonated on the side chain. Notably, no evidence for

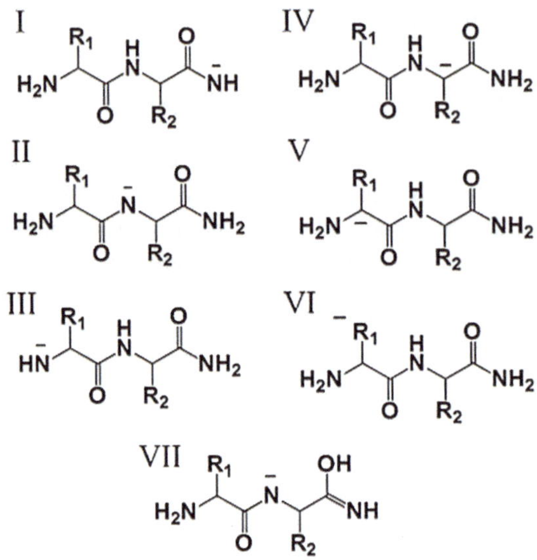

Fig. 14 Proposed sites of deprotonation for **c** ion structures. Adapted from Grzetic and Oomens [37]. Copyright 2012, with permission from Elsevier

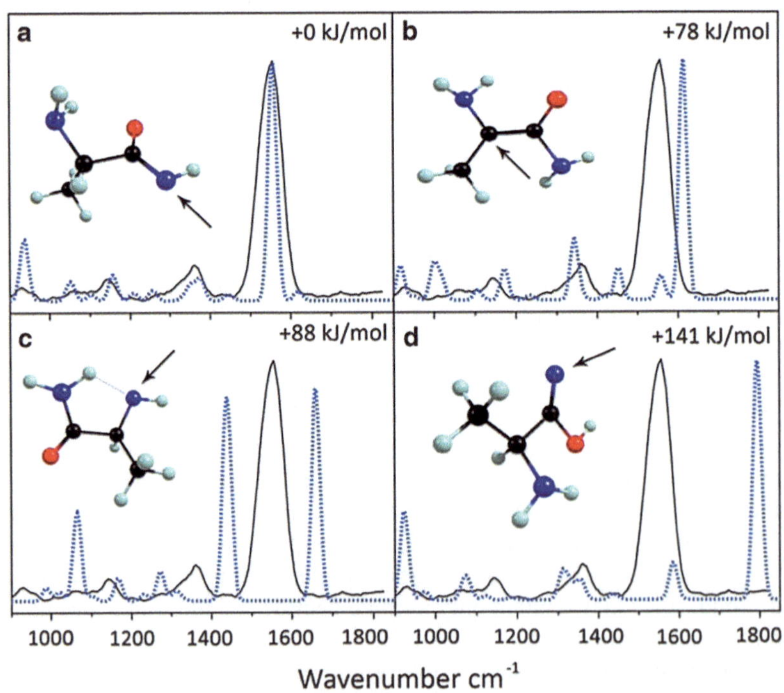

Fig. 15 The IRMPD spectrum (*solid black lines*) of the $\mathbf{c_1}$ anion from AlaPhe compared to the computed IR spectra (*blue dotted lines*) for putative structures deprotonated at (**a**) the C-terminal amide, (**b**) the α-carbon, (**c**) the N-terminus, and (**d**) the C-terminal amide with enol-imine tautomerization. Reprinted from Grzetic and Oomens [37]. Copyright 2012, with permission from Elsevier

the *enol-imide* motif was found for the deprotonated systems studied, though this motif is often suggested as the structure of **c**-type ions created from (positive-mode) ECD and ETD.

As with protonated peptides, **b**-type ions have proved to be, perhaps, the most compelling in terms of potential structures. Peptide motifs with and without nucleophilic side chains have been investigated with quite interesting results which are not completely unlike their protonated counterparts. The first two systems studied, AlaAlaAla and AlaTyrAla, both gave rise to $\mathbf{b_2}$ ions taking the deprotonated oxazolone structure [38]. It is notable that tyrosine was seemingly uncompetitive as a deprotonation site, in clear contrast to **c**-type ions.

The most intriguing effects are observed for systems containing nucleophilic side chains, as neither oxazolone nor diketopiperazine structures match the experimentally observed spectrum. Conversely, a nucleophilic attack by the side chain, similar to the mechanism provided for protonated counterparts in Scheme 1, pathway 4, seems to be consistent with the results [39]. In the case of asparagine, there are two potential nucleophiles on the side chain. Attack from the side-chain amide nitrogen on the peptide backbone carbonyl carbon would give rise to the cyclic *succinimide* structure, whereas attack by the side-chain oxygen would lead to an *imino-γ-butyrolactone* or *amino-γ-butyrolactone*, depending on the deprotonation site. Through comparison of the experimental spectrum with theoretical spectra for these various potential isomers, it became clear that the only possible match was with the succinimide structure deprotonated at the succinimide nitrogen.

Thus far, no evidence of a diketopiperazine **b** anion has been reported, possibly because of the role of basic residues (which are not amenable to negative mode ionization) in the formation of diketopiperazine structures.

Because of the scarcity of studies thus far, anionic peptide fragmentation is a fertile ground for future work. It will be especially interesting to see how sequence motif and chain length affect the propensity of anion fragments to take various structural forms. Furthermore, it will be useful for future applications to continue to delve into the structure of anions to see if there are ever cases where cyclization (and resultant sequence scrambling) occurs, which has been proposed as a potential pitfall in the sequencing of protonated peptides.

4 Radical Dissociation Chemistry of Peptides

Radical chemistry of peptides opens up different, and often fascinating, dissociation pathways. However, the gentle ionization methods suitable for peptides, such as electrospray ionization (ESI) and matrix-assisted laser desorption/ionization (MALDI), typically generate closed-shell ions. Producing open-shell peptide ions from closed-shell ions thus requires some gas-phase reaction chemistry, either involving redox chemistry or distonic cleavages of a preparative complex, attachment or removal of an electron, or electronic excitation by photons or particles.

4.1 Preparative Complex

Some amino acids, such as tryptophan, can be complexed with a redox-active complex, incorporating, for instance, Cu^{II}. When this complex is activated by collisions, the copper is reduced to Cu^{I} and the amino acid is oxidized to $Trp^{+\bullet}$. Another approach involves chemically altering the indole side chain from an indole NH to an N-nitrosylated group (i.e., N–NO); this complex exhibits distonic cleavage, so that two open-shell fragments are produced, neutral $NO^{\bullet}$ and charged $Trp^{+\bullet}$. The advantage of this chemistry is that the initial radical site is known, and thus radical migrations can be studied by spectroscopic means. A number of studies on molecular systems incorporating tryptophan, for example, have been reported in the literature [40, 41].

4.2 Electron Capture and Electron Transfer Dissociation

Radical dissociation chemistry of peptides can also be triggered by attaching or removing an electron. Historically, electron capture dissociation (ECD) is the first demonstrated process [42] where a peptide cation captures a slow electron, giving rise to an odd-electron, charge-reduced species. A similar phenomenon was later demonstrated in an ion–ion reaction by inducing an electron transfer from a radical anion to a peptide cation, in a process termed electron transfer dissociation (ETD) [43]. The reduction of a cation is an exothermic process, and a considerable portion of the charge-reduced peptide dissociates via cleavage of a backbone N–C_α bond, giving rise to so-called **c** and **z** ions. This is in contrast to the prevalent cleavage of the amide bond in CID, and thus illustrates the very different mechanism/s that underlies/underlie these radical dissociation processes. In fact, the mechanism of ECD/ETD remains a contentious area. Some observations are consistent with a fast mechanism, which may even be non-ergodic (i.e., faster than intermolecular vibrational redistribution). The fact that ECD/ETD is successful for large molecules, such as proteins, indicates that much of the energy in the electron capture event must remain localized in the vicinity where the bond cleavage takes place. Conversely, a complete randomization of the energy in such large molecules would lead to insignificant heating, resulting in no bond cleavage. A very useful aspect of ECD/ETD is the selective cleavage at backbone bonds over the loss of fragile side-chain groups in the molecule; this makes ECD/ETD extremely powerful for sequencing purposes, as labile post-translational modifications (PTMs) (e.g., glycosylations, phosphorylations, nitrosylations, etc.) can be conserved, and can hence be identified and localized [44].

One of the proposed mechanisms in ECD, the Utah–Washington mechanism [45, 46], involves fast cleavage of the N–C_α bond as a result of electron capture at the carbonyl π^* anti-bonding orbital, followed by slower proton transfer. The mechanism is intimately linked to the chemical structures of the fragments that

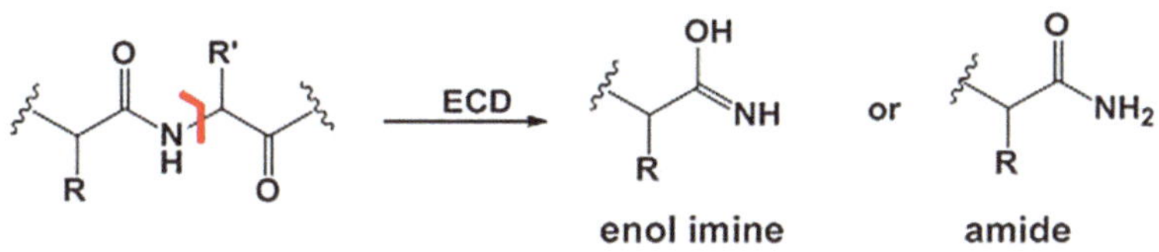

Scheme 4 Cleavage of N–C_α bond in ECD, leading to proposed enol imine or amide **c** ions

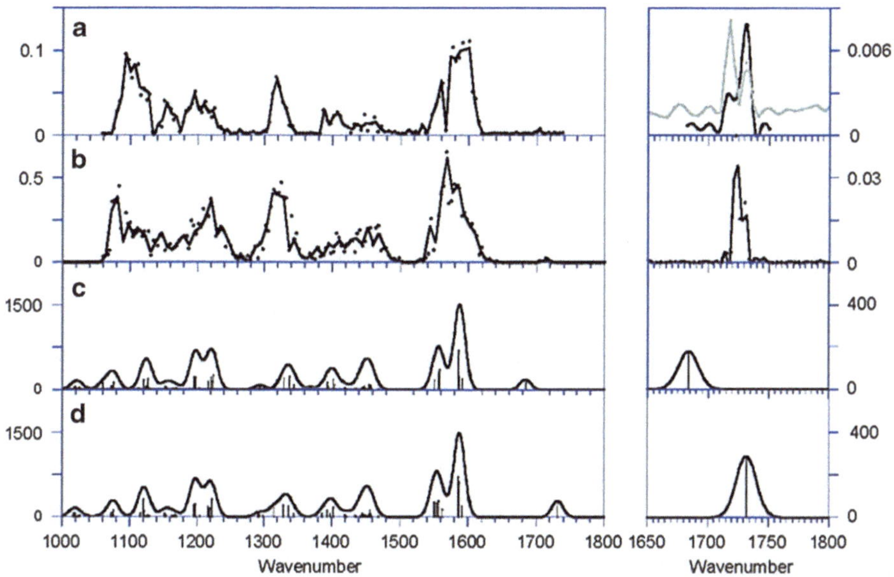

Fig. 16 (**a**) IRMPD spectrum of **c** ion made by ECD compared to (**b**) IRMPD spectrum of synthetically made amide structure and computed IR spectra for (**c**) enol imine and (**d**) amide structures. Adapted with permission from Frison et al. [47]. Copyright (2008) American Chemical Society

are produced. The N-terminal **c** ion can in principle adopt one of two isomeric structures, either an *enol–imine* or an *amide* structure (see Scheme 4). Although vibrational spectroscopy is an excellent tool to verify which structure is generated, thus far merely one IRMPD spectroscopy study has been carried out on an ECD fragment. The IRMPD spectrum of this **c** ion is depicted in Fig. 16, showing a convincing match with the reference spectrum for a synthetically made amide structure [47]. Notably, the amide C=O stretch at 1,730 cm^{-1} is key in assigning this structure. On the other hand, the enol-imine structure can be excluded based on the absence of the corresponding C=N stretch expected at 1,680 cm^{-1}. It remains to be seen whether the amide structure is generally observed for **c** ions. The particular **c** ion in Fig. 16 is rather special, as it is small and has a fixed-charge tris(2,4,6-trimethoxyphenyl)phosphonium group, as opposed to a site of proton attachment. Since the structures of the product ions in ECD/ETD are determined by proton or hydrogen atom migration, the absence of a remaining proton, after electron capture, is likely to affect the chemistry that takes place.

4.3 *Other Radical Processes*

Absorption of a high-energy ultraviolet (UV) photon leads to promotion of an electron into an excited orbital; the ion can then dissociate in an electronically excited state, or this energy can couple to vibrational energy (i.e., heat) by internal conversion. When peptides absorb photons in the vacuum UV range (e.g., 157 nm) or near-vacuum UV (e.g., 193 nm), a rich dissociation chemistry is observed, and many types of backbone fragments (i.e., *a*, *b*, *c*, *x*, *y*, *z*) are produced [48].

While most peptide dissociation is carried out in the positive ion mode, the negative ion mode is often better suited for acidic peptides, particularity those carrying acidic modifications (e.g., phosphorylations). There are a number of equivalent ion activation methods for peptide anions, involving ion-electron and ion-ion reactions, such as electron detachment dissociation (EDD) [49], negative electron transfer dissociation (NETD) [50, 51], and negative electron capture dissociation (nECD) [52].

Much less is known about the latter fragmentation mechanisms, and none of those dissociation products have yet been structurally elucidated by IR spectroscopy, leaving ample room for detailed mechanistic studies.

Conclusions and Outlook

In the past few years, IRMPD spectroscopy of peptide fragments has provided a number of critical insights into the sometimes surprising dissociation chemistries that take place in mass spectrometers. Characteristic diagnostic vibrations can verify the presence of proposed structures, such as oxazolones, macrocycles, imines, etc. The implications of some of these rearrangements on successfully deriving the sequences of peptides are now better understood, at least in terms of the CID chemistry of protonated peptides. However, much work still remains to be done in terms of characterizing various types of fragment ions, prepared by a range of ion activation methods.

The study of complex mixtures of fragment isomers also requires enhanced gas-phase structural tools. The advent of cold spectroscopy techniques, such as IR-UV ion dip [53] or IR pre-dissociation [54] spectroscopy, offers much higher resolution, and can hence provide much more structural detail. In cold spectroscopy techniques, the ions are cooled to cryogenic temperatures (i.e., 4–20 K), resulting in frozen conformations and narrower, better-defined spectral bands. Another key advantage of cold spectroscopy methods is that particular conformers can be selectively studied. In IR-UV

(continued)

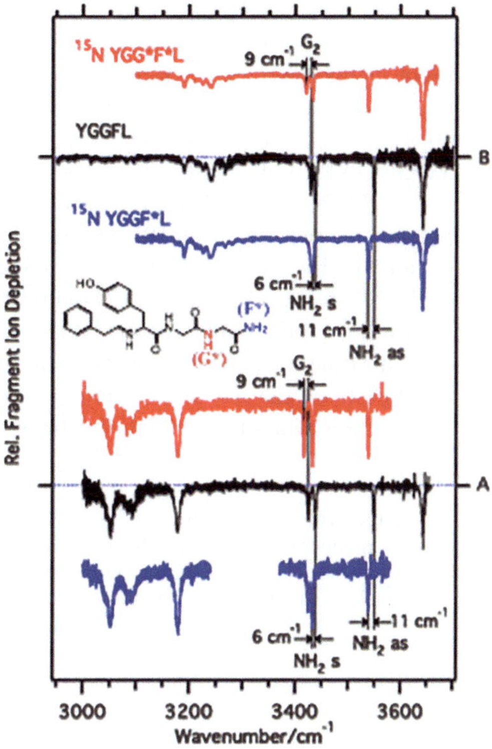

Fig. 17 IR-UV ion dip spectra for $\mathbf{a_4}$ ions from TyrGlyGlyPheLeu, as well as ^{15}N labeled variants of $\mathbf{a_4}$. Adapted from [55]. Copyright 2012, with kind permission from Springer Science and Business Media

ion dip spectroscopy, a UV photon selects a particular conformer so as to measure a conformer-resolved IR spectrum. As an example, in Fig. 17 the IR-UV dip spectra of two isomers of the $\mathbf{a_4}$ ion from TyrGlyGlyPheLeu are shown in the hydrogen stretching range [55]. In order to assign the various amide NH stretches, the peptides were also isotopically labeled with ^{15}N at various amide N sites (denoted with *). The heavier isotope results in a characteristic red shift of the amide NH stretch (caused by the larger reduced mass), allowing an unambiguous assignment of various amide bands. For instance, the NH stretch of the second glycine residue, G2, is selectively shifted upon ^{15}N substitution. In this particular example, the ^{15}N label at the phenylalanine residue reveals another very important insight: this group is observed to migrate to the C-terminus as an amide NH_2 group, as confirmed by the shifts in the symmetric and asymmetric NH_2 stretch modes. This very specific structural insight lends strong credence to the N-term imine structure in Scheme 3, which has an amide group at the C-terminus. It should be noted that these redshifts are only apparent as a result of the narrow bandwidths in these IR spectra, and that these shifts could not be observed by IRMPD spectroscopy.

(continued)

Another avenue for improving the structural elucidation of fragment ions involves the coupling of IR spectroscopy with other structural techniques in mass spectrometry. Ion mobility mass spectrometry is well suited to this approach. As demonstrated for the example of the $\mathbf{a_4}$ ion of TyrGlyGlyPheLeu, multiple structures could be separated by ion mobility, and could then be further structurally probed by IR spectroscopy. The coupling of ion mobility with IR spectroscopy presents a number of advantages: (1) ion mobility is capable of separating some isomers, allowing a 'clean-up' of the IR spectra of mixtures, (2) ion mobility allows a better quantitation of the various product ion structures that are generated, and (3) the complementary information from ion mobility and IR spectroscopy impose multiple constraints on the gas-phase structures of the fragment ions, thus allowing more detailed structural insights to be obtained.

Acknowledgements Some of this work was supported by the U.S. National Science Foundation under grant number CHE-0845450. Student travel support to the Netherlands for measurements with the free electron laser FELIX was provided by the NSF-PIRE (grant number OISE-0730072). This material is based upon work supported by the National Science Foundation Graduate Research Fellowship (A.P.) under Grant No. (DGE-1315138).

References

1. Aebersold R, Goodlett DR (2001) Chem Rev 101:269
2. Steen H, Mann M (2004) Nat Rev Mol Cell Biol 5:699
3. Roepstorff (1984) Biomed Mass Spectrom 11:601
4. Yu L, Tan Y, Tsai Y, Goodlett DR, Polfer NC (2011) J Proteome Res 10:2409
5. Dongré AR, Jones JL, Somogyi Á, Wysocki VH (1996) J Am Chem Soc 118:8365
6. Tsaprailis G, Somogyi Á, Nikolaev EN, Wysocki VH (2000) Int J Mass Spectrom 195/196:467
7. Polfer NC, Oomens J, Suhai S, Paizs B (2007) J Am Chem Soc 129:5887
8. Yalcin T, Csizmadia IG, Peterson MR, Harrison AG (1996) J Am Soc Mass Spectrom 7:233
9. Yalcin T, Khouw C, Csizmadia IG, Peterson MR, Harrison AG (1995) J Am Soc Mass Spectrom 6:1165
10. Chen X, Yu L, Steill JD, Oomens J, Polfer NC (2009) J Am Chem Soc 131:18272
11. Yoon SH, Chamot-Rooke J, Perkins BR, Hilderbrand AE, Poutsma JC, Wysocki VH (2008) J Am Chem Soc 130:17644
12. Oomens J, Young S, Molesworth S, Stipdonk M (2009) J Am Soc Mass Spectrom 20:334
13. Gucinski AC, Chamot-Rooke J, Steinmetz V, Somogyi Á, Wysocki VH (2013) J Phys Chem A 117:1291
14. Zou S, Oomens J, Polfer NC (2012) Int J Mass Spectrom 316–318:12
15. Gucinski AC, Chamot-Rooke J, Nicol E, Somogyi Á, Wysocki VH (2012) J Phys Chem A 116:4296
16. Grzetic J, Oomens J (2013) J Am Soc Mass Spectrom 24:1228
17. Perkins BR, Chamot-Rooke J, Yoon SH, Gucinski AC, Somogyi ARD, Wysocki VH (2009) J Am Chem Soc 131:17528
18. Paizs B, Suhai S (2005) Mass Spectrom Rev 24:508

19. Rodgers MT, Ervin KM, Armentrout PB (1997) J Chem Phys 106:4499
20. Polfer NC, Oomens J, Suhai S, Paizs B (2005) J Am Chem Soc 127:17154
21. Tirado M, Polfer NC (2012) Angew Chem 124:6542
22. Chen X, Steill JD, Oomens J, Polfer NC (2010) J Am Soc Mass Spectrom 21:1313
23. Tirado M, Rutters J, Chen X, Yeung A, Maarseveen J, Eyler J, Berden G, Oomens J, Polfer N (2012) J Am Soc Mass Spectrom 23:475
24. Gucinski A, Somogyi Á, Chamot-Rooke J, Wysocki V (2010) J Am Soc Mass Spectrom 21:1329
25. Durand S, Rossa M, Hernandez O, Paizs B, Maître P (2013) J Phys Chem A 117:2508
26. Harrison AG, Young AB, Bleiholder C, Suhai S, Paizs B (2006) J Am Chem Soc 128:10364
27. Bleiholder C, Osburn S, Williams TD, Suhai S, Van Stipdonk M, Harrison AG, Paizs B (2008) J Am Chem Soc 130:17774
28. Goloborodko A, Gorshkov M, Good D, Zubarev R (2011) J Am Soc Mass Spectrom 22:1121
29. Saminathan I, Wang XS, Guo Y, Krakovska O, Voisin S, Hopkinson A, Siu KWM (2010) J Am Soc Spectrom 21:2085
30. Bythell BJ, Dain RP, Curtice SS, Oomens J, Steill JD, Groenewold GS, Paizs B, Van Stipdonk MJ (2010) J Phys Chem A 114:5076
31. Verkerk UH, Zhao J, Van Stipdonk MJ, Bythell BJ, Oomens J, Hopkinson AC, Siu KWM (2011) J Phys Chem A 115:6683
32. Bythell BJ, Maître P, Paizs B (2010) J Am Chem Soc 132:14766
33. Verkerk UH, Zhao J, Lau JK-C, Lam T-W, Hao Q, Steill JD, Siu C-K, Oomens J, Hopkinson AC, Siu KWM (2012) Int J Mass Spectrom 330/332:254
34. Paizs B, Bythell B, Maître P (2012) J Am Soc Mass Spectrom 23:664
35. Polfer NC, Bohrer BC, Plasencia MD, Paizs B, Clemmer DE (2008) J Phys Chem A 112:1286
36. Oomens J, Steill JD (2010) J Am Soc Mass Spectrom 21:698
37. Grzetic J, Oomens J (2012) Int J Mass Spectrom 316/318:216
38. Grzetic J, Oomens J (2012) J Am Soc Mass Spectrom 23:290
39. Grzetic J, Oomens J (2013) Int J Mass Spectrom 354/355:70
40. Bagheri-Majdi E, Ke Y, Orlova G, Chu IK, Hopkinson AC, Siu KWM (2004) J Phys Chem B 108:11170
41. Piatkivskyi A, Osburn S, Jaderberg K, Grzetic J, Steill J, Oomens J, Zhao J, Lau J-C, Verkerk U, Hopkinson A, Siu KWM, Ryzhov V (2013) J Am Soc Mass Spectrom 24:513
42. Zubarev RA, Kelleher NL, McLafferty FW (1998) J Am Chem Soc 120:3265
43. Syka JEP, Coon JJ, Schroeder MJ, Shabanowitz J, Hunt DF (2004) Proc Natl Acad Sci U S A 101:9528
44. Mirgorodskaya E, Roepstorff P, Zubarev RA (1999) Anal Chem 71:4431
45. Syrstad E, Turecček F (2005) J Am Soc Mass Spectrom 16:208
46. Anusiewicz I, Berdys-Kochanska J, Simons J (2005) J Phys Chem A 109:5801
47. Frison G, van der Rest G, Tureček FE, Besson T, Lemaire Jl, Maître P, Chamot-Rooke J (2008) J Am Chem Soc 130:14916
48. Zhang L, Reilly JP (2010) Anal Chem 82:898
49. Budnik BA, Haselmann KF, Zubarev RA (2001) Chem Phys Lett 342:299
50. Coon JJ, Shabanowitz J, Hunt DF, Syka JEP (2005) J Am Soc Mass Spectrom 16:880
51. Huzarska M, Ugalde I, Kaplan DA, Hartmer R, Easterling ML, Polfer NC (2010) Anal Chem 82:2873
52. Yoo HJ, Wang N, Zhuang S, Song H, Håkansson K (2011) J Am Chem Soc 133:16790
53. Rizzo TR, Stearns JA, Boyarkin OV (2009) Int Rev Phys Chem 28:481
54. Robertson WH, Johnson MA (2003) Annu Rev Phys Chem 54:173
55. Wassermann T, Boyarkin O, Paizs B, Rizzo T (2012) J Am Soc Mass Spectrom 23:1029

Top Curr Chem (2015) 364: 183–224
DOI: 10.1007/128_2014_578

Published online: 12 March 2015

Spectroscopy of Metal-Ion Complexes with Peptide-Related Ligands

Robert C. Dunbar

Abstract With new experimental tools and techniques developing rapidly, spectroscopic approaches to characterizing gas-phase metal ion complexes have emerged as a lively area of current research, with particular emphasis on structural and conformational information. The present review gives detailed attention to the metal-ion complexes of amino acids (and simple derivatives), much of whose study has focused on the question of charge-solvation vs salt-bridge modes of complexation. Alkali metal ions have been most frequently examined, but work with other metal ions is discussed to the extent to which they have been studied. The majority of work has been with simple cationic metal ion complexes, while recent excursions into deprotonated complexes, anionic complexes, and dimer complexes are also of interest. Interest is growing in complexes of small peptides, which are discussed both in the context of possible zwitterion formation as a charge-solvation alternative, and of the alternative metal-ion bond formation to amide nitrogens in structures involving iminol tautomerization. The small amount of work on complexes of large peptides and proteins is considered, as are the structural consequences of solvation of the gas-phase complexes. Spectroscopy in the visible/UV wavelength region has seen less attention than the IR region for structure determination of gas-phase metal-ion complexes; the state of this field is briefly reviewed.

Keywords Action spectroscopy · Amino acids · Gas phase ion structures · Metal ion complexes · Peptides

R.C. Dunbar (✉)
Chemistry Department, Case Western Reserve University, Cleveland, OH 44106, USA
e-mail: rcd@po.cwru.edu

Contents

1 Introduction

Ionic complexes can be built by assembling one or more ligands around an ionic core, where the core may be a proton, a metal ion, or a more complicated ion (cluster ion, inorganic ion, etc.). Such structures can have biological interest and relevance when the ligand(s) wrapping around the ionic core comprises the stuff of living systems, and here we are interested specifically in the structures formed when the wrappings are essentially peptides and the core is a metal ion.

The convergence of remarkable technological developments on several fronts has quite suddenly made spectroscopic characterization of gas phase ions greatly more powerful, convenient and versatile (see Oomens [1]), with the result that the last decade has seen a flowering of ion spectroscopy. A major portion of the hundreds of publications in this area has addressed ionic complexes, and a large number of these reports in the last few years have had at least passing relevance to biological, biochemical and biophysical concerns. Such is the focus of the present survey. A number of reviews touching on this area have appeared in the last few years [2–9].

Interesting questions arise when the ligand is complex enough to offer a choice of binding sites and binding geometries as it arranges around the core. Answers to such questions can then reflect back to the native condensed-phase biological architecture to enhance understanding of the structure of metallic charge sites in living systems.

The flavor of the present chapter overviewing the field might be suggested by Fig. 1, which lays out in skeletal form the progression from spectroscopy of the simplest "peptide" complexes (amino acids), through small oligopeptides (here, a dipeptide), to large systems (proteins). The simplest systems show a few prominent spectroscopic features which can readily be identified (with computational help) with specific molecular surface groups (here, vibrations associated with the C-terminus). The dipeptide emphasizes the initial emergence of the spectroscopic signatures (still individually discrete) reflecting the body of the peptide (the amide linkages and the interior residue side chains); finally, the complex of the large cytochrome-c peptide [13] shows the complete submergence of the C-terminal features, leaving the spectrum dominated by the two amide bands of the 100 or

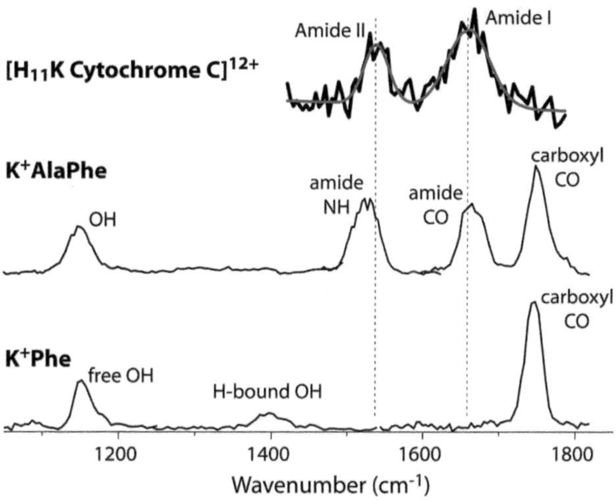

Fig. 1 Infrared photodissociation action spectra in the fingerprint infrared region from fairly early in the modern evolution of spectroscopy of biologically interesting complex ions. These spectra illustrate the use of electrospray ion production, ion trapping mass spectrometry (Fourier-transform ion cyclotron resonance in these examples) and the FELIX free electron laser light source. Lower spectrum originally derived from [10, 11]; middle spectrum from [12]; upper spectrum from [13]. Figure reproduced with permission from [12]

so amide linkages, just as with proteins in solution. Our survey follows the same path from smaller to larger ligands, first in some detail for the infrared (vibrational) spectral domain, and then more briefly for the less well developed domain of electronic (visible/ultraviolet) spectra.

Parallel to our survey of peptide metal complexes, which is encompassed by the present chapter, are a large number of spectroscopic structure studies of protonated amino acids and peptides (for example, amino acids and derivatives [14–29] and larger peptides [30–38]). We could consider the proton as a metal ion and weave the spectroscopy of protonated peptides into our discussion here, but we largely leave such a survey for another place. More broadly, spectroscopic information is becoming pervasive in the study of gas-phase biomolecules, and many further examples appear in other chapters in this volume. One of the exciting frontiers is the spectroscopy of cryogenic ions [39]. Not much has been reported yet in this latter domain about metal-ion complexes of biomolecules, but we can expect such cold ion spectroscopy to be a flourishing area in the near future. The present state of this area is well covered in the chapter of Boyarkin and Rizzo in this volume [40]. Yet another frontier area not explored in the present chapter is the observation and exploitation of gas-phase ion fluorescence [41], where some especially noteworthy recent work is bringing to bear the powerful tool of FRET (Fluorescence Resonant Energy Transfer) for characterization of large biomolecules [42, 43].

2 Infrared and Vibrational Spectroscopy

2.1 *Amino Acids and Derivatives*

2.1.1 Charge-Solvated vs Salt Bridge (Zwitterion)

Forming a bedrock for the understanding of metal-ion peptide interactions, great effort has gone into the study of metal-ion binding to gas-phase amino acids. A principal theme arises from the observation that the natural amino acids self-ionize to the zwitterionic form in aqueous solution (Scheme 1), while in gas phase their most stable form is the canonical conformation. It was natural to ask whether interaction with a metal ion in gas phase could preferentially stabilize the zwitterionic conformation of the amino acid ligand to make this the most favorable conformation. The answer to this question has been explored by now in considerable depth, revealing a delicately balanced and multifaceted interplay of factors which may or may not favor isomerization of the ligand to the zwitterionic form. Early on, the Bowers group [44] explored this question both experimentally, using largely inconclusive evidence from collision cross sections, and computationally. Starting in the mid-1990s, Williams' group pioneered the use of black-body infrared dissociation (BIRD) to address ion-structure investigation, including the present question (see [45] for one example). Near the beginning of this century, the Armentrout group began to use the high-quality thermochemical information about ion dissociation pathways coming from guided ion beam mass spectrometry (GBIMS) to address this question (for example, [46]), and they continue to use this approach extensively as a complementary technique to their spectroscopic studies. These and other experimental programs, along with widespread computational study, formed a backdrop of understanding of the issue. Realization that spectroscopy could provide more conclusive evidence than other experimental approaches [47] launched a decade of IR action spectroscopy centered around this theme.

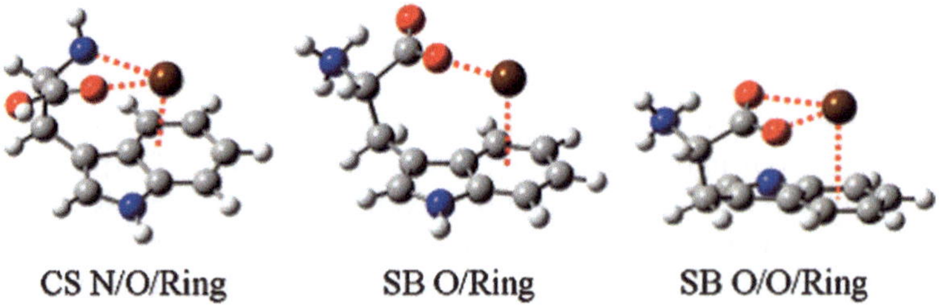

Scheme 1 Low-energy structures for the complex of Ba^{2+} and Trp. Structures can be classified as salt-bridge (SB: interaction between the positive metal ion and the negative carboxylate of the zwitterionic amino acid) or charge solvation (CS: interaction of the metal ion with Lewis-basic sites of the canonical amino acid). Nomenclature of the various structures further includes the main binding sites of the amino acid

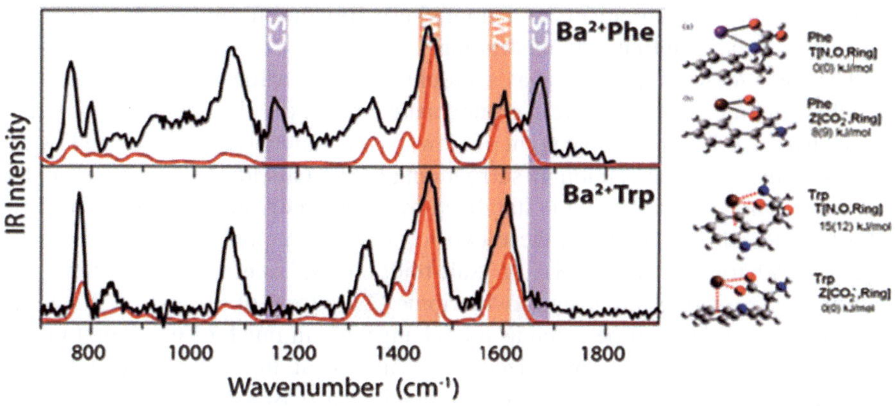

Fig. 2 Spectra of the trapped populations of Phe and Trp complexes of Ba^{2+} (figure adapted from [11]). Comparison of the IRMPD spectra (*upper trace* in each panel) of the Ba^{2+} complexes of Phe and Trp. The *lower* (*red*) trace in each panel shows the calculated spectrum of the SB conformation, and two expected diagnostic features for the Z[CO_2^-,Ring] (SB) conformation are designated by the (*orange*) bars labeled ZW. Two diagnostic features expected for the T[N,O,Ring] charge-solvated conformation are designated by the (*purple*) bars labeled CS, and are seen to be present in the observed spectra only in the Phe case

As an example of the strength of this tool for characterizing the populations of the electrosprayed, trapped ionic complexes, Fig. 2 contrasts spectra of the Ba^{2+} complexes of the two closely related amino acids Phe and Trp [11]. It is clear that the Trp complex is homogeneously SB, but for the Phe case the CS conformation has comparable stability leading to a mixture of both conformations with substantial abundance of each. Figures 1 and 2 underscore the strength of this structure-determining tool in circumstances where we can identify effective diagnostic features in the spectrum. In these examples of CS/SB identification for amino acid ligands, the diagnostic peaks for both conformations are clearly evident. Figure 1 (lowest trace) shows an uncontaminated CS complex with the characteristic markers of the carboxylic acid stretch band at short wavelength (above 1,700 cm^{-1} for singly charged metals) and OH bending mode at 1,150 cm^{-1}. These are absent for the uncontaminated SB complex in Fig. 2 (lower trace), and these spectral regions are quite empty in the latter spectrum. Conversely, the characteristic markers for SB amino acids, the asymmetric carboxylate COO^- stretch (1,600 cm^{-1}) and the $-NH_3^+$ bend (between 1,400 and 1,450 cm^{-1}), are completely absent in these empty spectral regions for K^+Phe in Fig. 1.

One would like to draw quantitative conclusions about the relative abundances of the conformations from the sizes of the peaks in Fig. 2 (upper trace), but this is unjustifiable for several reasons. For one thing, the one-photon optical absorption strengths of the different peaks as calculated quantum-mechanically do not correspond accurately to the effective multiphoton absorption cross sections appropriate to the IRMPD process [8]. For another, as clearly demonstrated by Prell et al. [48] among others, the time course of IRMPD dissociation can differ widely for different isomers, so that measurements at a single laser irradiation time (as is usually the practice) can be quantitatively misleading.

The early focus of this field was a simple binary decision of whether the given metal-ion/amino-acid pair gave a ground state complex with canonical or zwitter-ionic character. From a simple point of view, the canonical form is stabilized by the local solvation of the metal ion by surrounding ligand (shielding its charge from the vacuum), while the zwitterion is stabilized by electrostatic interactions in the form of a (+ − +) salt bridge. Thus, these two forms are commonly referred to respectively as "charge-solvated (CS)" or "salt-bridge (SB)" conformations. (Since the salt-bridge commonly involves a zwitterionic (ZW) form of the ligand, the designations SB and ZW can be taken as interchangeable in much of this literature). A review 4 years ago [5] summarized the evidence at that time, displayed in a comprehensive matrix of metal-ion/amino-acid pairs. Since then, it has become increasingly clear that a binary CS vs SB decision is too simple, and the combination of computation with spectroscopy has revealed trends governed by more subtle considerations. Table 1 presents an update of their matrix, displaying the spectroscopically determined evidence as of mid-2013. There is a useful update specifically of the computed CS/SB preferences for alkali metal ions in [62] (Fig. 7).

With the bias that spectroscopic assignments are the most definitive experimental evidence of which isomers are actually formed and observed under given instrumental conditions, we have focused our table primarily on structures assigned from spectroscopic data. Computation can now often predict the relative stabilities of different isomers with good confidence, but a full survey and evaluation of this extensive computational literature is beyond our scope. In order to clarify trends, we have included in our table some cases, indicated by square brackets, where theory or alternative experimental approaches are the only evidence so far available. We have ordered the metal ions within each charge state by a computed chelation-length parameter, which is akin to the standard ionic radius, but perhaps more appropriate for these chelation complexes. The amino acids are ordered according to their binding affinity for Na^+ [49], based on the observation that SB/CS preferences in tryptophan appeared to correlate well with this parameter [77].

Several basic structure types often compete to determine the ground state, as illustrated for Rb^+Ser in Scheme 2, where we have designated the isomers using the letter coding of Table 1, along with the square-bracket descriptive coding used by the Armentrout group. These basic structure themes can be summarized as follows:

1. T. Tridentate Chelating CS. T[CO,N,X] where the ligand chelates the metal through three Lewis-basic sites (X being a side-chain Lewis-basic atom)
2. B. Bidentate Chelating CS. B[CO,N] or B[CO,X] where the metal ion is chelated at only two sites
3. C. C-Terminal CS. Metal coordinated to COOH group, C[COOH] or C[COOH, X]. Metal ion coordinates both oxygens of the C-terminal COOH group without formation of a zwitterion
4. Z. Salt Bridge SB, (Zwitterion). COOH deprotonates to give zwitterion, Z [CO_2^-]

Table 1 Amino acid binding modes

AA	D‡	H^+	Li^+	Na^+	Ag^+	K^+	Rb^+	Cs^+	Mn^{2+}	Ca^{2+}	Ba^{2+}
R§			1.94	2.31	2.52	2.66	2.86	3.03	2.08^a	2.28	2.60
Gly	161	[B^{mm}]		[$B,X^{q,c}$]	[B^v]	[B^w]	[C^q]	[C^x]	[Z^a]	[$Z^{jj,kk}$]	Z^f
Ala	167	[B^{mm}]		[B^q]	[B^v]		[C^q]		[Z^a]		
Val	173	[B^{mm}]	[B^{ff}]	[B^{ff}]	[B^v]				[Z^a]		Z^s
Cys	175	B^g	T^g	T^g	[T^v]	C^g	C^g	C^g	[T^a]		
Leu	175	[B^{mm}]			[B^v]				[Z^a]		
Ile	176	[B^{mm}]			[B^v]				[Z^a]		
Pro	196	[B^{mm}]	Z^r	Z^c	[Z^v]	Z,C^r	Z,C^r	Z,C^r	[Z^a]	[Z^{ee}]	Z^s
Ser	192	[B^{mm}]	T^h	T^h	[T^v]	T,C^h	T,C^h	$T,C,Z^{h,t}$	[T^a]		Z^s
Thr	197	[B^{mm}]	T^l	T^l	[T^v]	T,C^l	T,C^l	T,C,Z^l	[T^a]		
Met	197	T^m	T^m	T^m	[T^v]	T,C^m	T,C^m	T,C,Z^m	[T^a]		
Phe	198	B^{cc}	T^n	T^n	T^n	$T^{hh,n}$	T,B^n	T,C^n	[T^a]		Z,T^n
Tyr	201	B^{cc}			[T^v]	T^{hh}			[T^a]		
Asp	203	[B^{mm}]	T^p		[T^v]			C'^p	[T^a]		Z^p
Glu	204	B^p	T^p		[T^v]			B^p	[T^a]	Z^p	Z^p
Trp	210	B^{dd}	T^o	T,B^o	$T^{o,ii}$	T,B^o	T,B^o	T,B^o	[T^a]		Z^{ll}
Asn	206	B^f	T^f	T^f	[T^v]	T,C^f	T,C^f	T,C^f	[T^a]		T^f
Gln	212	B^d	T^d	T^d	[T^v]	T,B^d		T,X^d	[T^a]	Z^p	Z^s
Lys	>213	B^i	T^i	T^i	[T^v]	B^i			[T^b]		
His	219	B^{bb}	T^j	$T^{j,k}$	[T^v]	B,T,C^j	B,T,C^j	B,T,C^j	[T^b]	$Z,T^{j,k}$	Z^k
Arg	>225	B^e	T^e	$Z,T^{e,t}$	B^e	Z^e	Z^e	Z^e	[T^a]		Z^s

‡ D = amino acid binding energy to Na^+ (kJ/mol) (Ref. [49]). Note that most of these kinetic-method binding energies have been remeasured subsequently by other techniques like guided beam dissociation, and the values listed here should not be taken as definitive. In particular, Asp [50] and Pro [51] are most likely lower by the order of 10 kJ mol^{-1}

§ R = B3LYP distance from metal to carbonyl O in M^+alanine (Å)

[Square brackets] = theory or non-spectroscopic experiment

Color coding:

Red=CS Blue=SB Purple=mixed CS and Z

Letter coding:

T=Tridentate CS B=Bidentate CS (CO,X) C=CS (COOH) C'=CS (COOH plus additional site) Z=SB X=unidentified CS ligand

a [52], b [53], c [46, 47], d [54], e [15, 20], f [21], g [18], h [55], i [14, 56], j [57], k [58], l [59], m [16], n [11], o [60], p [61], q [44], r [62, 63], s [64], t [48], u [65], v [66], w [67], x [68], aa [69], bb[17], cc [70], dd [39], ee [71], ff [72], gg [73], hh [10], ii [74], jj [75], kk [76], ll [77], mm [78]

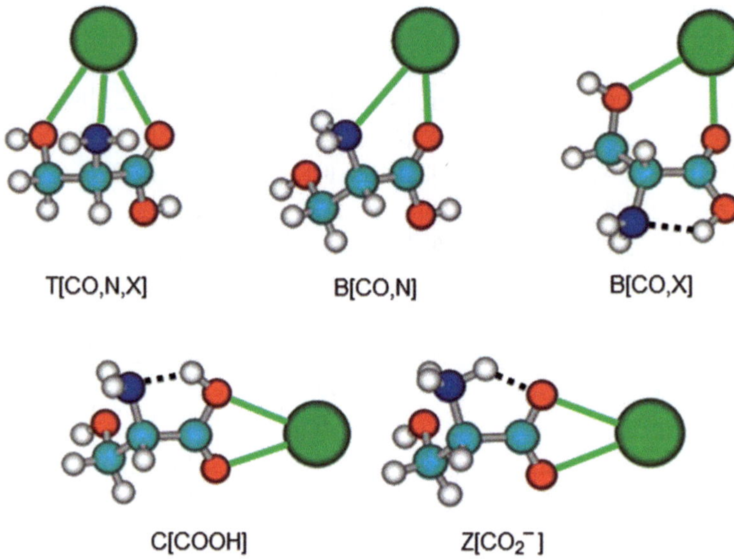

Scheme 2 Structures of the principal isomers of metal-ion serine complexes. Reproduced with permission from [55]

Some trends can be seen in Table 1, which are developed and explored extensively in the literature. Although there are occasional exceptions to most of these trends, the following principles seem to be mostly valid for understanding and predicting the relative stabilities of the different binding motifs described here:

1. Effect of metal size on choice among the three likely CS binding motifs B, T, and C. Small metals gain more from solvation, which enhances more highly solvated CS forms. Thus T and B are favored over C for small metals, and T is favored over B for small metals. (There is effectively no possibility of a T conformation for the aliphatic amino acids Gly, Ala, Val, Leu, Ile, and Pro.)
2. Effect of metal size on CS vs SB preferences. Other things being equal, small ion size is more favorable to CS stabilization (T, B, or C) by local micro-solvation of the bare metal ion, while large metal ion size reduces the micro-solvation energy gain to the point that SB stabilization (Z) becomes competitive. Thus it is among the larger metal ions that we find increasingly frequent cases of Z ions in the population. (Pro and Arg are exceptional cases often preferring Z, because of the availability of a nitrogen site with abnormally high proton affinity to attract the carboxyl proton.)
3. Effect of metal size, ion charge, and nitrogen-site basicity specifically for the choice between C[COOH] and Z[CO_2^-]. An important point is that C[COOH] converts into Z[CO_2^-] by simple displacement of the carboxyl proton through a short distance, which has the result of charge separation into the salt-bridge configuration. Some circumstances make the salt-bridge-forming position of the proton in Z better than the neutral proton position in C, namely (1) the metal ion is large (disfavoring CS microsolvation), (2) the charge is high (doubly charged

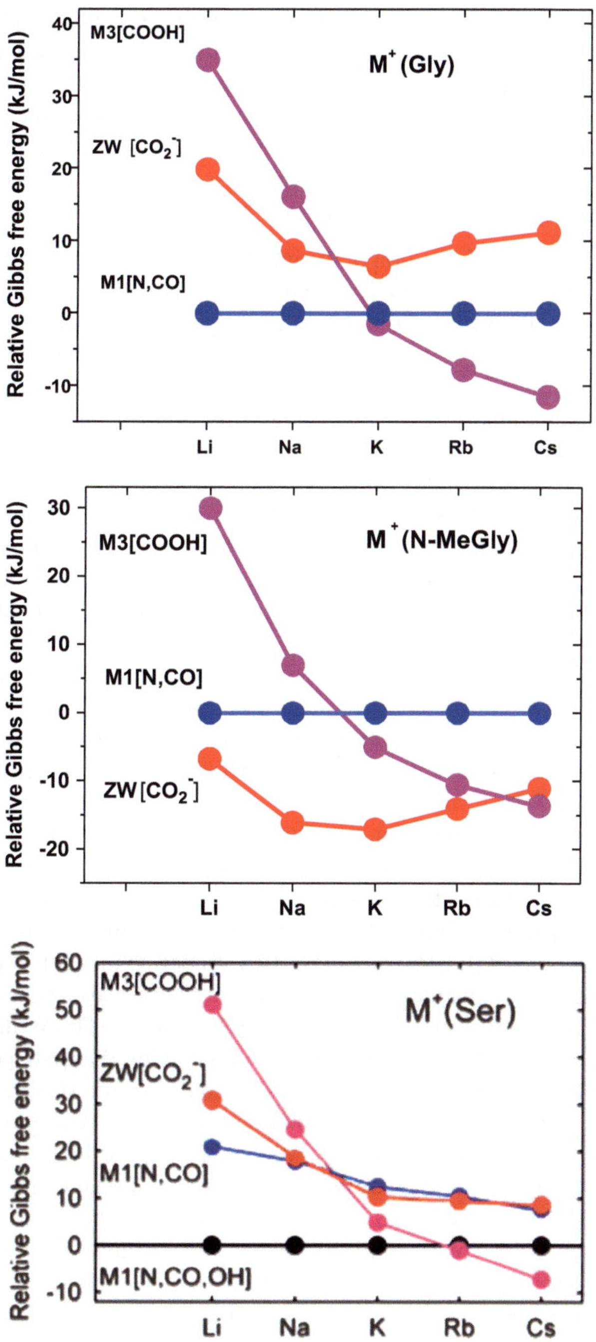

Fig. 3 Trends of the important isomers as a function of metal size for three different amino acids. Computational results plotted from [55] and [62]

metals), and (3) the proton-accepting site is strongly basic (Pro, Arg, N-methyl derivatives). A zwitterion ground state is more likely given any of these conditions.
4. Effect of ion charge on CS/SB preference. Higher charge on the metal enhances all binding interactions, but apparently favors the SB electrostatic stabilization associated with the salt bridge more strongly than the ion-solvation interactions leading to CS. Except for the special cases of Pro and Arg, all of the strongly favored SB ground states in Table 1 are doubly charged.

As an illustration of the interplay of factors across the alkali metals, Fig. 3 shows the shifting energies of the principal players for glycine [62] (where there is no side chain) and for serine [55] (as a representative amino acid where side-chain interactions play a role). Two effects are notable when the side chain is introduced in going from Gly to Ser: (1) the new CS tridentate structure T[N,CO,OH] appears and is dominant for smaller metal ions, and (2) the stability of Z[CO_2^-] zwitterion is greatly enhanced because the electronegative side chain can participate in extended salt bridge stabilization. However, the CS structure C[COOH] dominates in both cases for large metal ions. Moision and Armentrout showed by experiment and computation [67] that the small alkali metals favor binding to the NH_2 group, while the larger alkali metals switch to a preference for binding to the more extended carboxyl terminal COOH group. This principle rationalizes the switch from B or T toward C in going from Li^+ across to Cs^+ for many amino acids in the table.

Turning the nitrogen from primary to secondary amine increases its local basicity, and has been shown in many examples to enhance the tendency to separate charge and form an SB salt bridge by moving an acidic proton to this site. Methylation increases the stability of the SB form relative to CS forms by significant amounts (~15–20 kJ/mol) [14, 56, 62]. This often results in SB replacing CS as the ground state in N-methylated derivatives, and also results in the exceptional appearance of SB for most proline complexes. The effect has been frequently verified both computationally and spectroscopically through methylating this nitrogen in primary aliphatic amino acids, as in [54, 62]. The effect is nicely exemplified by the calculations of [62] shown in Fig. 3b. Comparing the plots for glycine vs *N*-methylglycine, it is seen that the effect of methylation is primarily to stabilize the zwitterion by about 20 kJ/mol relative to the other conformations. The Williams group noted a redoubled effect with two methyl substituents, giving the first spectroscopically characterized Li^+ amino acid complex with SB conformation (lysine) [56].

Trends comparing aliphatic amino acids with those having active side chains were originally noted by Wyttenbach et al. [44]) and were recently clarified by Drayss et al. [62]. Computationally, the aliphatic amino acids show a trend toward greater relative CS stability with increasing size of alkali metal ions, while the opposite trend, with SB being increasingly favored for larger metal ions, is observed in amino acids with active side chains. This apparent discrepancy is not real, because in the aliphatic cases, the stable CS structure which is favored in comparison with SB for large metal ions is the C-terminal C[COOH] or C[COOH,X] structure, whereas in the active side chain cases, it is a chelated bidentate B[CO,X] or tridentate T[CO,N,X] conformation which is in competition with the SB conformation. This situation is all illustrated computationally in Fig. 3. For the natural amino acids, these trends of CS vs SB character are not usually observable experimentally (not counting the exceptional cases of Pro and Arg), because one or another CS structure is usually more stable than SB so that CS is observed as the ground state in all cases, but various studies [14, 62, 80] have displayed IRMPD evidence for SB preferences using methyl-substituted amino acids, where the basicity of an amino nitrogen can be increased and adjusted by methylation to stabilize an SB form.

The exceptional case of arginine, with its extremely basic side chain available for metal-ion coordination, has long been interesting. The cationized molecule was one of the cases carefully studied by the Williams group in their groundbreaking applications of blackbody infrared radiative dissociation (BIRD) to many ion-structure questions including the CS vs SB distinction [81]. These complexes provided an early target for applying IRMPD spectroscopy to this problem using an OPO laser in the hydrogen-stretching region [14], and for application of the free-electron laser in the mid-infrared [20].

2.1.2 Transition Metals

Although transition metal ions are surely not unimportant partners for peptides and enzymes in nature, there is little to report of spectroscopic study of their amino acid interactions. As noted in Table 1, Ag^+ is the only one with a noticeable body of spectroscopic data for monomeric amino acid complexes. For comparison's sake, the comprehensive computational surveys of Mn^{2+} by the Ohanessian group [52, 53] and of Ag^+ by the York group [66] are displayed to fill some gaps in Table 1. Besides the entries in the table, we can note that the spectroscopy of Cd^{2+}His (CS tridentate) was reported [69], as was a computation [73] of Zn^+ (CS bidentate) and Zn^{2+} (SB) with glycine. Closely related is the spectroscopic characterization of the complex of $CdCl^+$ with histidine [69], where the $CdCl^+$ ion behaves in the same way as a singly-charged transition metal (such as Ag^+), and is chelated in a tridentate T[CO,N,N] complex. From these limited examples, we can conclude that the strong chelating interaction of the transition metal ions with Lewis-basic sites leads to maximally chelated CS structures (except for the exceptional case of Pro) always for the Ag^+ ion, and also for Zn^{2+}, Mn^{2+} and other metal ions complexing amino acids possessing Lewis-basic side-chain heteroatoms. If there is no active side chain and a doubly charged transition metal ion, it looks as if salt-bridge stabilization (SB) outweighs solvation interactions (CS), leading to ground state zwitterions for these transition metal ions with aliphatic amino acid ligands. Thus, for the example of Mn^{2+} at least, SB is predicted to be the favored ground state for the aliphatic amino acids. The transition metal ions considered here all have a filled (Ag^+, Zn^{2+}, Cd^{2+}) or half-filled (Mn^{2+}) *d*-shell, making their interactions relatively uncomplicated. Consideration of partially filled shells, giving the complications of non-zero spin and non-spherical electron distributions, is for the future.

2.1.3 Deprotonated Ligands

For doubly charged cations, the possibility is opened up of complexation of the metal ion with a deprotonated amino acid ligand, $[M^{2+}(AA\text{-}H)^{-}]^{+}$. It seems that it may actually be easier to produce these deprotonated complexes from an electrospray source than the intact $[M^{2+}(AA)]^{2+}$ complexes for transition metal ions, and reports about such complexes are appearing. The most straightforward outcome is for the metal ion to be simply chelated by the anionic ligand in the same pattern as the corresponding non-deprotonated complex. In these cases, the ligand is deprotonated to give the expected carboxylate anion, which then chelates to give an $[M^{2+}(AA\text{-}H)^{-}]^{+}$ tridentate complex T[CO^{-},N,X], or, in the case of lysine at least, a bidentate complex B[CO_2^{-},N]. This pattern seems natural for a strongly Lewis-basic side-chain group combined with a strongly interacting transition metal ion. The cases of $[Zn^{2+}(His\text{-}H)^{-}]^{+}$ [69], $[Cd^{2+}(His\text{-}H)^{-}]^{+}$ [69], $[Pb^{2+}(Phe\text{-}H)^{-}]^{+}$ [82], and $[Pb^{2+}(Lys\text{-}H)^{-}]^{+}$ [83] were all shown spectroscopically to follow this pattern. In a slight variation of this theme in the $[Pb^{2+}(Glu\text{-}H)^{-}]^{+}$ complex, the site of deprotonation is still carboxyl, but it is the side-chain COOH which deprotonates giving a tridentate complex which can be described as a T[CO_2^{-},N,CO] complex [82].

However, when the Fridgen group looked at Pb^{2+} complexes of the aliphatic amino acids, a different and unexpected pattern of deprotonated complexation was uncovered. Initially working with the complex of Pb^{2+} with deprotonated glycine [84], it was found that, rather than the expected C-terminal carboxyl deprotonation and metal binding by the carboxylate as diagramed in structure-types (ii) and (iii) in Scheme 3, IRMPD spectroscopy in the hydrogen stretching region firmly indicated the N-deprotonated structure of type (i) in Scheme 3. This bidentate conformation might be labeled B[N^{-},CO]. Their calculations convincingly supported this as the most stable isomer.

Follow-up experiments [83] showed this same pattern (Scheme 3 structure (i)) for all the aliphatic amino acids (Ala, Val, Leu, Ile, and Pro). This distinctive

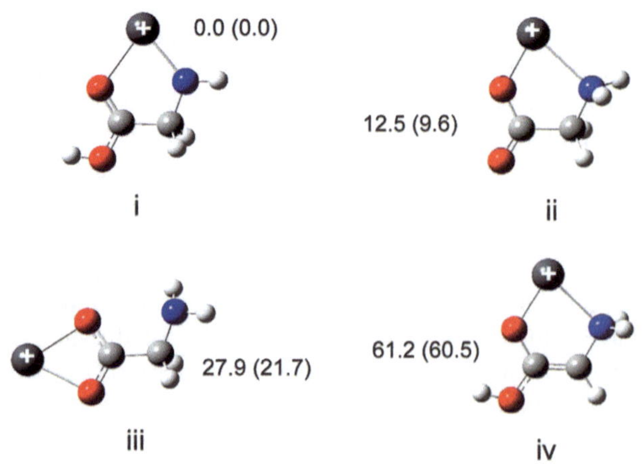

Scheme 3 Possible structures for the deprotonated Pb^{2+} glycine complex. The amine-deprotonated structure (i) is identified spectroscopically as the predominant form for glycine and the other aliphatic amino acids including Pro. Reproduced with permission from [84]

Scheme 4 The novel structure of the deprotonated zinc/proline complex. Adapted with permission from [85]

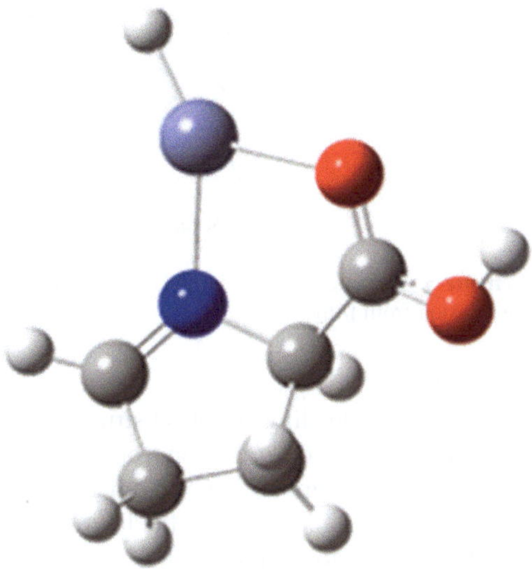

$[Zn^{2+}(Pro\text{-}H)^{-}]^{+}$

binding of Pb^{2+} to the deprotonated amino nitrogen of the amino acid was rationalized as reflecting a preference by the soft acid Pb^{2+} to bind the soft base NHR^{-} rather than the harder base COO^{-}.

Looking at complexes of the more active zinc ion [85], the proline complex revealed a further variation of this nitrogen/metal-ion binding theme for the deprotonated proline/zinc complex $[Zn^{2+}(Pro\text{-}H)^{-}]^{+}$. This was assigned the interesting structure shown in Scheme 4 having the same N-deprotonated structure as the Pb^{2+} complex, except that a proton has moved from the proline ring to the metal ion (formally an oxidative addition), giving the novel structure shown.

2.1.4 Anionic Complexes

In contrast with the dozens of studies of metal cation complexes with peptides, there have been only a small number of spectroscopic reports on complexes of anions. A study by the Williams group [86] of halide anion complexes with Glu, His, and Arg showed clear cut characteristic IRMPD spectra, indicating CS-type isomers favored for Glu and His, vs SB isomers for Arg (Scheme 5). As could be expected, the halide ion binds differently than metal cations, forming effectively hydrogen bonds to two acidic protons. Spectroscopically speaking, the halide ions give relatively small frequency shifts of the characteristic carbonyl (CS) or carboxylate (SB) stretching frequencies, because the perturbing anion is quite remote and isolated from the region of the CO or COO^{-} normal mode.

The Kass group [87] recently reported the IRMPD spectroscopy and calculations of the Cl^{-} complex of proline in a report reinforcing the cautionary note regarding

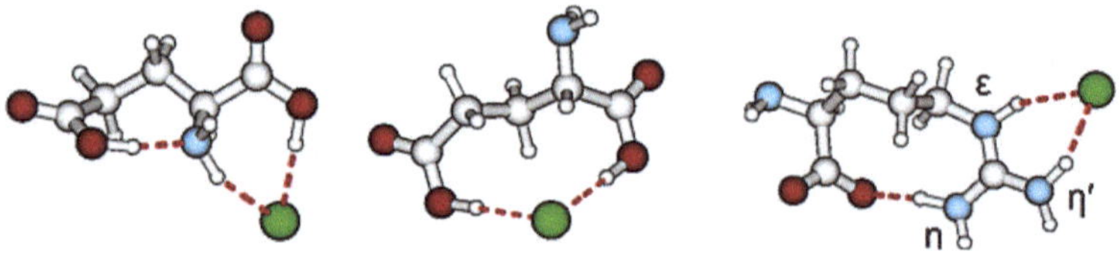

Glu X⁻ CS Isomers Arg X⁻ SB Isomer

Scheme 5 Anionic complexes of Glu and Arg amino acids with halide anions. Adapted with permission from [86]

simple but misleading, IRMPD structure results. The spectrum observed strongly indicated a CS conformation for this complex, giving an excellent fit to the calculated spectrum of a CS isomer, which is one of several isomers (both CS and SB) calculated to lie near the lowest energy. However, impelled by the fact that careful calculations indicated a very low-lying SB structure which should be populated, they looked more closely at the dissociation kinetics, finding a deviation from linear logarithmic decay of the parent ion near the point of complete dissociation (>79%). Noting the discussions by Prell et al. [48, 88, 89] of using photokinetic data to sort out misleading IRMPD results for mixed-isomer systems, they considered it most likely that the SB isomer was actually present at about 44% abundance, but was suppressed in the IRMPD spectrum because its dissociation is much slower than that of the strongly observed CS isomer.

2.1.5 Tripositive Complexes

The severe instability of a triply charged metal cation with a ligand as small as a single amino acid has made the production and characterization of such ions elusive. It was thus of considerable interest that a collaboration of York and FELIX groups was able to observe triply charged complexes of La(III) with a protected Trp derivative, with only innocuous modifications separating the complexes studied from the ideal complex of the bare amino acid [90]. The ligand (L) was the end-group derivatized *N*-acetyl-tryptophan-methyl ester. The two complexes for which they achieved IRMPD characterization were the solvent-stabilized La(III)L(ACN)$_2^{3+}$, where ACN is acetonitrile; and the dimer complex La(III)L$_2^{3+}$. In both cases, the IRMPD spectrum gave an excellent fit to a calculated spectrum for a conformation having CS character, and having chelation of the metal ion by all the available Lewis-basic sites for each amino acid ligand, namely the carbonyl oxygen, the amino nitrogen, and the π system of the Trp side chain.

2.1.6 Dimer Complexes

In the domain of ionic complexes, the term “dimer” is usually used for a composite system AXB, where A and B are ligands, which may be the same (“homogeneous” or “symmetric”) or different (“heterogeneous” or “asymmetric”), and X is a charged core group, usually a proton or a metal ion. These are often among the most abundant ions in the electrospray or molecular expansion products, and there has been some attention to the structures and spectroscopy of dimers involving peptide-related ligands.

There have been a number of studies of peptide proton-bound dimer spectroscopy in collaborations involving, among others, the Paris groups, the Fridgen group, the McMahon group, and the FELIX group, using the CLIO FEL [91–95], the FELIX FEL [96], and also benchtop OPOs in the H-stretching region [97]. These results are outside our present scope.

For metal-ion dimer complexes with amino acid ligands, it has been interesting to characterize the two ligands with respect to their CS or SB nature. In the mid-infrared this has proved to be difficult. In the case [98] of the $M^{2+}Trp_2$ dimers, where M was Ca, Sr, Ba, Zn, Cd, Mn, Co, and Ni, it was readily established from the mid-IR spectra that all dimers contain at least one CS ligand. Assigning the second ligand was more problematic because of the lack of clear positively diagnostic markers for the SB conformation, but the weight of evidence was for a CS/SB combination in all cases except Ni, which was considered to be CS/CS. A later further study [99] of the Mg, Ca, Sr, and Ba dimers of Trp in the H-stretching region using an OPO laser showed the H-stretching wavelength region (~3,000–4,000 cm^{-1}) to be much more clearly diagnostic, and gave a more definitive assignment of CS/CS for the Sr^{2+} and Ba^{2+} complexes, vs CS/SB for the Mg^{2+} and Ca^{2+} complexes. Finding a bias toward more CS binding specifically in the Ba^{2+} case presented an interesting contrast, in that much previous work with Ba^{2+} monomer complexes had always shown this metal's complexes to be more prone to SB conformations than the smaller alkaline earth metals.

The dimer complexes of histidine with Zn^{2+} and Cd^{2+} were also examined in the mid-IR [69], with a similar outcome to the mid-IR study of the Trp complexes. It was easily established that at least one ligand in each case was CS, and it was considered most likely that at least some of the population had an SB ligand forming CS/SB dimers. However, again the lack of definitive markers for possible SB ligands left uncertainty about the presence and extent of such ligands.

2.2 *Small Peptides*

2.2.1 CS and SB Binding

Among the earliest FEL work applying IRMPD spectroscopy to ion structures was the spectroscopy at CLIO of protonated dialanine [34]. A close match to the

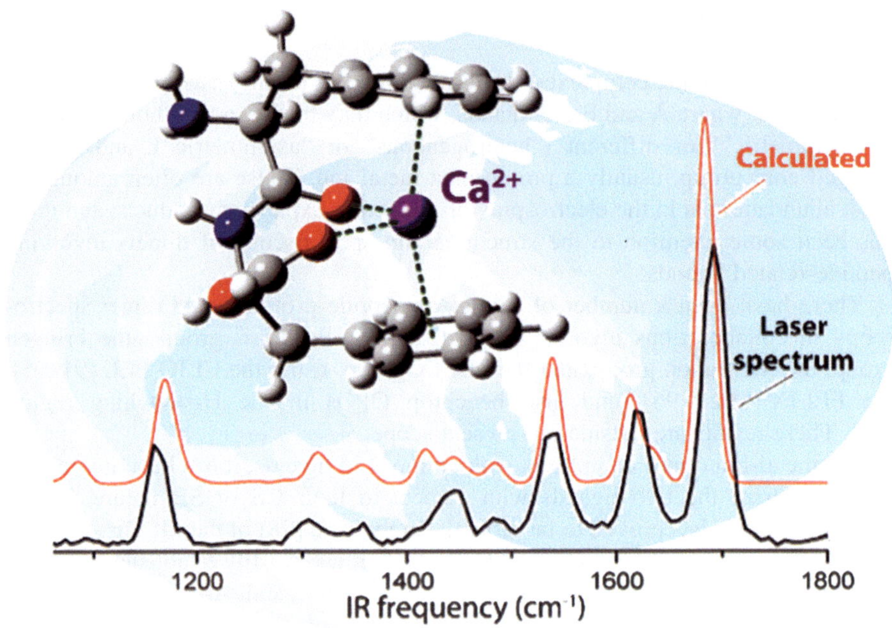

Fig. 4 The caging effect of the strong double cation-π interaction with the doubly charged calcium ion and the PhePhe ligand. Reproduced with permission from [101]

calculated spectrum was obtained, showing definitively that protonation was at the amino nitrogen. On the other hand, a report for protonated AlaHis soon followed [33], showing that in that case the site of protonation was at the imidazole side-chain nitrogen (that nitrogen being more basic than the amino nitrogen).

Further pioneering efforts at CLIO gave spectra for metal ion complexes of small peptides [100], namely the sodium ion complexes of di- and tri-glycines and -alanines. About the same time, the group at FELIX in collaboration with our group reported results for sodium and potassium complexes of PheAla and AlaPhe, giving clear evidence for cation-π coordination of the phenyl ring [12].

The cation-π theme was pursued in studies of the double-π system PhePhe, in which cations, especially the alkaline earths, are encapsulated in compact structures [101], as in Fig. 4. Chelation by both phenyl groups forming a cage in this way is strongly enforced for the doubly charged metal ions. For singly charged counterparts (alkali metals and Ag^+), binding of the N-terminal aromatic ring is still strong, but the cage is more likely to open up by giving up the binding of the C-terminal ring; the spectra of these latter complexes have broader peaks and more congestion, indicating conformational flexibility. The cage-structuring effect of the doubly charged cations is thus enhanced by the stronger micro-solvation experienced by the higher ion charge.

Accumulating experience indicates that salt-bridge formation in metal ion complexes of dipeptides and larger peptides is less common than for the mono-amino

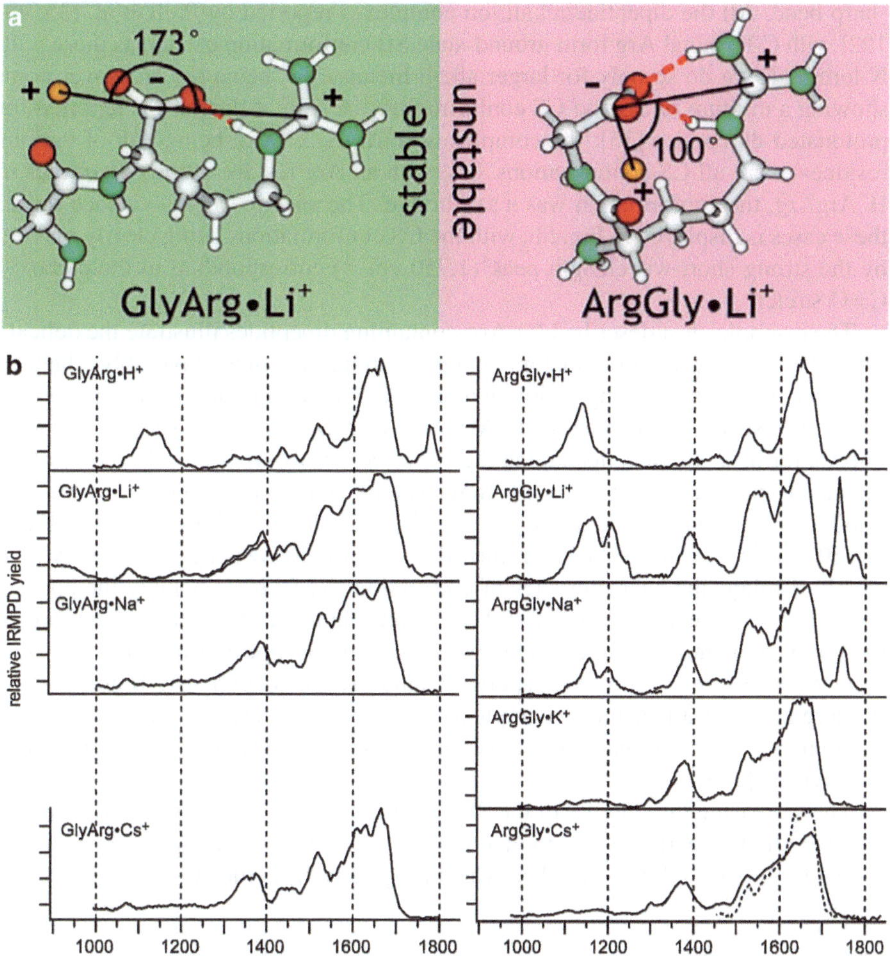

Fig. 5 Sequence dependence of the ability of an alkali cation to stabilize the salt bridge. (**a**) Structures of the most stable zwitterion form for the lithium ion complexes; (**b**) IRMPD spectra of the complexes. Arg at the C-terminus favors salt bridging, while the reverse sequence (in combination with a small ion such as Li^+ or Na^+) can give a stable CS ion as signaled by the strong peak near 1,750 cm^{-1} of the ArgGly complexes of those ions. Reproduced with permission from [102]

acid complexes, and so far all of the ground-state salt-bridge instances which have been characterized spectroscopically have been either singly charged dipeptide complexes containing at least one arginine residue [35, 48, 102] or dipeptide complexes of doubly charged metal ions [79].

Among Arg-containing dipeptides, the work of the Williams group [35, 48, 102] indicates that those with C-terminal Arg are more likely to favor SB formation, with the protonated arginine side chain forming an excellent linear salt bridge in combination with the deprotonated carboxyl and the metal ion (Fig. 5a, left). The salt bridge formed with N-terminal Arg is less favorable (Fig. 5a, right), having a sharp bend. All the dipeptide/alkali ion complexes reported by Prell et al. [35, 48, 102] with C-terminal Arg form ground-state SB conformations, whereas those with N-terminal Arg do so only for larger alkali metals, Na^+ being the crossover point showing a mixture of SB and CS conformations. A similar theme was reported for protonated dipeptides [35]: the complexes H^+ArgXxx (Xxx being any of various residues) were all CS conformations, but, with an Arg residue at the C-terminus in H^+ArgArg, the conformation was a salt bridge. The spectroscopic evidence for all these cases is displayed in Fig. 5b, with the CS conformations being clearly marked by the strong short-wavelength peak (1,750 cm^{-1}) corresponding to the carboxyl C=O stretch.

The examples just described for Arg-containing dipeptides illustrate the delicate role played by geometric factors in determining the most favorably chelated conformations for wrapping an ion in a peptide ligand. Similarly, sequence dependence was seen for complexes of (Ala,Phe) with Ba^{2+} [79], as shown in Fig. 6. PheAla forms a complex displaying the expected CS absorption peaks, comprising the metal-bound carboxyl carbonyl stretch (in the vicinity of 1,700 cm^{-1}), Amide I (1,620 cm^{-1}) and Amide II (1,540 cm^{-1}), and the free carboxyl COH bend (1,150 cm^{-1}). On the other hand, the AlaPhe complex switches completely, and clearly displays the characteristics of conversion to SB, most notably the appearance of a strong feature around 1,400 cm^{-1}, attributable to NH bending motions of the protonated amine nitrogen, along with apparent disappearance of the free COH bend at 1,150 cm^{-1}. An even subtler geometric distinction was found for the diastereomers of alkali metal complexes of PhePhe, where the DL and LL diastereomers were found to have substantially different conformational preferences [103].

The increasing favorability of the SB conformation for increasingly large doubly charged alkaline earth ions is pointed out in Fig. 7. The spectroscopic evidence is shown for the M^{2+}AlaAla complex, which progresses from mostly CS conformation

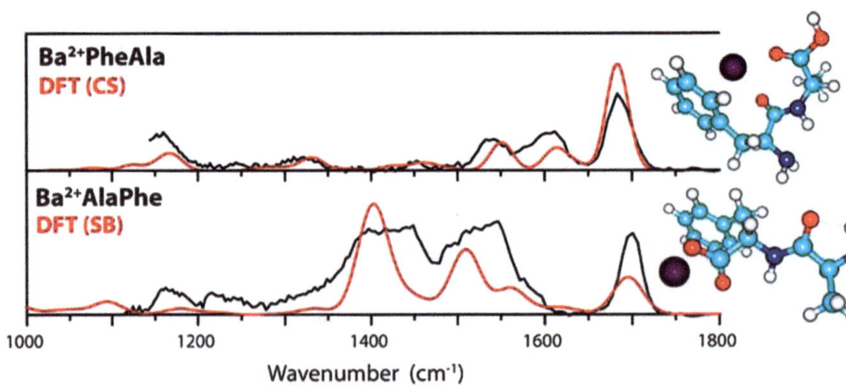

Fig. 6 Influence of sequence on the formation of CS vs SB structures. Complexes of Ba^{2+} with AlaPhe and PheAla are found to behave differently based on the comparison of experimental and theoretical spectra. Reproduced with permission from [79]

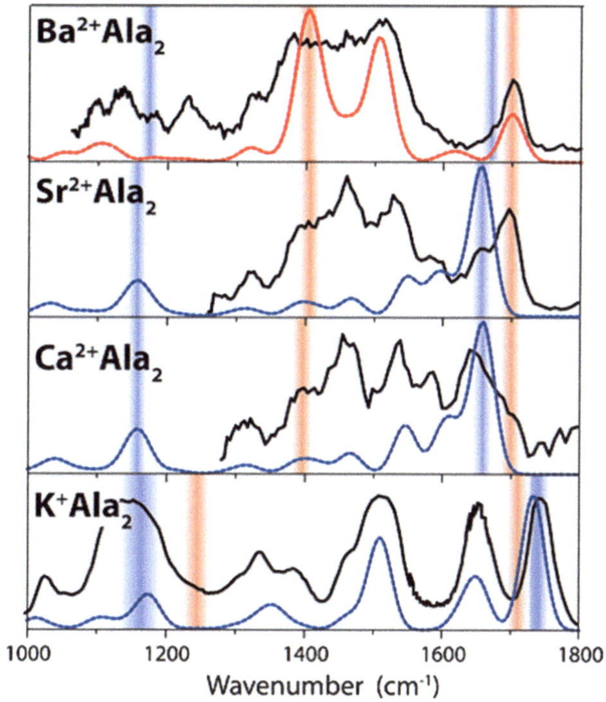

Fig. 7 Series of AlaAla metal ion complex spectra showing the relative prominence of SB vs CS signature features. Calculated positions of the characteristic SB peaks are indicated in *red*, and CS peaks are in *blue*. Traces of calculated spectra show the most stable SB conformation in the Ba complex (*red bars*) and the most stable CS conformations for the Sr, Ca, and K complexes (*blue bars*). Reproduced with permission from [79]

for Ca^{2+}, to mostly SB conformation for Ba^{2+}. The K^+ complex is fully CS, as expected.

Noting the interesting exceptions discussed above (as well as the iminol complexes described in the following section), most dipeptide and all tripeptide and higher complexes that have been characterized spectroscopically possess CS as opposed to SB ground states. Much detail has been obtained regarding the spectroscopic characteristics of CS-bound complexes. Taking trialanine as a typical set of spectra for a ligand without complicating side chains [104], Fig. 8 lays out the four characteristically strong and reliable features common to this spectroscopy. The two normal modes belonging to the COOH C-terminus (CO stretch and COH bend) are still prominent, even though they are outnumbered by the two sets of amide linkage modes (CO stretch, or Amide I, and NH bend, or Amide II). The modes follow very regular wavelength shifts correlated with the relative binding strengths of the series of metal ions. Thus the carboxylic acid CO stretch shows a systematically increasing red shift from 1,790 cm^{-1} in the neutral ligand, down to ~1,690 cm^{-1} for the divalent Ca^{2+} cation. The Amide I band (which, in the same way as the carboxylic acid band, has strong participation of a metal-bound oxygen in the normal mode) shows a similar progressive red shift, from ~1,700 to ~1,600 cm^{-1}. On the other hand, the Amide II band, which primarily involves atoms not directly bound to the metal, shows a progressive but smaller shift in the

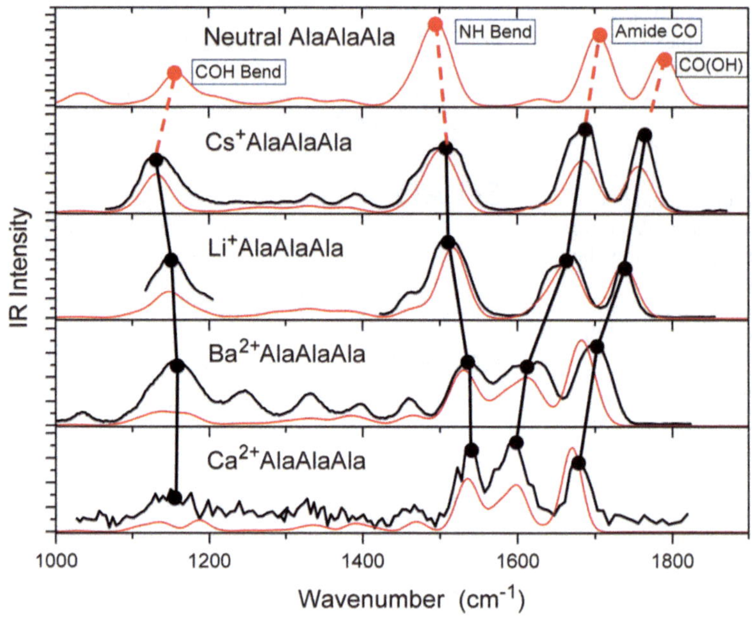

Fig. 8 IRMPD spectra of the trialanine complexes (*black traces*), along with the calculated CS (*red*) conformations considered most likely to dominate the ground state populations. The IRMPD plot for the Na^+ complex was extracted from [100]. Figure reproduced with permission from [104]

blue direction from ~1,500 to ~1,530 cm^{-1}, and the COH bend, also remote from metal-bound atoms, shows no discernible systematic shift.

2.2.2 Iminol Complexes

In nature, as in the model studies discussed thus far, metal ions of the main-group elements form complexes with the peptide backbone by interaction with the amide carbonyl oxygens, along with side-chain chelation sites which may be conformationally available. The Paris group laid out the guiding principle that the metal ion (Na^+) tends to coordinate as many amide carbonyls as possible in glycine peptide chains [105]; this interaction is so strong that it is reasonable to anticipate similar tendencies for more highly decorated chains as well. However, we recognized that peptide binding of highly charged transition metal ions in condensed phases often follows a fundamentally different pattern, primarily binding instead via strong metal bonds to deprotonated amide nitrogens, and such a pattern is possible also for gas-phase model systems involving more active metal ions than alkali metals [106].

For the intact peptide to bind in this way in a (non-deprotonated) gas-phase complex, the deprotonation of the amide nitrogen must transfer the proton to another site. In cases reported so far, this has meant tautomerism of the amide

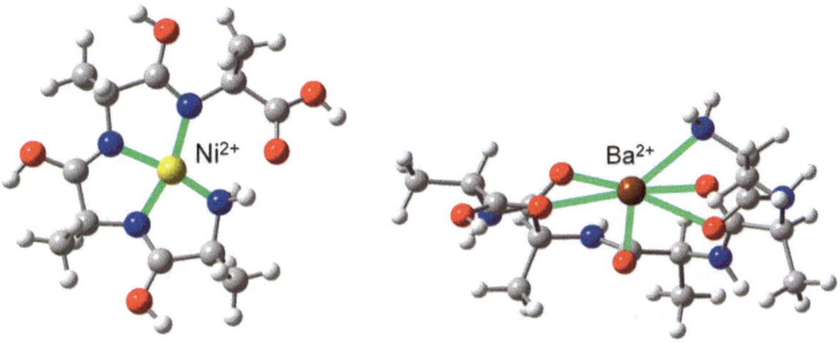

1 2

Scheme 6 Lowest-energy structures for complexes of Ni^{2+} and Ba^{2+} with poly-alanines. Formation of an iminol structure is observed for the Ni^{2+} complex. Reproduced with permission from [107]

linkage into the imine enol, or *iminol*, tautomer, analogous to an enol. Scheme 6 illustrates this theme by comparing a highly chelated barium ion having multiple carbonyl oxygen ligands [108], with a transition metal ion Ni^{2+} chelated by three amide nitrogens, each in the iminol configuration, along with the fourth nitrogen from the N-terminus (computed structure for $Ni^{2+}(Ala)_4$ shown for illustration) [107].

Solid spectroscopic evidence for iminol ground states (with computational support) was reported for several complexes of PhePhe [109]. As displayed in Fig. 9, iminol binding is preferred for Co^{2+}, Ni^{2+} and Mg^{2+}, while Ca^{2+} and Li^+ prefer CS binding. The tendency toward iminol binding appears to correlate with overall binding energy, so that the strongly binding Mg^{2+}, Ni^{2+}, and Co^{2+} ions clearly match the expected iminol spectral patterns, while the more weakly binding K^+ and Ca^{2+} ions show the familiar pattern of CS binding (as exemplified in Fig. 8).

A subsequent survey of HisGly complexes with most of these same metal ions gave similar results [110]. The strong-binding ions Mg^{2+} and Ni^{2+} gave clear spectra for iminol, while the weaker-binding ions K^+, Ca^{2+}, and Ba^{2+} gave CS spectra. Looking at larger peptides, a more extensive survey was reported with the aim of defining more broadly the scope of the iminol binding theme for gas-phase complexes [107]. Figure 10 compiles the results. A powerful discriminant between iminol and CS conformations was found to be the presence or absence of the Amide II band, which is consistently strong at 1,500–1,550 cm^{-1} for CS (and SB) conformations, but absent for the iminol forms. Unfortunately, no positive marker for the iminol form stands out in this mid-infrared wavelength span, so the presence of a weak but not vanishing band at the Amide II position can only be taken as an indication, but not a proof, that some of the population has taken the iminol form. Results are collected in the figure for six different peptides ranging from two to four residues (considering PheAla and PheGly to be equivalent). For the most part, the patterns described above were confirmed and strengthened, with weakly binding metals K^+ and Ba^{2+} being clearly CS or SB, and strongly binding Ni^{2+} being equally

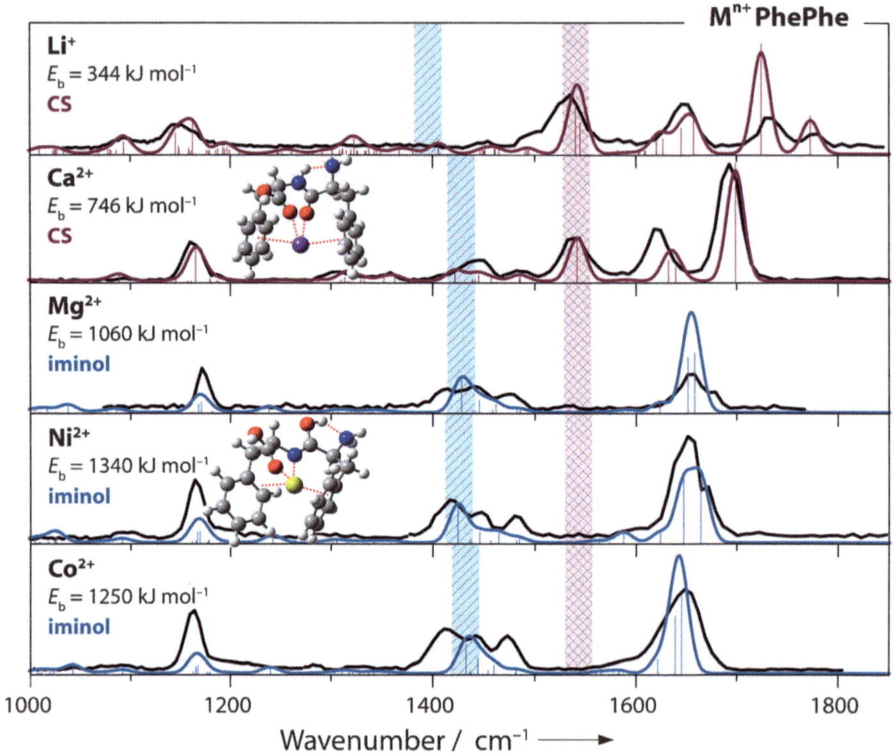

Fig. 9 IRMPD spectra (*black*) compared with calculated CS spectra (*red*) and calculated Iminol spectra (*blue*) for several M^{n+}PhePhe complexes. The *red bar* marks the expected position of the Amide II peak of the CS complex, while the *blue bar* marks that of the iminol COH bending mode. Reproduced with permission from [109]

clearly iminol. However, both Ca^{2+} and Mg^{2+} give indications of being transitional. The Ca^{2+} complexes of the two least strongly chelating ligands, AlaAla and PheAla (or PheGly), appear to have a relatively weak Amide II bands, suggesting partial iminol character, but become positively iminol in appearance for the more effective solvating ligands AlaAlaAla, PhePhe, PheGlyGly, and PheGlyGlyPhe. The Mg^{2+} complexes look positively iminol in character for the weakly solvating dipeptide ligands, but switch to an obvious CS character with the larger PheGlyGly and PheGlyGlyPhe ligands (which presumably give greater microsolvation of the metal ion in favor of CS). It looks as if both these latter metals are close to the point of transition between iminol and CS character, and tend to move to CS predominance as the ligand takes on a greater degree of micro-solvating capacity. The Mg^{2+} (and perhaps Ca^{2+}) results indicating iminol conformations are particularly interesting, because in condensed phase these ions are considered as strongly favoring CS-type coordination to amide CO oxygens [111, 112], while transition metal ions such as Ni^{2+} consistently favor deprotonated amide nitrogen binding (frequently square planar) [113–115].

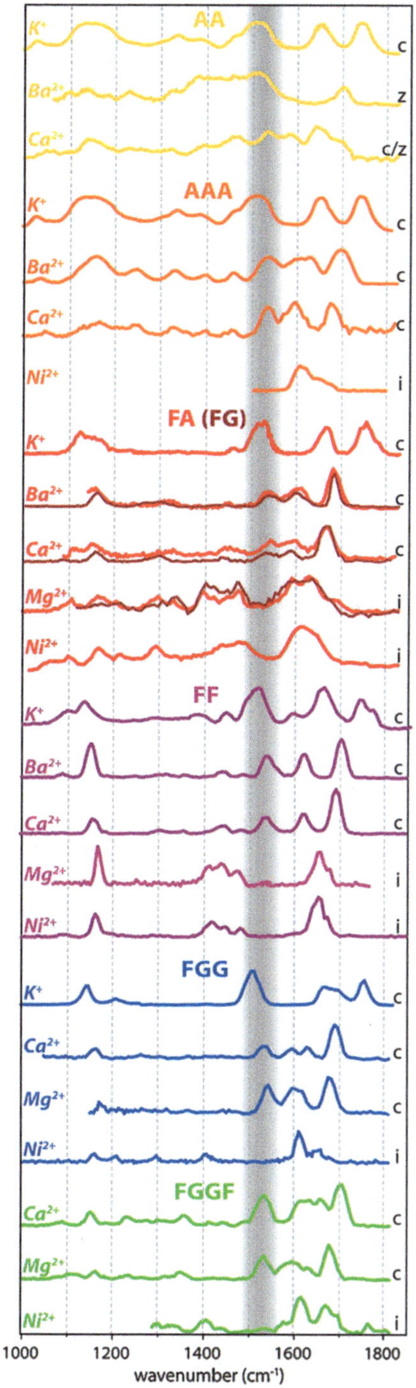

Fig. 10 IRMPD spectra of five metal ions complexed to various peptide ligands (identified by their one-letter amino acid codes). The spectra are grouped by ligand. The *gray shading* indicates the approximate position of the amide NH bending mode (Amide II mode). The presence of this band indicates that the ligand is in the 'normal' amide configuration, while absence of the band suggests a tautomerization to the iminol motif, in which the proton has moved from the amide nitrogen to the amide oxygen atom. The c, z,i-letter code to the right of each spectrum indicates our assignment as charge-solvated, **z**witterion or **i**minol, respectively. Note that the spectra for the FG (PheGly) ligand have been superimposed on the FA (PheAla) spectra in a *slightly darker color*, and that the spectra of this pair of ligands are consistently very similar. Reproduced with permission from [107]

2.3 *Infrared: Larger Peptides and Proteins*

IRMPD spectroscopy has shown itself to be well suited to pointing up a number of structural themes and transitions as one moves upward from amino acids, through di- and tri-peptides, to larger systems. An overview was offered in our Introduction by the correlation of the three gas-phase spectra juxtaposed in Fig. 1.

Moving out towards longer *n*-residue oligopeptides, a threshold is crossed between about $n = 4$ and $n = 6$, where it becomes no longer favorable to wrap the oligopeptide around a metal ion in a tightly chelated fashion with all carbonyl oxygens and the amine nitrogen coordinated (see Scheme 6, structure 2). The significance of this boundary as a function of different metal ions was explored by our group for $n = 5$ (specifically, for pentaalanine) [116] and by the Ohanessian group, whose studies of Na^+ complexes of Gly_n span this range of ligand sizes [105]. Useful spectroscopic markers of this size transition were identified in the mid-IR spectral region. We found that the tightly chelated class of complexes exhibits spectroscopic characteristics of a terminal COOH which is not perturbed by intramolecular H-bonding. Some or all of these features were prominent for smaller sodium ion complexes of Gly_n ($n \leq 5$) [105] and for the Ca^{2+} and Ba^{2+} complexes of Ala_5 [116]. (It can be seen in the $Ba^{2+}Ala_5$ structure shown in Scheme 6, structure 2, that the tightly wrapped chelation geometry of this complex allows no freedom for H-bonding of the COOH group.) On the other hand, larger alkali ion complexes show spectroscopic characteristics more indicative of a predominantly H-bonded carboxyl group, as expected when there is a surplus of Lewis-basic chelation sites exceeding the chelation-available space around the metal ion.

The Ohanessian group has led a systematic effort to apply spectroscopy and computation to complexes of sodium-complexed peptides, extending to larger sizes with six or more residues [100, 105, 117–119]. Calculations become challenging, and the conformation space becomes enormous, but they have discerned some key structural features from the spectra. Once there are metal-free carbonyl oxygens available, networks of hydrogen bonds can form, and it makes sense to organize the structural possibilities in terms of secondary peptide structures such as helices, sheets, and so on. Of recent interest is the study of the globular-to-helix transition for Na^+Ala_n, where *n* ranges from 8 to 12 [119]. For all peptide lengths, all levels of calculation predict a helical form as the lowest energy, but globular forms are not unreasonably high in free energy. Figure 11 displays the calculated low-energy structures for $n = 8$ for both helical and globular forms. The IRMPD spectra, while not giving very exact agreement with calculated spectra, are able to show that both helical and globular conformations contribute to the observed spectrum at $n = 8$. The IRMPD spectra are interpreted as indicating helical structures for $n = 9$ to 12, with the possibility of some globular contribution for $n = 9$.

The decapeptide gramicidin S has been a target of room temperature IRMPD [120], cold-ion IR/UV [121, 122], and computational [32] spectroscopic studies.

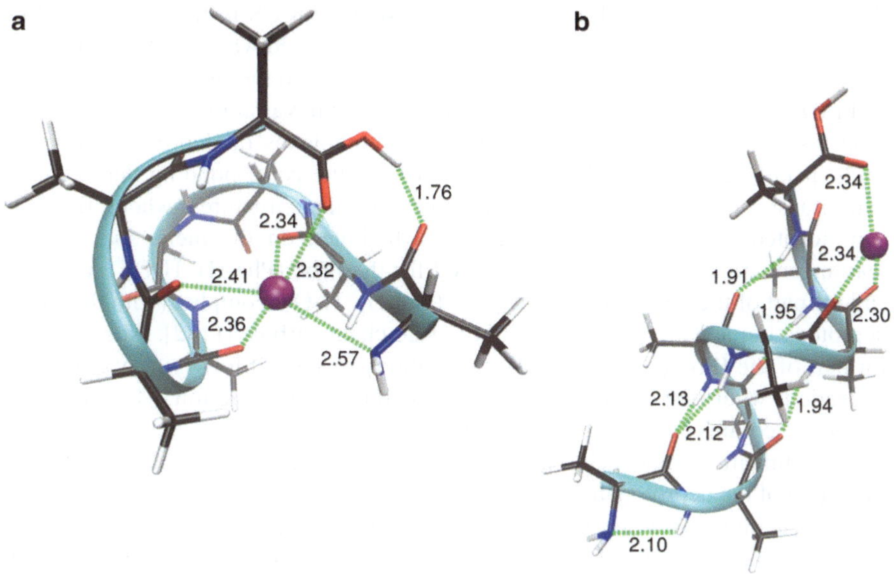

Fig. 11 Na^+Ala_8: (**a**) lowest energy globular structure; (**b**) α-helical structure (lowest energy). Calculations were performed at the M06/6-311+g(d,p)//M06/6-31g(*d,p*) level of theory. Reproduced with permission from [119]

Since the most incisive results have come from the cold-ion work, reports on progress there are appropriate to the Rizzo chapter [40].

2.4 Solvation Effects

The thermodynamically driven transition from solution-phase zwitterion ground states to gas-phase canonical tautomers can be probed by progressively adding water molecules to the bare gas-phase complex and characterizing the structure of the resulting complex. In their study of Li^+ and Na^+ complexes with Arg, Bush et al. [123] clearly stated the general principle at work in such progressive hydration studies: "Hydration of the metal ion in these clusters weakens the interactions between the metal ion and the amino acid, whereas hydrogen-bond strengths are largely unaffected. Thus, hydration preferentially stabilizes the zwitterionic structures, all of which contain strong hydrogen bonds." In this particular study, the transition from CS to SB for lithiated Arg was induced by attaching a single water molecule.

Spectroscopy in the hydrogen-stretching region was applied to characterize proton and lithium-ion complexed valine complexes as a function of hydration with up to four water molecules [22]. Although calculations and prior study by black-body infrared dissociation had suggested the possibility of zwitterions,

Kamariotis et al. observed no sign of conversion from CS to SB with up to four waters in either case. The only structural change observed was a switch from CS B [CO,N] to CS C[CO] upon addition of the third water to the lithiated complex.

Fridgen's group has uncovered several structural variations in their spectroscopic study of the formally mono-hydrated complexes of deprotonated amino acids, paralleling their work described above on the corresponding deprotonated bare transition metal complexes. Simplest are the structures where the bare deprotonated complex (see Sect. 2.1.3) is hydrated on the metal ion, without rearrangement. Such a pattern is assigned for the $[Pb^{2+}(Phe\text{-}H)^{-}H_2O]^{+}$ case [82], and for part of the population of the $[Pb^{2+}(Glu\text{-}H)^{-}H_2O]^{+}$ complex (with the ligand deprotonation actually coming off of the side-chain carboxyl) [82]. Also assigned for part of the $[Pb^{2+}(Glu\text{-}H)^{-}H_2O]^{+}$ population are several low-lying variant conformations where the complex switches to a carboxylate zwitterion, while the water molecule remains attached to the metal ion [82].

The aliphatic amino acids present more complicated structure changes upon hydration of the deprotonated complexes. $[Pb^{2+}(Pro\text{-}H)^{-}H_2O]^{+}$ is assigned a carboxylate structure with a proton transferred from the water, which can be viewed as a PbOH+ complexed to a Pro carboxylate zwitterion (Scheme 7a) [83]. For the hydrated Pb^{2+} complexes of the other aliphatic amino acids (Ala, Val, Leu, and Ileu), it was suggested that a structure type with intact carboxylate was most consistent with experiment, although not the lowest in energy. They suggest the structure shown in Scheme 7b, which, interestingly, amounts to a $PbOH^{+}$ cation complexed with the intact (un-deprotonated) amino acid [83].

The more active transition metal Zn^{2+} again differs from Pb^{2+} as in Sect. 2.1, again preferring a structure having a hydrogen atom covalently attached to the zinc atom. $[Zn^{2+}(Pro\text{-}H)^{-}H_2O]^{+}$ (the hydrated zinc-proline counterpart to the lead-proline complex) was assigned as the structure shown in Scheme 7c, which is the structure assigned in Scheme 4 with water attached to the metal ion [85].

Since the aqueous solution-phase structure of amino acids and peptides is normally zwitterionic, while that in the gas phase is frequently non-charge-separated, there is considerable interest in whether, and how completely, the complexes

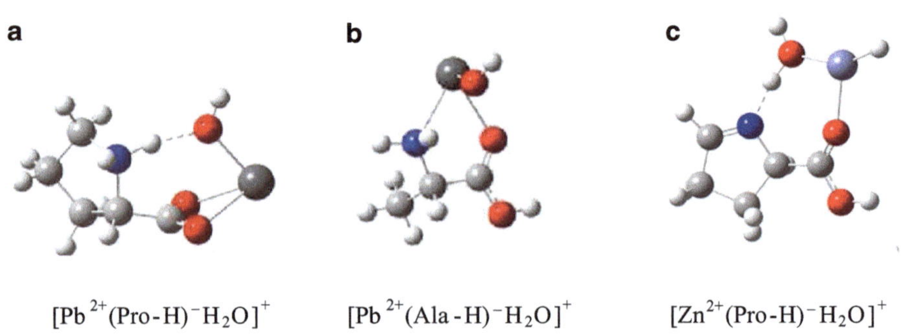

Scheme 7 Assigned structures for the observed hydrated deprotonated M^{2+} alanine complexes. (**a**) Pb^{2+} proline. (**b**) Pb^{2+} with aliphatic amino acids other than proline. (**c**) Zn^{2+} proline. Adapted from [83, 85]

of interest here re-equilibrate to the global minimum structure (more generally to the gas-phase equilibrium mixture of conformations) upon transfer by electrospray to the gas phase. For lack of better information, most spectroscopic studies have tacitly assumed that the structure mix characterized by gas-phase IRMPD represents an equilibrated population. A large number of the studies give this assumption indirect support, in that they show a gas-phase population mixture which is consistent with the best estimates available from computations of the relative free energies and the corresponding thermodynamically equilibrated populations in the gas phase. Concern about whether the gas-phase determinations actually reflect population equilibration is always lurking in the background of presentations of gas-phase experimental results of mixed populations, and there are in fact a number of reports of disagreement with the assumption of equilibration. For example, studies by the Kass group and others (both spectroscopic [124] and non-spectroscopic [87, 125, 126]) of the protonation site of electrosprayed *p*-aminobenzoic acid [124] and the deprotonation sites of tyrosine and *p*-hydroxybenzoic acid [127, 128] showed that the ratio of competing sites can be strongly affected by whether the electrospray is carried out using a protic or a non-protic solvent as well as other variables of the solution. A similar conclusion was affirmed in a following report by Steill and Oomens [129] for deprotonated *p*-hydroxybenzoic acid, where they used mid-infrared IRMPD spectra to show clearly that the site of deprotonation was the carboxyl for protic solvents, and the hydroxyl for aprotic solvents. Thus, the structure of the solution-phase ion was shown to be retained upon transfer to the gas phase. A CID study [130] showed differing fragmentation patterns of ions $[Gly\text{-}H\text{+}Zn]^+$, $[Asn\text{-}H\text{+}Zn]^+$, and $[Asp\text{-}H\text{+}Zn]^+$ from different solvents, which were attributed to structurally differing ion structures formed in the desolvation step, but these conclusions have not yet been confirmed by spectroscopic study.

The Williams group has been interested in the site of attachment of the initial water molecules during progressive solvation of peptides. IRMPD spectroscopy is well suited to characterizing these choices, which are likely to be closely balanced. For instance, at low temperature, the first water attachment to protonated proline is enthalpy controlled at the C-terminus, while at higher temperature the higher entropy of the N-terminal site induces a switch to water attachment there [88].

In an experiment of high technical sophistication and promise for the future, the Dopfer group has described the application of a transient three-color UV/UV/IR ion dip technique, which achieves the observation of the IR action spectrum with a time resolution of a few picoseconds after ion formation. A recent application [131] was to the mono-hydrated cation of acetanilide ($AA^+ \cdot H_2O$), which provides a model for a hydrated peptide amide linkage. As shown in Fig. 12, the ion is initially hydrated at the CO, but the water molecule migrates around to the NH site. The figure displays the time sequence of the evolution from the reactant IR spectrum to the product IR spectrum, revealing a time scale of about 5 ps for this reaction (at the cryogenic temperature of the supersonic beam expansion used to generate the original hydrated molecules).

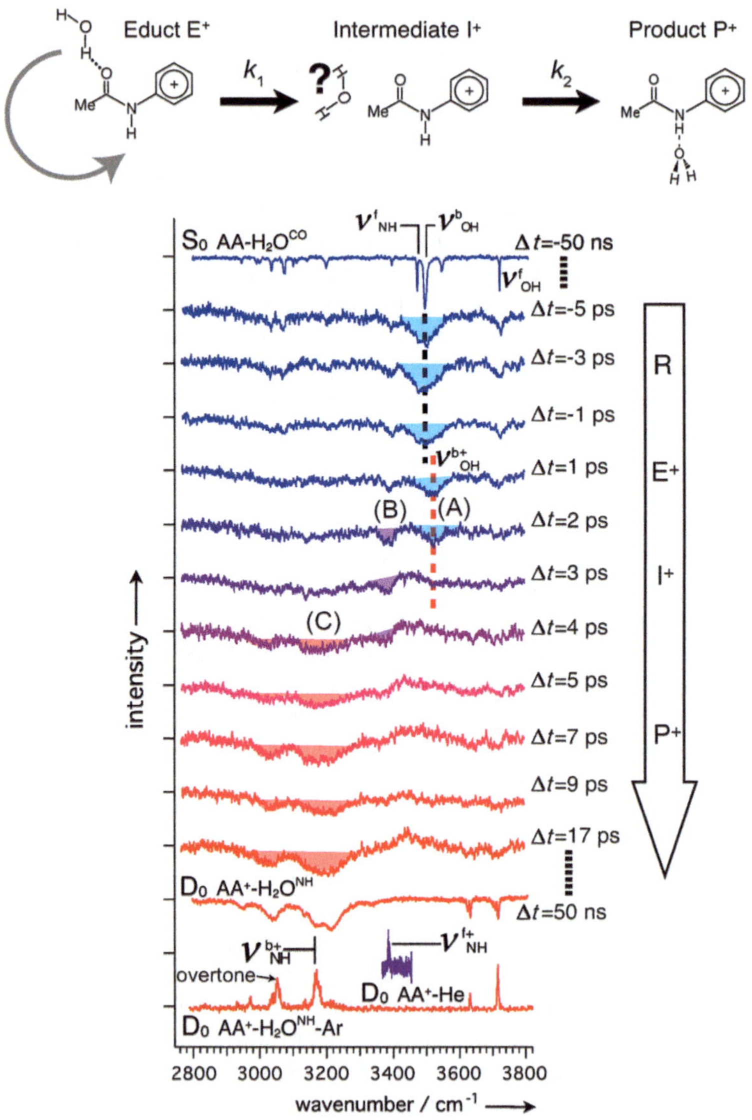

Fig. 12 Time-resolved IR dip spectra of acetanilide$^+$-H_2O. The spectra at $\Delta t = -50$ ns and 50 ns are "static" IR spectra of the CO-bound and NH-bound isomers in the neutral and cationic states measured by nanosecond lasers. The IR spectra of cold AA^+-He and AA^+-H_2O-Ar clusters generated by electron impact ionization are also shown for comparison. Reproduced with permission from [131]

3 Visible/UV and Electronic Spectroscopy

Compared with the recent ferment of activity in the infrared spectroscopy of ions, attention to the UV/visible spectroscopic region has been less active. Among the reasons for this lower interest are, first, the low information content of the typical (room temperature) spectra of molecules of chemical interest, which characteristically show a small number of broad peaks and shoulders, in which a few individual peaks occupy a substantial fraction of the total useful wavelength range, and, second, the greater difficulty and lower confidence of the calculations (most often time-dependent density functional theory (TDDFT) for systems large enough to be interesting) of electronic transitions compared with the high precision and confidence which has evolved for properly scaled vibrational calculations (both ab initio and DFT). It has up to now been unusual for interesting structural questions about larger ions to be answered definitively by correlation of UV/vis action spectra with quantum-chemical calculations.

Although considerations such as these have led to a less active field of UV/visible spectroscopic study compared with infrared study, this wavelength region offers some unique advantages, particularly for studies oriented toward kinetics, dynamics, thermodynamics, and mechanisms of gas-phase ion chemistry. As a particular example, we can mention the ability to initiate a chemical process with the insertion of a very accurately known (single-photon) increment of several eV of internal energy at a precisely determined time (for a view of some of these possibilities, see, for instance, [132]). Control of photon polarization is possible [133]. Moreover, excited electronic reaction and dissociation pathways, often involving radical ion chemistry, can be accessed in this way, which is not possible with vibrational activation.

A leading group in the exploration of the capabilities and limitations of UV/vis spectroscopy of metal complexes is the Nottingham group (Stace et al.), which has pioneered the use of novel ion-source technologies to produce interesting complexes for action-spectroscopic study. They reviewed the field in 2010 [6].

The Lyon group has, within the last few years, greatly expanded the UV/vis spectroscopy of bio-relevant ions (for example), including some metal-cationized peptides and proteins [134]. A thoroughly analyzed case [135] of a UV spectrum of a peptide system is the observation and comparison of the AcGlyGlyGlyTrpNH$_2$Ag$^+$ and AcHisGlyGlyGlyTrpNH$_2$Ag$^+$ complexes, which serve as models for the interesting prion octarepeat system (Fig. 13). The spectra have broad peaks with low information content, and it is hard to separate the features reflecting charge transfer vs intrachromophore transitions within the indole moiety of Trp. However, a distinctive influence of the introduction of His is observed in the form of a shift and enhancement of a feature at 280 nm in the latter peptide, and the authors analyze this peak as showing a close interaction of the imidazole lone-pair nitrogen of the His residue with the silver ion, along with cation-π binding to the indole ring of Trp. This wide-ranging study of the metal binding site in the prion octarepeat neighborhood also included ion mobility, IRMPD spectroscopy in the

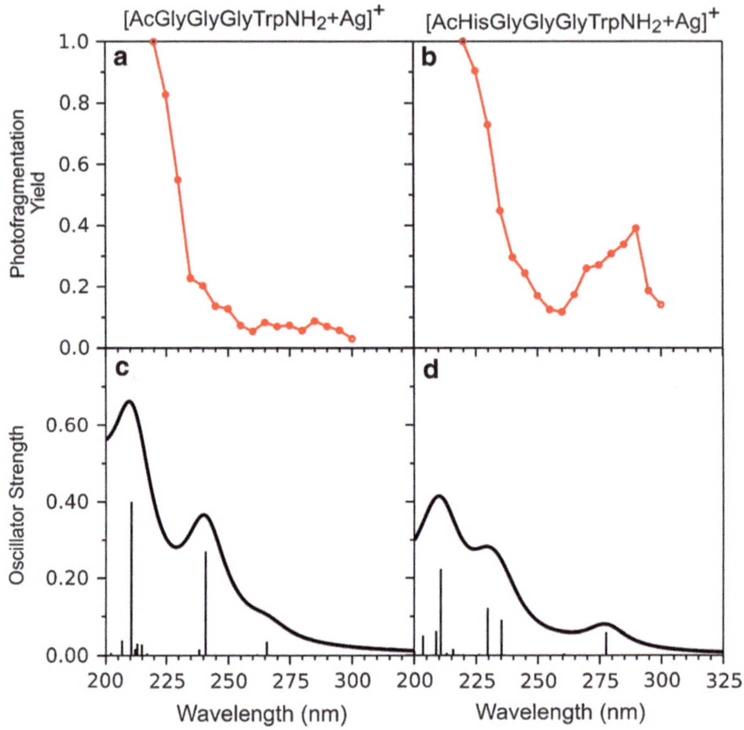

Fig. 13 Experimental photodissociation yield (*red*, arbitrary units) and calculated UV absorption spectra (*black*) for $[AcGlyGlyGlyTrpNH_2 + Ag]^+$ (**a**, **b**) and $[AcHisGlyGlyGlyTrpNH_2 + Ag]^+$ (**c**, **d**). Reproduced with permission from [135]

H-stretching region, and computations, so their conclusions were more broadly based than the UV spectrum alone would justify.

An illustration of the spectroscopic emergence of a strong isolated chromophore within the interior of a substantially large protein is given by this group's spectroscopy of the cytochrome-c system. The Soret band near 410 nm is nicely observed in the $[M\text{-}6H]^{6-}$ electron photodetachment action spectrum for cytochrome-c, containing the Fe(III) heme chromophore [136] (Fig. 14). The gas-phase peak is very close to the position observed in solution, suggesting that the chromophore is well shielded from solvent interactions in solution.

The UV spectroscopy of anions is apt to become complicated because of the competing possibilities of electron photodetachment and photodissociation as exit channels for the photoexcited anion. A case showing progress in sorting out such complexities was shown by the Lyon group. Complexes $[GSH + M - 2H]^-$ (where M is Ag or Au and GSH is the gamma tripeptide glutathione) were found to give a varying mix of detachment and dissociation at UV wavelengths [137]. The action spectra showed some intensity and structure at wavelengths below 300 nm. Time-dependent density functional theory (TDDFT) calculations gave general accord

Fig. 14 Laser-normalized electron photodetachment yield measured as a function of the laser wavelength for $[Prot - 6H]^{6-}$, where Prot is cytochrome-c. Reproduced with permission from [136]

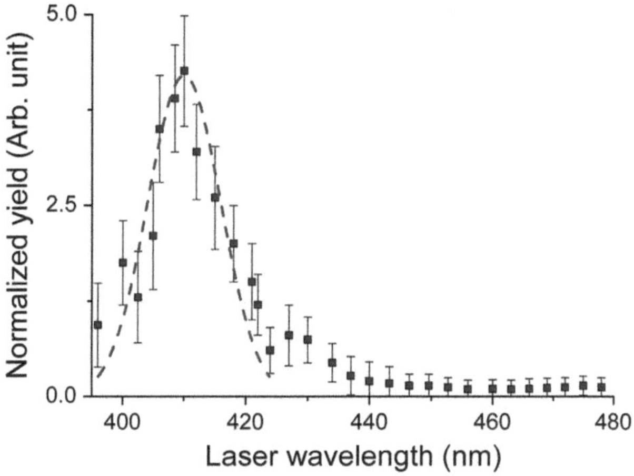

with the observed action spectra, and were useful in their interpretation. However, this work was not primarily a spectroscopy exercise, since structure assignments (predicting two-coordinate O-M-S metal binding) were made based on DFT structure calculations, not notably aided by the UV/vis spectra.

UV/vis spectra of silver and gold complexes with tryptophan both show strikingly strong long wavelength peaks at 330 nm for $AgTrp^+$ [74] and 400 nm for $AuTrp^+$ [138]. The peaks are attributed to indole-metal charge transfer, with the metal bound in a charge-solvated fashion to the six-membered ring of the indole group and the amino nitrogen.

Both IR and UV/vis action spectra were used [139] to characterize the radical cations of Trp-containing peptides, in particular $AcGly_3TrpNH_2$. Both types of spectra indicated a canonical π-radical cation structure, ruling out the possibilities of forming a zwitterion, or of an ionization site other than the indole π-system. The spectra are shown in Fig. 15 (compared with the IR spectrum of the protonated peptide). These spectroscopic probes are seen in general as useful structural diagnostics for ionized peptides containing a Trp residue.

The Lyon group made a comparison of the nonapeptide oxytocin (CysTyrIleGlnAsnCysProLeuGly-NH_2) complexed with Cu^{2+}, vs complexation with two protons, and also vs protonated tyrosine [140]. The action spectra in all three cases were considered to be dominated by the tyrosine chromophore, with only weak perturbations among its three environments. The spectroscopic evidence here is not very incisive, although, as Fig. 16 shows, copper complexation resulted in a substantial change in the UV spectrum. This, along with ion mobility and computational evidence, led to the main structural conclusion that the tyrosine side chain did not interact closely with the metal in the Cu(II) complex, while a major structural change was induced by 4-N chelation of the metal ion. The corresponding dianionic complex $[OT - 4H + Cu]^{2-}$ was also studied by photodetachment action spectroscopy [141]. Again, the spectroscopic evidence was not strong, but the arguments pointed toward a structure with copper bound to four deprotonated

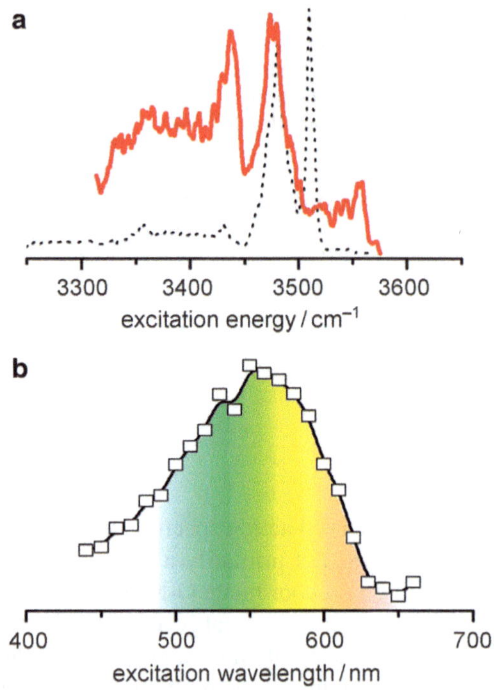

Fig. 15 (**a**) IRMPD spectra of $AcGly_3TrpNH_2^{\bullet+}$ (*solid line*) and $AcGly_3TrpNH_3^{+}$ (*dotted line*). (**b**) Visible photo-fragmentation spectrum of $AcGly_3TrpNH_2^{\bullet+}$. Reproduced with permission from [139]

nitrogens, and the neutral form of the tyrosine side chain. A contrast was drawn with the doubly deprotonated oxytocin ion $[OT\text{-}2H]^{2-}$, in which the tyrosine hydroxyl was considered to be deprotonated.

An interesting approach to probing peptide and protein structures via UV/vis action spectroscopy is the work of the group in Australia using iodotyrosine as a chromophoric label [142], as has been developed by Julian's group at Riverside [143, 144]. The local environment of the tyrosine is sampled by the perturbation of its UV spectrum, with the innocuous iodine substituent atom serving as a one-photon detachable label. The extent of folding to give microsolvation of the tyrosine site is reflected in coalescence of a split feature in the 290 nm region.

Turning to the extreme ultraviolet wavelengths now accessible with synchrotron light sources, VUV spectroscopy of metal complexed polysaccharide ions (photon energies ~8–20 eV) has not yet received much attention, but recent work at the high-energy photon source SOLEIL [145] on polysaccharides is suggestive for the future. In contrast to anionic species, where electron detachment is the primary decomposition pathway, the sodiated species show intense loss of the sodium ion. Attention has been mostly directed to differences in fragmentation patterns among different isomers, but the VUV spectrum of sodiated maltohexose, for example, shows interesting structure with several peaks between 12 and 18 eV, which suggest interesting future possibilities for useful spectroscopic structural information.

Fig. 16 Experimental fragmentation yields of $[OT+2H]^{2+}$ and $[OT+Cu]^{2+}$ ions measured as a function of the laser wavelength. Reproduced with permission from [140]

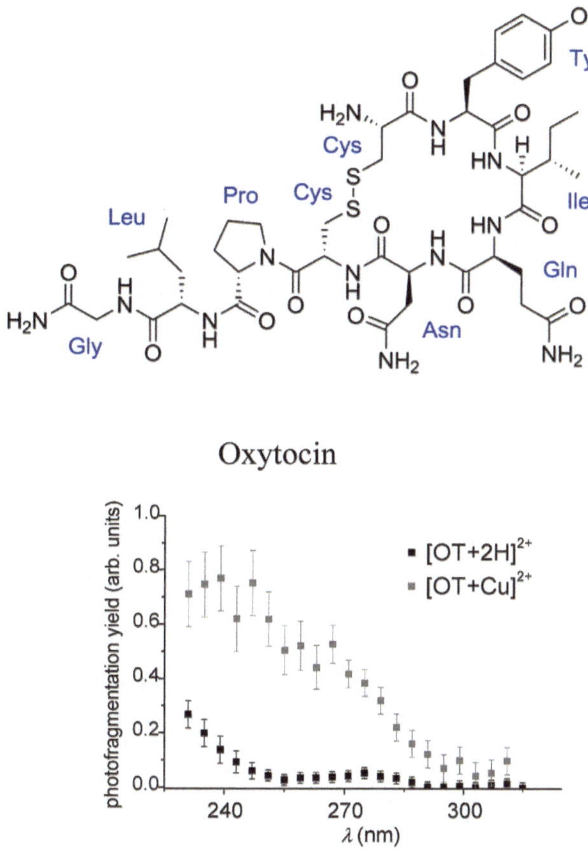

Acknowledgments This work was financially supported by the "Nederlandse Organisatie voor Wetenschappelijk Onderzoek" (NWO). R.C.D. acknowledges support from the National Science Foundation, Grant PIRE-0730072, and expresses gratitude to the FELIX facility for its continuing welcome. The FELIX staff, and particularly Dr. Lex van der Meer and Dr. Britta Redlich, are gratefully acknowledged for their assistance. We thank SURFsara Computing and Networking Services (www.surfsara.nl) for their support in using the Lisa Computer Cluster.

References

1. Rijs AM, Oomens J (2015) IR spectroscopic techniques to study isolated biomolecules. Top Curr Chem. doi:10.1007/128_2014_621
2. MacAleese L, Maître P (2007) Infrared spectroscopy of organometallic ions in the gas phase: from model to real world complexes. Mass Spectrom Rev 26:583–605
3. Eyler JR (2009) Infrared multiple photon dissociation spectroscopy of ions in Penning traps. Mass Spectrom Rev 28:448–467

4. Fridgen TD (2009) Infrared consequence spectroscopy of gaseous protonated and metal ion cationized complexes. Mass Spectrom Rev 28:586–607
5. Polfer NC, Oomens J (2009) Vibrational spectroscopy of bare and solvated ionic complexes of biological relevance. Mass Spectrom Rev 28:468–494
6. Cox H, Stace AJ (2010) Recent advances in the visible and UV spectroscopy of metal dication complexes. Int Rev Phys Chem 29:555–588
7. Burt MB, Fridgen TD (2012) Structures and physical properties of gaseous metal cationized biological ions. Eur J Mass Spectrom 18:235–250
8. Oomens J, Sartakov BG, Meijer G, Von Helden G (2006) Gas-phase infrared multiple photon dissociation spectroscopy of mass-selected molecular ions. Int J Mass Spectrom 254:1–19
9. Wu RH, McMahon TB (2009) Structures, energetics, and dynamics of gas-phase ions studied by FTICR and HPMS. Mass Spectrom Rev 28:546–585
10. Polfer NC, Paizs B, Snoek LC, Compagnon I, Suhai S, Meijer G, von Helden G, Oomens J (2005) Infrared fingerprint spectroscopy and theoretical studies of potassium ion tagged amino acids and peptides in the gas phase. J Am Chem Soc 127:8571–8579
11. Dunbar RC, Steill JD, Oomens J (2010) Cationized phenylalanine conformations characterized by IRMPD and computation for singly and doubly charged ions. PCCP 12:13383–13393
12. Polfer NC, Oomens J, Dunbar RC (2008) Alkali metal complexes of the dipeptides PheAla and AlaPhe: IRMPD spectroscopy. ChemPhysChem 9:579–589
13. Oomens J, Polfer N, Moore DT, van der Meer L, Marshall AG, Eyler JR, Meijer G, von Helden G (2005) Charge-state resolved mid infrared spectroscopy of a gas-phase protein. PCCP 7:1345–1348
14. Bush MF, Forbes MW, Jockusch RA, Oomens J, Polfer NC, Saykally RJ, Williams ER (2007) Infrared spectroscopy of cationized lysine and epsilon-N-methyllysine in the gas phase: effects of alkali-metal ion size and proton affinity on zwitterion stability. J Phys Chem A 111:7753–7760
15. Bush MF, O'Brien JT, Prell JS, Saykally RJ, Williams ER (2007) Infrared spectroscopy of cationized arginine in the gas phase: direct evidence for the transition from nonzwitterionic to zwitterionic structure. J Am Chem Soc 129:1612–1622
16. Carl DR, Cooper TE, Oomens J, Steill JD, Armentrout PB (2010) Infrared multiple photon dissociation spectroscopy of cationized methionine: effects of alkali-metal cation size on gas-phase conformation. PCCP 12:3384–3398
17. Citir M, Hinton CS, Oomens J, Steill JD, Armentrout PB (2012) Infrared multiple photon dissociation spectroscopy of protonated histidine and 4-phenyl imidazole. Int J Mass Spectrom 330:6–15
18. Citir M, Stennett EMS, Oomens J, Steill JD, Rodgers MT, Armentrout PB (2010) Infrared multiple photon dissociation spectroscopy of cationized cysteine: effects of metal cation size on gas-phase conformation. Int J Mass Spectrom 297:9–17
19. Correia CF, Balaj PO, Scuderi D, Maître P, Ohanessian G (2008) Vibrational signatures of protonated, phosphorylated amino acids in the gas phase. J Am Chem Soc 130:3359–3370
20. Forbes MW, Bush MF, Polfer NC, Oomens J, Dunbar RC, Williams ER, Jockusch RA (2007) Infrared spectroscopy of arginine cation complexes: direct observation of gas-phase zwitterions. J Phys Chem A 111:11759–11770
21. Heaton AL, Bowman VN, Oomens J, Steill JD, Armentrout PB (2009) Infrared multiple photon dissociation spectroscopy of cationized asparagine: effects of metal cation size on gas-phase conformation. J Phys Chem A 113:5519–5530
22. Kamariotis A, Boyarkin OV, Mercier SR, Beck RD, Bush MF, Williams ER, Rizzo TR (2006) Infrared spectroscopy of hydrated amino acids in the gas phase: protonated and lithiated valine. J Am Chem Soc 128:905–916
23. Mac Aleese L, Simon A, McMahon TB, Ortega JM, Scuderi D, Lemaire J, Maître P (2006) Mid-IR spectroscopy of protonated leucine methyl ester performed with an FTICR or a Paul type ion-trap. Int J Mass Spectrom 249:14–20

24. Maître P, Lemaire J, Scuderi D (2008) Structural characterization under tandem mass spectrometry conditions: infrared spectroscopy of gas phase ions. Phys Scr 78:058111
25. Scuderi D, Bakker JM, Durand S, Maître P, Sharma A, Martens JK, Nicol E, Clavaguera C, Ohanessian G (2011) Structure of singly hydrated, protonated phospho-tyrosine. Int J Mass Spectrom 308:338–347
26. Scuderi D, Correia CF, Balaj OP, Ohanessian G, Lemaire J, Maître P (2009) Structural characterization by IRMPD spectroscopy and DFT calculations of deprotonated phosphorylated amino acids in the gas phase. ChemPhysChem 10:1630–1641
27. Simon A, MacAleese L, Maître P, Lemaire J, McMahon TB (2007) Fingerprint vibrational spectra of protonated methyl esters of amino acids in the gas phase. J Am Chem Soc 129: 2829–2840
28. Talbot FO, Tabarin T, Antoine R, Broyer M, Dugourd P (2005) Photodissociation spectroscopy of trapped protonated tryptophan. J Chem Phys 2005:122
29. Wu RH, McMahon TB (2008) An investigation of protonation sites and conformations of protonated amino acids by IRMPD spectroscopy. ChemPhysChem 9:2826–2835
30. Correia CF, Clavaguera C, Erlekam U, Scuderi D, Ohanessian G (2008) IRMPD spectroscopy of a protonated, phosphorylated dipeptide. ChemPhysChem 9:2564–2573
31. Dunbar RC, Steill JD, Polfer NC, Oomens J (2009) Gas-phase infrared spectroscopy of the protonated dipeptides H^{+}PheAla and H^{+}AlaPhe compared to condensed-phase results. Int J Mass Spectrom 283:77–84
32. Joshi K, Semrouni D, Ohanessian G, Clavaguera C (2012) Structures and IR spectra of the gramicidin S peptide: pushing the quest for low-energy conformations. J Phys Chem B 116: 483–490
33. Lucas B, Gregoire G, Lemaire J, Maître P, Glotin F, Schermann JP, Desfrancois C (2005) Infrared multiphoton dissociation spectroscopy of protonated N-acetyl-alanine and alanyl-histidine. Int J Mass Spectrom 243:105–113
34. Lucas B, Gregoire G, Lemaire J, Maître P, Ortega JM, Rupenyan A, Reimann B, Schermann JP, Desfrancois C (2004) Investigation of the protonation site in the dialanine peptide by infrared multiphoton dissociation spectroscopy. PCCP 6:2659–2663
35. Prell JS, O'Brien JT, Steill JD, Oomens J, Williams ER (2009) Structures of protonated dipeptides: the role of arginine in stabilizing salt bridges. J Am Chem Soc 131:11442–11449
36. Wu RH, McMahon TB (2007) Infrared multiple photon dissociation spectroscopy as structural confirmation for GlyGlyGlyH(+) and AlaAlaAlaH(+) in the gas phase. Evidence for amide oxygen as the protonation site. J Am Chem Soc 129:11312
37. Wu RH, McMahon TB (2009) Protonation sites and conformations of peptides of glycine (Gly(1–5)H(+)) by IRMPD spectroscopy. J Phys Chem B 113:8767–8775
38. Kong XL, Tsai IA, Sabu S, Han CC, Lee YT, Chang HC, Tu SY, Kung AH, Wu CC (2006) Progressive stabilization of zwitterionic structures in [H(Ser)(2–8)](+) studied by infrared photodissociation spectroscopy. Angew Chem Int Ed 45:4130–4134
39. Rizzo TR, Stearns JA, Boyarkin OV (2009) Spectroscopic studies of cold, gas-phase biomolecular ions. Int Rev Phys Chem 28:481–515
40. Rizzo TR, Boyarkin OV (2014) Cryogenic methods for the spectroscopy of large, biomolecular ions. Top Curr Chem. doi:10.1007/128_2014_579
41. Iavarone AT, Patriksson A, van der Spoel D, Parks JH (2007) Fluorescence probe of Trp-cage protein conformation in solution and in gas phase. J Am Chem Soc 129:6726–6735
42. Baer T, Dunbar RC (2010) The Asilomar conference on ion spectroscopy, October 16–20, 2009. J Am Soc Mass Spectrom 21:I1–I3
43. Talbot FO, Rullo A, Yao H, Jockusch RA (2010) Fluorescence resonance energy transfer in gaseous, mass-selected polyproline peptides. J Am Chem Soc 132:16156–16164
44. Wyttenbach T, Witt M, Bowers MT (2000) On the stability of amino acid zwitterions in the gas phase: the influence of derivatization, proton affinity, and alkali ion addition. J Am Chem Soc 122:3458–3464

45. Jockusch RA, Lemoff AS, Williams ER (2001) Hydration of valine-cation complexes in the gas phase: on the number of water molecules necessary to form a zwitterion. J Phys Chem A 105:10929–10942
46. Moision RM, Armentrout PB (2002) Experimental and theoretical dissection of sodium cation/glycine interactions. J Phys Chem A 106:10350–10362
47. Kapota C, Lemaire J, Maître P, Ohanessian G (2004) Vibrational signature of charge solvation vs salt bridge isomers of sodiated amino acids in the gas phase. J Am Chem Soc 126:1836–1842
48. Prell JS, Chang TM, Biles JA, Berden G, Oomens J, Williams ER (2011) Isomer population analysis of gaseous ions from infrared multiple photon dissociation kinetics. J Phys Chem A 115:2745–2751
49. Kish MM, Ohanessian G, Wesdemiotis C (2003) The Na+ affinities of alpha-amino acids: side-chain substituent effects. Int J Mass Spectrom 227:509–524
50. Heaton AL, Moision RM, Armentrout PB (2008) Experimental and theoretical studies of sodium cation interactions with the acidic amino acids and their amide derivatives. J Phys Chem A 112:3319–3327
51. Moision RM, Armentrout PB (2006) The special five-membered ring of proline: an experimental and theoretical investigation of alkali metal cation interactions with proline and its four- and six-membered ring analogues. J Phys Chem A 110:3933–3946
52. Khodabandeh MH, Reisi H, Davari MD, Zare K, Zahedi M, Ohanessian G (2013) Interaction modes and absolute affinities of amino acids for Mn^{2+}: a comprehensive picture. ChemPhysChem 14:1733–1745
53. Khodabandeh MH, Reisi H, Zare K, Zahedi M (2012) A theoretical elucidation of coordination properties of histidine and lysine to Mn^{2+}. Int J Mass Spectrom 313:47–57
54. Bush MF, Oomens J, Saykally RJ, Williams ER (2008) Alkali metal ion binding to glutamine and glutamine derivatives investigated by infrared action spectroscopy and theory. J Phys Chem A 112:8578–8584
55. Armentrout PB, Rodgers MT, Oomens J, Steill JD (2008) Infrared multiphoton dissociation spectroscopy of cationized serine: effects of alkali-metal cation size on gas-phase conformation. J Phys Chem A 112:2248–2257
56. Bush MF, Oomens J, Williams ER (2009) Proton affinity and zwitterion stability: new results from infrared spectroscopy and theory of cationized lysine and analogues in the gas phase. J Phys Chem A 113:431–438
57. Citir M, Hinton CS, Oomens J, Steill JD, Armentrout PB (2012) Infrared multiple photon dissociation spectroscopy of cationized histidine: effects of metal cation size on gas-phase conformation. J Phys Chem A 116:1532–1541
58. Dunbar RC, Hopkinson AC, Oomens J, Siu CK, Siu KWM, Steill JD, Verkerk UH, Zhao JF (2009) Conformation switching in gas-phase complexes of histidine with alkaline earth ions. J Phys Chem B 113:10403–10408
59. Rodgers MT, Armentrout PB, Oomens J, Steill JD (2008) Infrared multiphoton dissociation spectroscopy of cationized threonine: effects of alkali-metal cation size on gas-phase conformation. J Phys Chem A 112:2258–2267
60. Polfer NC, Oomens J, Dunbar RC (2006) IRMPD spectroscopy of metal-ion/tryptophan complexes. PCCP 8:2744–2751
61. O'Brien JT, Prell JS, Steill JD, Oomens J, Williams ER (2008) Interactions of mono- and divalent metal ions with aspartic and glutamic acid investigated with IR photodissociation spectroscopy and theory. J Phys Chem A 112:10823–10830
62. Drayss MK, Armentrout PB, Oomens J, Schaefer M (2010) IR spectroscopy of cationized aliphatic amino acids: stability of charge-solvated structure increases with metal cation size. Int J Mass Spectrom 297:18–27
63. Drayss MK, Blunk D, Oomens J, Schafer M (2008) Infrared multiple photon dissociation spectroscopy of potassiated proline. J Phys Chem A 112:11972–11974

64. Bush MF, Oomens J, Saykally RJ, Williams ER (2008) Effects of alkaline earth metal ion complexation on amino acid zwitterion stability: results from infrared action spectroscopy. J Am Chem Soc 130:6463–6471
65. Wang P, Ohanessian G, Wesdemiotis C (2008) The sodium ion affinities of asparagine, glutamine, histidine and arginine. Int J Mass Spectrom 269:34–45
66. Shoeib T, Siu KWM, Hopkinson AC (2002) Silver ion binding energies of amino acids: use of theory to assess the validity of experimental silver ion basicities obtained from the kinetic method. J Phys Chem A 106:6121–6128
67. Moision RM, Armentrout PB (2004) An experimental and theoretical dissection of potassium cation/glycine interactions. PCCP 6:2588–2599
68. Armentrout PB, Chen Y, Rodgers MT (2012) Metal cation dependence of interactions with amino acids: bond energies of Cs + to Gly, Pro, Ser, Thr, and Cys. J Phys Chem A 116: 3989–3999
69. Hofstetter TE, Howder C, Berden G, Oomens J, Armentrout PB (2011) Structural elucidation of biological and toxicological complexes: investigation of monomeric and dimeric complexes of histidine with multiply charged transition metal (Zn and Cd) cations using IR action spectroscopy. J Phys Chem B 115:12648–12661
70. Stearns JA, Mercier S, Seaiby C, Guidi M, Boyarkin OV, Rizzo TR (2007) Conformation-specific spectroscopy and photodissociation of cold, protonated tyrosine and phenylalanine. J Am Chem Soc 129:11814–11820
71. Fleming GJ, McGill PR, Idriss H (2007) Gas phase interaction of L-proline with Be2, Mg2 and Ca2 ions: a computational study. J Phys Org Chem 20:1032–1042
72. Jockusch RA, Lemoff AS, Williams ER (2001) Effect of metal ion and water coordination on the structure of a gas-phase amino acid. J Am Chem Soc 123:12255–12265
73. Ai HQ, Bu YX, Han KL (2003) Glycine-Zn+/Zn2+ and their hydrates: on the number of water molecules necessary to stabilize the switterionic glycine-Zn+/Zn2+ over the nonzwitterionic ones. J Chem Phys 118:10973–10985
74. Antoine R, Tabarin T, Broyer M, Dugourd P, Mitric R, Bonacic-Koutecky V (2006) Optical properties of gas-phase tryptophan-silver cations: charge transfer from the indole ring to the silver atom. ChemPhysChem 7:524–528
75. Strittmatter EF, Lemoff AS, Williams ER (2000) Structure of cationized glycine, gly·M^{2+} (M = Be, Mg, Ca, Sr, Ba), in the gas phase: intrinsic effect of cation size on zwitterion stability. J Phys Chem A 104:9793–9796
76. Hoyau S, Pelicier JP, Rogalewicz F, Hoppilliard Y, Ohanessian G (2001) Complexation of glycine by atomic metal cations in the gas phase. Eur J Mass Spectrom 7:303–311
77. Dunbar RC, Polfer NC, Oomens J (2007) Gas-phase zwitterion stabilization by a metal dication. J Am Chem Soc 129:14562–14563
78. Gronert S, Simpson DC, Conner KM (2009) A reevaluation of computed proton affinities for the common alpha-amino acids. J Am Soc Mass Spectrom 20:2116–2123
79. Dunbar RC, Steill J, Polfer NC, Oomens J (2009) Peptide length, steric effects and ion solvation govern zwitterion stabilization in barium-chelated di- and tripeptides. J Phys Chem B 113:10552–10554
80. Drayss MK, Blunk D, Oomens J, Gao B, Wyttenbach T, Bowers MT, Schafer M (2009) Systematic study of the structures of potassiated tertiary amino acids: salt bridge structures dominate. J Phys Chem A 113:9543–9550
81. Jockusch RA, Price WD, Williams ER (1999) Structure of cationized arginine (Arg center dot M+, M = H, Li, Na, K, Rb, and Cs) in the gas phase: further evidence for zwitterionic arginine. J Phys Chem A 103:9266–9274
82. Burt MB, Fridgen TD (2013) Gas-phase structures of Pb2+—cationized phenylalanine and glutamic acid determined by infrared multiple photon dissociation spectroscopy and computational chemistry. J Phys Chem A 117:1283–1290

83. Burt MB, Decker SGA, Atkins CG, Rowsell M, Peremans A, Fridgen TD (2011) Structures of bare and hydrated [Pb(AminoAcid-H)](+) complexes using infrared multiple photon dissociation spectroscopy. J Phys Chem B 115:11506–11518
84. Atkins CG, Banu L, Rowsell M, Blagojevic V, Bohme DK, Fridgen TD (2009) Structure of [Pb(Gly-H)](+) and the monosolvated water and methanol solvated species by infrared multiple-photon dissociation spectroscopy, energy-resolved collision-induced dissociation, and electronic structure calculations. J Phys Chem B 113:14457–14464
85. Gholami A, Fridgen TD (2013) Structures and unimolecular reactivity of gas-phase [Zn (Proline-H)](+) and [Zn(Proline-H)(H_2O)](+). J Phys Chem B 117:8447–8456
86. O'Brien JT, Prell JS, Berden G, Oomens J, Williams ER (2010) Effects of anions on the zwitterion stability of Glu, His and Arg investigated by IRMPD spectroscopy and theory. Int J Mass Spectrom 297:116–123
87. Schmidt J, Kass SR (2013) Zwitterion vs neutral structures of amino acids stabilized by a negatively charged site: infrared photodissociation and computations of proline-chloride anion. J Phys Chem A 117:4863–4869
88. Prell JS, Correra TC, Chang TM, Biles JA, Williams ER (2010) Entropy drives an attached water molecule from the C- to N-terminus on protonated proline. J Am Chem Soc 132: 14733–14735
89. Prell JS, Chang TM, O'Brien JT, Williams ER (2010) Hydration isomers of protonated phenylalanine and derivatives: relative stabilities from infrared photodissociation. J Am Chem Soc 132:7811–7819
90. Verkerk UH, Zhao JF, Saminathan IS, Lau JKC, Oomens J, Hopkinson AC, Siu KWM (2012) Infrared multiple-photon dissociation spectroscopy of tripositive ions: lanthanum-tryptophan complexes. Inorg Chem 51:4707–4710
91. Fridgen TD, MacAleese L, Maître P, McMahon TB, Boissel P, Lemaire J (2005) Infrared spectra of homogeneous and heterogeneous proton-bound dimers in the gas phase. PCCP 7: 2747–2755
92. Fridgen TD, MacAleese L, McMahon TB, Lemaire J, Maître P (2006) Gas phase infrared multiple-photon dissociation spectra of methanol, ethanol and propanol proton-bound dimers, protonated propanol and the propanol/water proton-bound dimer. PCCP 8:955–966
93. Rajabi K, Fridgen TD (2008) Structures of aliphatic amino acid proton-bound dimers by infrared multiple photon dissociation spectroscopy in the 700–2000 cm(−1) region. J Phys Chem A 112:23–30
94. Wu RH, Marta RA, Martens JK, Eldridge KR, McMahon TB (2011) Experimental and theoretical investigation of the proton-bound dimer of lysine. J Am Soc Mass Spectrom 22: 1651–1659
95. Wu RH, McMahon TB (2007) Infrared multiple photon dissociation spectra of proline and glycine proton-bound homodimers. Evidence for zwitterionic structure. J Am Chem Soc 129: 4864–4865
96. Steill JD, Szczepanski J, Oomens J, Eyler JR, Brajter-Toth A (2011) Structural characterization by infrared multiple photon dissociation spectroscopy of protonated gas-phase ions obtained by electrospray ionization of cysteine and dopamine. Anal Bioanal Chem 399: 2463–2473
97. Atkins CG, Rajabi K, Gillis EAL, Fridgen TD (2008) Infrared multiple photon dissociation spectra of proton- and sodium ion-bound glycine dimers in the N-H and O-H stretching region. J Phys Chem A 112:10220–10225
98. Dunbar RC, Steill JD, Polfer NC, Oomens J (2009) Dimeric complexes of tryptophan with M^{2+} metal ions. J Phys Chem A 113:845–851
99. Mino WK, Szczepanski J, Pearson WL, Powell DH, Dunbar RC, Eyler JR, Polfer NC (2010) Vibrational signatures of zwitterionic and charge-solvated structures for alkaline earth-tryptophan dimer complexes in the gas phase. Int J Mass Spectrom 297:131–138

100. Balaj OP, Kapota C, Lemaire J, Ohanessian G (2008) Vibrational signatures of sodiated oligopeptides (GG-Na+, GGG-Na+, AA-Na + and AAA-Na+) in the gas phase. Int J Mass Spectrom 269:196–209
101. Dunbar RC, Steill JD, Oomens J (2011) Encapsulation of metal cations by the PhePhe ligand: a cation-π ion cage. J Am Chem Soc 133:9376–9386
102. Prell JS, Demireva M, Oomens J, Williams ER (2009) Role of sequence in salt-bridge formation for alkali metal cationized GlyArg and ArgGly investigated with IRMPD spectroscopy and theory. J Am Chem Soc 131:1232–1242
103. Dunbar RC, Steill JD, Oomens J (2011) Chirality-induced conformational preferences in peptide-metal ion binding revealed by IR spectroscopy. J Am Chem Soc 133:1212–1215
104. Dunbar RC, Steill JD, Oomens J (2010) Conformations and vibrational spectroscopy of metal ion/polylalanine complexes. Int J Mass Spectrom 297:107–115
105. Balaj OP, Semrouni D, Steinmetz V, Nicol E, Clavaguera C, Ohanessian G (2012) Structure of sodiated polyglycines. Chem Eur J 18:4583–4592
106. Dunbar RC, Steill JD, Polfer NC, Oomens J (2013) Metal cation binding to gas-phase pentaalanine: divalent ions restructure the complex. J Phys Chem A 117:1094–1101
107. Dunbar RC, Polfer NC, Berden G, Oomens J (2012) Metal ion binding to peptides: oxygen or nitrogen sites? Int J Mass Spectrom 330–332:71–77
108. Wyttenbach T, Liu D, Bowers MT (2008) Interactions of the hormone oxytocin with divalent metal ions. J Am Chem Soc 130:5993–6000
109. Dunbar RC, Steill JD, Polfer NC, Berden G, Oomens J (2012) Peptide bond tautomerization induced by divalent metal ions: characterization of the iminol configuration. Angew Chem Int Ed 51:4591–4593
110. Dunbar RC, Oomens J, Berden G, Lau JKC, Verkerk UH, Hopkinson AC, Siu KWM (2013) Metal ion complexes with HisGly: comparison with PhePhe and PheGly. J Phys Chem A 117: 5335–5343
111. Martin RB (1990) In: Sigel A, Sigel H (eds) Magnesium and its role in biology. Nutrition and physiology. Marcel Dekker, New York, pp 1–13
112. Martin RB (1984) In: Sigel A, Sigel H (eds) Metal ions in biological systems: probing of proteins by metal ions and their low-molecular-weight complexes. Marcel Dekker, New York, pp 1–49
113. Andrews RK, Blakely RL, Zerner B (1988) In: Sigel A, Sigel H (eds) Metal ions in biological systems: nickel and its role in biology. Marcel Dekker, New York, pp 165–284
114. Martin RB (1988) In: Sigel A, Sigel H (eds) Metal ions in biological systems: nickel and its role in biology. Marcel Dekker, New York, pp 123–164
115. Sovago I, Osz K (2006) Metal ion selectivity of oligopeptides. Dalton Trans 3841–3854
116. Dunbar RC, Steill JD, Polfer NC, Oomens J (2013) Metal cation binding to gas-phase pentaalanine: divalent ions restructure the complex. J Phys Chem A 117(6):1094–1101
117. Semrouni D, Balaj OP, Calvo F, Correia CF, Clavaguera C, Ohanessian G (2010) Structure of sodiated octa-glycine: IRMPD spectroscopy and molecular modeling. J Am Soc Mass Spectrom 21:728–738
118. Semrouni D, Clavaguera C, Dognon JP, Ohanessian G (2010) Assessment of density functionals for predicting the infrared spectrum of sodiated octa-glycine. Int J Mass Spectrom 297:152–161
119. Martens JK, Compagnon I, Nicol E, McMahon TB, Clavaguera C, Ohanessian G (2012) Globule to helix transition in sodiated polyalanines. J Phys Chem Lett 3:3320–3324
120. Kupser P, Pagel K, Oomens J, Polfer N, Koksch B, Meijer G, von Helden G (2010) Amide-I and -II vibrations of the cyclic beta-sheet model peptide gramicidin S in the gas phase. J Am Chem Soc 132:2085–2093
121. Nagornova NS, Rizzo TR, Boyarkin OV (2010) Highly resolved spectra of gas-phase gramicidin S: a benchmark for peptide structure calculations. J Am Chem Soc 132:4040

122. Nagornova NS, Guglielmi M, Doemer M, Tavernelli I, Rothlisberger U, Rizzo TR, Boyarkin OV (2011) Cold-ion spectroscopy reveals the intrinsic structure of a decapeptide. Angew Chem Int Ed 50:5383–5386
123. Bush MF, Prell JS, Saykally RJ, Williams ER (2007) One water molecule stabilizes the cationized arginine zwitterion. J Am Chem Soc 129:13544–13553
124. Schmidt J, Meyer MM, Spector I, Kass SR (2011) Infrared multiphoton dissociation spectroscopy study of protonated p-aminobenzoic acid: does electrospray ionization afford the amino- or carboxy-protonated ion? J Phys Chem A 115:7625–7632
125. Tian ZX, Kass SR (2009) Gas-phase versus liquid-phase structures by electrospray ionization mass spectrometry. Angew Chem Int Ed 48:1321–1323
126. Schroder D, Budesinsky M, Roithova J (2012) Deprotonation of p-hydroxybenzoic acid: does electrospray ionization sample solution or gas-phase structures? J Am Chem Soc 134: 15897–15905
127. Tian ZX, Kass SR (2008) Does electrospray ionization produce gas-phase or liquid-phase structures? J Am Chem Soc 130:10842–10843
128. Tian ZX, Wang XB, Wang LS, Kass SR (2009) Are carboxyl groups the most acidic sites in amino acids? Gas-phase acidities, photoelectron spectra, and computations on tyrosine, p-hydroxybenzoic acid, and their conjugate bases. J Am Chem Soc 131:1174–1181
129. Steill JD, Oomens J (2009) Gas-phase deprotonation of p-hydroxybenzoic acid investigated by IR spectroscopy: solution-phase structure is retained upon ESI. J Am Chem Soc 131: 13570
130. Rogalewicz F, Hoppilliard Y, Ohanessian G (2003) Structures and fragmentations of zinc(II) complexes of amino acids in the gas phase – IV. Solvent effect on the structure of electrosprayed ions. Int J Mass Spectrom 227:439–451
131. Tanabe K, Miyazaki M, Schmies M, Patzer A, Schutz M, Sekiya H, Sakai M, Dopfer O, Fujii M (2012) Watching water migration around a peptide bond. Angew Chem Int Ed 51: 6604–6607
132. Lifshitz C (1997) Energetics and dynamics through time-resolved measurements in mass spectrometry: aromatic hydrocarbons, polycyclic aromatic hydrocarbons and fullerenes. Int Rev Phys Chem 16:113–139
133. Orth R, Dunbar RC, Riggin M (1977) Measurement of angular-distribution and energy of ionic fragments from photodissociation of molecular-ions. Chem Phys 19:279–288
134. Bellina B, Compagnon I, Joly L, Albrieux F, Allouche AR, Bertorelle F, Lemoine J, Antoine R, Dugourd P (2010) UV spectroscopy of entire proteins in the gas phase. Int J Mass Spectrom 297:36–40
135. Bellina B, Compagnon I, MacAleese L, Chirot F, Lemoine J, Maître P, Broyer M, Antoine R, Kulesza A, Mitric R, Bonacic-Koutecky V, Dugourd P (2012) Binding motifs of silver in prion octarepeat model peptides: a joint ion mobility, IR and UV spectroscopies, and theoretical approach. PCCP 14:11433–11440
136. Brunet C, Antoine R, Lemoine J, Dugourd P (2012) Soret band of the gas-phase ferri-cytochrome c. J Phys Chem Lett 3:698–702
137. Bellina B, Compagnon I, Bertorelle F, Broyer M, Antoine R, Dugourd P, Gell L, Kulesza A, Mitric R, Bonacic-Koutecky V (2011) Structural and optical properties of isolated noble metal-glutathione complexes: insight into the chemistry of liganded nanoclusters. J Phys Chem C 115:24549–24554
138. Antoine R, Bertorelle F, Broyer M, Compagnon I, Dugourd P, Kulesza A, Mitric R, Bonacic-Koutecky V (2009) Gas-phase synthesis and intense visible absorption of tryptophan-gold cations. Angew Chem Int Ed 48:7829–7832
139. Bellina B, Compagnon I, Houver S, Maître P, Allouche AR, Antoine R, Dugourd P (2011) Spectroscopic signatures of peptides containing tryptophan radical cations. Angew Chem Int Ed 50:11430–11432

140. Joly L, Antoine R, Albrieux F, Ballivian R, Broyer M, Chirot F, Lemoine J, Dugourd P, Greco C, Mitric R, Bonacic-Koutecky V (2009) Optical and structural properties of copper-oxytocin dications in the gas phase. J Phys Chem B 113:11293–11300
141. Joly L, Antoine R, Allouche AR, Broyer M, Lemoine J, Dugourd P (2009) Optical properties of isolated hormone oxytocin dianions: ionization, reduction, and copper complexation effects. J Phys Chem A 113:6607–6611
142. Kirk BB, Trevitt AJ, Blanksby SJ, Tao Y, Moore BN, Julian RR (2013) Ultraviolet action spectroscopy of iodine labeled peptides and proteins in the gas phase. J Phys Chem A 117: 1228–1232
143. Ly T, Julian RR (2010) Elucidating the tertiary structure of protein ions in vacuo with site specific photoinitiated radical reactions. J Am Chem Soc 132:8602–8609
144. Ly T, Julian RR (2009) Ultraviolet photodissociation: developments towards applications for mass-spectrometry-based proteomics. Angew Chem Int Ed 48:7130–7137
145. Enjalbert Q, Brunet C, Vernier A, Allouche AR, Antoine R, Dugourd P, Lemoine J, Giuliani A, Nahon L (2013) Vacuum ultraviolet action spectroscopy of polysaccharides. J Am Soc Mass Spectrom 24:1271–1279

Top Curr Chem (2015) 364: 225–270
DOI: 10.1007/128_2014_580

Published online: 25 March 2015

Isolated Neutral Peptides

Eric Gloaguen and Michel Mons

Abstract This chapter examines the structural characterisation of isolated neutral amino-acids and peptides. After a presentation of the experimental and theoretical state-of-the-art in the field, a review of the major structures and shaping interactions is presented. Special focus is made on conformationally-resolved studies which enable one to go beyond simple structural characterisation; probing flexibility and excited-state photophysics are given as examples of promising future directions.

Keywords Amide • Amino-acid • Backbone-side chain interactions • Conformation-selective IR spectroscopy • Gas phase laser spectroscopy • Hydrates • Secondary structures • Supersonic expansion

Contents

E. Gloaguen (✉)
CNRS, INC and INP, Lab. Francis Perrin, 91191 Gif-sur-Yvette, France

CEA, IRAMIS, Laboratoire Interactions, Dynamique et Lasers, 91191 Gif-sur-Yvette, France
e-mail: eric.gloaguen@cea.fr

M. Mons (✉)
CEA, IRAMIS, Laboratoire Interactions, Dynamique et Lasers, 91191 Gif-sur-Yvette, France

CNRS, INC and INP, Lab. Francis Perrin, 91191 Gif-sur-Yvette, France
e-mail: michel.mons@cea.fr

Abbreviations

Ac	Acetyl
Aib	Aminoisobutyric acid
BB	Backbone
Bn	Benzyl
CI	Conical intersection
CT	Charge transfer
DFT	Density functional theory
FC	Franck–Condon
FEL	Free electron laser
FTIR	Fourier transform infrared
IR	Infrared
LE	Locally excited
Me	Methyl
NCI	Non-covalent interactions
OPO	Optical parametric oscillator
PES	Potential energy surface
SC	Side-chain
SEP	Stimulated emission pumping
UV	Ultraviolet
VUV	Vacuum ultraviolet
Z	Benzyloxycarbonyl

1 Introduction

Peptides and proteins are complex systems, whose functionalities in the living world ensue from a subtle compromise between structuration and flexibility [1]. On one hand, their backbone has its own steric limitations caused by the rigidity of the peptide bonds which link residues together, restricting the set of conformations and therefore the types of possible secondary structures [2, 3]. On the other hand, the large diversity brought by the side chains of the 20 natural amino-acid residues explains the tremendous variety of tertiary and quaternary structures, together with the various tasks performed by these molecules, including enzymatic properties, molecular recognition, mechanical tasks, etc.

Besides the classical techniques for structural determination of proteins, namely X-ray diffraction or nuclear magnetic resonance, molecular modelling has become a complementary approach, providing refined structural details [4–7]. This view on the atomic scale paves the way to a comprehensive study of the correlations between protein structure and function, but a realistic description relies strongly on the performance of the theoretical tools. Nowadays, a full size protein is treated by force fields models [7–10], and smaller motifs, such as an active site of an enzyme, by multiscale approaches involving both quantum chemistry methods for local description, and molecular mechanics for its environment [11]. However, none of these methods are ab initio: force fields require a parameterisation based on experimental data of model systems; DFT quantum methods need to be assessed by comparison against high level ab initio calculations on small systems.

In such a context, isolated model systems can play the role of benchmarks, on which theory and experiment can be confronted at a very high level of detail. At the turn of the century, this rationale has led gas phase spectroscopists to devote experimental effort towards simple systems of biological relevance [12, 13]. Small building blocks of proteins were first targeted: conformational distributions of these flexible systems, as well as the interactions that shape molecules, in particular their H-bonding networks, have been the focus of these pioneering studies [14–20]. In this spirit, the classical gas-phase methods of the physicist, inherited from atomic or small molecule spectroscopy, namely optical absorption, either in the microwave, IR/Raman or UV spectral range, have been mobilised to tackle these issues. First, microwave experiments, carried out in supersonic expansions, remain devoted to relatively small molecules, from models of biomolecules to more recently investigated natural and capped amino-acids and sugars [21]. The advantage of this approach stems from its direct sensitivity to the structure, through the rotational constant, allowing it to provide stringent tests to the theoretical structures calculated. However, similar to X-ray crystallography or NMR studies, these experiments remain only indirectly sensitive to the intramolecular interactions which shape the molecule. In contrast, vibrational probes, which are more directly sensitive to their immediate environment, provide an invaluable opportunity to document the interactions occurring in these systems. In particular, NH, OH, and CO stretching modes appeared to be promising diagnostics, leading to an accurate H-bond network characterisation, and have therefore been explored by several groups. Although Raman absorption [22–25], FTIR [26–31] or IR/VUV [32, 33] spectra can be successfully carried out in an expansion, these techniques suffer from their lack of conformer selectivity. Conversely, the selectivity of double resonance IR/UV techniques has raised considerable interest, leading laser spectroscopists to provide an accurate picture of the structure of each conformer through the IR spectra, as well as information on the conformational populations through UV spectroscopy. This latter approach has proved to be quite successful [34–39]. It has been applied to numerous biosystems, either neutral or charged, isolated or in small clusters. Owing to the extent of this field, the present review focuses on globally neutral peptides, isolated or microsolvated [40, 41].

Characterisation of isolated neutral peptides through their vibrational signature resulted from the synergy between two complementary approaches:

- Gas phase spectroscopy of jet-cooled neutral molecules benefited from the development of laser-vaporisation and laser-desorption techniques as well as their coupling with a supersonic expansion. Improvements of spectroscopic procedures involving several lasers, e.g. the IR/UV double resonance spectroscopy, helped to collect information on weakly populated conformers.
- Quantum chemistry methods have been constantly improved, gaining in accuracy and transferability. Helped by the computer "Moore's law" during the past two decades, theoretical studies have been able to treat larger and larger species at an increasingly better level of theory.

The first studies focused on the identification of small species by comparison between their theoretical and experimental vibrational fingerprints. Starting from aromatic natural amino acids which possess a convenient UV chromophore (see Sect. 2), many seminal studies have been devoted to these natural model peptides. Their biological relevance can be questioned, in particular because the N and C natural terminals adopt a zwitterionic structure under physiological conditions which is usually not the structure observed in the gas phase. These natural peptides nevertheless constitute handy model systems on which experimental techniques and theoretical methodologies have been refined. This approach became increasingly difficult as the targeted systems were larger and larger. Among these, capped peptides, whose N and C terminals were made of complete peptide bonds, enabled progress to be made towards greater biological relevance. Successful investigation of systems as large as tripeptides stemmed from a gradual knowledge of species of increasing size, following the so-called bottom-up approach. A set of essential secondary structures has been identified, demonstrating the biological relevance of such gas-phase studies. By demonstrating that relevant models mimicking the backbone features of proteins could be studied in the gas phase at an unprecedented level of accuracy, these pioneering works opened new horizons and stimulated a large community. Taking advantage of decisive quantum chemistry progress, a more thorough approach appeared where a global theoretical study of the PES was conducted. Close comparison between the experimentally observed landscape and its theoretical counterpart provides much richer outcomes, allowing a fruitful interplay between experiment and theory. A comprehensive and precise PES enables a reliable assignment of the species observed, including all minor conformers. In return, failures of quantum chemistry methods to reproduce experimental observations could be spotted, guiding further theoretical developments.

The aim of this review is first to describe the parallel advancements in both experimental and theoretical methods and procedures, together with the cross-fertilisation that resulted. Examples of striking secondary structures of proteins isolated in the gas phase are given. Highlights on specific shaping interactions, including intermolecular interactions, e.g. with the solvent, provide an in-depth understanding of protein structure. In a last part, it is shown how conformation-selective spectroscopy is able to document the diversity of the PES landscape of these species, thus revealing their flexibility. It is also shown how coupling IR and UV information enables spectroscopists to target conformer-selective photophysics in peptides. Finally, future prospects of gas phase conformation-selective spectroscopy are presented.

2 Two Synergetic Approaches

2.1 Experiment

Vibrational spectroscopy, based on IR or Raman absorption, is an efficient technique to identify the conformational structure of small molecules, including model peptide systems [22–33]. It is, however, confronted by spectral congestion when dealing with molecules of increasing flexibility with a pool of conformations which interconvert at room temperature. Advanced procedures are required to gain valuable conformation-specific spectroscopic information. An efficient cooling usually provided by a supersonic expansion is needed to remove the room temperature internal energy, typically 3.10^3 cm^{-1} in a capped amino-acid. In addition, techniques such as temperature- or concentration-dependence studies are needed to identify the contributions of conformers and clusters [23, 27, 28, 30]. Several conformer-selective experiments were proposed, such as those based on a strong field deflection [42].

In this context, IR/UV double resonance experiments [43, 44] became quite popular. When conformer-selective excitation of a UV chromophore is possible, these experiments provide spectroscopists with an elegant way to record conformer-selective IR spectra [45]. Combined with classical thermal vaporisation techniques [46, 47] or with more sophisticated laser-desorption-based techniques [48–50] coupled to a supersonic expansion, IR/UV experiments have dramatically boosted the research field pioneered by Levy's group [48]. They made possible the observation of vibrationally resolved IR spectra (Fig. 1), first from the NH stretch spectroscopy permitted by the development of table-top OPOs in this range [45]. From the basic principles of IR spectroscopy, H-bonding in jet-cooled peptides can be reliably identified, providing a first basis to conformational assignment, further refined by comparison with quantum chemistry calculations. Rapidly, several conformers of the three natural amino-acids Phe [16], Trp [20] and Tyr [56], and sometimes their hydrates [57, 58], were identified through their vibrational fingerprint, refining former pioneering studies based on UV spectroscopy only [48, 59–62]. Numerous studies on peptides containing one of the three natural aromatic residues followed [46, 47, 51–55, 63–118], the presence of a near-UV chromophore being the tribute to pay to gain the conformer-selectivity brought by the IR/UV double resonance technique. Non-natural residues derived from the natural UV chromophores have also been widely employed [52, 111, 119–129]. Beyond this constraining aspect, the UV frequency and FC pattern being sensitive to the interactions involving the aromatic ring, they can provide very useful structural clues (Sect. 4.2). Of course, the principle of the experiment relies on the conformational selectivity of the UV chromophore. This is made possible by the intrinsic sensitivity of the UV transition to the chromophore environment, the cooling achieved in the expansion enabling the detection of frequency differences as small as a few cm^{-1}, in practice limited by the rotational contour of the UV bands. This selectivity can in general be achieved for small species, apart from spectral coincidence cases, which require a more sophisticated treatment [130]. In larger species, the spectral broadening caused by the difficulty in evacuating the internal energy results in a lack of UV resolution which affects the conformational selectivity [54, 89, 107, 108]. In these cases, one has to rely on the interpretation of the IR spectrum to figure out whether conformational selectivity is achieved or whether several species are simultaneously present. It should be noted that sequences deprived of UV chromophore can also been investigated by introducing so-called UV-tags which can be covalently [105, 107, 127, 131–145] or non-covalently bound, as in peptide-toluene complexes (Gloaguen E, Mons M, unpublished results). The assumption that the desired UV chromophore (often a phenyl ring) is weakly interacting with the rest of the molecule turned out to be a relatively seldom situation, owing to the huge flexibility of the peptides. Biases are introduced by what can be seen as parasitic interactions, e.g. NH-π [131], which prevent to extract any intrinsic data about the untagged system, especially in terms of conformational energetics.

Gas phase peptides of various sizes have been investigated in several spectral ranges (Fig. 1a–e), the first of them, in both importance (number of studies), reliability and interpretative potentialities is probably the so-called *amide A* NH-stretch region (Fig. 1c, d). The NH stretch oscillator exhibits an exquisite sensitivity to its environment, and is virtually uncoupled with the potentially numerous other NH groups of the molecules. Then, each individual IR band bears qualitative information about the strength of the H-bond in which the corresponding NH group is engaged [34], apart from very specific cases where strong hyperconjugation effects involve the σ* orbital of the vibrational probe [146]. Moreover, other vibrational probes lie in the *amide A* NH-stretch region, and contribute to make this spectral range very popular: amine NH and carboxylic acid OH for natural peptides, indole NH fot Trp, imidazole NH for His, phenol OH for Tyr, alcohol OH for Ser and Thr as well as water OH in microhydrated peptides. For the simplest systems, the H-bonding content can be deduced just by "reading" the conformer-selective amide A spectrum. Spectroscopic tricks, such as spectral correlations between physically coupled NH oscillators in NH_2-capped C-terminal groups [34, 70], together with the analysis of UV spectroscopic features (see Sect. 4.2), provide additional structural clues which have enabled the tackling of more complex species. Phe-containing dipeptides have been confidently assigned without relying on quantum calculations, and a ranking of the H-bonds according to their strength has been proposed (C5<NH-π<C10<C7 [34]). This pioneering work paved the way to systematic tests of quantum chemical methods in their ability to reproduce the IR spectra of these species as well as the conformer

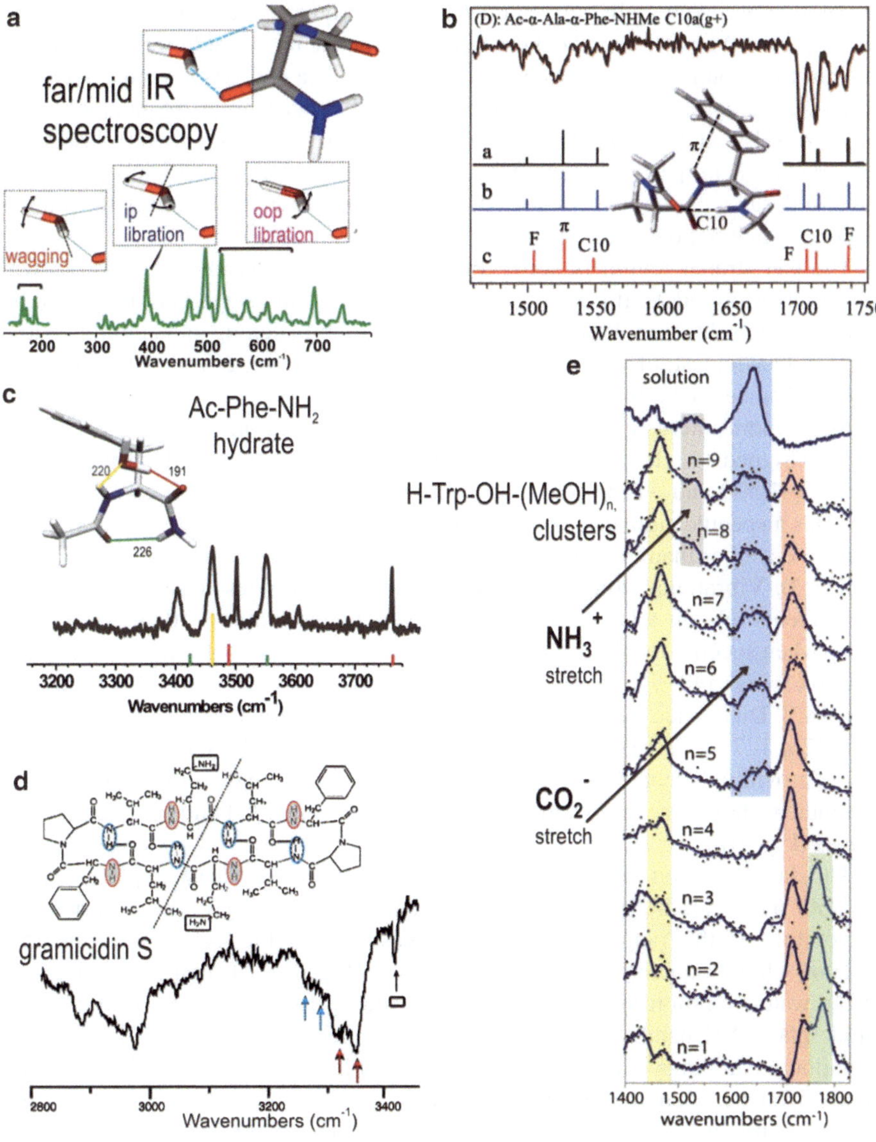

Fig. 1 Selected IR/UV spectra of isolated peptides, hydrates and clusters in various spectral ranges. Far-IR (**a**) conformer X of the Ac-Phe-NH_2 monohydrate [51]; amide I and II. (**b**) Ac-Ala-Phe-NH-Me dipeptide (adapted from [52] with permission of AIP), amide A NH stretch region. (**c**) Ac-Phe-NH_2 H_2O (X) (adapted from [53] with the permission of ACS). (**d**) Gramicidin S (adapted from [54] with the permission of Wiley). (**e**) H-Trp-OH (methanol)$_n$ clusters. Adapted from [55] with the permission of ACS

energetic order. Now, reliability and limitations of quantum chemistry methods are better known. Most studies use them first to derive the most stable forms, which are thought to account for experiment, and then propose a conformational assignment based on the agreement between theoretical and experimental frequencies in the *amide A* NH-stretch region. This strategy has proven to be quite powerful provided the structure is properly accounted for, which has been quite a critical issue during the past decade, owing to the real challenge that represented a proper description of dispersive interactions (see Sect. 2.2).

The CH stretch region is active in peptides and increasingly accessible with recent table-top OPOs. It could, in principle, be used for assignment purposes, but examples are seldom in the literature because band broadening and couplings often make a decisive use of this region difficult [26, 28, 31, 46, 63, 95, 113], although recent advances seem to be promising [147, 148].

Much more interesting are the amide I, II and III regions (Fig. 1b, e), respectively assigned to: (I) CO stretch (~1,700 cm^{-1}); (II and III) NH bend and CN stretch (~1,500 and ~1,200 cm^{-1} respectively). This so-called fingerprint region provides a specific signature of the backbone. It is however much less straightforward to decipher than the amide A region (some exceptions are discussed later in Sect. 3.3). Vibrational couplings make the interpretation less easy, and confident assignments rely on a large spectral range (the whole amide I–III regions), especially when spectral resolution is limited, usually 1% of the IR frequency for a free-electron source [48, 54, 55, 66, 69, 79, 80, 82, 83, 95, 97, 105, 108, 131, 132, 136, 141, 142]. Difference frequency generation from table-top OPOs [52, 64, 90, 104, 120, 126, 127, 137, 139, 144, 145] has become a valuable alternative source to FELs, providing spectral resolutions in the cm^{-1} range, and solid conformational assignments have emerged from studies combining amide I/II and amide A spectral ranges (Fig. 1b).

The far-IR region (wavelength >20 μm) remains the least explored so far [51, 66, 69, 117]. IR/UV double resonance experiments have been carried out successfully in this range, illustrating the feasibility of the technique in the 150–700 cm^{-1} low frequency range at a FEL facility (Fig. 1a). Clear specific signatures of the water libration motions in peptide hydrates, together with specific out-of-amide-plane NH bends for the monomer, have been obtained [51]. Inclusion of vibrational couplings in the theoretical treatment has recently enabled a successful comparison between theory and experiment [117], which should foster further developments in the far-IR. However, the advantage of this range for conformational assignment purposes remains to be demonstrated. One should note that SEP constitutes an alternative to FEL excitation in this range [149], but has not yet been employed for peptides.

Apart from vibrational spectroscopy, other techniques may also provide interesting diagnostics for structural characterisation of gas phase neutral amino-acids and peptides. Ionisation and fragment appearance energies [150–153] or photoelectron spectra [154–156] are indeed sensitive to conformation and provide additional clues for conformational assignment.

2.2 *Theory*

The role of theory is two-fold in the quest for conformational identification. First, peptides, being flexible, exist under many different conformations. The first theoretical task is to get a rough idea of the potential energy landscape by listing the minima, and sorting them according to their relative energy: only the lowest-energy conformations are expected to contribute to the experimental conformational mixture. Once this task is achieved, the next step for theory is to provide quantitative values of a few observables for assignment by comparison with experimental data. These two features therefore dictate the nature of the theoretical approaches required: a thorough exploration of the potential energy surface, reliable energy calculations, and frequency calculations as far as vibrational spectroscopy is concerned.

2.2.1 Explorations

They are typically carried out using force fields such as Gromacs, Amber or Charmm [157–159]. However, these force fields have been parameterised for condensed phases, and have some biaises which impede them in properly accounting for specific gas phase features. As an example, C7 H-bonded structures are not found, or sometimes found strongly distorted with Amber. Similarly, the poor energetics cannot pretend to describe the relative stabilities accurately enough for the trial procedure which follows the exploration. This step should not be too limiting so that no stable structure is eliminated at this stage [160]. To reduce the risk of missing potentially relevant structures in complex systems, theoretical explorations should ideally be carried out using different force fields, and eventually backed up by a critical analysis.

2.2.2 Quantum Chemistry Calculations of Structures

Gas phase studies of flexible systems owe their success to the high sensitivity of the observables to the structure, but, in return, this approach is very demanding for theoretical methods. As an illustration, an H-bond distance mispredicted by 1 pm may lead to an error of 2 cm^{-1} on the calculated vibrational frequency. The main issue of this step is then to get a theoretical structure accurate enough to predict observables quantitatively at a reasonable computational cost. In this quest, it has been rapidly realised how important it was to take dispersive interactions into account in these systems. Early Hartree–Fock calculations or DFT calculations dominated by the popular B3LYP functional with post-Hartree–Fock methods rapidly reached their limits because of the underestimation of dispersion interactions. In contrast, the Möller–Plesset approach treated as a second-order perturbation (MP2) overestimated dispersion at an additional computational cost, making its

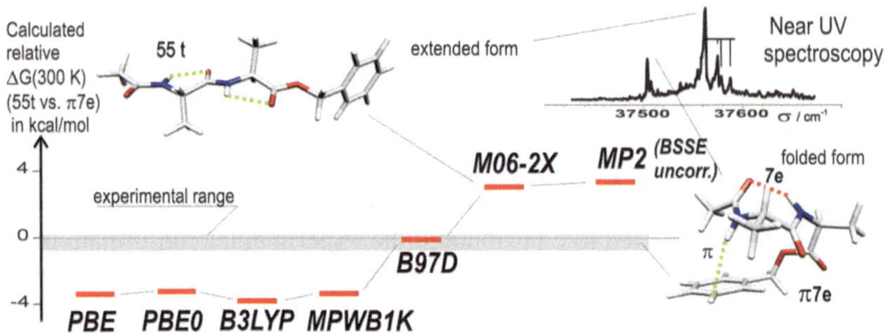

Fig. 2 Calculated relative Gibbs energies between the folded and extended form of the Ac-Ala-Ala-O-Bn model peptide as a function of the method. Adapted from [135] with the permission of ACS

benefits questionable. Many careful studies [99, 103, 135, 160–166] on model peptide systems document this issue. Failures to reproduce conformer energy ordering or specific structural features are often taken as a clue for a poor description of dispersive interactions.

Recent advances in DFT have reduced this problem either though the form of new hybrid functionals, M05-2X [167, 168] and M06-2X [169] being among the most popular, or by adding an explicit semi-empirical correction [170–173] to a dedicated functional, e.g. B97-D [171], both at a reasonable calculation cost. Several assessments carried out on model peptides [103, 135, 174] (Fig. 2) have shown the interest of these alternative approaches, which are now considered the highest standards in the community. Comparison with benchmark experimental data suggests that real improvements of both the structure description and the energetics are obtained. Errors of the order of a few kJ mol^{-1} for the 0 K enthalpy and up to 5 kJ mol^{-1} for the Gibbs energies can be anticipated on systems as large as dipeptides, including those where dispersion plays a decisive shaping role, as in aromatic-containing peptide chains [95, 96, 99, 103, 111, 112, 135, 160]. Correlatively, the development of these functionals provides faster calculations, making possible the investigation of even larger systems [53, 107, 128, 134, 139].

2.2.3 Vibrational Spectra

The accuracy of the calculated frequencies determines the quality of the assignment of an IR spectrum to the correct conformation among the potentially numerous candidates of energy compatible with experimental observation. Providing that the structure is correctly predicted, frequency calculations in the harmonic approximation are, surprisingly, sufficient to provide reliable vibrational spectra in the *amide A* region. Usually the disagreement between experimental and harmonic frequencies is systematic and can be overcome by scaling these harmonic frequencies to fit the experimental ones for benchmark systems, which allows one to account for both an approximate description of the vibration and systematic anharmonicity effects.

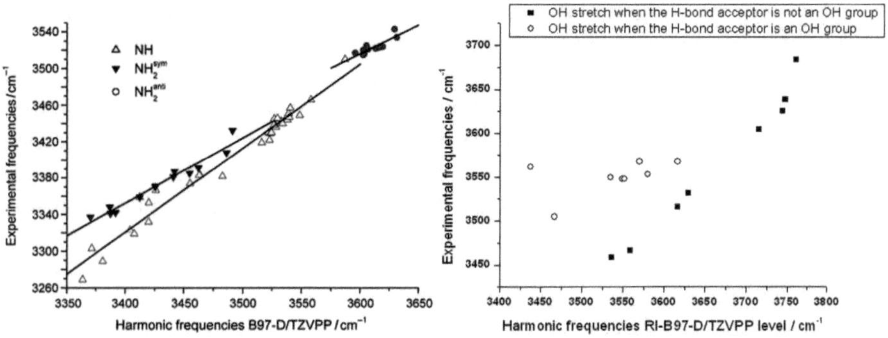

Fig. 3 Experimental vs B97-D/TZVPP calculated harmonic frequencies of previously assigned peptides. (**a**) For the NH stretch for which a satisfactory correlation can be observed whatever the nature of the H-bond acceptor, [107]. Reprinted with permission of ACS. (**b**) For the OH(alcohol) stretch for which the nature of the H-bond acceptor must be taken into account when a scaling procedure of the theoretical frequencies is considered [Gloaguen E, Mons M, unpublished results]

Different scaling procedures have been followed depending on the size of the set of benchmark systems available. When few benchmark data exist, a method-dependent global scaling factor is applied to the harmonic frequencies, providing a rough correction which can be sufficient for assignment in the simplest cases, but which do not render the subtleties captured by the level of calculation used. To avoid such a loss of information in the scaling process, a more refined strategy was needed. The availability of a large set of peptide conformations enabled the determination of mode-dependent scaling factors [175, 176]. In contrast to global scaling factors, this correction mode by mode reshapes the raw theoretical vibrational pattern for better comparison with the experimental spectra. Mode-dependent linear ($\nu_{\text{th scaled}} = a\,\nu_{\text{raw}} + b$) scaling functions constitute the finest variation of this pragmatic approach. Particularly well-suited for systems with vibrational probes of different nature, it reached an unsurpassed precision in the 20 cm^{-1} range when applied to peptide hydrates in the *amide A* region [53, 107] (Fig. 3a). The level of precision reached enables one to make critical use of automatic assignment procedures, where the experimental spectra are checked against the series of conformations issued from the exploration/optimisation procedure; the best candidates are ranked on the basis of an agreement factor based on both frequencies and energetics [112].

This mode-specific approach, however, can fail when vibrational modes are coupled. Therefore, in the amide I–III region where couplings are more frequent, a lesser precision is generally achieved, even if domain-selective scaling factors provide fair predictions [52, 80]. A further theoretical effort towards more accurate frequencies is needed to extract all the valuable information contained in this spectral region, especially the mid-IR region, involving backbone motions. This can be achieved by calculating anharmonic corrections at the second-order of perturbation [177]. However, this approach is often disappointing and is hampered by the size of the systems. More accurate approaches [178, 179] have also been tried, with some success, but have been confronted by the same large size issue.

Finally, one can mention the far-IR region where the first experimental/theoretical combined study on a microhydrated peptide demonstrated that couplings between backbone, side chain and solvent modes are ubiquitous [51]. Excellent theoretical treatment of the vibrational levels was indeed needed to exploit this spectral range fully, which has recently been achieved [117], opening interesting perspectives for this range.

2.3 *Lessons from the Confrontation...*

Interfaces between experiment and theory are of critical importance: agreement between both datasets suggest that an issue is solved, whereas discrepancies have fostered methodological theoretical developments and/or deeper understanding of the physical phenomena involved. For isolated peptides, two major interfaces between experiment and theory have been extensively exploited – vibrational frequencies and energetics – and both raise questions, some of which are discussed in this section.

A good correlation between experimental and harmonic vibrational frequencies is at the heart of the conformational assignment process. However, such a correlation is not obvious, and exceptions exist as illustrated by the OH(alcohol) stretch mode (Fig. 3b). At first sight, the correlation is quite poor, which is disappointing in the perspective of predicting the experimental frequency from harmonic calculations. A more detailed analysis reveals that a reasonable correlation is achieved, provided the OH–OH H-bond patterns are not included. This example illustrates the interest of such an investigation of the vibrational frequency interface between experiment and theory on a set of benchmark systems: (1) limits of applicability of the scaling procedure is better known; (2) the error made between scaled harmonic and experimental frequencies can be anticipated; and (3) solutions to study systems outside the correlation can be considered, e.g. by taking into account the coupling between OH modes in order to improve the OH–OH H-bond cases.

The energetics interface is also very instructive, although its investigation is complicated by the qualitative nature of most of the experimental results in the field. Conformational energetics are indeed not measured *per se*, but more intuitively deduced from the apparent conformational distribution revealed by UV spectroscopy. Yet, experimental techniques exist to measure conformational distribution [47], and even conformational energetics [149]. Even though conformational distributions have indeed been measured for a few peptides [47, 71, 72, 125, 126], conformational energetics have been measured only for slightly simpler systems [149, 180, 181]. Conclusions about conformational energetics from the majority of experiments have thus been limited so far to identifying major and minor conformations of the distribution.

Despite this experimental context, relevant information about the parameters controlling the conformational population in expansions could be obtained. One of the striking features shared by the majority of the peptides studied is the small number of conformations observed in the expansion [107, 139], despite the significant, if not large, number of energetically low-lying calculated conformations. This clearly points to important conformational relaxation effects during the expansion caused by collisions with the carrier gas, in line with the spectacular energy-controlled isomerisation experiments [47, 71, 72, 119, 180]. Laser-desorbed species are often coupled to an expansion of argon because of its capability to remove efficiently the large internal energy resulting from the vaporisation process and therefore eventually achieve a better vibrational cooling. Correlatively, its large mass and polarisability make it an active solvent, which favours interconversion between conformations, leading to a better conformational relaxation. Such an effect has been revealed in the presence of solvent molecules in the expansion, leading to the concept of "solvent-assisted conformational isomerisation" [182, 183]. As a matter of fact, using He or Ne instead of Ar as carrier gas enables the observation of additional minor species calculated at high energy [58, 112], in agreement with the usual picture of the supersonic expansion being a rapid event, which freezes the conformational distributions at high temperature (kinetic trapping) [184].

Such carrier-gas effects pointing to kinetically-sensitive conformational distributions do raise a few questions. How to interpret the simultaneous observation of several conformations of comparable apparent intensities? What are the theoretical parameters relevant for comparison with the experimental conformational distributions? In thermally-vaporised molecules, the reservoir and valve temperatures are typically in the $T_0 = 450$ K range, with a temperature at the nozzle throat already decreased by a quarter in the ~340 K range, assuming $\gamma = 5/3$ [184]. In laser-desorption or laser-ablation devices, the molecules are vaporised as relatively hot species. Rough estimates of the starting temperature are in the 600 K range [185], which suggests a total loss of memory of the molecular structure in the solid state. Once in the gas phase, peptides already begin to cool down along the desorption plume. Then collisions with the carrier gas leads to a sudden freeze below 225 K (at the nozzle, the carrier gas is at around three-quarters of its initial temperature, which is usually 300 K). During this complex cooling process following vaporisation, conformational isomerisations take place until the isomerisation rates fall down below the characteristic time of the expansion, freezing de facto the populations [184]. These final populations resulting from a high temperature conformational equilibrium suggest that Gibbs energies are more suitable than enthalpy at 0 K to compare with experiment, which has been supported by various examples on small molecules [186] where the distribution is correctly predicted by a room temperature equilibrium, as well as by recent work on peptides and flexible molecules [88, 107, 135, 187, 188]. These systems existing under conformations of very different entropic behaviours, e.g. one stretch and one compact form, can be used as "conformational thermometers": the population ratio between the two species depending on the conformational temperature, its measurement provides an estimate of the temperature at which interconversions become infrequent in the expansion [135]. It should be noted that this temperature is clearly linked to the height of isomerisation barriers, and thus varies with the molecules and conformations considered. At worst, in systems where more than two conformers are observed, isomerisation paths may close at very different temperatures, leading to a final distribution totally out of equilibrium, raising the question of comparison with theory: a careful and quantitative analysis should therefore take into account not only the thermodynamic, but also the kinetic parameters governing the isomerisations taking place in the expansion. Contributions to the final distribution of all the conformers initially populated at high temperature are difficult to anticipate theoretically, given the possible population transfers during the expansion. This would require the knowledge of the interconversion history, which is in practice out of reach and therefore always disregarded, even in the most careful analyses of the experimental populations [88]. In this respect, it is also worth mentioning the specific case of complexes for which the point in the expansion where the partners finally stick together is critical for the final conformational distribution. Early formation enables the complex to rearrange and find the most stable structures, whereas an aggregation in the late stages of the expansion favours the easiest way to assemble the partners from their own initial distribution. This kinetic trapping has been observed to play a predominant role by hampering a full exploration of peptide hydrates during the expansion [53]. In particular, it can explain that a few conformers, for which water needs energy to insert in an already formed H-bond between a donor and an acceptor site of the peptide, are not observed, despite their predicted low energy.

Finally, other apparent discrepancies between the predicted conformational distribution and the observation may simply come from the way the distributions are estimated. Experimental populations are subjected to flaws when based on the assumption of similar detectivity for each conformation. This hypothesis is indeed questionable. In IR/UV experiments, detection uses either fluorescence or resonant ionisation detection, which both depend drastically upon the excited-state lifetime. In DNA bases, for instance, it falls in the sub-picosecond regime preventing any successful detection with nanosecond lasers [189]. Such a situation has not yet been revealed on peptides, only lifetimes strongly varying with conformation in the nanosecond regime having been encountered. These measurements are scarce and are limited to Phe-based peptides [75, 91, 110, 115] and Tyr [56]. In a few peptide-related compounds, presenting a phenyl chromophore and an amide group, short monomer lifetimes are suspected [18]. For Trp-based peptides, indirect clues about sub-nanosecond excited-state dynamics are derived from apparently "missing" conformers, i.e. the absence in the spectra of conformations predicted to be very stable [160]. This point is further discussed in Sect. 4.2. Apart from excited-state lifetimes, other conformation-dependent effects are likely to impact the estimate of the conformational distribution: the photoionisation cross section, which depends on the structural changes between neutral and ion [190], or the existence of fast relaxation of the ion into dark states, can yield an additional source of imprecision. The former, however, can be qualitatively overcome by ensuring a comfortable excess energy above the photoionisation threshold [135]. Alternatively, elegant IR/UV experiments, based on IR-induced population transfers in the expansion, have been recently demonstrated as an efficient way to determine the relative conformational populations, provided that these species are detected in the UV spectrum [47, 71, 72, 125, 126].

In summary, the interface between experiment and theory on energetics is instructive but full of potential pitfalls. Without significant theoretical developments to make it more reliable, vibrational frequencies remain the preferred interface in the conformational assignment process.

3 Peptide Structures Identified in the Gas Phase

3.1 *Secondary Structures of Protein Backbones*

One of the first goals of spectroscopic studies of isolated neutral peptides was to check whether gas phase folding in the supersonic expansion was capable of forming the secondary structures typical of peptides and proteins spontaneously, i.e. using the sole intramolecular interactions as driving force. The results, as an outcome of the past decade, were beyond initial expectations, even if it took some time to cross-check the assignments. They revealed a wealth of backbone structural features, centred on the formation of intrabackbone H-bonds, linking more or less remote CO and NH groups of the backbone, labelled as Cn, where n stands for the number of atoms present in the ring formed by the H-bond. In natural proteins, four typical intramolecular interactions are found, C5 in β-strands, C7 in γ-turns, C10 in β-turns and 3_{10} helices, and C13 in α-helices. Apart from the latter, all these features have been observed in isolated neutral peptides, together with β-sheet-like structures and mixed helices. Figure 4 summarizes the typical spectral signatures in the *amide A* range for these specific structures.

3.1.1 C5 Interactions, a Signature of Extended β-Strand-Like Backbone Structures

Short-range C5 H-bonds occur on the amino-acid scale. They have been observed in model peptides based on Trp [63], Phe [70] and Tyr [114], but their occurrence requires specific stabilisation conditions because of the intrinsic weakness of the H-bond caused by its nonlinearity, in spite of a short H–O distance (228 pm). They are usually challenged by the other short range, but much stronger, C7 H-bonds, which lead in turn to folded backbone structures [63, 70, 73, 83, 114, 188, 191]. In esterified C-terminal models [46, 87], the absence of an NH group forbids the folded structures. The same reason holds for longer molecules such as Ac-Ala-Ala-O-Bn where both NH groups are engaged in C5 H-bonds, leading to a 5–5

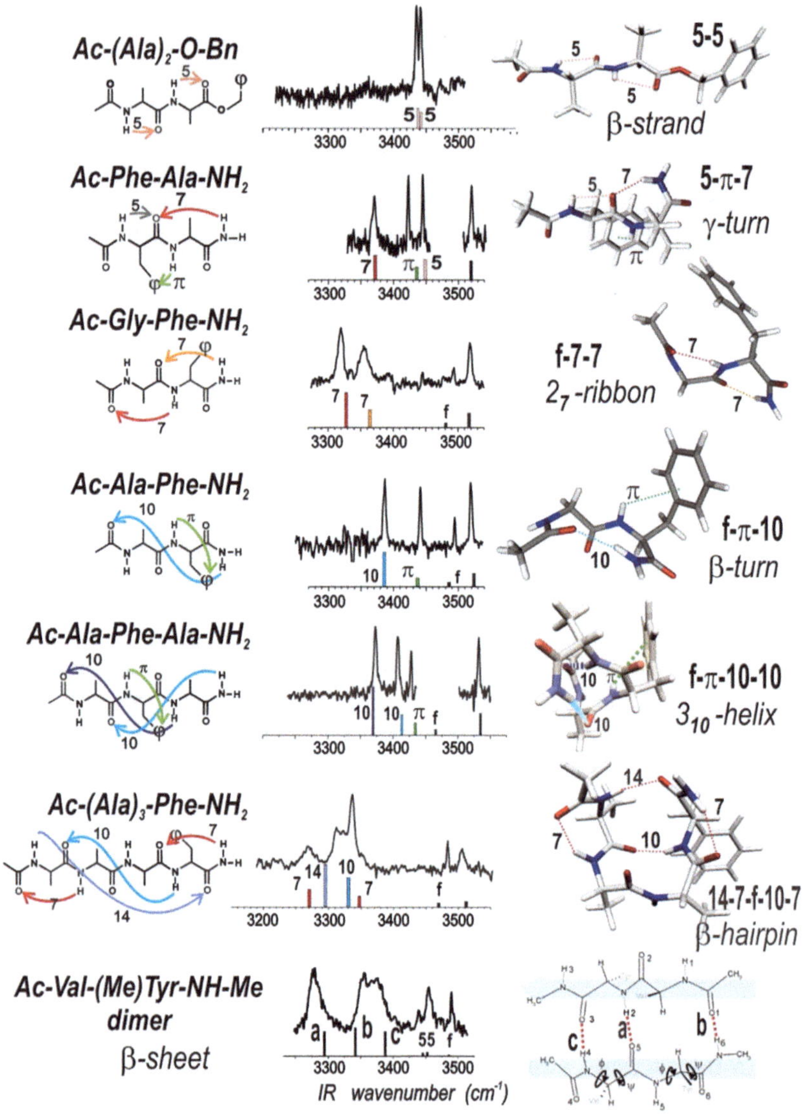

Fig. 4 Secondary structure gallery, from gas phase IR/UV laser spectroscopy of a series of model capped peptides: β-strand [135], γ-turn (in combination with a nearby β-strand N-terminal [84, 94]), 2_7 ribbon [82], β-turn [81], incipient 3_{10} helix [85], β-hairpin [107] and antiparallel β-sheet model [124]. Spectra of Ac-(Ala)$_2$-O-Bn and the Ac-Val-(Me)Tyr-NH-Me dimer are adapted from [135] and [124] respectively, with the permission of ACS. Spectrum of Ac-(Ala)$_3$-Phe-NH$_2$ is adapted from [107] with the permission of Wiley

secondary structure [135] (Fig. 4, top). In addition, the folding from the NH of the second Ala residue into a C7 H-bond is hampered by significant entropic effects which favour extended forms because of their multiple low frequency modes.

Amidated models, [63, 70, 73, 83, 114] as well as di- or tri-peptides with Phe or Trp as first residue [84, 85, 94, 104, 112], have also revealed C5-containing secondary structures. In these conformations, the C5 H-bond occurring at the aromatic residue is additionally stabilised by an interaction between the π system of the aromatic ring and the NH of the following residue along the chain , making the 5-π motif very common among these secondary structures. C5s are characterised by a usually weak, narrow (~2 cm^{-1}) NH stretch band usually located in the narrow range 3,430–3,455 cm^{-1} (see Fig. 4, the two narrow lines of the extended Ac-Ala-Ala-O-Bn), which can sometimes be mistaken for a NH-π interaction (see Sect. 3.2).

3.1.2 C7 H-Bonds, at the Heart of γ-Turns

These H-bonds, which are centred on amino-acid residue *i*), actually span over three residues, linking the NH(*i* +*1*) and the CO(*i*-*1*) moieties of the backbone. They are quite easily observed in the gas phase, in capped amino-acids [63, 70, 73, 83, 114, 131, 188], or often in combination with other secondary structures, for instance following an extended Phe residue with a C5 interactions leading to 5-π-7 structures [70, 84, 85, 94, 112] (Fig. 4). Their formation is made possible by the flexibility of the backbone around residue *i*, or, when less flexible, as in the case of Pro, by convenient structural constraints. The interatomic H–O H-bond distance is in the 200 pm range, depending upon the residue, testifying a significant interaction strength. The corresponding signatures are much broader than the C5 (typically 10 cm^{-1}), leading to apparently weaker depletion signals, in spite of having a significantly more intense oscillator strength according to calculations. C7 features are chiral, existing under both direct and indirect enantiomers, nearly isoenergetic in achiral residues. For chiral residues, one form becomes more stable and is labelled γ_D or γ_L according to the residue chirality (D vs L [94]). In residues of natural chirality (L), these forms differ by the position of the side chain relative to the C7 H-bond ring, with an equatorial or axial position for γ_L or γ_D, respectively. The usual large energy difference between these forms explains why only γ_L is generally observed in the gas phase. Their NH stretch frequency is sensitive to the nature of the residue, i.e. the nature of the side chain in the equatorial position; steric constrains in bulky residues and Pro giving rise to larger red shifts [84]. When not combined with other strong H-bonds, γ_L frequencies range typically from 3,380 cm^{-1} for Gly [84] down to 3,330 cm^{-1} for Pro. Their γ_D counterparts are red-shifted by 30–40 cm^{-1}, as shown by IR spectroscopy of the α-amino-isobutyric acid (α-dimethylglycine), an achiral non-natural residue, which revealed that both γ_L and γ_D structures of Aib (which both exhibit a side chain in an axial position) have a frequency of ~3,340 cm^{-1}, suggesting that the C7 H-bond frequencies are also sensitive to the (equatorial vs axial) location of the side chain [94].

3.1.3 C10 H-bonds, the Stabilizing Interactions of β-turns and 3_{10} Helix Structures

These H-bonds are strong and relatively short, with H–O distances in the 210–220 pm range. Their characteristic NH stretch signature consists of medium intensity, moderately broadened (~5 cm^{-1}) bands, typically located in the 3,405 cm^{-1} region, or 3,390 cm^{-1} in the case of NH_2 C-terminals, with little dependence on the residue content. C10 examples are provided by β-turns, within a series of Phe-containing dipeptides [81, 82, 92, 103, 109, 111] (see Figs. 4 and 8). As β-turns are centred over two residues, they are penalised in short peptides by the weak number of H-bonds per residue: only one of the three backbone NHs of a capped β-turn-folded dipeptide is stabilised by an H-bond. In order to compete efficiently with multiple H-bond motifs such as f-7-7 or 5-π-7, additional interactions are required. This can be achieved either by H-bond acceptor groups on the side chains, which stabilize the backbone NH groups, or by bulky residues, enabling strong side chain/side chain interactions (see Sect. 3.3). The β-turns belong to four main types, corresponding to different relative orientations of the three peptide bonds of the turn. Their relative stability is strongly dependent upon the chirality of the backbone residues at the turn. Whereas the amide A region provides direct information of the H-bond strength, it is insensitive to the type of turn. Amide I and II regions, in contrast, appear much more discriminating for the turn type assignment [52, 192].

3.1.4 Combinations of Elementary Motifs

These combinations are quite common in most of the di- and tri-peptides studied. The following repetitions of local features have been reported: short 2_7 ribbons [80–82, 85, 121], 10–10 motifs including 3_{10} helix [85, 93] as well as successive intertwined C10 turns [80] and mixed combinations of different local structures 10–7 or 7–10 in tripeptides [85, 121, 138, 139].

When the chain goes longer, head-to-tail interactions can take place, giving rise to additional H-bonding features [98, 107, 134, 139]. Noticeable are β-hairpins exhibiting antiparallel C10 and C14 H-bonds stabilising the chain reversal [107, 134]. C13 H-bonds could also be formed in tripeptides and larger species, but have not yet been observed because of the intrinsically high energy of this conformation in short peptides. Observation of C13 H-bonds and α-helices in neutral species is still pending until a short sequence stabilising these features can be found.

The comparison between isolated features and the same feature embedded in a more or less complex H-bond network enables one to document the existence of cooperative electronic effects, presumably caused by an electronic polarisation of the amide groups favoured by the H-bond network, especially when a circular chain is formed [82, 107, 121, 134, 139]. As expected for a vibrational motion coupled to the rest of the molecules, the linewidth associated with these coupled oscillators is often significantly broadened compared to the poorly coupled oscillators such as those engaged in a C5 interaction.

Finally, peptide dimers characterised in the supersonic expansion also provide evidence for the stabilisation of β-strands by interstrand H-bonds [31, 46, 64, 73, 90, 124]. This is interesting when isolated monomers are found to adopt folded structures, thus illustrating the intrinsic flexibility of peptides and their sensitivity upon the environment.

3.2 Backbone-Side Chain Interactions in Proteins

The H-bond donor (NH_{bb}) and acceptor (CO_{bb}) groups along the backbone peptide chain may alternatively form H-bonds with donor or acceptor sites located on side chains instead of taking part in the H-bond network of the secondary structure. Often established between close lying sites, these backbone-side chain interactions are capable of orientating side chains and influencing backbone local folding. They can also take place between more remote parts of the chains or between different chains contributing to tertiary or quaternary contacts, respectively.

NH_{bb}-π interactions are nearly ubiquitous in the species studied by IR/UV spectroscopy. The necessary UV chromophore in these studies is either one of the natural aromatic residue (Trp, Tyr, Phe) or a specific tag, most often based on a phenyl ring. These interactions, sometimes referred to as "π H-bonds", give rise to narrow bands, usually in the 3,420–3,440 cm^{-1} range, depending upon the interaction strength [63, 70, 73, 81–85, 94, 109, 111, 114, 135]. The NH moiety involved usually points towards the π-electron cloud, with a relative arrangement depending upon the degrees of freedom existing along the chain between the two partners (Fig. 5). Various situations have been observed, ranging from strongly constrained intraresidue NH_{bb}-π bonds to an almost unconstrained interaction between remote partners, enabling an NH orientation normal to the aromatic ring [135]. Stronger π H-bonds (in the 3,380–3,420 cm^{-1} region) can be encountered in aromatic-rich peptides when the same NH interacts with two aromatic clouds in a sandwich-type motif [112] (Fig. 5).

These interactions are quite good candidates for a visualisation using recent theoretical tools based on topological analyses of electron density. Especially appealing is the recently developed NCI (for non-covalent interactions) procedure [193], which provides an elegant picture of the points of close contacts for both inter- and intra-molecular interactions [194] deduced from isosurfaces of reduced electron density gradient. Figure 5 (bottom right) shows the isosurfaces associated with the NH-π interactions, which makes more tangible the close contact between the NHs involved and π clouds. In this capped Trp-Tyr dipeptide, NCI supports the picture of a potential general shaping motif between three distinct partners where the NH_{bb} is sandwiched between the Trp and Tyr π clouds [112].

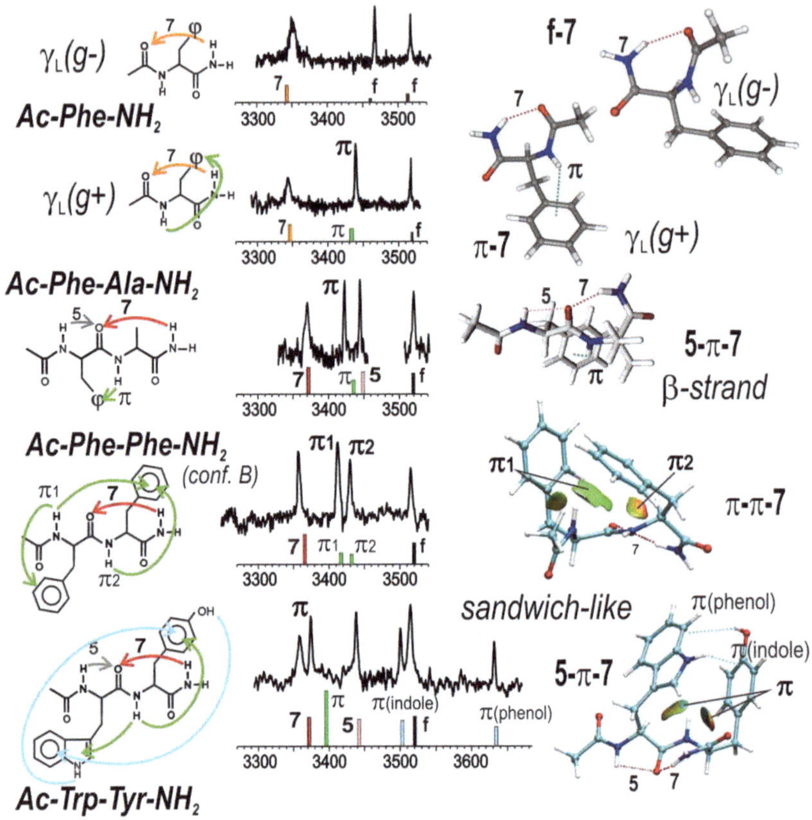

Fig. 5 Gallery of NH-π interactions between an aromatic residue (Phe, Tyr or Trp) and a backbone amide NH bond, from gas phase IR/UV laser spectroscopy of a series of model capped peptides: a folded backbone (γ-turn), without and with a NH-π interaction within the Phe residue [70], an extended backbone on Phe, with the interaction of Phe with NH of the next residue [84, 94]) and various aryl-rich dipeptides, leading to sandwich-like conformations of the NH embedded in the aryl environment and to significantly red shifted IR NH stretch signatures [112]. In the latter cases, the simultaneous presence of two π H-bonds originating from the same NH moiety (made possible through a direct interaction between aromatic side chains) is indicated by the 3D-NCI plot of these interactions [193, 194]

Interestingly, gas-phase studies of NH_{bb}-π bonds enable one to tackle general trends which are usually deduced from data mining. In proteins of the Protein Data Bank, the phenyl ring of the Phe residue (i), binds twice as frequently to the NH ($i + 1$) as to the NH(i) group. As a matter of fact, the conformation of the isolated Ac-Phe-NH_2 residue characterised by a C5 interaction stabilised by such a NH ($i + 1$)–Phe interaction is by far the most stable [83].

Other backbone-side chain interactions have been documented for specific residues. A C6 H-bond linking the S atom of methionine to its NH_{bb} group has been observed in model capped dipeptides [109], mimicking a folding pattern also found in proteins [195]. The IR spectral signature of the NH–S(Met) interaction is significantly red-shifted relative to the C10 features for instance, pointing out its potential strength and relevance as a major interaction, deserving consideration in protein modelling.

In turn, the much shorter side chain of cysteine and serine residues (which differ in the nature of their heteroatom) only enables C5 H-bonds between the $NH_{bb}(i)$ group of the residue and the heteroatom (sulphur or oxygen, respectively), leading to stronger interactions than observed for intrabackbone C5 NH_{bb}–OC_{bb} H-bonds [118]. In parallel, the SH and OH groups can also establish a C5 H-bond with the carbonyl atom $CO(i)$ of the residue [116, 118].

3.3 Side Chain–Side Chain Interactions in Proteins

Side chain–side chain (SC–SC) interactions also play an important role in the backbone folding. In the presence of a workable vibrational probe in the side-chain (NH in Trp, OH in Tyr, OH in Ser or Thr, SH in Cys, $CONH_2$ in Asp or Glu, etc.), SC–SC interactions can be directly monitored. For instance, the moderate red-shifts simultaneously observed for both OH(phenol) and NH(indole) bands in the capped Trp-Tyr dipeptide revealed, respectively, the OH-π and NH-π interactions characteristic of the face-to-face arrangement of the aromatic side-chains [112]. Beyond H-bonding, IR spectroscopy is also sensitive to other interactions. A spectacular one is the formation of an intramolecular salt bridge, corresponding to an intramolecular zwitterion formation after an intramolecular proton transfer between an acidic (Glu) and a basic side chain (Arg) [105, 141], revealed by the spectral carbonyl signatures.

Although some SC–SC interactions are probably amenable to IR spectroscopic investigations, others are more difficult to prove. Namely, aromatic–aromatic [103, 111, 112] or aromatic–alkyl [82, 94] interactions not only shape the molecule but also impact the conformational landscape by significantly changing the relative conformational energetics. Unfortunately, because of the lack of workable vibrational probe on these groups, these major interactions are only remotely detected through the constraints they bring to the regions where the vibrational probes lie [103]. The frequent and stabilizing CH–aromatic interactions [194] should provide useful evidence, but are in practice difficult to identify from the sole CH stretch spectroscopy, particularly because of the large coupling between the numerous CH oscillators of a peptide.

3.4 Terminal-Controlled Structures

Natural terminations ($-NH_2$ and $-COOH$) of peptides have been well documented in the gas phase, and have been quite popular, particularly because of their typical

NH and OH stretch signatures [16, 20, 26, 28, 56, 74, 77, 79, 88, 89, 95, 96, 101, 106, 113, 137, 138, 140]. Other studies [22–25, 66, 69, 75, 97, 142] have also illustrated the flexibility of these terminations, giving rise to several OH–N or NH–O bonds. This extreme flexibility generally leads to a high number of low-energy lying conformations [99, 160, 161, 166], rendering the structure determination quite cumbersome for large species, since it requires an extensive structural exploration and a subsequent structural optimisation, taking dispersive interactions into account. In dipeptides, the natural terminal groups establish H-bonds with backbone amide groups, leading to specific structures far from those observed in the analogous capped dipeptides. In this respect, the edge effects caused by the terminal overcome and mask the intrinsic propensities of the backbone, disqualifying these species as models for protein segments.

3.5 Microsolvation Structures and Complexation

Hydration is a classical topic of quantum chemistry, and studying hydrates of increasing size isolated in the gas phase is an elegant way to assess the gradual effects of hydration as a function of the molecular environment. Beyond the characterisation of the hydration sites of model peptide bonds, whose structure [15], energetics [17] and barriers between them [180] have been determined, the approach is especially appealing for flexible molecules, such as peptides, for which dramatic effects can be anticipated, even for a modest environment. Theoretically predicted some time ago [196], experimental illustrations of these effects have been published only quite recently. Several groups have shown that peptides or related flexible molecules undergo a significant structural change upon solvation, as indicated by changes in their H-bonding network [53, 55, 102, 129, 136].

One of the most dramatic studies, carried out on a capped amino-acid solvated by one water molecule, documented the effect of microhydration and its dependence on the solvation site occupied by the water molecule for each conformation. It provided evidence of a complete conformational distribution reversal where a folded backbone form overcomes an extended structure as the most stable when solvated by a single water molecule [53]. The presence of this first molecule, which bridges the two ends of the protected peptide (Fig. 1c), enables the establishment of a daisy chain of H-bonds within the molecule, contributing to an enthalpic stabilisation, induced by cooperative effects. This type of experiment on hydrates is, however, difficult to interpret and requires heavy reliance on vibrational spectral data to be successful, at least in terms of an unambiguous assignment. Experimental results beyond a single solvent molecule remain seldom; the interpretation being complicated [55, 102, 136] by vibrational coupling between solvent molecules.

Beyond microsolvation, peptide–peptide interactions [31, 46, 73, 87, 90, 120, 124] and complexation with larger molecules have also been considered by several groups, especially to describe specifically interesting interactions between a peptide and its biological environment: peptides have been complexed with pyrazole to favour β-sheets [65, 100], with sugars to describe the interactions responsible for cellular recognition in glycopeptides [114, 197], or to document the anomeric effect [198].

3.6 Beyond Short Linear α-Peptides

Natural and capped α-peptides have played the role of test molecules on which to assess the potentiality of IR gas phase spectroscopy. It has quickly been extended to other species, beyond the simple canonical structures. Short cyclic peptides have been investigated, documenting the *cis* conformation of the peptide bonds [76, 78, 129], which is usually too high in energy to be observed in linear peptides. More recently, other peptides, with longer spacers between amide bonds, i.e. β- and γ-peptides, having respectively two or three methylene spacers between their peptide bonds, have been investigated, either alone or in combination between them or with alpha peptides [52, 104, 122, 123, 125–128, 143–145]. More flexible C11, C8 and C6 H-bonded structures in β-peptides as compared to their C10, C7 and C5 equivalents in the regular α-peptide are a general characteristic of these species. As another consequence of the increased flexibility provided by flexible spacers between amide groups, amide staking became possible and has indeed been revealed as a new shaping interaction on γ-peptides [125–128]. In turn, cyclic spacers bring more constraints which leave fewer folding choices for the peptide chain [143–145], paving the way for a controlled folding of larger species.

3.7 Pushing Gas Phase Investigation to Its Limits

Going further in terms of size is, of course, a challenge, first because both UV and IR spectroscopy become barely resolved for a series of reasons. First, experiments using thermal vaporisation are hampered by the need to heat up the carrier gas; alternatively, a laser-desorbed set-up can use room temperature or cooled carrier gas, but start from much hotter molecule conditions, even if desorption is accompanied by a significant expansion of the desorbed matter [185, 199]. These hot conditions do not favour the cooling of the molecules, inasmuch as they exhibit a large flexibility, measured in terms of the number of low frequency modes, and hence a large internal energy content. The number of collisions scaling as P_0D/T_0, where P_0, T_0 and D are the generating pressure, temperature and nozzle diameter, respectively [184], means that the cooling capabilities of the expansion are fixed for given generating conditions. This consideration leads to the notion of critical system size, for which the internal energy content of the vaporised molecules is no longer efficiently removed. In addition, species which are laser-desorbed downstream of the nozzle meet an already partially expanded carrier gas, and thus experience relatively fewer collisions, limiting the cooling efficiency. The spectroscopic consequence of incomplete vibrational cooling is congestion of the UV spectral features, leading at best to a decrease of the UV absorption and more likely to a lack of conformational selectivity in the optical excitation. Together with the broadening of the UV feature, a decrease in the IR/UV double resonance signal should also be anticipated. Indeed the depletion signal, which is essentially due to the difference in probe intensity between cold and hot molecules, will therefore drastically be affected in presence of an incomplete cooling. From the IR point of view, vibrational spectroscopy is relatively more robust relative to temperature effects; however, since the number of IR chromophores increases linearly with the molecular size, vibrational congestion is always an issue.

Few examples of large molecules for which resolved or partly resolved features have been observed. One can cite tri- [80, 85, 88, 106, 112, 121, 124, 133, 138–140], tetra- [107, 129, 134] and pentapeptides [89, 98, 101, 107, 139], and larger systems up to gramicidins [54, 79, 89, 108]. In all cases, in addition to the potential lack of conformational selectivity, the theoretical investigation required for the assessment is a cumbersome task, which is not always successful. In most cases, the interpretation of IR spectra consists of providing clues about the general typology of the forms observed, i.e. helical vs extended or coiled (Fig. 1d).

4 Added Value of Conformation-Resolved Studies

4.1 *Flexibility*

4.1.1 Probing the Flexibility

Flexible systems, by definition, have at least two energetically low-lying conformations (thermodynamic criterion) separated by an energy barrier that can be easily crossed (kinetic criterion). Focusing first on the necessary conditions to have quasi isoenergetic minima, evidence for the flexibility of isolated peptides lies in UV spectroscopy: conformationally-resolved UV signatures properly identified from IR/UV or UV/UV double resonance experiments provide apparent conformer populations which can be regarded as reflecting the conformational distribution in the molecular beam. Despite the questionable character of this approach (see Sect. 2.3), energetics are often qualitatively in line with the observed abundances. This suggests that the employed theoretical method gets the physics of the interactions, and that experimental artefacts do not occur systematically. Practically, the coexistence of conformers in the expansion generally turns out to be related to neighbouring Gibbs energies, namely within ~5–10 kJ mol^{-1}. The conformations differ either by the type of peptide backbone (folding) and/or, more trivially, by the orientation of the side chain relative to the peptide backbone. In all cases, the number of conformations reported remains small, usually two or three, partly because of ignored weak signals resulting from a low signal-to-noise ratio, especially with laser-desorbed samples. Figure 6 shows the remarkable example of a

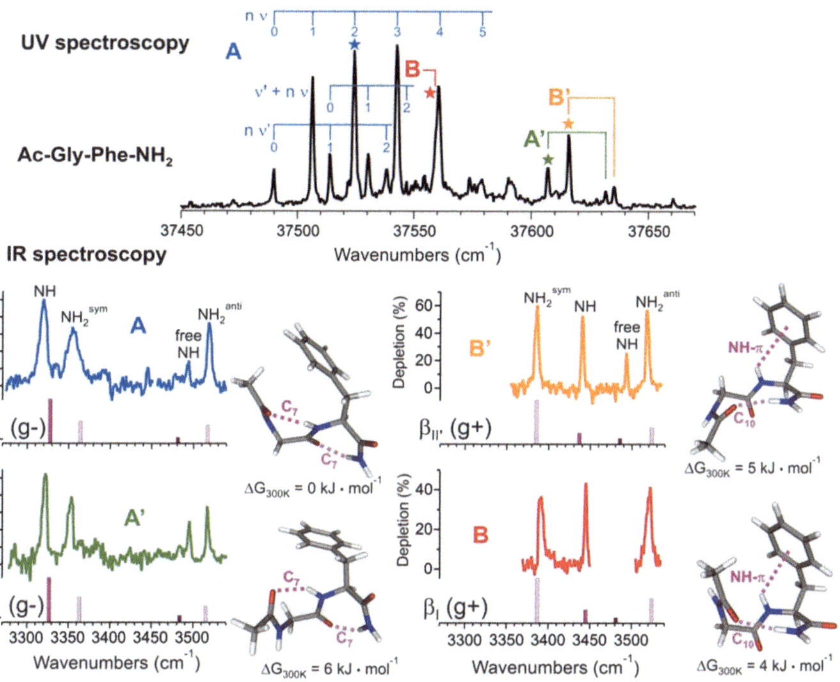

Fig. 6 Conformational diversity in the Ac-Gly-Phe-NH_2 capped dipeptide. The R2PI spectrum (*top panel*), combined with fours IR/UV experiments (*below*) carried out on the UV bands marked with an *asterisk* provides evidence for the existence of four conformers in this system, whose apparent intensities in the UV spectrum (one major conformer A and three minor ones) are fairly correlated to the calculated Gibbs energies (A is the most stable by more than 4 kJ mol^{-1}) [200]

capped Gly-Phe sequence, where four conformations with different types of backbones are found: two f-7-7 (double γ-turn) extended conformations, differing by the chirality L or D of the turns, and two β-turn folded structures of types I and II'. The latter differ only by the backbone folding on the Gly residue, and share the same backbone conformation and side chain orientation on Phe [200].

This conformational diversity is clearly caused by the specific flexibility of the Gly conformation, which does not suffer from any steric hindrance produced by the absence of a side chain. The same study on a capped Ala-Phe sequence (see Fig. 8) exhibits only two conformers; those two in which a steric hindrance appears when substituting Gly by Ala are rejected at energies too high to be observed. Put in other terms, γ_D-turns and type II' β-turns are specific to Gly residues and are not observed in other natural residues because of the presence of a side chain [200].

Flexibility can also be elegantly addressed by more sophisticated experimental techniques focusing on photoinduced isomerisation in the early stages of a supersonic expansion. The pioneering work of the Zwier group demonstrated how IR or UV/UV pump-dump excitations (SEP, Fig. 7) in the early expansion can be used to measure accurately conformational distribution together with isomerisation

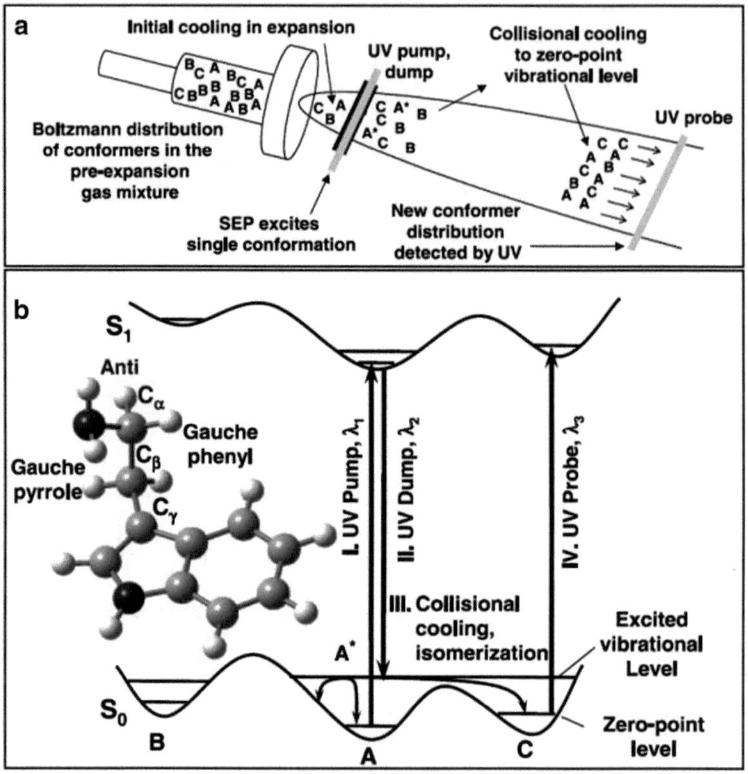

Fig. 7 Set-up (*top panel*) and principle (*lower panel*) of population transfer experiments: SEP or IR excitation causes a conformational selective excitation in the expansion (Steps I and II), isomerisation take place (step III) and products are ultimately probed by UV spectroscopy in the late expansion (step IV). Extracted from [149]

quantum yields on isolated peptides [47, 71, 72, 125, 126]. The principle (Fig. 7) stems from the fact that sufficient cooling is already achieved in the early expansion, allowing optical excitation to take place under cold spectroscopic conditions. Then, in a modest collisional regime, conformational isomerisations can take place and the new conformational populations are eventually cooled down. The usual late expansion UV spectroscopy can detect the population transfers, through "hole burning" or "hole filling" effects. A variety of interesting information can be extracted from these experiments. First, they document qualitatively the topological connectivity linking the several conformers observed and provide an experimental counterpart to the theoretical disconnectivity diagrams describing these conformational basins [71]. Furthermore, these experiments can result in a quantitative analysis of the population transfers between conformers following IR photoexcitation, by comparing experimental results with a statistical model of conformational cooling in the expansion [72].

Despite the lack of any spectacular selectivity with the vibrational mode excited [47, 71], this approach provided a quite interesting by-product: a measurement of conformations populations. Once measured, the photo-induced population transfers to each of the conformers detected is worked out, and a detailed balance analysis of all the contributions enables one to estimate precisely the conformational abundances, thus avoiding the bias produced by a conformer-dependence of the probe (ionisation or fluorescence) efficiency [47, 71, 72, 125, 126]. It should even be noted that this strategy should also allow one to detect the presence of a major dark conformer: in this case, even if IR photoexcitation only causes a small change in the population of the major form, this would lead to a significant increase in the population of other forms, easily detectable as an significant increase in their probe signals. Although a quantitative measurement of their abundance is not feasible, a significant population of the minor form (hole filling) would reveal the existence of the dark species, together with their IR spectroscopy.

The same type of experiments can also be used to provide quantitative energetic information. The control of the excess energy deposited in the system, especially in UV/UV pump-dump experiments, together with the detection of the isomerisation (or population transfer) threshold, ultimately led to a convincing measurement of the relative energies of the conformers observed, together with the height of the barriers between them. This has been carried out for small biomolecules [149, 181], but not yet for molecules of the size of a peptide.

4.1.2 When Side Chains Operate a Fine-Tuning of the Backbone Structure

A survey of the published spectroscopy of Xxx-Phe and Phe-Xxx capped isolated sequences provides a rich illustration of the peptide backbone flexibility (Fig. 8). While the coexistence of a major conformer with an extended backbone (5-π-7 or f-7-7) vs a minor conformer with a folded backbone (β-turns) appears to be a general trend for the simplest residues Gly and Ala [82, 84], this picture starts to change when more complex residues are considered. Val and Phe are more bulky residues capable of forming aromatic–alkyl or aromatic–aromatic interactions between side chains which further stabilises the β-turn enough to make it the major conformation observed [82, 84, 103, 111]. In Phe-Phe or Phe-Tyr sequences, this effect is still enhanced by the existence of additional NH-π interactions, which lead to the exclusive observation of the β-turn [103, 111]. More generally, H-bond acceptor groups within the flexible side chain, as in Met [109], Cys or Ser [116, 118], bind to nearby backbone NH and stabilise the β-turn, which again turns out to be the only/major conformation observed. In this context where analysis of the conformal distribution provides a deeper insight into peptide backbone diversity and its sensitivity to nearby interactions, study of high energy conformers is then of special interest. They are produced in the supersonic expansion and observed as minor forms when performing a relatively faster cooling which traps these conformers populated at high temperature. Recent experiments carried out on the

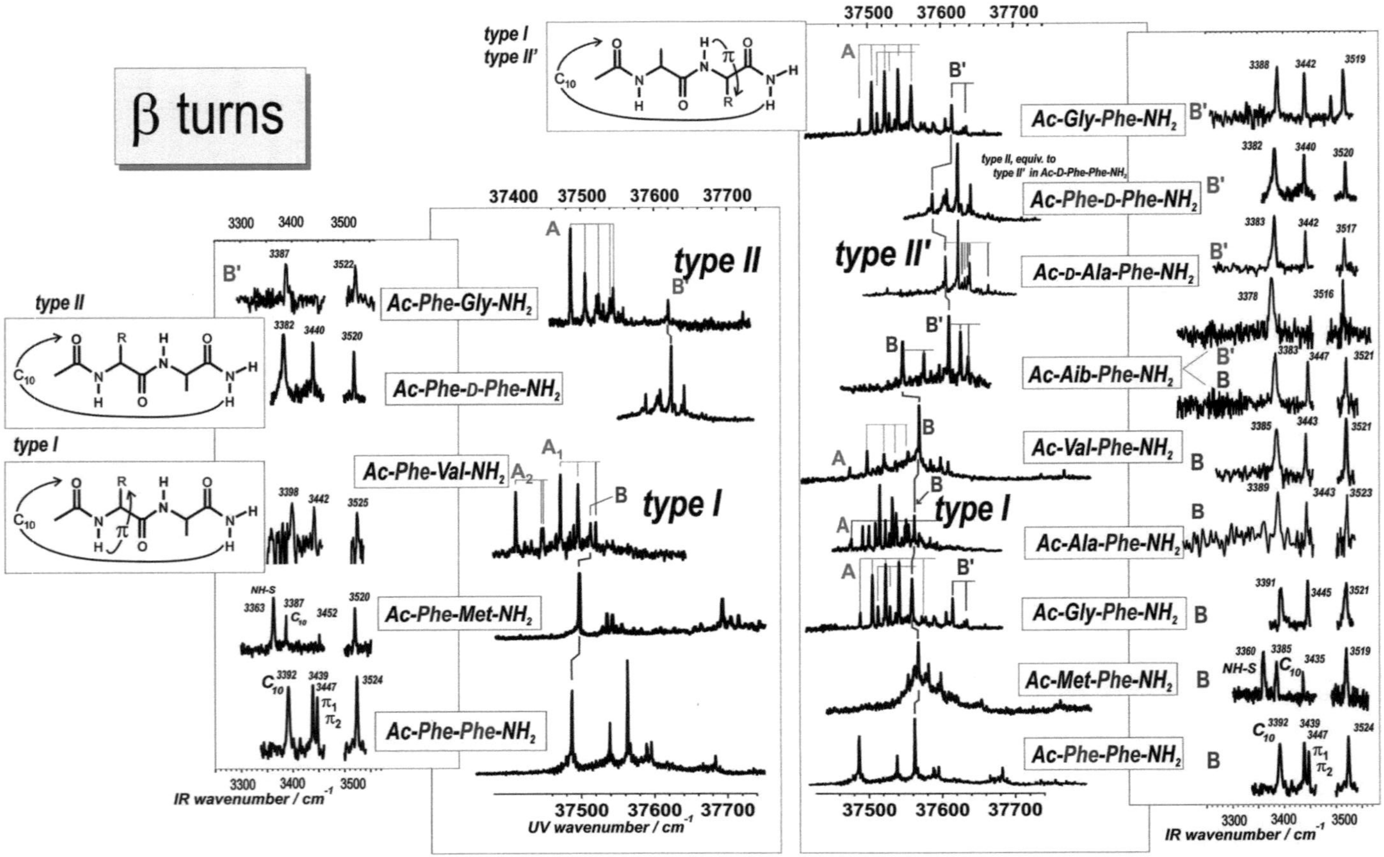

Fig. 8 β-Turn gallery: near UV and amide A IR region signatures of β-turn conformers in a series of Phe-containing model dipeptides (chiral residues are of L-chirality, unless indicated as D-). Type I turns are observed when Phe is in the first or in second position. Type II and II′ are characteristics of Gly (and of achiral residues, such as Aib, the iso-amino-butyric acid) in second and first position respectively, but can also be encountered in the presence of residues of D-chirality. These spectra reveal that the UV spectroscopy (band origin position) is barely sensitive to the type of residue, and only depends on the type of turn, as expected since the type controls the Phe chromophore environment [82, 84, 92, 103, 109]

Phe-Phe sequence in a He/Ne carrier gas mixture, together with an extensive theoretical conformational analysis, have revealed how specific NH-π and π-π interactions stabilise other structures than β-turns, namely π-π-7 and 5-π-7 backbone forms (Fig. 9) [112]. This picture of ancillary interactions governing conformational distribution seems to hold on larger species, as testified by the systematic study of capped tripeptides, Phe-Ala-Ala, Ala-Phe-Ala and Ala-Ala-Phe, which shows how the presence of an NH-π interaction can guide the backbone overall folding, and in this case control the formation of an incipient helix [85].

Interestingly, residue chirality can also contribute to backbone flexibility [92] through slight modifications of hindrance effects and NH-π interactions between homochiral and heterochiral sequences. Thus the type (I vs II′) of β-turn in Xxx-Phe sequences is controlled by the chirality of Xxx: while both forms coexist in the

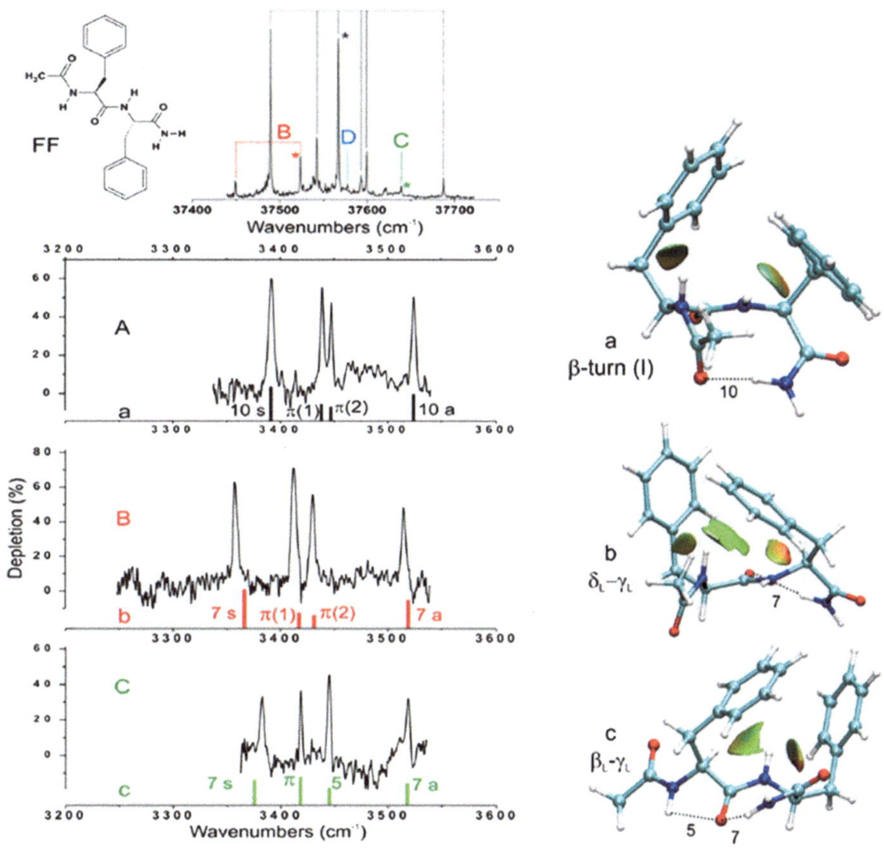

Fig. 9 Diversity of possible backbones in an aromatic-rich dipeptide, Ac-Phe-Phe-NH_2: *Top left*: UV spectrum from R2PI experiment; *bottom left*: the three IR spectra obtained on the main bands of the three conformers, and, *right*, corresponding structural assignment, with the emphasis put on the NH-π interactions, indicated by partial 3D-NCI plots, pointing out the close contacts between the H atom and its aromatic cloud environment. Adapted from [112]

presence of the achiral Aib residue, only type I β-turns are found for Ala. For D-Ala, not only are type II′ turns found, but no other conformation is observed, unlike the Ala and Aib cases. The formation of β-turns induced by heterochiral sequences is a property currently used for purpose-synthesised molecules, including drugs.

4.1.3 Environment and Flexibility

Interactions with species external to the peptide chain (solvent, ion, another peptide chain or any other type of biomolecular environment) are also expected to guide backbone folding and impact peptide flexibility. In this scope, complexation of model peptides [46, 53, 55, 65, 73, 87, 90, 100, 102, 120, 124, 129, 136] shed light on the interface between a protein and its environment.

In this respect, microhydrated peptides have been found to be quite a fruitful study-case. Monohydrates observed in a supersonic expansion are mainly folded in a slightly distorted γ-turn-like form, where a water molecule bridges the free CO and NH of the two consecutive amide groups involved in the γ-turn (Fig. 10) [53]. The distortion consists essentially of a ~30 pm elongation of the intramolecular H-bond, which illustrates the necessary structural compromise found by this flexible system to satisfy both inter- and intra-molecular H-bonds. This propensity of a γ-turn to be distorted to accommodate the water molecule is the very same process expected from a protein binding to a ligand within the induced-fit theory. It documents a slightly different aspect of molecular flexibility which does not characterise the ability to switch from one secondary structure to the other, but which consists of moving around an average geometry along soft vibrational modes, typically backbone motions. Interestingly, isolated and monohydrated γ-turns can be interpreted as two extreme situations where respectively zero or maximum constraint is brought to the peptide chain by the solvent. In a fully solvated γ-turn, intermediate situations can then be anticipated leading to a range of possible intramolecular H-bond distances, which testify to the increased flexibility brought by the environment.

Additionally, conformational population analysis revealed that peptide chain folding could be triggered by putting together one extended form and one water molecule [53]. This microsolvation-induced isomerisation on the residue scale

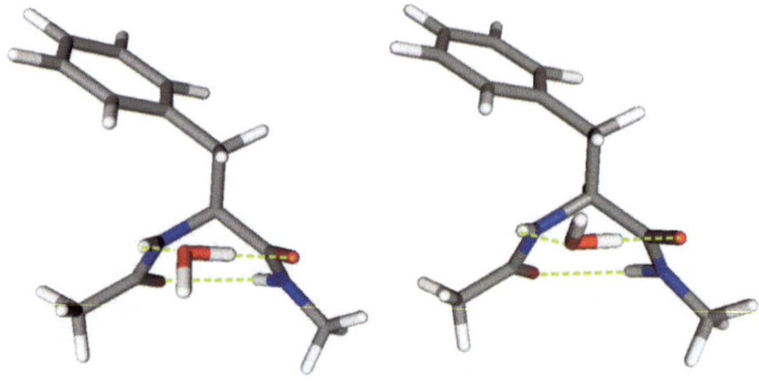

Fig. 10 Structure of the main forms Ac-Phe-NH_2 monohydrates observed in the expansion, as deduced from IR/UV double resonance spectroscopy (see [53])

illustrates the strong contribution of solvent in protein flexibility. One should note that the predicted most stable hydrate corresponds to a water molecule inserted in the intramolecular H-bond of the γ-turn, a structure which has not been observed. Kinetic trapping does not favour water insertion because of a too high barrier under the expansion conditions. However, one cannot exclude the possibility that these complexes, although being formed, cannot be detected because of a too short-lived excited state (cf. Sects. 2.3 and 4.2). The sub-nanosecond lifetime estimated for the hydrates [200] suggests that photophysics effects deserve to be taken into consideration.

4.2 *Excited States*

4.2.1 UV Spectroscopy: A Source of Assignment Information Still Too Often Neglected

The major interest of the IR/UV double resonance approach stems from the dependence of UV spectroscopy on the chromophore environment. If one examines it, it provides useful indication for structural assignment. Figure 11 shows the

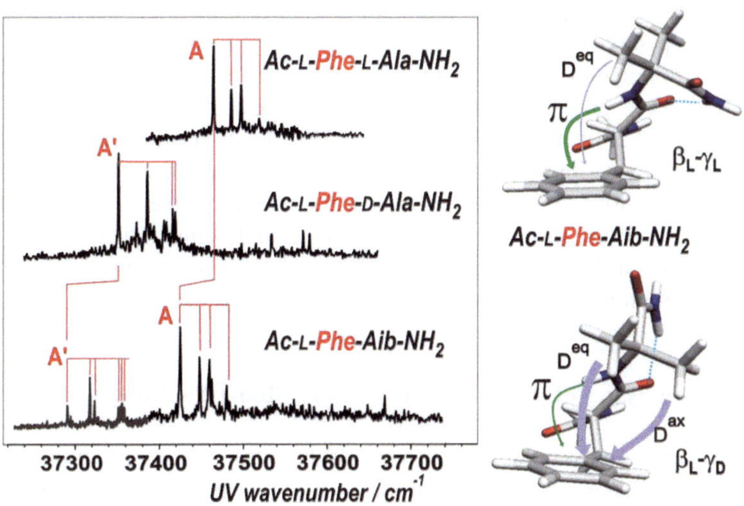

Fig. 11 Interresidue chiral recognition in Ac-Phe-Xxx-NH_2 (Xxx=L-Ala, D-Ala and Aib) capped dipeptides. All these species possess the same β_L-γ feature, with an extended backbone over the phenyl ring and a γ-turn on the Ala-based residue. The R2PI spectra (*left*) shows that, with the achiral Aib residue, two spectral features (A and A′) are observed, produced by the simultaneous presence of conformations with γ_L and γ_D turns; only one remains in chiral residues, illustrating the preference of these chiral species for a well-defined fold (γ_L in L residues and γ_D in D ones; A and A′ features, respectively). These two forms (*right*) differ by the interactions between the turn and the phenyl ring (NH-π and dispersive interactions D^{ax} and D^{eq} between the axial and equatorial methyl groups respectively and the Phe ring), leading eventually to a spectral recognition of the turn chirality by the neighbouring Phe residue [94]

spectroscopic features of the lowest energy $\pi - \pi^*$ transitions of the Phe chromophore within a series of Phe-containing dipeptides. It provides a good example of the information collected, namely the frequency of the UV transitions and the FC vibronic pattern measured. Both are quite characteristic of the Phe environment. A Phe residue in a β_L(a) conformation, provides a characteristic progression of two to four bands separated by ~20 cm^{-1}, assigned to the side chain torsional motion. The absolute position of the intense origin band depends slightly upon the nature and number of the following residues within the sequence. This property has, for instance, been used as an indicator for chiral recognition of γ_L or γ_D turns on L- or D-Ala residues by the preceding Phe residue within the sequence (Fig. 11) [94].

A Phe residue with a γ_L(g-) side chain conformation in an Ac-Xxx-Phe sequence gives rise to a long progression caused by the dispersive interaction between the Phe side chain and the N-terminal acetyl group (Fig. 6). Besides these easily identifiable FC signatures, β-turns exhibit more discrete signatures, with a very modest FC pattern. However, they lie in well-defined spectral regions, characteristic of the type of turn and of the position of the chromophore within the turn. Figure 8 shows examples of these characteristics, which help a great deal to find the turn signature when hidden in a congested region dominated by another conformer (e.g. type I β-turns in Ala-Phe and Gly-Phe sequences [92, 200]). Phe-Phe type I β-turn also provides a nice illustration, where the origin positions of each chromophore match well the typical positions of Phe transitions found for all the other type I β-turns: ~35,500 and ~35,560 cm^{-1} for Phe in first and second position of the sequence, respectively.

The development of excited-state calculations for molecules of the size of dipeptides, made possible through the use of quantitative quantum chemistry methods (e.g. CC2 [201]), should foster the use of UV spectroscopy as a complementary assignment source. However, these methods still need to be validated on flexible systems, for which making reliable predictions of the spectroscopic variations with the conformation can be a challenge. Model systems such as Phe-Cys, for which large spectral shifts of typically 200 cm^{-1} have been found [116, 118], are trial models of choice to assess these excited-state methods.

4.2.2 Photophysics of Aromatic Residues in Model Peptides

Although important from the point of view of the intrinsic stability of proteins upon UV irradiation, this issue has been only seldom addressed in gas phase experiments. Attention has been drawn to the field by pioneering excited-state calculations in the past decade conducted on peptides [177, 202–209], following an approach already successful for showing the role of $\pi\sigma^*$ excited states in the deactivation of aromatic compounds, such as phenol or indole [210]. These authors emphasised the possible role of several excited states, namely a locally excited (LE) state and an intrabackbone charge transfer (CT) state to convert efficiently the photodeposited electronic energy into vibrational energy. The proposed deactivation mechanism involves a series of conical intersections (CI), which enables sequential non-radiative transitions from the optically excited $\pi\pi^*$ state, the LE and CT states down to the ground state. In this model, the LE-CT transition occurs thanks to an intra-backbone C7 H-bond, necessary for the electron-driven proton transfer to take place.

Beyond its fundamental importance, fast relaxation processes may also seriously impair the assignment strategies based on ns-UV excitation. Despite this threat, this issue is seldom discussed or addressed directly. Only a few experimental studies suggest that the excited-state lifetime could be much shorter than the radiative lifetime. Thus conformer-dependent lifetimes in the 1–80 ns range have been reported at the origin of the first $\pi\pi^*$ transition of isolated natural Phe [75, 91] as well as of a more biologically relevant capped amino-acid [110, 115, 200], for which quantum chemical calculations have shown that peptide bond $n\pi^*$ states are responsible for the lifetime shortening observed. Still shorter (sub-nanosecond) lifetimes are out of reach of the techniques employed, but are suspected to occur in the hydrates of capped Phe [200], suggesting that detection by ns-UV excitation might become a concern. With other chromophores, the most stable conformer of natural Trp was shown to exhibit specific excited fluorescence behaviour, interpreted as a conformer-selective coupling between L_b and L_a states [59, 211], although this rationalisation was later disputed [20]. Confirmation of this excited-state dynamics was obtained from time-resolved experiments [212]. To our knowledge, no conformer-selective lifetime measurements have been reported for Trp- or Tyr-related capped compounds. Only suspicion of "missing" conformers has been reported for Trp-based natural dipeptides, but this kind of alert remains rare, since chasing missing conformers requires a full and reliable reconstruction of the PES, together with a correct thermal correction of the energetics, for a close comparison with the experimental populations detected [160].

All these studies suggest that a more universal detection, insensitive to ultrafast relaxation, should be used in these difficult cases. One possibility is to rely on femtosecond excitation and ionisation in order to counterbalance the ultrafast relaxation. The resulting loss of conformational selectivity of the excitation step caused by the large spectral linewidth must be compensated for by additional diagnostics. Coupling of ns-IR excitation with femtosecond detection is a way to fix the issue: this has been done on DNA bases, their hydrates [213] or on water dimer and complexes of flexible systems [214]. However, in addition to the selective ns-IR excitation, distinguishing several conformers also requires a selective detection. This can be carried out either from fragmentation channels, photoelectron spectroscopy, or through lifetime measurements in a pump-probe fs-UV detection scheme.

Alternative detection schemes can be used. IR spectroscopy of biomolecules embedded in He droplets has already been proved an efficient way of characterising conformations. The lack of selectivity, however, required the use of additional tricks (alignment in a strong electric field) to identify and disentangle the several spectral contributions [215]. IR/VUV experiments also provide interesting perspectives [32, 33].

Finally, a variant of the population transfer experiment (see Sect. 4.1) can be proposed to monitor the presence of dark conformers, “missing” in the UV spectra. Although not detected by R2PI in the nanosecond regime, the presence of such dark conformers could be revealed through their IR excitation in the supersonic expansion.

4.2.3 IR Spectroscopy of Excited States

IR spectroscopy of excited states has been conducted on several biologically relevant systems [18, 216–218], as well as very recently on peptides [200], in order to provide a structural characterisation of the excited state. A first approach to record such spectra involves the use of a two-laser UV/IR detection scheme [18, 216, 217]. A scan of the IR wavelength reveals the long-lived species present in the UV-irradiated part of the molecular beam, i.e. excited or ionised molecules, and unexcited molecules in their ground state. The resulting overlapped IR signatures [18, 216] can be disentangled by performing similar scans when the IR pulse is sent before or after the UV pulse. A detection scheme involving three sequential pulses UV/IR/UV has also been used [218], but still leads to the overlapped signature of the ground and excited states. When UV does not provide enough conformational selectivity, an IR pulse can precede the whole detection scheme to warm up the undesired conformation selectively, leading to a quadruple resonance method IR/UV/IR/UV which has been successfully applied to record a conformation-selective vibrational spectrum of the S_1 excited state of a dihydrated 3-hydroxyflavone [219]. Given the size of the systems investigated, all these techniques are applicable to isolated neutral peptides [200], and should bring interesting developments in the near future.

5 Conclusion and Future Prospects

Conformer-selective IR spectroscopy is a powerful tool for determining the conformational population present in a low temperature medium, e.g. in a supersonic expansion. Despite the existence of pitfalls under specific conditions, mostly linked to the occurrence of extremely short excited-state lifetimes for specific conformations, it remains a very efficient way to characterize the conformers. The development of conformer-selective precise experimental data together with high level theoretical modelling, namely quantum chemistry calculations, made possible a fruitful cross-fertilised comparison between these two approaches. A series of emblematic structures could be detected in the gas phase and libraries of several types of peptides structures are being constructed. This provides benchmark experimental data to assess, calibrate, or parameterise theoretical methods. In this respect, the parameterisation of the vibrational couplings between CO stretches and NH bends [52] is an inspiring example, showing how gas-phase nanosecond data can be of benefit to time-resolved 2D IR spectroscopy [220]. Besides this remarkable application in a widely visited spectral range, more exotic frequency domains are still to be documented in more detail. This can be done thanks to the laser facilities available nowadays, down to the THz range. Even if size is an issue in studies of neutrals, it might be interesting to keep extending the range of structures investigated to the size domain where helical become prominent and can be reached. This would help confirm the helical assignments proposed for large ionic species [221, 222]. It would also enable one to provide a clear signature for complete 3_{10} and α-helices and to document the transition between these two secondary structures.

Beyond α-peptides, gas phase investigations of more complex but comparable systems are also possible. In this case, the motivations are on another level: benchmark data on individual building blocks as well as on the small oligomers provide guidelines to the organic chemist towards the design of synthetic foldamers, having specific folding propensities. The large variety of possible building blocks, with extended flexibility (such as β- or γ-peptides) or, in contrast, specific constraints (β-peptides based on a cyclobutane spacer for example [223]), endow these foldamers with a wealth of specific properties, making them suitable for, for example, catalytic or therapeutic applications.

Finally, last but not least, the possibility to detect "dark" missing conformers, as well as to collect conformer-selective excited-state data from gas phase experiment combining IR spectroscopy and pump-probe detection provides a great opportunity to document the elementary deactivation processes taking place in the protein photophysics from the investigation of neutral models, bringing a necessary counterpart to the difficult quantum modelling of the excited states. IR spectroscopy of the excited states, IR spectroscopy in He droplets and IR-induced population transfer experiments are among the most promising techniques, based on IR spectroscopy, which should allow spectroscopists to shed light on dark conformations.

References

1. Tzeng SR, Kalodimos CG (2012) Protein activity regulation by conformational entropy. Nature 488:236
2. Perczel A, Angyán JG, Kajtar M, Viviani W, Rivail JL, Marcoccia JF, Csizmadia IG (1991) Peptide models. 1. Topology of selected peptide conformational potential-energy surfaces (glycine and alanine derivatives). J Am Chem Soc 113:6256
3. Berg JM, Tymoczko JL, Stryer L (2002) Biochemistry. W.H. Freeman, New York
4. Scheraga HA, Khalili M, Liwo A (2007) Protein-folding dynamics: overview of molecular simulation techniques. Annu Rev Phys Chem 58:57
5. Shea JE, Brooks CL (2001) From folding theories to folding proteins: a review and assessment of simulation studies of protein folding and unfolding. Annu Rev Phys Chem 52:499
6. Chen YW, Ding F, Nie HF, Serohijos AW, Sharma S, Wilcox KC, Yin SY, Dokholyan NV (2008) Protein folding: then and now. Arch Biochem Biophys 469:4

7. Karplus M, McCammon JA (2002) Molecular dynamics simulations of biomolecules. Nat Struct Biol 9:646
8. Gordon MS, Slipchenko L, Li H, Jensen JH (2007) The effective fragment potential: a general method for predicting intermolecular interactions. Annu Rep Comp Chem 3:177
9. Gresh N (2006) Development, validation, and applications of anisotropic polarizable molecular mechanics to study ligand and drug-receptor interactions. Curr Pharm Design 12:2121
10. Gresh N, Cisneros GA, Darden TA, Piquemal J-P (2007) Anisotropic, polarizable molecular mechanics studies of inter- and intramolecular interactions and ligand-macromolecule complexes. A bottom-up strategy. J Chem Theory Comput 3:1960
11. Senn HM, Thiel W (2009) QM/MM methods for biomolecular systems. Angew Chem Int Ed 48:1198
12. Zwier TS (2001) Laser spectrosocpy of jet-cooled biomolecules and their water-containing clusters: water bridges and molecular conformation. J Phys Chem A 105:8827
13. Weinkauf R, Schermann JP, de Vries MS, Kleinermanns K (2002) Molecular physics of building blocks of life under isolated or defined conditions. Eur Phys J D 20:309
14. Robertson EG, Hockridge MR, Jelfs PD, Simons JP (2000) IR-UV ion-dip spectroscopy of N-benzylformamide clusters: stepwise hydration of a model peptide. J Phys Chem A 104:11714
15. Robertson EG (2000) IR-UV ion-dip spectroscopy of N-phenyl formamide, and its hydrated clusters. Chem Phys Lett 325:299
16. Snoek LC, Robertson EG, Kroemer RT, Simons JP (2000) Conformational landscapes in amino acids: infrared and ultraviolet ion-dip spectroscopy of phenylalanine in the gas phase. Chem Phys Lett 321:49
17. Mons M, Dimicoli I, Tardivel B, Piuzzi F, Robertson EG, Simons JP (2001) Energetics of the gas phase hydrates of trans-formanilide: a microscopic approach to the hydration sites of the peptide bond. J Phys Chem A 105:969
18. Robertson EG, Hockridge MR, Jelfs PD, Simons JP (2001) IR-UV ion-depletion and fluorescence spectroscopy of 2-phenylacetamide clusters: hydration of a primary amide. Phys Chem Chem Phys 3:786
19. Robertson EG, Simons JP (2001) Getting into shape: conformational and supramolecular landscapes in small biomolecules and their hydrated clusters. Phys Chem Chem Phys 3:1
20. Snoek LC, Kroemer RT, Hockridge MR, Simons JP (2001) Conformational landscapes of aromatic amino acids in the gas phase: infrared and ultraviolet ion dip spectroscopy of tryptophan. Phys Chem Chem Phys 3:1819
21. Alonso JL, López JC (2014) Microwave spectroscopy of biomolecular building blocks. Top Curr Chem. doi:10.1007/128_2014_601
22. Balabin RM (2010) Conformational equilibrium in glycine: experimental jet-cooled Raman spectrum. J Phys Chem Lett 1:20
23. Balabin RM (2010) The identification of the two missing conformers of gas-phase alanine: a jet-cooled Raman spectroscopy study. Phys Chem Chem Phys 12:5980
24. Balabin RM (2010) The first step in glycine solvation: the glycine-water complex. J Phys Chem B 114:15075
25. Balabin RM (2012) Experimental thermodynamics of free glycine conformations: the first Raman experiment after twenty years of calculations. Phys Chem Chem Phys 14:99
26. Linder R, Nispel M, Häber T, Kleinermanns K (2005) Gas-phase FT-IR-spectra of natural amino acids. Chem Phys Lett 409:260
27. Albrecht M, Rice CA, Suhm MA (2008) Elementary peptide motifs in the gas phase: FTIR aggregation study of formamide, acetamide, N-methylformamide, and N-methylacetamide. J Phys Chem A 112:7530
28. Linder R, Seefeld K, Vavra A, Kleinermanns K (2008) Gas phase infrared spectra of nonaromatic amino acids. Chem Phys Lett 453:1
29. Hesse S, Suhm MA (2009) Conformation and aggregation of proline esters and their aromatic homologs: pyramidal vs. planar RR' N-H in hydrogen bonds. Z Phys Chem 223:579

30. Otto KE, Hesse S, Wassermann TN, Rice CA, Suhm MA, Stafforstz T, Diederichsen U (2011) Temperature-dependent intensity anomalies in amino acid esters: weak hydrogen bonds in protected glycine, alanine and valine. Phys Chem Chem Phys 13:14119
31. Lee JJ, Albrecht M, Rice CA, Suhm MA (2013) Adaptive aggregation of peptide model systems. J Phys Chem A 117:7050
32. Hu YJ, Bernstein ER (2009) Vibrational and photoionization spectroscopy of neutral valine clusters. J Phys Chem A 113:8454
33. Hu Y, Bernstein ER (2008) Vibrational and photoionization spectroscopy of biomolecules: aliphatic amino acid structures. J Chem Phys 128:164311
34. Chin W, Piuzzi F, Dimicoli I, Mons M (2006) Probing the competition between secondary structures and local preferences in gas phase isolated peptide backbones. Phys Chem Chem Phys 8:1033
35. Zwier TS (2006) Laser probes of conformational isomerization in flexible molecules and complexes. J Phys Chem A 110:4133
36. Gerhards M (2006) In: Laskin J, Lifshitz C (eds) Principles of mass spectrometry applied to biomolecules. Wiley, Hoboken, p. 3
37. de Vries MS, Hobza P (2007) Gas-phase spectroscopy of biomolecular building blocks. Annu Rev Phys Chem 58:585
38. Schermann JP (2008) Spectroscopy and modelling of biomolecular building blocks. Elsevier, Amsterdam
39. Simons JP (2009) Good vibrations: probing biomolecular structure and interactions through spectroscopy in the gas phase. Mol Phys 107:2435
40. Patrick AL, Polfer NC (2014) Peptide fragmentation products in mass spectrometry probed by infrared spectroscopy. Top Curr Chem. doi:10.1007/128_2014_576
41. Dunbar RC (2014) Spectroscopy of metal-ion complexes with peptide-related ligands. Top Curr Chem. doi:10.1007/128_2014_578
42. Filsinger F, Erlekam U, von Helden G, Küpper J, Meijer G (2008) Selector for structural isomers of neutral molecules. Phys Rev Lett 100:133003
43. Page RH, Shen YR, Lee YT (1988) Infrared–ultraviolet double resonance studies of benzene molecules in a supersonic beam. J Chem Phys 88:5362
44. Rijs AM, Oomens J (2014) IR spectroscopic techniques to study isolated biomolecules. Top Curr Chem. doi:10.1007/128_2014_621
45. Pribble RN, Zwier TS (1994) Size-specific infrared-spectra of benzene-$(H_2O)_n$ clusters ($n = 1$ through 7): evidence for noncyclic $(H_2O)_n$ structures. Science 265:75
46. Gerhards M, Unterberg C (2002) Structures of the protected amino acid Ac-Phe-OMe and its dimer: a β-sheet model system in the gas phase. Phys Chem Chem Phys 4:1760
47. Dian BC, Longarte A, Zwier TS (2002) Conformational dynamics in a dipeptide after single-mode vibrational excitation. Science 296:2369
48. Cable JR, Tubergen MJ, Levy DH (1987) Laser desorption molecular beam spectroscopy: the electronic spectra of tryptophan peptides in the gas phase. J Am Chem Soc 109:6198
49. Meijer G, de Vries M, Hunziker HE, Wendt HR (1990) Laser desorption jet-cooling of organic-molecules – cooling characteristics and detection sensitivity. Appl Phys B 51:395
50. Piuzzi F, Dimicoli I, Mons M, Tardivel B, Zhao Q (2000) A simple laser vaporization source for thermally fragile molecules coupled to a supersonic expansion: application to the spectroscopy of tryptophan. Chem Phys Lett 320:282
51. Cirtog M, Rijs AM, Loquais Y, Brenner V, Tardivel B, Gloaguen E, Mons M (2012) Far/mid-infrared signatures of solvent solute interactions in a microhydrated model peptide chain. J Phys Chem Lett 3:3307
52. Buchanan EG, James WH, Choi SH, Guo L, Gellman SH, Muller CW, Zwier TS (2012) Single-conformation infrared spectra of model peptides in the amide I and amide II regions: experiment-based determination of local mode frequencies and inter-mode coupling. J Chem Phys 137:094301

53. Biswal HS, Loquais Y, Tardivel B, Gloaguen E, Mons M (2011) Isolated monohydrates of a model peptide chain: effect of a first water molecule on the secondary structure of a capped phenylalanine. J Am Chem Soc 133:3931
54. Abo-Riziq A, Crews BO, Callahan MP, Grace L, de Vries MS (2006) Spectroscopy of isolated gramicidin peptides. Angew Chem Int Ed 45:5166
55. Blom MN, Compagnon I, Polfer NC, von Helden G, Meijer G, Suhai S, Paizs B, Oomens J (2007) Stepwise solvation of an amino acid: the appearance of zwitterionic structures. J Phys Chem A 111:7309
56. Inokuchi Y, Kobayashi Y, Ito T, Ebata T (2007) Conformation of L-tyrosine studied by fluorescence-detected UV-UV and IR-UV double-resonance spectroscopy. J Phys Chem A 111:3209
57. Snoek LC, Kroemer RT, Simons JP (2002) A spectroscopic and computational exploration of tryptophan-water cluster structures in the gas phase. Phys Chem Chem Phys 4:2130
58. Ebata T, Hashimoto T, Ito T, Inokuchi Y, Altunsu F, Brutschy B, Tarakeshwar P (2006) Hydration profiles of aromatic amino acids: conformations and vibrations of L-phenylalanine-$(H_2O)_n$ clusters. Phys Chem Chem Phys 8:4783
59. Cable JR, Tubergen MJ, Levy DH (1988) The electronic-spectra of small peptides in the gas-phase. Faraday Discuss 86:143
60. Sipior J, Sulkes M (1988) Spectroscopy of tryptophan derivatives in supersonic expansions – addition of solvent molecules. J Chem Phys 88:6146
61. Cable JR, Tubergen MJ, Levy DH (1989) Fluorescence spectroscopy of jet cooled tryptophan peptides. J Am Chem Soc 111:9032
62. Martinez SJ III, Alfano JC, Levy DH (1992) The electronic spectroscopy of the amino acids tyrosine and phenylalanine in a supersonic jet. J Mol Spectrosc 156:421
63. Dian BC, Longarte A, Mercier S, Evans DA, Wales DJ, Zwier TS (2002) The infrared and ultraviolet spectra of single conformations of methyl-capped dipeptides: N-acetyl tryptophan amide and N-acetyl tryptophan methyl amide. J Chem Phys 117:10688
64. Gerhards M, Unterberg C, Gerlach A (2002) Structure of a β-sheet model system in the gas phase: analysis of the C = O stretching vibrations. Phys Chem Chem Phys 4:5563
65. Unterberg C, Gerlach A, Schrader T, Gerhards M (2002) Clusters of a protected amino acid with pyrazole derivatives: β-sheet model systems in the gas phase. Eur Phys J D 20:543
66. Bakker JM, Aleese LM, Meijer G, von Helden G (2003) Fingerprint IR spectroscopy to probe amino acid conformations in the gas phase. Phys Rev Lett 91:203003
67. Hünig I, Seefeld KA, Kleinermanns K (2003) REMPI and UV–UV double resonance spectroscopy of tryptophan ethylester and the dipeptides tryptophan-serine, glycine- tryptophan and proline-tryptophan. Chem Phys Lett 369:173
68. Unterberg C, Gerlach A, Schrader T, Gerhards M (2003) Structure of the protected dipeptide Ac-Val-Phe-OMe in the gas phase: towards a β-sheet model system. J Chem Phys 118:8296
69. Çarçabal P, Kroemer RT, Snoek LC, Simons JP, Bakker JM, Compagnon I, Meijer G, von Helden G (2004) Hydrated complexes of tryptophan: ion dip infrared spectroscopy in the "molecular fingerprint" region, 100–2,000 cm^{-1}. Phys Chem Chem Phys 6:4546
70. Chin W, Mons M, Dognon J-P, Piuzzi F, Tardivel B, Dimicoli I (2004) Competition between local conformational preferences and secondary structures in gas-phase model tripeptides as revealed by laser spectroscopy and theoretical chemistry. Phys Chem Chem Phys 6:2700
71. Dian BC, Longarte A, Winter PR, Zwier TS (2004) The dynamics of conformational isomerization in flexible biomolecules. I. Hole-filling spectroscopy of N-acetyl tryptophan methyl amide and N-acetyl tryptophan amide. J Chem Phys 120:133
72. Evans DA, Wales DJ, Dian BC, Zwier TS (2004) The dynamics of conformational isomerization in flexible biomolecules. II. Simulating isomerizations in a supersonic free jet with master equation dynamics. J Chem Phys 120:148
73. Gerhards M, Unterberg C, Gerlach A, Jansen A (2004) β-Sheet model systems in the gas phase: structures and vibrations of Ac-Phe-NHMe and its dimer $(Ac\text{-}Phe\text{-}NHMe)_2$. Phys Chem Chem Phys. 6:2682

74. Hünig I, Kleinermanns K (2004) Conformers of the peptides glycine-tryptophan, tryptophan-glycine and tryptophan-glycine-glycine as revealed by double resonance laser spectroscopy. Phys Chem Chem Phys 6:2650
75. Lee YH, Jung JW, Kim B, Butz P, Snoek LC, Kroemer RT, Simons JP (2004) Alanyl side chain folding in phenylalanine: conformational assignments through ultraviolet rotational band contour analysis. J Phys Chem A 108:69
76. Wiedemann S, Metsala A, Nolting D, Weinkauf R (2004) The dipeptide cyclic(glycyltryptophanyl) in the gas phase: a concerted action of density functional calculations, S_0-S_1 two-photon ionization, spectral UV/UV hole burning and laser photoelectron spectroscopy. Phys Chem Chem Phys 6:2641
77. Abo-Riziq AG, Bushnell JE, Crews B, Callahan MP, Grace L, De Vries MS (2005) Discrimination between diastereoisomeric dipeptides by IR-UV double resonance spectroscopy and ab initio calculations. Int J Quantum Chem 105:437
78. Abo-Riziq AG, Crews B, Bushnell JE, Callahan MP, De Vries MS (2005) Conformational analysis of cyclo(Phe-Ser) by UV-UV and IR-UV double resonance spectroscopy and ab initio calculations. Mol Phys 103:1491
79. Bakker JM, Plützer C, Hünig I, Häber T, Compagnon I, von Helden G, Meijer G, Kleinermanns K (2005) Folding structures of isolated peptides as revealed by gas-phase mid-infrared spectroscopy. ChemPhysChem 6:120
80. Chin W, Compagnon I, Dognon JP, Canuel C, Piuzzi F, Dimicoli I, von Helden G, Meijer G, Mons M (2005) Spectroscopic evidence for gas-phase formation of successive β-turns in a three-residue peptide chain. J Am Chem Soc 127:1388
81. Chin W, Dognon JP, Piuzzi F, Tardivel B, Dimicoli I, Mons M (2005) Intrinsic folding of small peptide chains: spectroscopic evidence for the formation of beta-turns in the gas phase. J Am Chem Soc 127:707
82. Chin W, Dognon J-P, Canuel C, Piuzzi F, Dimicoli I, Mons M, Compagnon I, von Helden G, Meijer G (2005) Secondary structures of short peptide chains in the gas phase: double resonance spectroscopy of protected dipeptides. J Chem Phys 122:054317
83. Chin W, Mons M, Dognon J-P, Mirasol R, Chass G, Dimicoli I, Piuzzi F, Butz P, Tardivel B, Compagnon I, von Helden G, Meijer G (2005) The gas-phase dipeptide analogue acetyl-phenylalanyl-amide: a model for the study of side chain/backbone interactions in proteins. J Phys Chem A 109:5281
84. Chin W, Piuzzi F, Dognon J-P, Dimicoli I, Mons M (2005) Gas phase models of γ-turns: effects of side-chain/backbone interactions investigated by IR/UV spectroscopy and quantum chemistry. J Chem Phys 123:084301
85. Chin W, Piuzzi F, Dognon J-P, Dimicoli I, Tardivel B, Mons M (2005) Gas phase formation of a 3_{10}-helix in a three-residue peptide chain: role of side chain-backbone interactions as evidenced by IR-UV double resonance experiments. J Am Chem Soc 127:11900
86. Chin W, Dognon JP, Piuzzi F, Dimicoli I, Mons M (2005) Secondary structures of Val-Phe and Val-Tyr(Me) peptide chains in the gas phase: effect of the nature of the protecting groups. Mol Phys 103:1579
87. Gerlach A, Unterberg C, Fricke H, Gerhards M (2005) Structures of Ac-Trp-OMe and its dimer $(Ac\text{-}Trp\text{-}OMe)_2$ in the gas phase: influence of a polar group in the side-chain. Mol Phys 103:1521
88. Řeha D, Valdés H, Vondrášek J, Hobza P, Abu-Riziq A, Crews B, de Vries MS (2005) Structure and IR spectrum of phenylalanyl-glycyl-glycine tripetide in the gas-phase: IR/UV experiments, ab initio quantum chemical calculations, and molecular dynamic simulations. Chem-Eur J 11:6803
89. Abo-Riziq A, Bushnell JE, Crews B, Callahan M, Grace L, De Vries MS (2006) Gas phase spectroscopy of the pentapeptide FDASV. Chem Phys Lett 431:227
90. Fricke H, Gerlach A, Gerhards M (2006) Structure of a β-sheet model system in the gas phase: analysis of the fingerprint region up to 10 μm. Phys Chem Chem Phys 8:1660

91. Hashimoto T, Takasu Y, Yamada Y, Ebata T (2006) Anomalous conformer dependent S_1 lifetime of L-phenylalanine. Chem Phys Lett 421:227
92. Brenner V, Piuzzi F, Dimicoli I, Tardivel B, Mons M (2007) Chirality-controlled formation of β-turn secondary structures in short peptide chains: gas-phase experiment versus quantum chemistry. Angew Chem Int Ed 46:2463
93. Brenner V, Piuzzi F, Dimicoli I, Tardivel B, Mons M (2007) Spectroscopic evidence for the formation of helical structures in gas-phase short peptide chains. J Phys Chem A 111:7347
94. Gloaguen E, Pagliarulo F, Brenner V, Chin W, Piuzzi F, Tardivel B, Mons M (2007) Intramolecular recognition in a jet-cooled short peptide chain: γ-turn helicity probed by a neighbouring residue. Phys Chem Chem Phys 9:4491
95. Häber T, Seefeld K, Kleinermanns K (2007) Mid- and near-infrared spectra of conformers of H-Pro-Trp-OH. J Phys Chem A 111:3038
96. Häber T, Seefeld K, Engler G, Grimme S, Kleinermanns K (2008) IR/UV spectra and quantum chemical calculations of Trp-Ser: stacking interactions between backbone and indole side-chain. Phys Chem Chem Phys 10:2844
97. von Helden G, Compagnon I, Blom MN, Frankowski M, Erlekam U, Oomens J, Brauer B, Gerber RB, Meijer G (2008) Mid-IR spectra of different conformers of phenylalanine in the gas phase. Phys Chem Chem Phys 10:1248
98. Vaden TD, Gowers SAN, de Boer T, Steill JD, Oomens J, Snoek LC (2008) Conformational preferences of an amyloidogenic peptide: IR spectroscopy of Ac-VQIVYK-NHMe. J Am Chem Soc 130:14640
99. Valdés H, Spirko V, Rezac J, Řeha D, Abo-Riziq AG, de Vries MS, Hobza P (2008) Potential-energy and free-energy surfaces of glycyl-phenylalanyl-alanine (GFA) tripeptide: experiment and theory. Chem-Eur J 14:4886
100. Fricke H, Gerlach A, Unterberg C, Wehner M, Schrader T, Gerhards M (2009) Interactions of small protected peptides with aminopyrazole derivatives: the efficiency of blocking a beta-sheet model in the gas phase. Angew Chem Int Ed 48:900
101. Vaden TD, Gowers SAN, Snoek LC (2009) Infrared spectroscopy of 'forbidden' peptide sequences. Phys Chem Chem Phys 11:5843
102. Fricke H, Schwing K, Gerlach A, Unterberg C, Gerhards M (2010) Investigations of the water clusters of the protected amino acid Ac-Phe-OMe by applying IR/UV double resonance spectroscopy: microsolvation of the backbone. Phys Chem Chem Phys 12:3511
103. Gloaguen E, Valdes H, Pagliarulo F, Pollet R, Tardivel B, Hobza P, Piuzzi F, Mons M (2010) Experimental and theoretical investigation of the aromatic-aromatic interaction in isolated capped dipeptides. J Phys Chem A 114:2973
104. James WH, Baquero EE, Choi SH, Gellman SH, Zwier TS (2010) Laser spectroscopy of conformationally constrained alpha/beta-peptides: Ac-ACPC-Phe-NHMe and Ac-Phe-ACPC-NHMe. J Phys Chem A 114:1581
105. Rijs AM, Ohanessian G, Oomens J, Meijer G, von Helden G, Compagnon I (2010) Internal proton transfer leading to stable zwitterionic structures in a neutral isolated peptide. Angew Chem Int Ed 49:2332
106. Abo-Riziq A, Grace L, Crews B, Callahan MP, van Mourik T, de Vries MS (2011) Conformational structure of tyrosine, tyrosyl-glycine, and tyrosyl-glycyl-glycine by double resonance spectroscopy. J Phys Chem A 115:6077
107. Plowright RJ, Gloaguen E, Mons M (2011) Compact folding of isolated four-residue neutral peptide chains: H-bonding patterns and entropy effects. ChemPhysChem 12:1889
108. Rijs AM, Kabeláč M, Abo-Riziq A, Hobza P, de Vries MS (2011) Isolated gramicidin peptides probed by IR spectroscopy. ChemPhysChem 12:1816
109. Biswal HS, Gloaguen E, Loquais Y, Tardivel B, Mons M (2012) Strength of NH···S hydrogen bonds in methionine residues revealed by gas-phase IR/UV spectroscopy. J Phys Chem Lett 3:755
110. Mališ M, Loquais Y, Gloaguen E, Biswal HS, Piuzzi F, Tardivel B, Brenner V, Broquier M, Jouvet C, Mons M, Došlić N, Ljubić I (2012) Unraveling the mechanisms of nonradiative

deactivation in model peptides following photoexcitation of a phenylalanine residue. J Am Chem Soc 134:20340

111. Schwing K, Fricke H, Bartl K, Polkowska J, Schrader T, Gerhards M (2012) Isolated β-turn model systems investigated by combined IR/UV spectroscopy. ChemPhysChem 13:1576
112. Gloaguen E, Loquais Y, Thomas JA, Pratt DW, Mons M (2013) Spontaneous formation of hydrophobic domains in isolated peptides. J Phys Chem B 117:4945
113. Shimozono Y, Yamada K, S-i I, Tsukiyama K, Fujii M (2013) Revised conformational assignments and conformational evolution of tyrosine by laser desorption supersonic jet laser spectroscopy. Phys Chem Chem Phys 15:5163
114. Stanca-Kaposta EC, Çarçabal P, Cocinero EJ, Hurtado P, Simons JP (2013) Carbohydrate-aromatic interactions: vibrational spectroscopy and structural assignment of isolated mono-saccharide complexes with p-hydroxy toluene and N-acetyl L-tyrosine Methylamide. J Phys Chem B 117:8135
115. Mališ M, Loquais Y, Gloaguen E, Jouvet C, Brenner V, Mons M, Ljubić I, Došlić N (2014) Non-radiative deactivation of electronically excited phenylalanine in model peptides: quenching properties of a primary amide group. Phys Chem Chem Phys 16:2285
116. Yan B, Jaeqx S, van der Zande WJ, Rijs AM (2014) A conformation-selective IR-UV study of the dipeptides Ac-Phe-Ser-NH_2 and Ac-Phe-Cys-NH_2: probing the SH···O and OH···O hydrogen bond interactions. Phys Chem Chem Phys 16:10770
117. Jaeqx S, Oomens J, Cimas A, Gaigeot MP, Rijs AM (2014) Gas-phase peptide structures unraveled by far-IR spectroscopy: combining IR-UV ion-dip experiments with Born-Oppenheimer molecular dynamics simulations. Angew Chem Int Ed 53:3663
118. Alauddin M, Biswal HS, Gloaguen E, Mons M (2014) Intra-residue interactions in proteins: interplay between serine or cysteine side chains and backbone conformations, revealed by laser spectroscopy of isolated model peptides. Phys Chem Chem Phys. doi:10.1039/c4cp04449e
119. Dian BC, Florio GM, Clarkson JR, Longarte A, Zwier TS (2004) Infrared-induced conformational isomerization and vibrational relaxation dynamics in melatonin and 5-methoxy-*N*-acetyl tryptophan methyl amide. J Chem Phys 120:9033
120. Fricke H, Gerlach A, Unterberg C, Rzepecki P, Schrader T, Gerhards M (2004) Structure of the tripeptide model Ac-Val-Tyr(Me)-NHMe and its cluster with water investigated by IR/UV double resonance spectroscopy. Phys Chem Chem Phys 6:4636
121. Fricke H, Schäfer G, Schrader T, Gerhards M (2007) Secondary structure binding motifs of the jet cooled tetrapeptide model Ac-Leu-Val-Tyr(Me)-NHMe. Phys Chem Chem Phys 9:4592
122. Baquero EE, James WH, Choi SH, Gellman SH, Zwier TS (2008) Single-conformation ultraviolet and infrared spectroscopy of model synthetic foldamers: β-peptides Ac-β^3-hPhe-NHMe and Ac-β^3-hTyr-NHMe. J Am Chem Soc 130:4784
123. Baquero EE, James WH, Choi SH, Gellman SH, Zwier TS (2008) Single-conformation ultraviolet and infrared spectroscopy of model synthetic foldamers: β-peptides Ac-β^3-hPhe-β^3-hAla-NHMe and Ac-β^3-hAla-β^3-hPhe-NHMe. J Am Chem Soc 130:4795
124. Fricke H, Funk A, Schrader T, Gerhards M (2008) Investigation of secondary structure elements by IR/UV double resonance spectroscopy: analysis of an isolated β-sheet model system. J Am Chem Soc 130:4692
125. James WH, Müller CW, Buchanan EG, Nix MGD, Guo L, Roskop L, Gordon MS, Slipchenko LV, Gellman SH, Zwier TS (2009) Intramolecular amide stacking and its competition with hydrogen bonding in a small foldamer. J Am Chem Soc 131:14243
126. Buchanan EG, James WH, Gutberlet A, Dean JC, Guo L, Gellman SH, Zwier TS (2011) Single-conformation spectroscopy and population analysis of model gamma-peptides: new tests of amide stacking. Faraday Discuss 150:209
127. James WH, Buchanan EG, Guo L, Geman SH, Zwier TS (2011) Competition between amide stacking and intramolecular H bonds in γ-peptide derivatives: controlling nearest-neighbor preferences. J Phys Chem A 115:11960

128. James WH, Buchanan EG, Muller CW, Dean JC, Kosenkov D, Slipchenko LV, Guo L, Reidenbach AG, Gellman SH, Zwier TS (2011) Evolution of amide stacking in larger gamma-peptides: triamide H-bonded cycles. J Phys Chem A 115:13783
129. Schwing K, Reyheller C, Schaly A, Kubik S, Gerhards M (2011) Structural analysis of an isolated cyclic tetrapeptide and its monohydrate by combined IR/UV spectroscopy. ChemPhysChem 12:1981
130. Shubert VA, Zwier TS (2007) IR-IR-UV hole-burning: conformation specific IR spectra in the face of UV spectral overlap. J Phys Chem A 111:13283
131. Compagnon I, Oomens J, Bakker J, Meijer G, von Helden G (2005) Vibrational spectroscopy of a non-aromatic amino acid-based model peptide: identification of the gamma-turn motif of the peptide backbone. Phys Chem Chem Phys 7:13
132. Compagnon I, Oomens J, Meijer G, von Helden G (2006) Mid-infrared spectroscopy of protected peptides in the gas phase: a probe of the backbone conformation. J Am Chem Soc 128:3592
133. Cocinero EJ, Stanca-Kaposta EC, Gamblin DP, Davis BG, Simons JP (2009) Peptide secondary structures in the gas phase: consensus motif of N-linked glycoproteins. J Am Chem Soc 131:1282
134. Gloaguen E, Pollet R, Piuzzi F, Tardivel B, Mons M (2009) Gas phase folding of an $(Ala)_4$ neutral peptide chain: spectroscopic evidence for the formation of a β-hairpin H-bonding pattern. Phys Chem Chem Phys 11:11385
135. Gloaguen E, de Courcy B, Piquemal JP, Pilmé J, Parisel O, Pollet R, Biswal HS, Piuzzi F, Tardivel B, Broquier M, Mons M (2010) Gas-phase folding of a two-residue model peptide chain: on the importance of an interplay between experiment and theory. J Am Chem Soc 132:11860
136. Zhu H, Blom M, Compagnon I, Rijs AM, Roy S, von Helden G, Schmidt B (2010) Conformations and vibrational spectra of a model tripeptide: change of secondary structure upon micro-solvation. Phys Chem Chem Phys 12:3415
137. Miyazaki M, Makara K, Ishiuchi S, Fujii M (2011) Gas-phase infrared spectroscopy of monopeptides from 10 to 3 μm. Chem Lett 40:1157
138. Chakraborty S, Yamada K, Ishiuchi S, Fujii M (2012) Gas phase IR spectra of tri-peptide Z-Pro-Leu-Gly: effect of C-terminal amide capping on secondary structure. Chem Phys Lett 531:41
139. Dean JC, Buchanan EG, Zwier TS (2012) Mixed 14/16 helices in the gas phase: conformation-specific spectroscopy of Z-$(Gly)_n$, $n = 1, 3, 5$. J Am Chem Soc 134:17186
140. Ishiuchi S, Yamada K, Chakraborty S, Yagi K, Fujii M (2013) Gas-phase spectroscopy and anharmonic vibrational analysis of the 3-residue peptide Z-Pro-Leu-Gly-NH_2 by the laser desorption supersonic jet technique. Chem Phys 419:145
141. Jaeqx S, Oomens J, Rijs AM (2013) Gas-phase salt bridge interactions between glutamic acid and arginine. Phys Chem Chem Phys 15:16341
142. Jaeqx S, Du WN, Meijer EJ, Oomens J, Rijs AM (2013) Conformational study of Z-Glu-OH and Z-Arg-OH: dispersion interactions versus conventional hydrogen bonding. J Phys Chem A 117:1216
143. Kusaka R, Zhang D, Walsh PS, Gord JR, Fisher BF, Gellman SH, Zwier TS (2013) Role of ring-constrained γ-amino acid residues in α/γ-peptide folding: single-conformation UV and IR spectroscopy. J Phys Chem A 117:10847
144. Walsh PS, Kusaka R, Buchanan EG, James WH III, Fisher BF, Gellman SH, Zwier TS (2013) Cyclic constraints on conformational flexibility in γ-peptides: conformation specific IR and UV spectroscopy. J Phys Chem A 117:12350
145. Gord JR, Walsh PS, Fisher BF, Gellman SH, Zwier TS (2014) Mimicking the first turn of an α-helix with an unnatural backbone: conformation-specific IR and UV spectroscopy of cyclically constrained β/γ-peptides. J Phys Chem B 118:8246

146. Gloaguen E, Brenner V, Alauddin M, Tardivel B, Mons M, Zehnacker-Rentien A, Declerck V, Aitken DJ (2014) Direct spectroscopic evidence of hyperconjugation unveils the conformational landscape of hydrazides. Angew Ch Int Ed. doi:10.1002/anie.201407801
147. Buchanan EG, Sibert EL, Zwier TS (2013) Ground state conformational preferences and CH stretch-bend coupling in a model alkoxy chain: 1,2-diphenoxyethane. J Phys Chem A 117:2800
148. Buchanan EG, Dean JC, Zwier TS, Sibert EL (2013) Towards a first-principles model of Fermi resonance in the alkyl CH stretch region: application to 1,2-diphenylethane and 2,2,2-paracyclophane. J Chem Phys 138:064308
149. Dian BC, Clarkson JR, Zwier TS (2004) Direct measurement of energy thresholds to conformational isomerization in tryptamine. Science 303:1169
150. Wilson KR, Belau L, Nicolas C, Jimenez-Cruz M, Leone SR, Ahmed M (2006) Direct determination of the ionization energy of histidine with VUV synchrotron radiation. Int J Mass Spectrom 249:155
151. Wilson KR, Jimenez-Cruz M, Nicolas C, Belau L, Leone SR, Ahmed M (2006) Thermal vaporization of biological nanoparticles: fragment-free vacuum ultraviolet photoionization mass spectra of tryptophan, phenylalanine-glycine-glycine, and, beta-carotene. J Phys Chem A 110:2106
152. Lee KT, Sung J, Lee KJ, Park YD, Kim SK (2002) Conformation-dependent ionization energies of L-phenylalanine. Angew Chem Int Ed 41:4114
153. Jochims HW, Schwell M, Chotin JL, Clemino M, Dulieu F, Baumgärtel H, Leach S (2004) Photoion mass spectrometry of five amino acids in the 6–22 eV photon energy range. Chem Phys 298:279
154. Plekan O, Feyer V, Richter R, Coreno M, de Simone M, Prince KC, Carravetta V (2007) Investigation of the amino acids glycine, proline, and methionine by photoemission spectroscopy. J Phys Chem A 111:10998
155. Tia M, de Miranda BC, Daly S, Gaie-Levrel F, Garcia GA, Powis I, Nahon L (2013) Chiral asymmetry in the photoionization of gas-phase amino-acid alanine at Lyman-alpha radiation wavelength. J Phys Chem Lett 4:2698
156. Powis I, Rennie EE, Hergenhahn U, Kugeler O, Bussy-Socrate R (2003) Investigation of the gas-phase amino acid alanine by synchrotron radiation photoelectron spectroscopy. J Phys Chem A 107:25
157. Van der Spoel D, Lindahl E, Hess B, Groenhof G, Mark AE, Berendsen HJC (2005) GROMACS: fast, flexible, and free. J Comput Chem 26:1701
158. Case DA, Cheatham TE, Darden T, Gohlke H, Luo R, Merz KM, Onufriev A, Simmerling C, Wang B, Woods RJ (2005) The Amber biomolecular simulation programs. J Comput Chem 26:1668
159. Brooks BR, Brooks CL, Mackerell AD, Nilsson L, Petrella RJ, Roux B, Won Y, Archontis G, Bartels C, Boresch S, Caflisch A, Caves L, Cui Q, Dinner AR, Feig M, Fischer S, Gao J, Hodoscek M, Im W, Kuczera K, Lazaridis T, Ma J, Ovchinnikov V, Paci E, Pastor RW, Post CB, Pu JZ, Schaefer M, Tidor B, Venable RM, Woodcock HL, Wu X, Yang W, York DM, Karplus M (2009) CHARMM: the biomolecular simulation program. J Comput Chem 30:1545
160. Valdés H, Řeha D, Hobza P (2006) Structure of isolated tryptophyl-glycine dipeptide and tryptophyl-glycyl-glycine tripeptide: ab initio SCC-DFTB-D molecular dynamics simulations and high-level correlated ab initio quantum chemical calculations. J Phys Chem B 110:6385
161. Toroz D, Van Mourik T (2006) The structure of the gas-phase tyrosine-glycine dipeptide. Mol Phys 104:559
162. Černý J, Hobza P (2007) Non-covalent interactions in biomacromolecules. Phys Chem Chem Phys 9:5291

163. Holroyd LF, van Mourik T (2007) Insufficient description of dispersion in B3LYP and large basis set superposition errors in MP2 calculations can hide peptide conformers. Chem Phys Lett 442:42
164. Shields AE, van Mourik T (2007) Comparison of ab initio and DFT electronic structure methods for peptides containing an aromatic ring: effect of dispersion and BSSE. J Phys Chem A 111:13272
165. van Mourik T (2008) Assessment of density functionals for intramolecular dispersion-rich interactions. J Chem Theory Comput 4:1610
166. Toroz D, van Mourik T (2010) Structure of the gas-phase glycine tripeptide. Phys Chem Chem Phys 12:3463
167. Zhao Y, Truhlar DG (2007) Density functionals for noncovalent interaction energies of biological importance. J Chem Theory Comput 3:289
168. Zhao Y, Truhlar DG (2006) Assessment of model chemistries for noncovalent interactions. J Chem Theory Comput 2:1009
169. Zhao Y, Truhlar DG (2008) Density functionals with broad applicability in chemistry. Acc Chem Res 41:157
170. Grimme S (2004) Accurate description of van der Waals complexes by density functional theory including empirical corrections. J Comput Chem 25:1463
171. Grimme S (2006) Semiempirical GGA-type density functional constructed with a long-range dispersion correction. J Comput Chem 27:1787
172. Grimme S, Antony J, Ehrlich S, Krieg H (2010) A consistent and accurate ab initio parametrization of density functional dispersion correction (DFT-D) for the 94 elements H-Pu. J Chem Phys 132:154104
173. Mackie ID, DiLabio GA (2008) Interactions in large, polyaromatic hydrocarbon dimers: application of density functional theory with dispersion corrections. J Phys Chem A 112:10968
174. Lill SON (2010) Evaluation of dispersion-corrected density functional theory (B3LYP-DCP) for compounds of biochemical interest. J Mol Graph 29:178
175. Bouteiller Y, Poully JC, Desfrançois C, Grégoire G (2009) Evaluation of MP2, DFT, and DFT-D methods for the prediction of infrared spectra of peptides. J Phys Chem A 113:6301
176. Bouteiller Y, Gillet JC, Grégoire G, Schermann JP (2008) Transferable specific scaling factors for interpretation of infrared spectra of biomolecules from density functional theory. J Phys Chem A 112:11656
177. Došlić N, Kovačević G, Ljubić I (2007) Signature of the conformational preferences of small peptides: a theoretical investigation. J Phys Chem A 111:8650
178. Neff M, Rauhut G (2009) Toward large scale vibrational configuration interaction calculations. J Chem Phys 131:124129
179. Scribano Y, Lauvergnat DM, Benoit DM (2010) Fast vibrational configuration interaction using generalized curvilinear coordinates and self-consistent basis. J Chem Phys 133:094103
180. Clarkson JR, Baquero E, Shubert VA, Myshakin EM, Jordan KD, Zwier TS (2005) Laser-initiated shuttling of a water molecule between H-bonding sites. Science 307:1443
181. LeGreve TA, Clarkson JR, Zwier TS (2008) Experimental determination of conformational isomerization energy thresholds in serotonin. J Phys Chem A 112:3911
182. Sohn WY, Cho K-J, Lee SY, Kang SS, Park YD, Kang H (2012) Solvent-assisted conformational isomerization (SACI) of meta-substituted phenols: tuning relative stability, isomerization barrier, and IVR rate. Chem Phys Lett 525–26:37
183. Sohn WY, Kim M, Kim S-S, Park YD, Kang H (2011) Solvent-assisted conformational isomerization and the conformationally-pure REMPI spectrum of 3-aminophenol. Phys Chem Chem Phys 13:7006
184. Miller RD (1988) In: Scoles G (ed) Atomic and molecular beam methods, vol 1. Oxford University Press, New York, p 14
185. Handschuh M, Nettesheim S, Zenobi R (1999) Is infrared laser-induced desorption a thermal process? The case of aniline. J Phys Chem B 103:1719

186. Godfrey PD, Brown RD (1998) Proportions of species observed in jet spectroscopy vibrational energy effects: histamine tautomers and conformers. J Am Chem Soc 120:10724
187. Shubert VA, Baquero EE, Clarkson JR, James WH, Turk JA, Hare AA, Worrel K, Lipton MA, Schofield DP, Jordan KD, Zwier TS (2007) Entropy-driven population distributions in a prototypical molecule with two flexible side chains: O-(2-acetamidoethyl)-N-acetyltyramine. J Chem Phys 127:234315
188. Cabezas C, Varela M, Cortijo V, Jiménez AI, Peña I, Daly AM, López JC, Cativiela C, Alonso JL (2013) The alanine model dipeptide Ac-Ala-NH_2 exists as a mixture of C_7^{eq} and C_5 conformers. Phys Chem Chem Phys 15:2580
189. Mons M, Piuzzi F, Dimicoli I, Gorb L, Leszczynski J (2006) Near-UV resonant two-photon ionization spectroscopy of gas phase guanine: evidence for the observation of three rare tautomers. J Phys Chem A 110:10921
190. Mons M, Dimicoli I, Piuzzi F (2002) Gas phase hydrogen-bonded complexes of aromatic molecules: photoionization and energetics. Int Rev Phys Chem 21:101
191. Head-Gordon T, Head-Gordon M, Frisch MJ, Brooks CL, Pople JA (1991) Theoretical-study of blocked glycine and alanine peptide analogs. J Am Chem Soc 113:5989
192. Vass E, Hollósi M, Besson F, Buchet R (2003) Vibrational spectroscopic detection of beta- and gamma-turns in synthetic and natural peptides and proteins. Chem Rev 103:1917
193. Johnson ER, Keinan S, Mori-Sánchez P, Contreras-García J, Cohen AJ, Yang W (2010) Revealing noncovalent interactions. J Am Chem Soc 132:6498
194. Chaudret R, de Courcy B, Contreras-García J, Gloaguen E, Zehnacker-Rentien A, Mons M, Piquemal J-P (2014) Unraveling non covalent interactions within flexible biomolecules: from electron density topology to gas phase spectroscopy. Phys Chem Chem Phys 16:2285
195. Zhou P, Tian F, Lv F, Shang Z (2009) Geometric characteristics of hydrogen bonds involving sulfur atoms in proteins. Proteins Struct Funct Bioinf 76:151
196. Han WG, Jalkanen KJ, Elstner M, Suhai S (1998) Theoretical study of aqueous N-acetyl-L-alanine N′-methylamide: structures and Raman, VCD, and ROA spectra. J Phys Chem B 102:2587
197. Cocinero EJ, Çarçabal P, Vaden TD, Davis BG, Simons JP (2011) Exploring carbohydrate-peptide interactions in the gas phase: structure and selectivity in complexes of pyranosides with N-acetylphenylalanine methylamide. J Am Chem Soc 133:4548
198. Cocinero EJ, Çarçabal P, Vaden TD, Simons JP, Davis BG (2011) Sensing the anomeric effect in a solvent-free environment. Nature 469:76
199. Zhigilei LV, Leveugle E, Garrison BJ, Yingling YG, Zeifman MI (2003) Computer simulations of laser ablation of molecular substrates. Chem Rev 103:321
200. Loquais Y, Gloaguen E, Habka S, Vaquero-Vara V, Brenner V, Tardivel B, Mons M (2014) Secondary structures in phe-containing isolated dipeptide chains: laser spectroscopy vs quantum chemistry. J Phys Chem A. doi:10.1021/jp509494c
201. Christiansen O, Koch H, Jorgensen P (1995) The 2nd-order approximate coupled-cluster singles and doubles model CC2. Chem Phys Lett 243:409
202. Shemesh D, Domcke W (2011) Effect of the chirality of residues and gamma-turns on the electronic excitation spectra, excited-state reaction paths and conical intersections of capped phenylalanine-alanine dipeptides. ChemPhysChem 12:1833
203. Shemesh D, Hättig C, Domcke W (2009) Photophysics of the Trp-Gly dipeptide: role of electron and proton transfer processes for efficient excited-state deactivation. Chem Phys Lett 482:38
204. Shemesh D, Sobolewski AL, Domcke W (2009) Efficient excited-state deactivation of the Gly-Phe-Ala tripeptide via an electron-driven proton-transfer process. J Am Chem Soc 131:1374
205. Shemesh D, Sobolewski AL, Domcke W (2010) Role of excited-state hydrogen detachment and hydrogen-transfer processes for the excited-state deactivation of an aromatic dipeptide: N-acetyl tryptophan methyl amide. Phys Chem Chem Phys 12:4899

206. Sobolewski AL, Domcke W (2006) Relevance of electron-driven proton-transfer processes for the photostability of proteins. ChemPhysChem 7:561
207. Sobolewski AL, Shemesh D, Domcke W (2009) Computational studies of the photophysics of neutral and zwitterionic amino acids in an aqueous environment: tyrosine-$(H_2O)_2$ and tryptophan-$(H_2O)_2$ clusters. J Phys Chem A 113:542
208. Clavaguéra C, Piuzzi F, Dognon JP (2009) Electronic spectrum of tryptophan-phenylalanine. A correlated ab initio and time-dependent density functional theory study. J Phys Chem B 113:16443
209. Pollet R, Brenner V (2008) Assessment of time-dependent density functional theory for predicting excitation energies of bichromophoric peptides: case of tryptophan-phenylalanine. Theor Chem Acc 121:307
210. Sobolewski AL, Domcke W, Dedonder-Lardeux C, Jouvet C (2002) Excited-state hydrogen detachment and hydrogen transfer driven by repulsive $^1\pi\sigma^*$ states: a new paradigm for nonradiative decay in aromatic biomolecules. Phys Chem Chem Phys 4:1093
211. Tubergen MJ, Cable JR, Levy DH (1990) Substituent effects on the electronic spectroscopy of tryptophan derivatives in jet expansions. J Chem Phys 92:51
212. Ovejas V, Fernández-Fernández M, Montero R, Castaño F, Longarte A (2013) Ultrafast nonradiative relaxation channels of tryptophan. J Phys Chem Lett 4:1928
213. Nosenko Y, Kunitski M, Riehn C, Harbach PHP, Dreuw A, Brutschy B (2010) The structure of adenine monohydrates studied by femtosecond multiphoton ionization detected IR spectroscopy and quantum chemical calculations. Phys Chem Chem Phys 12:863
214. León I, Montero R, Castaño F, Longarte A, Fernández JA (2012) Mass-resolved infrared spectroscopy of complexes without chromophore by nonresonant femtosecond ionization detection. J Phys Chem A 116:6798
215. Choi MY, Miller RE (2006) Four tautomers of isolated guanine from infrared laser spectroscopy in helium nanodroplets. J Am Chem Soc 128:7320
216. Seurre N, Le Barbu-Debus K, Lahmani F, Zehnacker-Rentien A, Sepiol J (2003) Electronic and vibrational spectroscopy of jet-cooled m-cyanophenol and its dimer: laser-induced fluorescence and fluorescence-dip IR spectra in the S_0 and S_1 states. Chem Phys 295:21
217. Dian BC, Longarte A, Zwier TS (2003) Hydride stretch infrared spectra in the excited electronic states of indole and its derivatives: direct evidence for the $^1\pi\sigma^*$ state. J Chem Phys 118:2696
218. Bartl K, Funk A, Gerhards M (2008) IR/UV spectroscopy on jet cooled 3-hydroxyflavone $(H_2O)_n$ $(n = 1{,}2)$ clusters along proton transfer coordinates in the electronic ground and excited states. J Chem Phys 129:234306
219. Weiler M, Bartl K, Gerhards M (2012) Infrared/ultraviolet quadruple resonance spectroscopy to investigate structures of electronically excited states. J Chem Phys 136:114202
220. Asplund MC, Zanni MT, Hochstrasser RM (2000) Two-dimensional infrared spectroscopy of peptides by phase-controlled femtosecond vibrational photon echoes. Proc Natl Acad Sci U S A 97:8219
221. Xie YM, Schaefer HF, Silaghi-Dumitrescu R, Peng B, Li QS, Stearns JA, Rizzo TR (2012) Conformational preferences of gas-phase helices: experiment and theory struggle to agree: the seven-residue peptide Ac-Phe-$(Ala)_5$-Lys-H^+. Chem Eur J 18:12941
222. Stearns JA, Seaiby C, Boyarkin OV, Rizzo TR (2009) Spectroscopy and conformational preferences of gas-phase helices. Phys Chem Chem Phys 11:125
223. Altmayer-Henzien A, Declerck V, Merlet D, Baltaze JP, Farjon J, Guillot R, Aitken DJ (2013) Solution state conformational preferences of dipeptides derived from n-aminoazetidinecarboxylic acid: an assessment of the hydrazino turn. J Org Chem 78:6031

Top Curr Chem (2015) 364: 271–298
DOI: 10.1007/128_2014_577

Published online: 7 February 2015

Gas-Phase IR Spectroscopy of Nucleobases

Mattanjah S. de Vries

Abstract IR spectroscopy of nucleobases in the gas phase reflects simultaneous advances in both experimental and computational techniques. Important properties, such as excited state dynamics, depend in subtle ways on structure variations, which can be followed by their infrared signatures. Isomer specific spectroscopy is a particularly powerful tool for studying the effects of nucleobase tautomeric form and base pair hydrogen-bonding patterns.

Keywords Clusters · Gas phase · Hole burning · IR spectroscopy · Nucleobases · R2PI · REMPI

Contents

M.S. de Vries (✉)
Department of Chemistry and Biochemistry, University of California, Santa Barbara, CA 93106, USA
e-mail: devries@chem.ucsb.edu

Abbreviations

A	Adenine
C	Cytosine
DRS	Double resonant spectroscopy
G	Guanine
T	Thymine
U	Uracil

1 Introduction

Gas-phase techniques provide a reductionist approach to the study of nucleobases and nucleotides. Collision-free conditions in vacuo provide insight into the intrinsic properties of individual molecules, free of any interactions. Without such isolation, many fundamental properties can be masked by the biological environment, such as the macromolecular structure of the double helix, base pairing interactions, and the role of the solvent. Among the properties of interest are conformational preferences, tautomeric population distributions, hydrogen bonded and pi-stacking structures, inter- and intramolecular interactions, and excited state dynamics. Once such properties are mapped out for isolated bases, one can hope to extrapolate to more complex systems, including larger DNA segments and solvent contributions. Many aspects of these properties can be probed with the help of IR spectroscopy, which is particularly diagnostic for structural variation, especially when hydrogen bonding is involved.

In order to derive structural information from infrared frequencies, input is required from quantum chemical calculations at computational levels which match the experimental resolution. Experimentally, gas-phase conditions imply extremely low sample densities, requiring special techniques in order to acquire infrared data. Some of those techniques involve double resonance approaches which provide unique opportunities for isomer selective IR spectroscopy. This facet is among the advantages of gas-phase experiments, making it possible to follow certain properties, such as excited state dynamics, as a function of molecular structure. At the same time, the availability of gas-phase data provides opportunities to calibrate computational methods, force fields, and functionals.

2 Techniques

Gas-phase spectroscopy of neutral molecules, as opposed to ions, usually involves the use of supersonic molecular beams [1–4]. For smaller compounds this can be achieved by seeding in the inert drive gas. This limitation excludes the study of neutral nucleosides or larger compounds while even some of the bare nucleobases, such as guanine, cannot be sufficiently heated without thermal degradation. Some work with bases and base mimics has been done in seeded beams [5–10]. Larger compounds can now be vaporized successfully by pulsed laser desorption, followed by entrainment in a supersonic jet [11–14]. This experimental advance has opened up the field of study of nucleobases and nucleosides in isolation in the gas phase, especially by IR spectroscopy. The cooling in molecular beams makes this approach particularly attractive for spectroscopy. Although temperatures are not as low as in ion traps or helium droplets, molecular beams can achieve internal temperatures typically of the order of 10–20 K, which provides very useful optical resolution.

In a supersonic beam, typical densities are of the order of 10^{12} molecules/cm^3. Therefore, at a typical absorption cross section of 10^{-18} cm^2, a 1-mm supersonic beam would absorb a fraction of 10^{-7} of incident photons. The low densities inherent in gas-phase experiments have led to the development and application of a number of suitable techniques in order to acquire infrared spectra.

2.1 *IR-UV Double Resonance*

Most of the work described in this chapter involves action spectroscopy in the form of IR-UV double resonance spectroscopy (DRS). In this approach, a secondary step reports on the absorption of the IR photon. A disadvantage is that the resulting spectrum is not a pure absorption spectrum but rather the composite result of two processes. An advantage can be that the secondary step can provide additional information. In the case of IR-UV double resonance, one combines direct IR absorption with either resonant two-photon ionization spectroscopy (R2PI) or laser induced fluorescence (LIF). A promising new variant is Ionization Loss Stimulated Raman spectroscopy [15]. The optical selection of the second step makes these techniques isomer-selective, which is their greatest strength. Comparison is sometimes possible with direct absorption techniques, although those generally lack isomer specificity, such as the following.

2.2 *Fourier Transform Microwave Spectroscopy*

In Fourier transform microwave spectroscopy, isomeric analysis is derived from rotational spectra; see [16]. Alonso and coworkers have combined this approach with laser desorption jet cooling for the study of tautomeric forms of nucleobases.

This technique identified tautomeric forms for all the canonical bases [17–20], as well as for several of their complexes with water [21], allowing comparisons with tautomer identifications from IR techniques.

2.3 Helium Droplets

In helium droplets, spectroscopy is usually performed in the infrared [22–25]. Resonant absorption by specific vibrational modes implies heating, resulting in helium atoms boiling off the droplets, recorded in a mass spectrometer or with a bolometer. The temperature in helium droplets is lower than in supersonic expansions and the cooling is so fast as essentially to freeze the starting population distribution. The starting temperatures need not be elevated much above room temperature because "pick-up" sources do not require very high gas densities.

2.4 Cavity Ringdown Spectroscopy and Multipass Absorption

Direct absorption probabilities can be enhanced by multipass arrangements and a particularly elegant and sensitive form of this principle is cavity ring down spectroscopy, where signal damping is recorded, rather than direct absorption. However, even at the largest ringdown times achievable with the highest quality mirrors, it is still difficult to measure IR absorption in a molecular beam. Saykally and co-workers obtained IR spectra of uracil and of nucleobase clusters with water by combining cavity ringdown and multipass spectroscopy with a slit nozzle, which produced an unusually wide molecular beam [26–28].

3 Monomers and Tautomeric Forms

The nucleobase monomers can adopt a variety of tautomeric forms and IR spectroscopy in the near IR is diagnostic for this property, especially with the NH and OH stretch frequencies as fingerprints of keto, imino, or enol forms.

3.1 Purines

For guanine, four species have been observed in R2PI spectra, labeled A–D, distinguished by UV–UV hole burning and characterized by IR-UV DRS

[29–35]. Figure 1e, g and h show experimental spectra from Nir et al., obtained by IR-UV DRS. Originally these tautomers were assigned as the four lowest energy forms, with A and D as enol forms and B and C as keto forms. The N7 vs N9 forms of each were subsequently distinguished by Mons et al. by selective methyl substitution [33]. However, the B, C assignment to the keto forms turned out to be incorrect. Surprisingly, these are imino forms, even though these tautomers are significantly higher in energy, and the three lowest energy tautomers are thus absent in the R2PI spectra, as described by Mons et al. and by Marian [36, 37]. This absence is explained by short, sub-picosecond, excited state lifetimes, to be discussed below, which render nanosecond timescales R2PI blind for these tautomers. The three "missing" tautomeric forms are shown in red in Fig. 1. Evidence that these tautomers are in fact present in the gas phase comes from two other

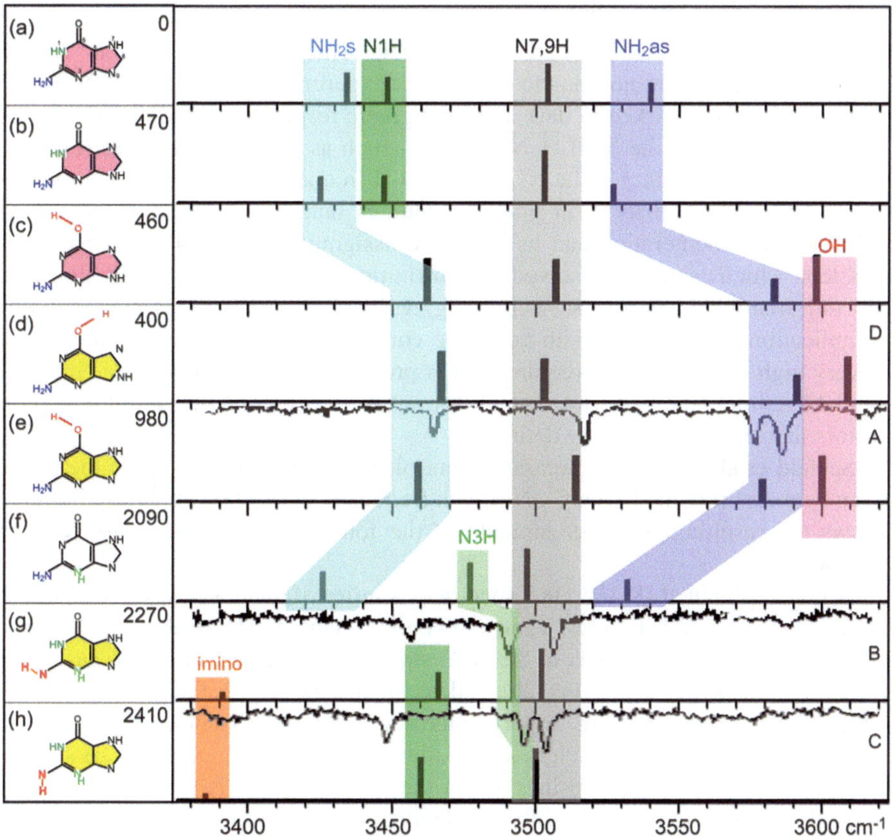

Fig. 1 a-h: IR frequencies for the eight lowest energy guanine tautomers, as calculated by Marian [36]. Modes are color-coded as shown in the structures on the *left*. Color coding of the structures is detailed in the text. Numbers are relative energies in cm^{-1}. **e**, **g**, and **h** show experimental IR-UV double resonance spectra from Nir et al. [29]. **A–D** refer to the four tautomers observed in R2PI spectroscopy

experiments which are direct absorption measurements, as opposed to action spectroscopy, and thus do not involve the excited state. One is FT microwave spectroscopy by Alonso et al., who observe the four lowest energy tautomeric forms (a)→(d) [19]. The second evidence comes from a helium droplet experiment by Choi and Miller who found the same four lowest energy tautomers [25]. These experiments are not isomer selective, so a large number of vibrational bands are observed in the infrared spectrum. The assignment was aided by aligning the molecules in an electric field and recording absorption as a function of the polarization angle of the light relative to the field. This elegant approach provides vibrational transition moment angles as additional data for comparison with ab initio theory.

Figure 1 shows IR frequencies for guanine as calculated by Marian [36]. The enol forms are characterized by the OH stretch at about 3,600 cm^{-1}, marked in red, and a red shift of the symmetric and antisymmetric NH_2 stretches, marked in blue and purple, respectively. The imino forms are characterized by the imino NH stretch below 3,400 cm^{-1}, marked in orange, which is, however, very low in intensity and therefore not diagnostic in most experimental spectra. The N1H and N3H stretches are marked in dark and light green, respectively. Pairs of tautomers which only differ in the N7H vs N9H forms, such as (a) vs (b) or (c) vs (e), have very similar IR signatures. The N7-H and N9-H frequencies are almost identical at about 3,500 cm^{-1}, marked in gray. Sometimes tautomeric blocking by methyl substitution or deuteration can help resolve assignments [33, 34]. An analogous problem, which cannot be resolved by substitution, is the similarity in IR signature of conformational tautomers, such as (g) vs (h) or (c) vs (d). Because of this complication, assignments with complete confidence call for computational data at very high accuracy. This requirement is problematic, especially since currently available computations do not account for anharmonicity and thus employ scaling factors that are not rigorously defined.

Seefeld et al. confirmed the assignment of the B and C tautomers to the imino form by measuring the imino C=N stretch frequency at ~1,700 cm^{-1} [38]. Figure 1 shows the resulting final assignments of the four long-lived tautomers A–D in yellow.

Nir et al. obtained IR-UV DRS spectra of a series of guanosines, some of which appear in Fig. 2 [29]. Figure 2a shows 9H enol guanine. The peak at 3,525 cm^{-1} marked in yellow corresponds to the N9H stretch and is absent in all the guanosine traces: 2′,3′-deoxyguanosine (Fig, 2b), 3′-deoxyguanosine (Fig. 2c), 2′-deoxyguanosine (Fig. 2d), and guanosine (Fig. 2e). The blue and purple bands denote the symmetric and antisymmetric NH_2 stretch modes, while the red band denotes the OH stretch, showing that all these guanosines are in the enol form. As is the case with bare nucleobase, the keto form is not observed with nanosecond timescale R2PI. The red peaks correlate with the 2′-OH and the blue peaks correlate with the 3′-OH modes. The small red shift of the 3′-OH mode in Fig. 2e compared to Fig. 2d indicates a small amount of hydrogen bonding, consistent with a "windshield wiper" orientation with the 3′-OH pointing towards the 2'O. The 5′-OH frequency is absent in this frequency range, presumably because of strong red shifting, suggesting a strong

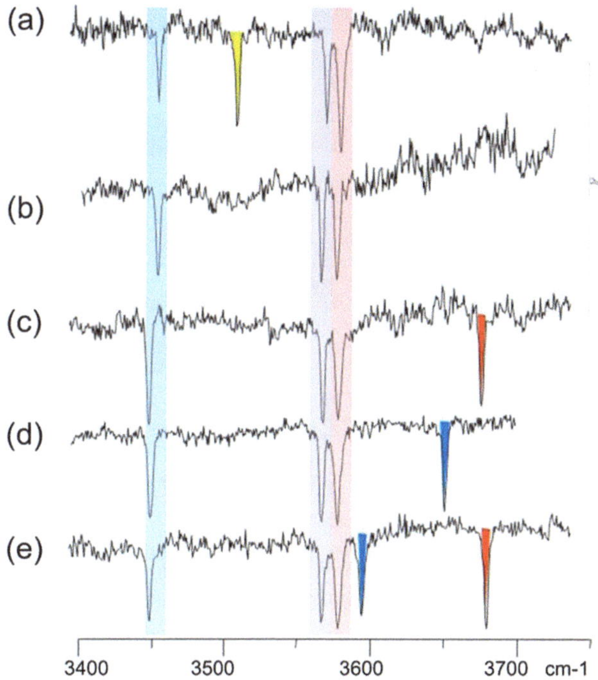

Fig. 2 IR-UV DRS spectra of (**a**) guanine, (**b**) 2′,3′-deoxyguanosine, (**c**) 3′-deoxyguanosine, (**d**) 2′- deoxyguanosine, and (**e**) guanosine

internal hydrogen bond consistent with a *cis* orientation of the sugar. In the biological context this hydrogen bond is absent. Asami et al. have shown that, while indeed the *cis* form is most stable, by contrast, in the case of 5′-*O*-ethylguanosine, the anti conformer is stabilized by the formation of hydrogen bonding involving the 2′-OH group [39, 40].

The near IR range between 3,000 and 4,000 cm^{-1} is very diagnostic for structure since it contains the N–H and O–H stretches. Inclusion of the C=O stretch frequency around 1,800 cm^{-1}, which now becomes possible with table-top laser systems, extends the useful range further. The mid-IR range, typically 500–2,000 cm^{-1}, attainable with a free electron laser, usually does not add much extra structural information. Those lower frequency modes can add further detail, however, especially for the sugar in nucleosides. Figure 3 shows IR-UV DRS spectra obtained at the FELIX free electron laser facility of guanosine (Fig. 3a) and 2′-deoxyguanosine (Fig. 3b) [41]. The red peaks correspond to modes in the guanine, while blue colored peaks denote modes of the sugar moiety. Figure 3c shows the spectrum of 9-ethyl-guanine for comparison, with the main ethyl modes marked in yellow.

The tautomeric landscape of adenine is somewhat less varied than in the case of guanine because of the absence of the oxygen. Plützer and Kleinermanns reported IR-UV double resonance and observed two tautomers, for which Fig. 4 shows the

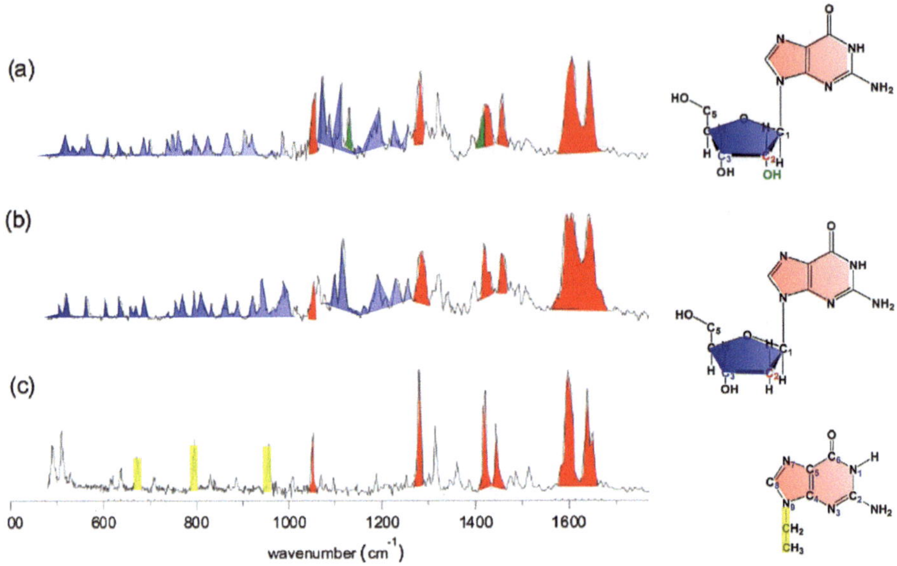

Fig. 3 IR-UV DRS spectra of (**a**) guanosine, (**b**) 2′-deoxyguanosine, and (**c**) 9-ethyl-guanine

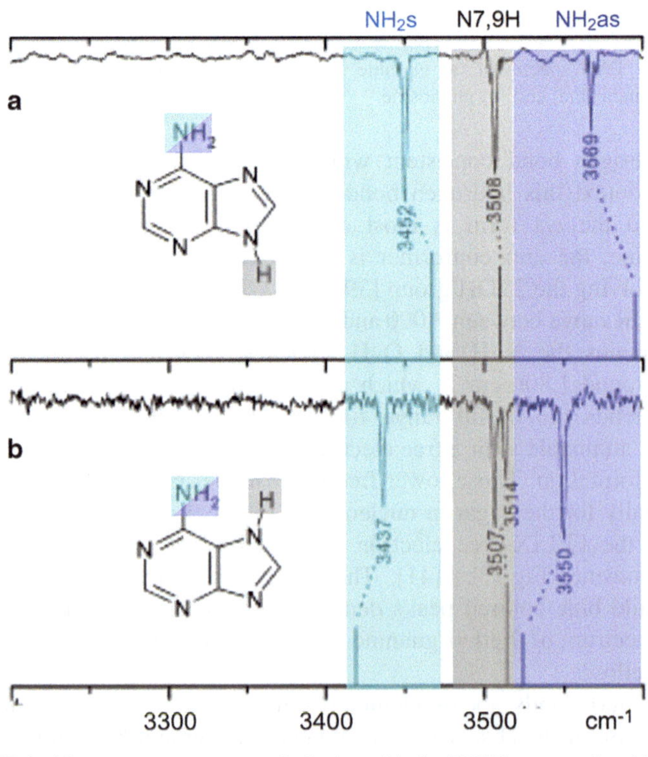

Fig. 4 IR-UV double resonance spectrum of adenine with the R2PI laser tuned to (**a**) 36,105 cm^{-1} and (**b**) 35,824 cm^{-1}. The *stick spectrum* shows the vibrational frequencies calculated at the B3LYP/6-311G(d,p) level. Data from Plützer and Kleinermanns (reprinted with permission, color coding added) [42]. Stick spectra were computed at the B3LYP/6-31G(d) level with a scaling factor of 0.9613

IR spectra [5, 42]. Both tautomers are of the amino form, with the *9H* form by far the most abundant relative to the *7H* form. This finding is consistent with the microwave measurements by Brown et al. [43]. At the conditions of jet cooling, the imino form appears to be absent, although in the gas phase at elevated temperature the IR spectra seem to comprise multiple tautomers, including imino. The analysis is somewhat complicated by the fact that the UV spectra contain contributions to two excited states, of $\pi\pi^*$ and $n\pi^*$ character, respectively [44, 45].

3.2 Pyrimidines

Figure 5 displays the main tautomers observed for the pyrimidine bases. Cytosine appears in both the keto and enol form, while uracil and thymine each appear almost exclusively in the diketo form. The IR-UV DRS spectrum of keto-cytosine appears in Fig. 6a, with 1-methyl and 5-methyl derivatives for comparison in Fig. 6b, c, respectively. The two enol forms for cytosine cannot be distinguished realistically by available IR spectra. The enol tautomer is slightly more stable than the keto form by about 0.03 eV [46–48]. The keto form is the biologically important one, with Watson–Crick base pairing in DNA, and predominant in solution. The keto-imino form in cytosine is much higher in energy. In matrix isolation, Szczesniak et al. observed both keto and enol forms with higher abundances for the latter and small contributions from the imino form [49]. Brown et al. have obtained rotational constants for all three tautomeric forms by microwave spectroscopy [50].

While the energy difference between the enol and keto forms is very small, some of their other properties, such as their vertical excitation energies, are very different [51]. Nir et al. reported R2PI and hole burning spectra and concluded that the keto and enol tautomers have band origins which differ by a remarkable half an electron volt at 314 and 278 nm, respectively [52, 53]. The R2PI spectra of U and T exhibit a very broad structure with an onset in the frequency range of the origin of the enol cytosine first reported by Brady et al. [4]. No spectroscopic detail could be extracted

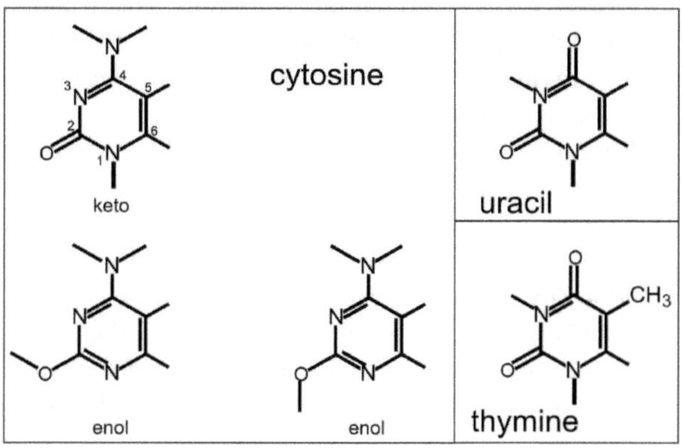

Fig. 5 Structures of lowest energy tautomers of the canonical pyrimidine bases

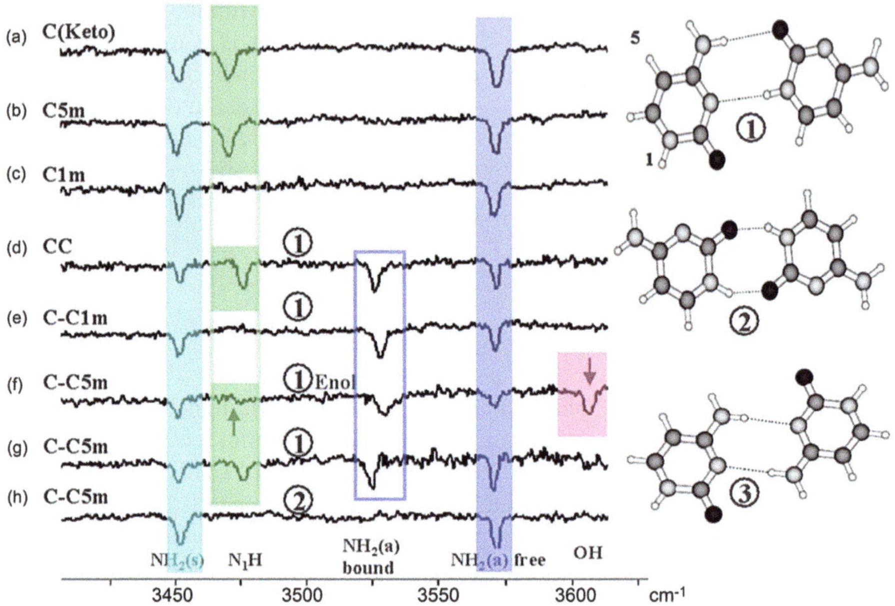

Fig. 6 IR-UV DRS spectra of cytosine monomers (**a–c**) and dimers (**d–h**). C1m and C5m denote 1-methylcytosine and 5-methylcytosine, respectively. *Circled numbers* on the dimer traces refer to the cluster structures shown on the right. (**f–h**) were recorded at three different UV wavelengths, corresponding to three different structures of the C-C5m cluster

from these broad spectra and tautomeric information was limited for a long time to bulk measurements.

Microwave measurements of uracil in a heated cell suggested the diketo form as the most abundant [54]. Brown et al. reported the first microwave measurements in a seeded molecular beam and also concluded that the diketo form was predominant [55]. Viant et al. reported the first rotationally resolved gas phase IR spectra of uracil [28]. This work employed a slit nozzle, an IR diode laser, and a multipass arrangement to obtain high resolution IR absorption spectra of the out-of-phase $\nu_6(C_2{=}O, C_4{=}O)$ stretching vibration. The rotational analysis unambiguously assigned the species to the diketo tautomer. Brown et al. also observed the diketo form of thymine in a seeded molecular beam, based on the hyperfine structure in the $14_{4,10}{-}13_{3,11}$ transition [56].

Recently, Ligare et al. confirmed the diketo character of the broad absorption of both U and T by IR-UV DRS. Their spectra showed ion gain rather than dips upon IR excitation. Double resonant spectroscopy relies on the fact that the burn laser changes the ground state vibrational distribution and thus the overall Franck–Condon (FC) factors. When the probe laser is tuned to a strong resonance, this invariably leads to ion dip. However, in the case of a broad absorption, the consequences of modified FC factors are not always predictable and an ion dip may not necessarily occur. In the case of U and T, it is probable that there is a large geometry change between the ground state and the excited electronic state, which leads to more favorable FC factors and thus ion gain. Such a scenario would also help explain the gradual onset of the R2PI spectrum and the absence of a strong 0–0 transition.

3.3 Excited State IR

An additional reason for excited state broadening may be lifetime broadening. As discussed below, most of the $\pi\pi^*$ excited state returns to the ground state on a picosecond timescale by internal conversion. However, there is a small quantum yield for a process leading to a "dark" excited state of $n\pi^*$ character or a triplet state ($^3\pi\pi^*$ or $^3n\pi^*$). Kunitski et al. have characterized this dark state for 1-methylthymine by performing IR-UV DRS on the excited state, rather than the ground state [57]. Figure 7 shows the spectra obtained for the NH stretch. The bottom trace is the usual ground state IR-UV DRS spectrum. The top trace results from a different pulse sequence in which the molecule is first excited by the pump laser and partly relaxes into the dark state, which is subsequently subjected to be IR burn pulse and ionized by the probe pulse. The result is a small red shift, interpreted as characteristic for a triplet state, whereas an nπ^* state would have produced a blue shift. Ligare et al. have recently obtained similar results for thymine itself.

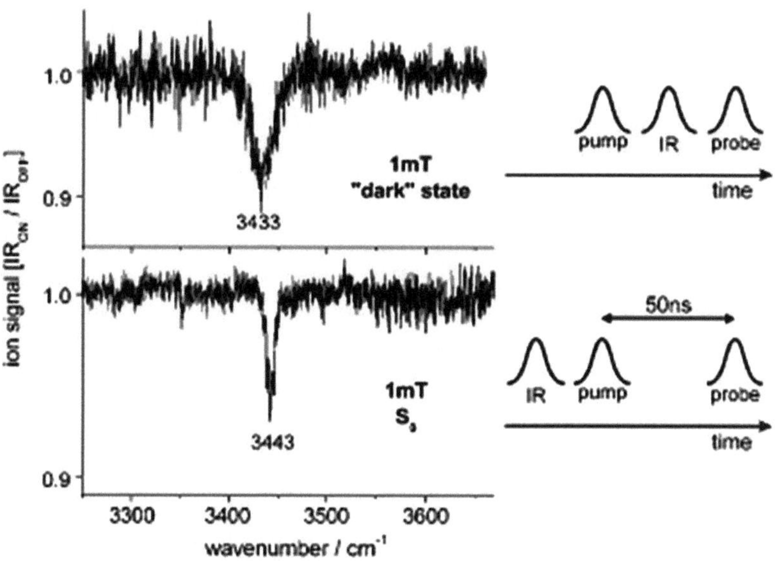

Fig. 7 Ionization-detected IR spectra of 1-methylthymine in the ground state (*bottom*) and long lived transient excited state (*top*) in the NH/OH stretch region. The corresponding laser pulse sequences are shown on the *right* [57]

3.4 Ionic Nucleobases and Nucleotides

For nucleotides, the charge on the phosphate group generally precludes the use of the IR-R2PI hole burning technique. Instead, it is possible to study ions in a trap by IR multiphoton dissociation (IRMPD). The characteristics of a free electron laser, such as FELIX, with its macro and micro pulses, are very suitable for this type of multiphoton IR spectroscopy [58]. Since there is no isomer selection in this case, the interplay with theory is especially important and the occurrence of multiple structural forms could complicate interpretation. van Zundert et al. compared results for neutral (by DRS) and protonated (by IRMPD) adenine and 9-methyladenine in the same mid-IR frequency range of 525–1,750 cm^{-1} [59]. They found the 9H tautomer to be dominant for both neutrals and ions. Salpin et al. studied protonated uracil, thymine, and cytosine by IRMPD spectroscopy [60]. Calculated infrared frequencies of the different low-lying isomers (computed at the B3LYP/6-31++GACHTUNGTRENUNG(d,p) level) predict the global energy minimum for an enolic tautomer in each case, whose infrared absorption spectra matched very well with the experimental IRMPD spectra. An additional very weak IRMPD signal observed at about 1,800 cm^{-1} suggests the presence of the second lowest energy oxo tautomer. Oomens et al. studied protonated cytosine dimers, concluding that the proton moves from one basic atom to another when the dimer ion dissociates [61]. Yang et al. studied the effect of C5 substituents on the cytosine base pairing motifs and concluded that the substitution does not alter the lowest energy base pairing structures [62, 63]. Yang et al. found a single tautomer for alkali metal cation cytosine complexes [64], as did Nei et al. for sodiated uracil and thiouracil complexes [65, 66]. On the other hand, Crampton et al. found that protonation preferentially stabilizes minor tautomers of halouracils [67].

Nei et al. have reported the IR spectra of all deprotonated canonical nucleotides, trapped in an FTICR instrument and subjected to IRMPD in the mid-IR [68, 69]. The measured IRMPD action spectra were compared to the linear IR spectra calculated at the B3LYP/6-311 + G(d,p) level of theory and the most stable conformations of the deprotonated forms of dA5′p, dC5′p, and T5′p were found to be conformers where the ribose moiety adopts a C3′ *endo* conformation and the nucleobase is in an *anti* conformation. By contrast, the most stable conformations of the deprotonated form of dG5’p are conformers where the ribose adopts a C3′ *endo* conformation and the nucleobase is in a *syn* conformation. In addition to the ground-state conformers, several stable low-energy excited conformers that differ slightly in the orientation of the phosphate ester moiety were also accessed in the experiments. Comparison of the conformations found by these authors for DNA nucleotides vs RNA nucleotides would suggest that the intrinsic difference between the DNA and RNA mononucleotides is probably not caused by their relative conformations but by the change in their chemical properties induced by the different substituents at the C2′ position [68]. By the same technique, Ligare et al. studied deprotonated clusters of nucleotides and found that, unlike in multimer double stranded DNA structures, the hydrogen bonds in these isolated

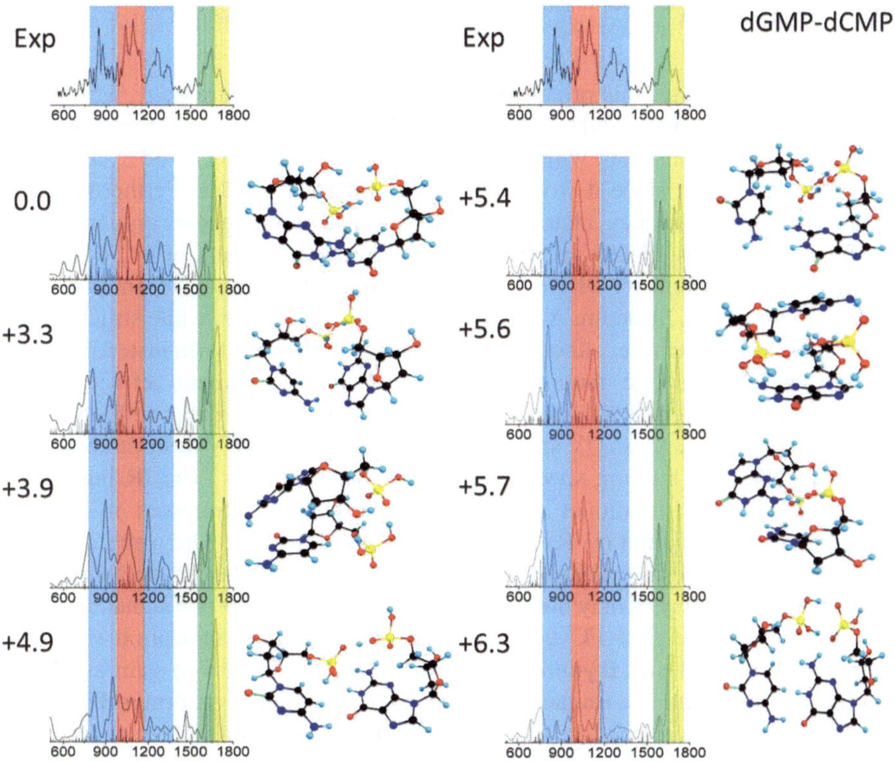

Fig. 8 The six lowest calculated energy spectra for deprotonated [dGMP-dCMP]- compared with IRMPD experiment. The energies listed are Gibbs free energies in kcal/mol relative to the ground state structure [70]

nucleotide pairs are predominantly formed between the phosphate groups, as shown in Fig. 8.

4 Cluster Structures

IR spectroscopy is also very diagnostic for the structure of hydrogen bonded clusters. Upon hydrogen bonding, NH and OH stretch frequencies in the near IR generally shift to the red, often by hundreds of wavenumbers, while also broadening. The extent of the red shift is correlated with the strength of the hydrogen bond. The red shift and broadening cannot be computationally predicted with much accuracy, although they are very clear indicators, showing which hydrogen bonding sites are implicated in bonds, while frequencies that do not shift mark the free sites. Therefore near IR spectra can provide insight into the cluster structure, often to the point of complete structural assignment.

4.1 Base Pairs

Figure 6 demonstrates this principle for cytosine dimers. Figure 6c–g, with both a free and a bound (shifted) NH_2 vibration clearly point to structure 1. Figure 6h with only free NH_2 is consistent with structure 2, while structure 3 can be excluded for all of these spectra, since it would contain no free NH_2 modes. Furthermore, the 3,600 cm^{-1} OH frequency, characteristic for the enol form, appears only in Fig. 6f.

Figure 12 shows IR-UV spectra for three structural isomers of GC base pairs [72]. The stick spectra are DFT calculated vibrational frequencies for the structures shown in the insets. Structure A is the Watson–Crick structure, while structure C is almost the same structure, although the cytosine is in the enol form instead of in the keto form. This subtle tautomeric difference has a dramatic effect on the photophysical properties as discussed below. The IR-UV spectrum in F correlates with a broad, structureless UV spectrum. The fact that broad UV absorption yields a sharp IR-UV hole burning spectrum is somewhat fortuitous. The IR burn pulse excites a specific vibrational mode around 3,000 cm^{-1}, followed by internal vibrational redistribution (IVR) on the picosecond timescale. The resulting vibrational state populations produce a different Franck–Condon landscape, which for a broad absorption may or may not lead to depletion of the UV probe signal. In fact, it is possible that the new FC landscape leads to a gain in ionization signal, as observed for uracil and thymine. Mayorkas et al. also observed gain signals in ionization loss stimulated Raman spectroscopy of tryptamine conformers [73]. Similarly, gain signals are possible following excitation in the mid-IR, which imparts a small amount of energy and thus provide less opportunity of vibrational redistribution.

A particular computational challenge is the incorporation of anharmonicity of the potentials. A common fix is the application of scaling factors but this approach, while useful, is clearly fraught with uncertainty. While it is good practice always to use the same empirical factor for consistency, it is also reasonable to assume that different modes could require different factors, leaving one with a somewhat uncomfortable number of empirically adjustable parameters. In the quest to develop more comprehensive computational strategies, it is useful to be able to compare with both near- and mid-IR data as sensitive tests. Figure 9 shows such a comparison for the GC base pair with the cytosine in the enol form, as depicted in Fig. 12c [74]. The mid-IR part of this spectrum (left panel) was obtained at FELIX and the lower frequencies provide an especially sensitive test for theory. Comparison of the theoretical frequencies with the experimental results indicates that the average absolute percentage deviation for the methods is 2.6% for harmonic RI-MP2/cc-pVDZ (3.0% with the inclusion of a 0.956 scaling factor which compensates for anharmonicity), 2.5% for harmonic RI-MP2/TZVPP (2.9% with a 0.956 anharmonicity factor included), and 2.3% for adapted PM3 CC-VSCF; the empirical scaling factor for the ab initio harmonic calculations improves the stretching frequencies but decreases the accuracy of the other mode frequencies.

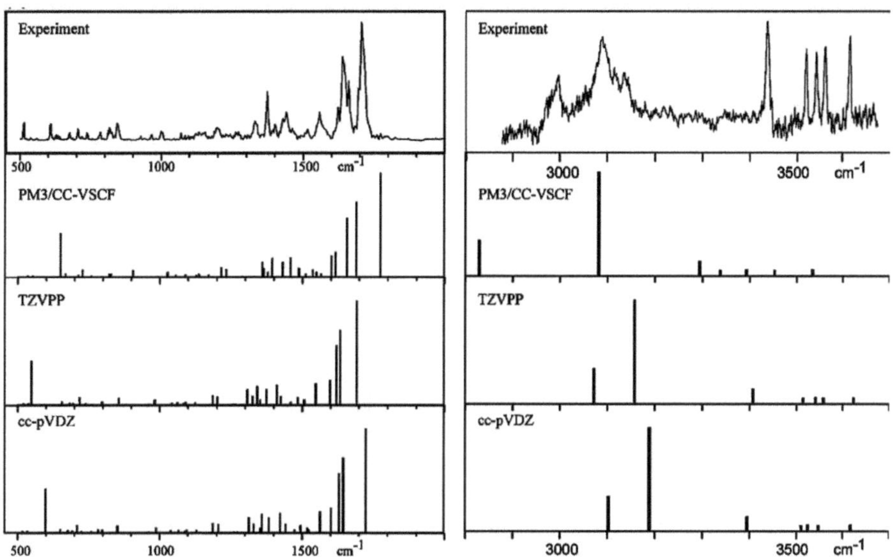

Fig. 9 Comparison of experimental data with calculations according to different models aimed at correcting for anharmonicity [74]

This work is an example of the ongoing cross fertilization of theory and experiment, in which gas-phase data serve as benchmarks for computational method development, while the computations provide the analysis of the experimental results.

For AT base pairs, Plützer et al. reported the IR-UV spectrum, which they assigned to cluster structures with (HNHO)-O-· · ·=C/(NHN)-H-· · · hydrogen bonding by comparison with ab initio calculated vibrational spectra of the most stable A–T isomers [75]. The Watson–Crick A–T base pair is not the most stable base-pair structure at different levels of ab initio theory, and its vibrational spectrum is not in agreement with the observed experimental spectrum. This is directly shown by the free N3H vibration in the IR spectrum of AT. This group is involved in hydrogen bonding in the Watson–Crick pair. Experiments with methylated A and T further support the structure assignment.

4.2 Clusters with Water

IR-UV DRS also lends itself to analysis of clusters with water in order to study the details of hydration, one water molecule at a time. Figure 10 shows IR-UV spectra of guanine water clusters as an example [76]. Figure 10a–c shows spectra from three observed structures while Fig. 10d–f show spectra from corresponding non-hydrated structures for comparison. The hydrated spectra contain broad, red-shifted peaks caused by hydrogen bonded modes, but the free modes suffice

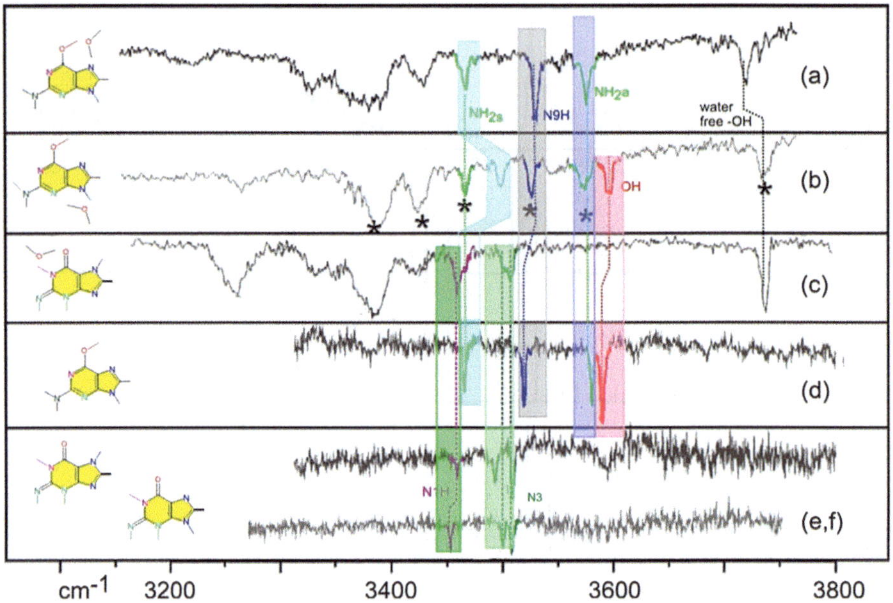

Fig. 10 IR-UV hole burning spectra of guanine with (**a–c**) and without (**d–f**) water. Corresponding modes for corresponding structures with and without water are indicated by *dotted lines* and color coding. Structures are schematic only. *Asterisks* denote peaks in **b** which also show up in **a**, probably caused by overlap in the probe spectra. Adapted from [76]

for analysis. They each contain a free water OH stretch, excluding bridged structures. Figure 10a, b parallel Fig. 10d. Figure 10b contains the free enol OH mode. Figure 10a is almost identical but without the enol OH mode indicating an enol with the water bound to the OH. Figure 10c parallels Fig. 10e, f, missing both NH_2 modes as well as the enol mode, but showing the N3H and N1H mode characteristic for the imino form. Once again, the keto form is not observed in these experiments. Chin et al. also reach that conclusion in their detailed study of rotamers of 9-methylguanine with water [32, 35].

Saigusa et al. reported IR-UV DRS spectra of hydrated clusters of nucleosides [77, 78] and of GG and GC base pairs [79]. For the monohydrated cluster of Gs they found multiple structural isomers. For one of the monohydrates they assigned a hydration structure involving the 5′-OH group of the sugar and the amino group of the guanine moiety. From the IR spectrum of the dehydrated clusters they inferred that the 2′-OH group is significantly influenced by the addition of the second water, which suggests the possibility of specific dihydrate structures for Gs. Their results also suggest that the amino-keto forms of Gs and 9MG, which are missing in the R2PI spectrum, can be observed upon hydration.

4.3 *Stacking vs H-Bonding Structures*

Two structural motifs dominate in DNA: π-stacking provides most of the structural stability of the helix while hydrogen bonding provides most of the recognition properties. For pairs of single bases H-bonding is the exclusive motif in the gas phase, while stacking dominates in solution. Water molecules stabilize the stacked structure both by competing with hydrogen bonding sites on the bases and by stabilizing the structure by forming bridges. Hobza and coworkers have calculated how many water molecules are needed to lead to preferential stacking and concluded that, depending on the base pair, this number can range from two to five [80, 81]. Clusters of that size have not yet been characterized in the gas phase, so this transition has not yet been experimentally verified. Another way to encourage stacking is blocking hydrogen bonding sites by methylation and several examples have been reported.

Callahan et al. have used xanthine and its methylated derivatives as models for studying the two motifs [82, 83]. For the 7-methylxanthine dimer, they observed hydrogen bonding on the N3H position, suggesting three possible combinations, one that is reverse Watson–Crick type and two that are reverse Hoogsteen type. For the 3–7-dimethylxanthine dimer, they observed a stacked structure, as determined by the free N1H stretch frequency. For trimethylxanthine dimers they inferred a stacked structure as well.

Plützer et al. reported a stacked structure for 9-methyladenine-adenine (9 mA-A) clusters based on one free NH_2 group and one weakly interacting NH_2 and N9H group in the IR-UV spectrum [84]. They found no evidence of stacking for A-A, 7 mA-A, or 9 mA-9 mA clusters. The latter shows only broad vibronic structure and is probably symmetrically hydrogen bound via the NH_2 groups. Interestingly, the symmetrically bound A-A cluster structure was not observed, although it is predicted to be the lowest in energy. Smith et al. studied adenine and microhydrated $A_m(H_2O)_n$ clusters by femtosecond pump–probe mass and photoelectron spectroscopy [85]. For the predominantly hydrogen-bonded adenine dimer, excited state relaxation is dominated by monomer-like processes. However, when the adenine dimer is clustered with several water molecules, they observed a nanosecond lifetime ascribed to excimer states in π-stacked clusters.

Asami et al. reported IR-UV DRS spectra for base pairs of adenine nucleosides, adenosine (Ado) and *N*6,*N*6-dimethyladenosine (DMAdo) [77]. They found that the dimer possesses a stacked structure, stabilized by the formation of a hydrogen-bonding network involving the two sugar groups. The occurrence of the frequency shift and broadening is explained satisfactorily based on the anharmonic coupling of the OH stretching modes with specific bending modes and low-frequency modes of base and sugar moieties. By contrast, Saigusa et al. performed experiments with GG and GC pairs with methyl substitutions in the sugar position, which for GC forces the Watson–Crick structure. The IR markers for this structure remained essentially unchanged upon hydration, suggesting no significant structural change nor stabilization of π stacking [79].

5 Excited State Dynamics

With the ability to perform isomer-selective IR spectroscopy has come the opportunity to study effects of subtle structural variations. One of the remarkable outcomes is that certain low energy isomers have not been observed in IR-UV DRS. Notable examples are the keto tautomer of guanine and the Watson–Crick structure of the GC base pair. Similarly, the lowest energy symmetrically hydrogen bonded structures of homodimers, such as GG and AA, are missing in these IR-UV measurements. Intriguingly, these "missing" isomers tend to be the biologically most relevant forms. Inevitably, double resonant spectroscopy involves a form of action spectroscopy, usually two-photon ionization and sometimes laser induced fluorescence. Both techniques detect the excited state. In a number of cases, where direct absorption measurements are available, the "missing" isomers are in fact observed, implying that they do exist in the gas phase. A prime example is the keto tautomer of guanine, which is identified both in microwave spectroscopy in a molecular beam [19] and in helium droplets [25].

The failure to observe certain isomers can be attributed to short-lived electronically excited states. An excited-state lifetime of the order of picoseconds or less is in fact four or more orders of magnitude shorter than the typical laser pulses of several nanoseconds, routinely used for two-photon ionization. This fact renders this form of action spectroscopy blind for the species with the shortest excited state lifetimes. A picture is now emerging in which the electronic excited state can undergo rapid internal conversion to the ground state. The key is the occurrence of conical intersections (CIs) which connect the excited state potential energy surface, reached by photon absorption, to the ground state potential energy surface. The dramatic lifetime differences between isomers and derivatives of nucleobases appear to be caused by variations in the excited state potential surfaces which restrict or slow access to these conical intersections [86]. As illustrated schematically in Fig. 11, conical intersections are the crossings of multidimensional potential surfaces. Therefore, these features can only occur in regions of the potential energy landscape which represent a deformation of the molecular frame from the ground state equilibrium geometry. For example, Fig. 11 shows the geometry at the conical intersection for the keto tautomer of guanine, calculated by Marian [36]. The CI involves strong out-of-plane bending of the C2 coordinates and different tautomeric arrangements lead to different potential surfaces along those coordinates. As a result, other tautomers do not lead to the same trajectories on the excited state potential surface and only the keto form exhibits the barrierless pathway through the CI, producing its sub-picosecond internal conversion. This mechanism also explains the strong dependence of excited state lifetimes on derivative structure [87–89]. When a subpicosecond internal conversion pathway is available (indicated schematically in red in Fig. 11), it can compete favorably with other processes, such as fluorescence (green) or other photochemical processes (orange).

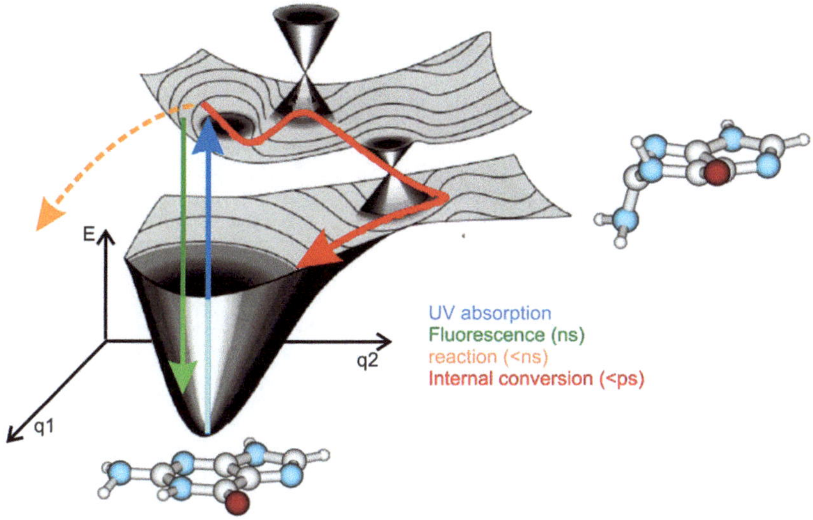

Fig. 11 Schematic potential energy diagram as a function of two generic internuclear coordinates, q1 and q2, showing the concept of conical intersections connecting different electronic states. *Insets* show the geometry of guanine at the ground state equilibrium and at the conical intersection between the $\pi\pi^*$ and the SO state, as calculated by Marian [36]

In the case of the pyrimidine bases, the major coordinates forming the CI are torsion and stretching of the C5=C6 bond. For 4-aminopyrimidine, surface hopping calculations identified two conical intersections [5, 42, 45, 90–101]: deformation at the C2 position leads to deactivation of the excited state with a lifetime, τ^*, of 1 ps and deformation at the C5=C6 bond with a τ^* of 400 fs. Immobilizing the latter with a 5-membered ring forms adenine with a conical intersection produced by the C2 deformation, τ^*, of 1 ps and dynamics similar to that in guanine. C2 substitution changes the potential energy landscape along this coordinate, such that adenine's isomer 2-aminopurine has a nanosecond excited state lifetime and strongly fluoresces. Similarly, for the pyrimidine bases the oxygen substitution at the C2 position eliminates the CI associated with the C2 puckering but leaves the CI associated with the C5=C6 coordinates available for internal conversion. C5 substituents in pyrimidines further alter excited state lifetimes over a range of picoseconds to nanoseconds by modification of the topography of the potential energy surfaces around C5=C6 torsion and stretching coordinates [89, 102, 103]. Interestingly, the same coordinates are found to play a role in thymine photo-dimerization in DNA [104].

We can now rationalize why, for GC clusters, the one structure that is prevalent in DNA, the Watson–Crick structure, has not yet been observed by two-photon ionization with nanosecond laser pulses. In the work of Nir et al. only a substituted version (9-ethyl-G-1-methyl-C) was reported (Fig. 12a), and its corresponding two-photon UV spectrum was very broad [72, 105]. Presumably, the Watson–Crick structure has a sub-picosecond excited state lifetime, while other structural

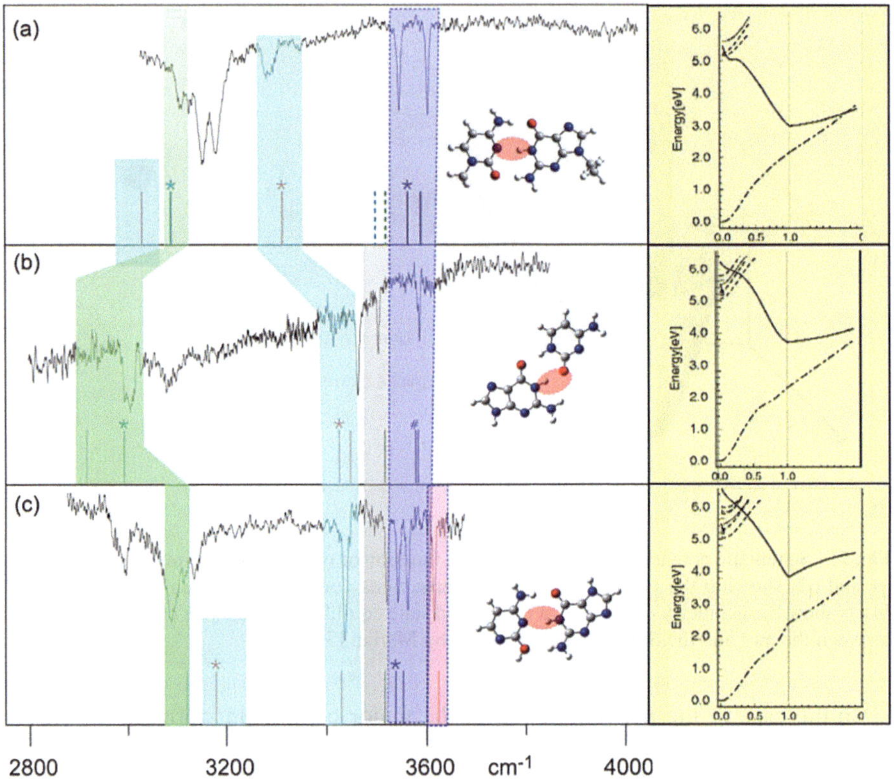

Fig. 12 *Left column* shows three structures observed for isolated GC base pairs, as identified by IR-UV spectroscopy. Stick spectra are DFT calculated vibrational frequencies. Structure A is the Watson–Crick structure. *Right column* shows calculated potential curves [71]. A charge transfer state (CT in *red*) connects the S1 (*green*) and S0 state (*blue*) by two conical intersections in (**a**) but not in (**b**) and (**c**). The reaction coordinate is middle hydrogen motion, N1-H, indicated with a *red ellipse* in the structures. Color coding of the modes is the same as in Fig. 1

arrangements of the same base pair have sharp spectra, consistent with much longer excited state lifetimes [106]. To explain the effect, the right column of Fig. 12 shows calculated potential curves from Sobolewski and Domcke [71]. The reaction coordinate is hydrogen motion, N1-H, indicated with a red circle in the structures. A charge transfer state connects the S1 and S0 states by two conical intersections in A but not in B and C. The difference in tautomeric form between the A and the C case induces a subtle change in the excited state potential energy landscape. The charge transfer state is just slightly higher in energy relative to S0 and S1, creating a barrier for reaching the first conical intersection. Thus a small change in potential energies upon tautomerization results in a difference of orders of magnitude in excited state lifetime.

It has been argued that a short excited state lifetime, by virtue of rapid internal conversion, is nature's strategy to protect the building blocks of life against UV

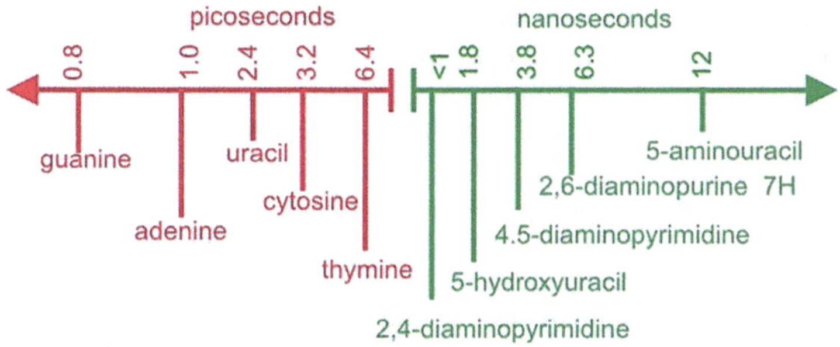

Fig. 13 Representative excited state lifetimes in the gas phase

induced damage which would otherwise result from slower photochemical processes. Figure 13 shows excited state lifetimes for the canonical bases in comparison with those for several of their derivatives. It appears that sub-picosecond lifetimes are a fairly unique property of the canonical bases and, as outlined above, this property is even specific to tautomers and base pair structures encountered in DNA. This would, however, not be a strategy that could have been adopted by biological evolution. For evolution to proceed, the replication machinery has to be in place first. Therefore, any photochemical properties the bases have, must have originated from prebiotic times. Assuming that a primordial soup would have contained many possible variations of the heterocyclic compounds, it is conceivable that the building blocks of life underwent photochemical selection on an early Earth. If so, these properties would be relics from chemistry that took place some four billion years ago.

In this context, it needs to be remembered that gas-phase spectroscopy constitutes a reductionist approach to the study of basic chemistry. Extrapolation to bulk conditions requires consideration of hydrogen bonding and stacking interactions. These interactions modify the photochemistry. For example, π stacking can lead to exciplex formation, which opens up additional deexcitation channels [107].

6 Summary and Outlook

The prime objective of gas-phase studies is the investigation of intrinsic properties in *isolated* molecules. For a long time this approach has been limited to small molecules. The application to larger molecules is made possible by simultaneous progress in both experimental and computational technique. Laser desorption has made it possible to vaporize low vapor pressure neutral compounds intact and progress in computing power and theoretical treatments have made larger systems amenable to high level computation. The resulting interplay between theory and experiment has been a powerful driver of this field of study. Theory has been instrumental in analyzing and guiding experiments, while at the same time gas-phase data are valuable for calibrating computational strategies. IR frequencies are among the best experimental data, available for this purpose. IR-UV DRS has emerged as a particularly fruitful meeting ground between theory and experiment, given its isomer selectivity. At this point, even very high level computations still lack precision, particularly in predicting hydrogen bond shifts and accounting for anharmonicity, so progress on this front may still be expected. It is still unclear what the molecular size limit is for this approach. The extrapolation from isolated molecules to bulk is probably aided by studies of clusters. In addition to base pair clusters, work on several small clusters with water has been reported, but data on larger clusters, at least approaching a first solvation shell, are still desirable.

A major use of IR data is structure determination, which for nucleobases includes tautomeric forms and cluster structures to model base pair interactions and microhydration. This capability also makes it possible to follow other properties as a function of structure. For the nucleobases, it turns out that self-protection against UV photodamage by internal conversion depends dramatically on molecular structure. Comparison with solution experiments probably further elucidates these findings and stimulates further research.

Acknowledgement This material is based upon work supported by the National Science Foundation under CHE-1301305 and by NASA under Grant No. NNX12AG77G.

References

1. Kim SK, Lee W, Herschbach DH (1996) Cluster beam chemistry: hydration of nucleic acid bases; ionization potentials of hydrated adenine and thymine. J Phys Chem 100:7933–7937
2. Anderson JB, Fenn JB (1965) Velocity distributions in molecular beams from nozzle sources. Phys Fluids 8:780
3. Levy DH (1980) Laser spectroscopy of cold gas-phase molecules. Annu Rev Phys Chem 31:197–225
4. Brady BB, Peteanu LA, Levy DH (1988) The electronic spectra of the pyrimidine bases uracil and thymine in a supersonic molecular beam. Chem Phys Lett 147:538–543
5. Plützer C, Nir E, de Vries MS, Kleinermanns K (2001) IR-UV double-resonance spectroscopy of the nucleobase adenine. PCCP 3:5466–5469
6. Muller A, Talbot F, Leutwyler S (2000) Intermolecular vibrations of jet-cooled (2-pyridone)/sub 2/: a model for the uracil dimer. J Chem Phys 112:3717–3725
7. Muller A, Talbot F, Leutwyler S (2001) Intermolecular vibrations of the jet-cooled 2-pyridone center dot 2-hydroxypyridine mixed dimer, a model for tautomeric nucleic acid base pairs. J Chem Phys 115:5192–5202
8. Muller A, Talbot F, Leutwyler S (2002) Hydrogen bond vibrations of 2-aminopyridine center dot 2-pyridone, a Watson–Crick analogue of adenine center dot uracil. J Am Chem Soc 124:14486–14494
9. Frey JA, Muller A, Frey HM, Leutwyler S (2004) Infrared depletion spectra of 2-aminopyridine center dot 2-pyridone, a Watson–Crick mimic of adenine center dot uracil. J Chem Phys 121:8237–8245
10. Held A, Pratt DW (1992) Hydrogen bonding in the symmetry-equivalent C/sub 2h/ dimer of 2-pyridone in its S/sub 0/ and S/sub 2/ electronic states. Effect of deuterium substitution. J Chem Phys 96:4869–4876

11. Arrowsmith P, de Vries MS, Hunziker HE, Wendt HR (1988) Pulsed laser desorption near a jet orifice: concentration profiles of entrained perylene vapor. Appl Phys B 46:165–173
12. Meijer G, de Vries MS, Hunziker HE, Wendt HR (1990) Laser desorption jet-cooling of organic molecules. Cooling characteristics and detection sensitivity. Appl Phys B 51:395–403
13. Tembreull R, Lubman DM (1987) Resonant two-photon ionization of small peptides using pulsed laser desorption in supersonic beam mass spectrometry. Anal Chem 59:1003–1006
14. Weyssenhoff HV, Selzle HL, Schlag EW (1985) Laser-desorbed large molecules in a supersonic jet. Zeitschrift fur Naturforschung Teil A 40a:674–676
15. Mayorkas N, Malka I, Bar I (2011) Ionization-loss stimulated Raman spectroscopy for conformational probing of flexible molecules. PCCP 13:6808–6815
16. Alonso JL, López JC (2014) Microwave spectroscopy of biomolecular building blocks. Topcis Curr Chem. doi:10.1007/128_2014_601
17. Lopez JC, Pena MI, Sanz ME, Alonso JL (2007) Probing thymine with laser ablation molecular beam Fourier transform microwave spectroscopy. J Chem Phys 2007:126
18. Vaquero V, Sanz ME, Lopez JC, Alonso JL (2007) The structure of uracil: a laser ablation rotational study. J Phys Chem A 111:3443–3445
19. Alonso JL, Pena I, Lopez JC, Vaquero V (2009) Rotational spectral signatures of four tautomers of guanine. Angew Chem Int Ed 48:6141–6143
20. Alonso JL, Vaquero V, Pena I, Lopez JC, Mata S, Caminati W (2013) All five forms of cytosine revealed in the gas phase. Angew Chem Int Ed 52:2331–2334
21. Lopez JC, Alonso JL, Pena I, Vaquero V (2010) Hydrogen bonding and structure of uracil-water and thymine-water complexes. PCCP 12:14128–14134
22. Merritt JM, Douberly GE, Miller RE (2004) Infrared-infrared-double resonance spectroscopy of cyanoacetylene in helium nanodroplets. J Chem Phys 121:1309–1316
23. Choi MY, Dong F, Miller RE (2005) Multiple tautomers of cytosine identified and characterized by infrared laser spectroscopy in helium nanodroplets: probing structure using vibrational transition moment angles. Philos Trans R Soc Lond Ser A Mathemat Phys Eng Sci 363:393–412
24. Smolarek S, Rijs AM, Buma WJ, Drabbels M (2010) Absorption spectroscopy of adenine, 9-methyladenine, and 2-aminopurine in helium nanodroplets. PCCP 12:15600–15606
25. Choi MY, Miller RE (2006) Four tautomers of isolated guanine from infrared laser spectroscopy in helium nanodroplets. J Am Chem Soc 128:7320–7328
26. Amirav A, Even U, Jortner J (1981) Absorption-spectroscopy of ultracold large molecules in planar supersonic expansions. Chem Phys Lett 83:1–4
27. Liu K, Fellers RS, Viant MR, McLaughlin RP, Brown MG, Saykally RJ (1996) A long path length pulsed slit valve appropriate for high temperature operation: infrared spectroscopy of jet-cooled large water clusters and nucleotide bases. Rev Sci Instrum 67:410–416
28. Viant MR, Fellers RS, Mclaughlin RP, Saykally RJ (1995) Infrared-laser spectroscopy of uracil in a pulsed slit jet. J Chem Phys 103:9502–9505
29. Nir E, Janzen C, Imhof P, Kleinermanns K, de Vries MS (2001) Guanine tautomerism revealed by UV-UV and IR-UV hole burning spectroscopy. J Chem Phys 115:4604–4611
30. Nir E, Grace L, Brauer B, de Vries MS (1999) REMPI spectroscopy of jet-cooled guanine. J Am Chem Soc 121:4896–4897
31. Cerny J, Spirko V, Mons M, Hobza P, Nachtigallova D (2006) Theoretical study of the ground and excited states of 7-methyl guanine and 9-methyl guanine: comparison with experiment. PCCP 8:3059–3065
32. Chin W, Mons M, Piuzzi F, Tardivel B, Dimicoli I, Gorb L, Leszczynski J (2004) Gas phase rotamers of the nucleobase 9-methylguanine enol and its monohydrate: optical spectroscopy and quantum mechanical calculations. J Phys Chem A 108:8237–8243
33. Mons M, Dimicoli I, Piuzzi F, Tardivel B, Elhanine M (2002) Tautomerism of the DNA base guanine and its methylated derivatives as studied by gas-phase infrared and ultraviolet spectroscopy. J Phys Chem A 106:5088–5094

34. Chin W, Mons M, Dimicoli I, Piuzzi F, Tardivel B, Elhanine M (2002) Tautomer contribution's to the near UV spectrum of guanine: towards a refined picture for the spectroscopy of purine molecules. Eur Phys J D 20:347–355
35. Piuzzi F, Mons M, Dimicoli I, Tardivel B, Zhao Q (2001) Ultraviolet spectroscopy and tautomerism of the DNA base guanine and its hydrate formed in a supersonic jet. Chem Phys 270:205–214
36. Marian CM (2007) The guanine tautomer puzzle: quantum chemical investigation of ground and excited states. J Phys Chem A 111:1545–1553
37. Mons M, Piuzzi F, Dimicoli I, Gorb L, Leszczynski J (2006) Near-UV resonant two-photon ionization spectroscopy of gas phase guanine: evidence for the observation of three rare tautomers. J Phys Chem A 110:10921–10924
38. Seefeld K, Brause R, Häber T, Kleinermanns K (2007) Imino tautomers of gas-phase guanine from mid-infrared laser spectroscopy. J Phys Chem A 111:6217–6221
39. Asami H, Urashima S, Tsukamoto M, Motoda A, Hayakawa Y, Saigusa H (2012) Controlling glycosyl bond conformation of guanine nucleosides: stabilization of the anti conformer in 5′-O-ethylguanosine. J Phys Chem Lett 3:571–575
40. Asami H, Tsukamoto M, Hayakawa Y, Saigusa H (2010) Gas-phase isolation of diethyl guanosine 5′-monophosphate and its conformational assignment. PCCP 12:13918–13921
41. Abo-Riziq A, Crews BO, Compagnon I, Oomens J, Meijer G, Von Helden G, Kabelac M, Hobza P, de Vries MS (2007) The mid-IR spectra of 9-ethyl guanine, guanosine, and 2-deoxyguanosine. J Phys Chem A 111:7529–7536
42. Plützer C, Kleinermanns K (2002) Tautomers and electronic states of jet-cooled adenine investigated by double resonance spectroscopy. PCCP 4:4877–4882
43. Brown RD, Godfrey PD, McNaughton D, Pierlot A (1989) A study of the major gas-phase tatomer of adenine by microwave spectroscopy. Chem Phys Lett 156:61–63
44. Lee Y, Schmitt M, Kleinermanns K, Kim B (2006) Observation of ultraviolet rotational band contours of the DNA base adenine: determination of the transition moment. J Phys Chem A 110:11819–11823
45. Kim NJ, Jeong G, Kim YS, Sung J, Kim SK, Park YD (2000) Resonant two-photon ionization and laser induced fluorescence spectroscopy of jet-cooled adenine. J Chem Phys 113:10051–10055
46. Tomic K, Tatchen J, Marian CM (2005) Quantum chemical investigation of the electronic spectra of the keto, enol, and keto-imine tautomers of cytosine. J Phys Chem A 109:8410–8418
47. Trygubenko SA, Bogdan TV, Rueda M, Orozco M, Luque FJ, Sponer J, Slavicek P, Hobza P (2002) Correlated ab initio study of nucleic acid bases and their tautomers in the gas phase, in a microhydrated environment and in aqueous solution – Part 1. Cytosine PCCP 4:4192–4203
48. Kobayashi R (1998) A CCSD(T) study of the relative stabilities of cytosine tautomers. J Phys Chem A 102:10813–10817
49. Szczesniak M, Szczepaniak K, Kwiatkowski JS, Kubulat K, Person WB (1988) Matrix-isolation infrared studies of nucleic-acid constituents. 5. Experimental matrix-isolation and theoretical ab initio SCF molecular-orbital studies of the infrared-spectra of cytosine monomers. J Am Chem Soc 110:8319–8330
50. Brown RD, Godfrey PD, McNaughton D, Pierlot A (1989) Tautomers of cytosine by microwave spectroscopy. J Am Chem Soc 111:2308–2310
51. Trygubenko SA, Bogdan TV, Rueda M, Orozco M, Luque FJ, Sponer J, Slavíek P, Hobza P (2002) Correlated ab initio study of nucleic acid bases and their tautomers in the gas phase, in a microhydrated environment and in aqueous solution. Part 1. Cytosine. PCCP 4:4192–4204
52. Nir E, Muller M, Grace LI, de Vries MS (2002) REMPI spectroscopy of cytosine. Chem Phys Lett 355:59–64
53. Nir E, Hünig I, Kleinermanns K, de Vries MS (2003) The nucleobase cytosine and the cytosine dimer investigated by double resonance laser spectroscopy and ab initio calculations. PCCP 5:4780–4785

54. Nowak MJ, Szczepaniak K, Barski A, Shugar D (1978) Spectroscopic studies on vapor-phase tautomerism of natural bases found in nucleic-acids. Z Naturforsch C 33:876–883
55. Brown RD, Godfrey PD, McNaughton D, Pierlot A (1988) The microwave spectrum of uracil. J Am Chem Soc 110:2329–2330
56. Brown RD, Godfrey PD, Mcnaughton D, Pierlot AP (1989) Microwave-spectrum of the major gas-phase tautomer of thymine. J Chem Soc Chem Comm 1989:37–38
57. Kunitski M, Nosenko Y, Brutschy B (2011) On the nature of the long-lived "dark" state of isolated 1-methylthymine. ChemPhysChem 12:2024–2030
58. Oepts D, van der Meer AFG, van Amersfoort PW (1995) The free-electron-laser user facility FELIX. Infrared Phys Technol 36:297–308
59. van Zundert GCP, Jaeqx S, Berden G, Bakker JM, Kleinermanns K, Oomens J, Rijs AM (2011) IR spectroscopy of isolated neutral and protonated adenine and 9-methyladenine. ChemPhysChem 12:1921–1927
60. Salpin JY, Guillaumont S, Tortajada J, MacAleese L, Lemaire J, Maitre P (2007) Infrared spectra of protonated uracil, thymine and cytosine. ChemPhysChem 8:2235–2244
61. Oomens J, Moehlig AR, Morton TH (2010) Infrared multiple photon dissociation (IRMPD) spectroscopy of the proton-bound dimer of 1-methylcytosine in the gas phase. J Phys Chem Lett 1:2891–2897
62. Yang B, Wu RR, Berden G, Oomens J, Rodgers MT (2013) Infrared multiple photon dissociation action spectroscopy of proton-bound dimers of cytosine and modified cytosines: effects of modifications on gas-phase conformations. J Phys Chem B 117:14191–14201
63. Ung HU, Moehlig AR, Kudla RA, Mueller LJ, Oomens J, Berden G, Morton TH (2013) Proton-bound dimers of 1-methylcytosine and its derivatives: vibrational and NMR spectroscopy. PCCP 15:19001–19012
64. Yang B, Wu RR, Polfer NC, Berden G, Oomens J, Rodgers MT (2013) IRMPD action spectroscopy of alkali metal cation-cytosine complexes: effects of alkali metal cation size on gas phase conformation. J Am Soc Mass Spectrom 24:1523–1533
65. Nei YW, Akinyemi TE, Kaczan CM, Steill JD, Berden G, Oomens J, Rodgers MT (2011) Infrared multiple photon dissociation action spectroscopy of sodiated uracil and thiouracils: effects of thioketo-substitution on gas-phase conformation. Int J Mass spectrom 308:191–202
66. Nei YW, Akinyemi TE, Steill JD, Oomens J, Rodgers MT (2010) Infrared multiple photon dissociation action spectroscopy of protonated uracil and thiouracils: effects of thioketo-substitution on gas-phase conformation. Int J Mass spectrom 297:139–151
67. Crampton KT, Rathur AI, Nei YW, Berden G, Oomens J, Rodgers MT (2012) Protonation preferentially stabilizes minor tautomers of the halouracils: IRMPD action spectroscopy and theoretical studies. J Am Soc Mass Spectrom 23:1469–1478
68. Nei YW, Crampton KT, Berden G, Oomens J, Rodgers MT (2013) Infrared multiple photon dissociation action spectroscopy of deprotonated RNA mononucleotides: gas-phase conformations and energetics. J Phys Chem A 117:10634–10649
69. Nei YW, Hallowita N, Steill JD, Oomens J, Rodgers MT (2013) Infrared multiple photon dissociation action spectroscopy of deprotonated DNA mononucleotides: gas-phase conformations and energetics. J Phys Chem A 117:1319–1335
70. Ligare M, Rijs AM, Berden G, Kabelac M, Nachtigallova D, Oomens J, de Vries MS (2014) Resonant IRMPD of nucleotide monophosphate anionic clusters. Submitted
71. Sobolewski AL, Domcke W, Hättig C (2005) Tautomeric selectivity of the excited-state lifetime of guanine/cytosine base pairs: the role of electron-driven proton-transfer processes. Proc Natl Acad Sci U S A 102:17903–17906
72. Nir E, Janzen C, Imhof P, Kleinermanns K, de Vries MS (2002) Pairing of the nucleobases guanine and cytosine in the gas phase studied by IR-UV double-resonance spectroscopy and ab initio calculations. PCCP 4:732–739
73. Mayorkas N, Izbitski S, Bernat A, Bar I (2012) Simultaneous ionization-detected stimulated raman and visible-visible-ultraviolet hole-burning spectra of two tryptamine conformers. J Phys Chem Lett 3:603–607

74. Brauer B, Gerber RB, Kabelac M, Hobza P, Bakker JM, Riziq AGA, de Vries MS (2005) Vibrational spectroscopy of the G center dot center dot center dot C base pair: experiment, harmonic and anharmonic calculations, and the nature of the anharmonic couplings. J Phys Chem A 109:6974–6984
75. Plützer C, Hünig I, Kleinermanns K, Nir E, de Vries MS (2003) Pairing of isolated nucleobases: double resonance laser spectroscopy of adenine-thymine. ChemPhysChem 4:838–842
76. Crews B, Abo-Riziq A, Grace L, Callahan M, Kabelac M, Hobza P, de Vries MS (2005) IR-UV double resonance spectroscopy of guanine-H_2O clusters. PCCP 7:3015–3020
77. Asami H, Yagi K, Ohba M, Urashima S, Saigusa H (2013) Stacked base-pair structures of adenine nucleosides stabilized by the formation of hydrogen-bonding network involving the two sugar groups. Chem Phys 419:84–89
78. Saigusa H, Urashima S, Asami H (2009) IR-UV double resonance spectroscopy of the hydrated clusters of guanosine and 9-methylguanine: evidence for hydration structures involving the sugar group. J Phys Chem A 113:3455–3462
79. Saigusa H, Urashima S, Asami H, Ohba M (2010) Microhydration of the guanine–guanine and guanine–cytosine base pairs. J Phys Chem A 114:11231–11237
80. Zeleny T, Hobza P, Kabelac M (2009) Microhydration of guanine center dot center dot center dot cytosine base pairs, a theoretical study on the role of water in stability, structure and tautomeric equilibrium. PCCP 11:3430–3435
81. Kabelac M, Ryjacek F, Hobza P (2000) Already two water molecules change planar H-bonded structures of the adenine ... thymine base pair to the stacked ones: a molecular dynamics simulations study. PCCP 2:4906–4909
82. Callahan MP, Gengeliczki Z, Svadlenak N, Valdes H, Hobza P, de Vries MS (2008) Non-standard base pairing and stacked structures in methyl xanthine clusters. PCCP 10:2819–2826
83. Callahan MP, Crews B, Abo-Riziq A, Grace L, de Vries MS, Gengeliczki Z, Holmes TM, Hill GA (2007) IR-UV double resonance spectroscopy of xanthine. PCCP 9:4587–4591
84. Plützer C, Hünig I, Kleinermanns K (2003) Pairing of the nucleobase adenine studied by IR-UV double-resonance spectroscopy and ab initio calculations. PCCP 5:1158–1163
85. Smith VR, Samoylova E, Ritze HH, Radloff W, Schultz T (2010) Excimer states in microhydrated adenine clusters. PCCP 12:9632–9636
86. Malone RJ, Miller AM, Kohler B (2003) Singlet excited-state lifetimes of cytosine derivatives measured by femtosecond transient absorption. Photochem Photobiol 77:158–164
87. Nachtigallova D, Lischka H, Szymczak JJ, Barbatti M, Hobza P, Gengeliczki Z, Pino G, Callahan MP, de Vries MS (2010) The effect of C5 substitution on the photochemistry of uracil. PCCP 12:4924–4933
88. Mburu E, Matsika S (2008) An ab initio study of substituent effects on the excited states of purine derivatives. J Phys Chem A 112:12485–12491
89. Kistler KA, Matsika S (2007) Cytosine in context: a theoretical study of substituent effects on the excitation energies of 2-pyrimidinone derivatives. J Phys Chem A 111:8708–8716
90. Broo A (1998) A theoretical investigation of the physical reason for the very different luminescence properties of the two isomers adenine and 2-aminopurine. J Phys Chem A 102:526–531
91. Andreasson J, Holmén A, Albinsson B (1999) The photophysical properties of the adenine chromophore. J Phys Chem B 103:9782–9789
92. Mishra SK, Shukla MK, Mishra PC (2000) Electronic spectra of adenine and 2-aminopurine: an ab initio study of energy level diagrams of different tautomers in gas phase and aqueous solution. Spectrochim Acta Part A 56:1355–1384
93. Lührs DC, Viallon J, Fischer I (2001) Excited state spectroscopy and dynamics of isolated adenine and 9-methyladenine. Phys Chem Chem Phys 3:1827–1831
94. Kang H, Jung B, Kim SK (2003) Mechanism for ultrafast internal conversion of adenine. J Chem Phys 118:6717–6719

95. Sobolewski AL, Domcke W (2002) On the mechanism of nonradiative decay of DNA bases: ab initio and TDDFT results for the excited states of 9H-adenine. Eur Phys J D 20:369–374
96. Barbatti M, Lischka H (2007) Can the nonadiabatic photodynamics of aminopyrimidine be a model for the ultrafast deactivation of adenine? J Phys Chem A 111:2852–2858
97. Marian CM (2005) A new pathway for the rapid decay of electronically excited adenine. J Chem Phys 122:104314
98. Hünig I, Plützer C, Seefeld KA, Löwenich D, Nispel M, Kleinermanns K (2004) Photostability of isolated and paired nucleobases: N–H dissociation of adenine and hydrogen transfer in its base pairs examined by laser spectroscopy. ChemPhysChem 5:1427–1431
99. Crespo-Hernández CE, Cohen B, Hare PM, Kohler B (2004) Ultrafast excited-state dynamics in nucleic acids. Chem Rev 104:1977–2019
100. Zierhut M, Roth W, Fischer I (2004) Dynamics of H-atom loss in adenine. PCCP 6:5178–5183
101. Zechmann G, Barbatti M (2008) Ab initio study of the photochemistry of aminopyrimidine. Int J Quantum Chem 108:1266–1276
102. Kistler KA, Matsika S (2008) Three-state conical intersections in cytosine and pyrimidinone bases. J Chem Phys 128
103. Kistler KA, Matsika S (2007) Radiationless decay mechanism of cytosine: an ab initio study with comparisons to the fluorescent analogue 5-methyl-2-pyrimidinone. J Phys Chem A 111:2650–2661
104. Yarasi S, Brost P, Loppnow GR (2007) Initial excited-state structural dynamics of thymine are coincident with the expected photochemical dynamics. J Phys Chem A 111:5130–5135
105. Nir E, Kleinermanns K, de Vries MS (2000) Pairing of isolated nucleic-acid bases in the absence of the DNA backbone. Nature 408:949–951
106. Abo-Riziq A, Grace L, Nir E, Kabelac M, Hobza P, de Vries MS (2005) Photochemical selectivity in guanine-cytosine base-pair structures. Proc Natl Acad Sci U S A 102:20–23
107. Middleton CT, de La Harpe K, Su C, Law YK, Crespo-Hernandez CE, Kohler B (2009) DNA excited-state dynamics: from single bases to the double helix. Annu Rev Phys Chem 60:217–239

Top Curr Chem (2015) 364: 299–334
DOI: 10.1007/128_2014_596

Published online: 11 February 2015

Carbohydrates

Emilio J. Cocinero and Pierre Çarçabal

Abstract Although carbohydrates represent one of the most important families of biomolecules, they remain under-studied in comparison to the other biomolecular families (peptides, nucleobases). Beyond their best-known function of energy source in living systems, they act as mediator of molecular recognition processes, carrying molecular information in the so-called "sugar code," just to name one of their countless functions. Owing to their high conformational flexibility, they encode extremely rich information conveyed via the non-covalent hydrogen bonds within the carbohydrate and with other biomolecular assemblies, such as peptide subunits of proteins. Over the last decade there has been tremendous progress in the study of the conformational preferences of neutral oligosaccharides, and of the interactions between carbohydrates and various molecular partners (water, aromatic models, and peptide models), using vibrational spectroscopy as a sensitive probe. In parallel, other spectroscopic techniques have recently become available to the study of carbohydrates in the gas phase (microwave spectroscopy, IRMPD on charged species).

Keywords Carbohydrates · Sugars · Glycans · Biomolecular interactions · Non-covalent interactions · Hydrogen bonding · Conformations · Micro-hydration · Gas-phase spectroscopy

---

E.J. Cocinero
Departamento de Química Física, Facultad de Ciencia y Tecnología, Universidad del País Vasco (UPV – EHU), Apartado 644, 48940 Bilbao, Spain
e-mail: emiliojose.cocinero@ehu.es

P. Çarçabal (✉)
Institut des Sciences Moléculaire d'Orsay-CNRS, Université Paris Sud, Bât. 210, 91405 Orsay Cedex, France
e-mail: pierre.carcabal@u-psud.fr

Contents

1 Introduction

Carbohydrates (CBHs) are found virtually everywhere in living systems and play key roles in countless biological processes. Their best known functions are as energy resources, as a backbone element in the DNA and RNA genetic molecules, and as builders of structural bio-scaffoldings in plants and insects. However, their biological activity is way more diverse. CBHs are central actors in cell signaling, molecular recognition, immunity, and inflammation, just to name a few. The diversity of their functions is well reflected by the slightly provocative title of a 1993 review by Varki: "Biological roles of oligosaccharides: all of the theories are correct"! [1], followed by no less than 11 pages of tables listing the effects and roles of oligosaccharides in living systems.

There exist several resources in the literature where glycoscience is introduced and reviewed [2–6]. In particular, an important document is the very complete report entitled "Transforming Glycoscience: A Roadmap for the Future", edited in 2012 by the National Research Council of National Academies (USA) [6]. It illustrates how glycoscience impacts on many diverse scientific fields, while remaining a vast under-explored territory, in comparison to the more mature fields of proteo-science and genetic science. To elucidate the structural fundamentals underlying glycoscience, all approaches have their role to play. These include mass spectrometry, NMR, and X-ray crystallography, which are not discussed here. Gas-phase spectroscopy, the topic this chapter focuses on, has the potential to determine accurate structures of complex molecular systems under controlled conditions.

The spectroscopic investigation of CBH conformational landscapes in the gas phase was pioneered by the group of John P. Simons in Oxford in the early 2000s. This work began by solving the structural preferences of the most naturally abundant monosaccharides – the building blocks of oligosaccharides, and was rapidly extended to the study of their micro-hydrated complexes and to larger oligosaccharides. Most of the early results on the vibrational spectroscopy of neutral CBHs obtained in Oxford have already been reviewed [7, 8]. Since then, the studies in vibrational spectroscopy of neutral CBHs have evolved towards the structural characterization of oligosaccharides and to the study of the interaction between CBHs and other molecules such as water, aromatic models, or other biomolecules, simulating the interactions taking place in the active sites in biological systems under controlled conditions. A major part of this chapter is dedicated to these two aspects. We also report progress in microwave spectroscopy and in vibrational spectroscopy of ions, as the applications of these two methods to CBHs will surely grow in the years to come.

The results surveyed in this chapter have provided key information on the structural properties of CBHs and on the nature of non-covalent interactions which rigidify their preferred conformations and govern the interplay between CBHs and other molecular partners. These results can be connected to two fundamental aspects of CBHs in biology:

1. Certain recurring CBH structural motifs are conserved across many biological systems. Their structural and conformational properties must be central to guarantee their functions and may be at the origin of their prevalence. However, the underlying reasons for their selection by nature remain to be understood fully [9]. Such recurrent motifs are often found in glycoproteins where the peptide is chemically bound to oligosaccharide units [10–12]. They have been the subject of NMR studies, complemented by molecular dynamics simulations [12–17]. In such studies, conducted in solution, it is difficult to disentangle the contributions made by intramolecular and solvent interactions. Gas-phase spectroscopy, following a bottom-up strategy, provides an alternative and complementary approach which can be used to probe the intrinsic structural preferences of the constituents of these motifs and, subsequently, the role of water in shaping or supporting them.
2. The ability to create, from a limited set of building blocks, extremely diverse and complex structures allows CBHs to act as subtle mediators of molecular recognition processes: they can encode rich and precise molecular information. Their high conformational flexibility also ensures that they can effectively adapt to the environment and optimize the association with their targets and, in consequence, the recognition process. This implies that lectins (CBH-binding proteins) have the ability to decipher the so-called "glycocode," or "sugar code" [18–20]. Glycoscientists are still working on breaking this code, aiming at exploiting its full potential for improved CBH-based therapeutics, with the emerging fields of glycomics and glycomimetics [18–25]. In the gas phase, the interrogation of the interaction between CBHs and model molecular systems (water, aromatic molecules, small peptides, metal cations) in size-controlled molecular complexes and clusters has helped to identify the different non-covalent intermolecular forces which may govern the structural basis of CBH–lectin interactions [26, 27].

To date, only a few experimental groups have dedicated their activity to the study of this molecular family, but it is hoped (and believed) that more gas-phase spectroscopy players will join in. The review is designed to inform non-specialists new to these systems, first by establishing the few conventions necessary to follow the work presented thereafter, and then by surveying the work carried out within the framework already outlined. Finally, we present some concluding remarks and signposts for the future.

2 Toolbox for Gas-Phase Spectroscopy of Carbohydrates

2.1 *Carbohydrate "Zoology": Structural Diversity and Conformational Flexibility*

Complete introductions to the chemistry and to the structural diversity of CBHs, glycoconjugates, and glycans can be found in the literature [2, 3, 28]. Here we present only the basic properties of mono- and oligosaccharides relevant for gas-phase spectroscopy and necessary to follow the discussions which follow. We begin by introducing the main actors of this chapter, their naming, atomic numbering, and conformational notation.

Strictly speaking, CBHs are defined as aldehydes or ketones with the empirical formula of a "carbon hydrate" $(C–H_2O)_{n\geq 3}$. The building blocks for biologically relevant saccharides are monosaccharides with $n \geq 5$. The best known monosaccharide Glucose (Glc) is shown in Fig. 1 in several representations.

Biological CBHs are usually pentoses (five carbon units) or hexoses (six carbon units) which can be found either in linear forms or as cyclic five-membered (furanose) or six-membered (pyranose) rings. Pyranoses have a relatively rigid chair skeleton with dominant $^1C_4$ or $^4C_1$ conformations connected by ring inversion. On the other hand, furanoses are more flexible because of the possibility of

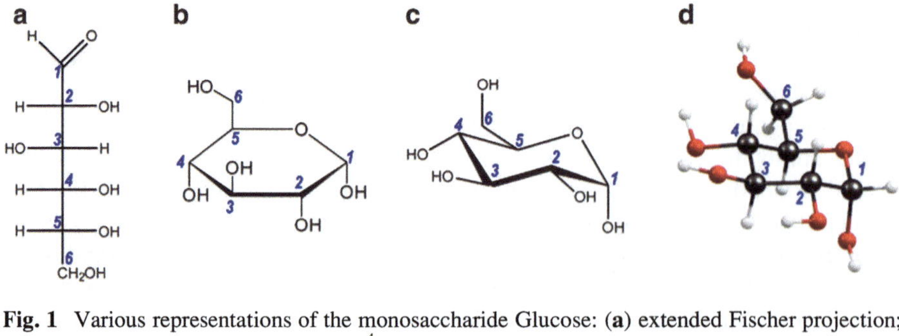

Fig. 1 Various representations of the monosaccharide Glucose: (**a**) extended Fischer projection; (**b**) Haworth projection; (**c**) pyranoside $^4C_1$ chair representation; (**d**) 3D conformational representation. The *numbers* indicate the standard designations of each carbon atom in the monosaccharide. This numbering is also used to identify the OH groups, OHn being the OH group bound to Cn

interconverting different envelope forms such as C2-*endo* or C3-*endo* through non-planar structures ("pseudoration").

The hexoses Glc, Galactose (Gal), and Mannose (Man) are the most abundant monosaccharides in natural oligosaccharides. They have the same chemical composition ($C_6H_{12}O_6$) but different configurations of the OH2 or OH4 groups. In their D-pyranose $^4C_1$-chair ring structure, as shown in Fig. 1c, d, all the OH groups of Glc are equatorial to the ring, but the OH4 (respectively, OH2) group is axial for Gal (respectively, Man). Monosaccharides, differing only by the position of a single OH group, are called epimers; see Fig. 2.

The carbon in position 1 is called the anomeric carbon. The location (equatorial or axial) of the oxygen linked to this carbon is determined upon cyclization of the monosaccharides and all monosaccharides can exist in two anomeric forms. In the D-pyranoside ring, the α anomer is formed if O1 is axial, and if O1 is equatorial the β anomer is formed; both are illustrated in Fig. 2. When an OH group is bound to C1, the anomers may co-exist and interconvert (mutarotation), in relative

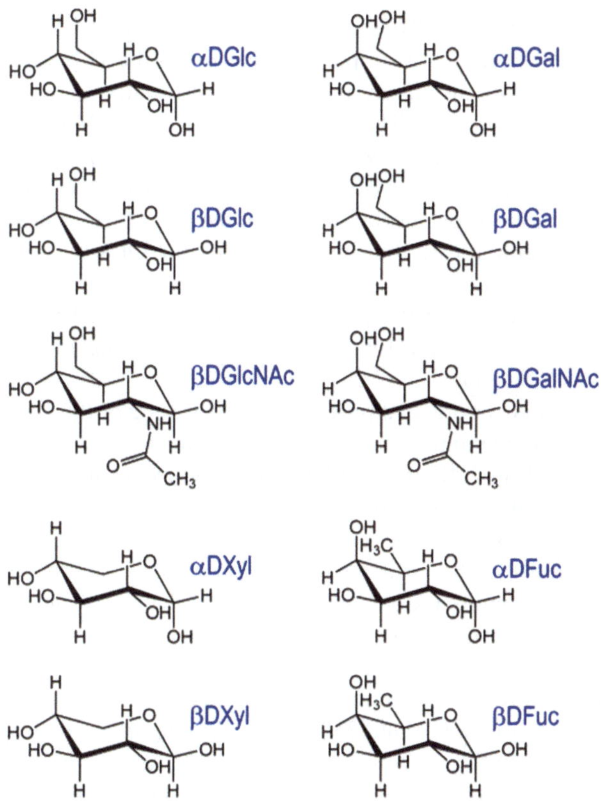

Fig. 2 Pyranose $^4C_1$ chair representation of some common monosaccharides. The standard orientation of the pyranose ring, with the endocyclic oxygen on the *top right corner*, has been adopted

abundances governed by the anomeric effect and other steric, electrostatic and solvent mediated interactions. The "anomericity" of the monosaccharides can be synthetically defined if the OH1 group is replaced by a "blocking" group (*O*-methyl or *O*-phenyl for instance).

Fucose (Fuc, shown in Fig. 2) is a deoxy-monosaccharide, formed by replacing the hydroxymethyl group of Gal with a methyl group; it can be termed, 6-deoxy-galactose. Xylose (Xyl) is a pentose, without the flexible hydroxymethyl group; in its $^4C_1$ pyranoside chair configuration, the OH2,3,4 groups have the same orientation (all equatorial) as Glc.

In terms of chirality, Fuc can be considered as the "odd monosaccharide out" – it is the only natural monosaccharide found in the L rather than the D form. The reason for this singularity is unknown; likewise the prebiotic origin of homochirality of all the other D-monosaccharides or, more generally, of all biomolecular building blocks, and the debate is still open [29, 30]. All the monosaccharides shown in Fig. 2 are drawn in their D chiral form apart from Fuc, which is drawn in both its L and D forms.

Monosaccharides provide ideal model systems for interrogating the non-covalent interactions at play within CBHs and with other molecules, using gas-phase spectroscopy. It is possible, for example, to "fine tune" the H-bonded interactions between neighboring OH groups by modifying their relative orientations. The flexible hydroxymethyl group (CH_2OH) can modulate the H-bonded networks by optimizing the interactions at play. The effect of this group can be isolated and "switched off" by replacing Glc by Xyl or Gal by Fuc, (replacements that retain the respective configurations equatorial or axial of the other OH groups). It is possible to modulate the interaction further by substituting the OHs by other functional groups. Possible substitutions of monosaccharides in biology are virtually infinite but one important example in the present context is provided by *N*-acetylation of Glc and Gal at C-2 to create two key biological building blocks, GlcNAc and GalNAc (shown in Fig. 2).

Starting from these few monosaccharide examples – the lowest rung of the CBH ladder – it is possible to appreciate the high flexibility and conformational diversity of the CBH family, illustrated in Fig. 3. Since each OH group can rotate around its connecting C–O bond (Fig. 3a), the number of possible different conformations (rotamers) for even a single monosaccharide is very high. Considering all the possible OH relative orientations in the pentose CBH, ribose, for example, as well as all the carbon skeletal configurations ($^4C_1$ and $^1C_4$ chairs, pyranose, furanose, boats...), no less than 70 different structures are found with relative energies in the range 0–40 kJ/mol [31].

A detailed notation of the conformation of a monosaccharide defines the *gauche* (g+/g−) or *trans* (t) configuration of each rotatable bond. Following this notation, the conformation of βDGlc shown in Fig. 3a would be written as tg+g-g+G-g+ (the capital letter refers to the C5–C6 torsion angle). In the gas phase, OH2, OH3, and OH4 often adopt concerted orientations to create cooperative chains of H-bonded intramolecular interactions. It is then possible to simplify the notation by defining the overall orientation of such chains of hydrogen bonds as clockwise (noted c) or

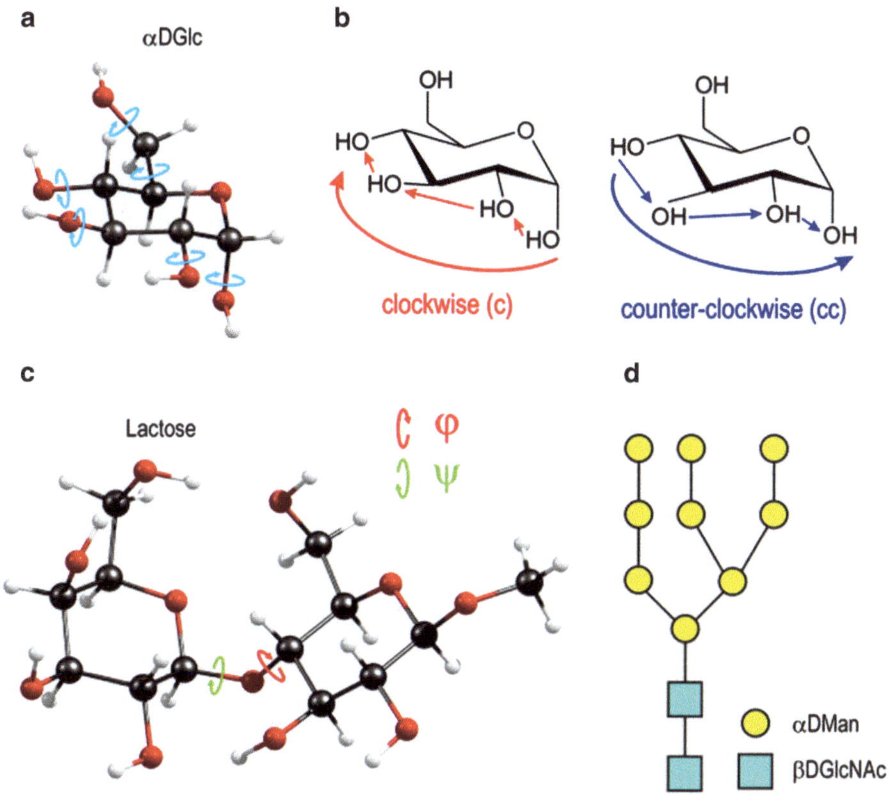

Fig. 3 (**a–d**) The origin of structural diversity and conformational flexibility of CBHs

counter-clockwise (noted cc), taking the standard orientation of the pyranose ring (with the endocyclic oxygen on the top right corner of the ring, as illustrated in Fig. 3b) as reference. In the βDGlc conformation shown in Figs. 1c and 3a, OH2–OH3–OH4 form a clockwise H-bond chain and the notation becomes tcG-g+. When the anomeric group is blocked, the first letter (t in this example) is often omitted. Besides the simplification of the notation, the c and cc convention has two advantages: it is more informative for the non-specialist and it is also more descriptive of the non-covalent interactions governing the molecular conformation.

In disaccharides, two monosaccharide units are bound via the O-glycosidic linkage, identified by the numbering of the connected groups. For instance, the lactose (Fig. 3c) is formed by Gal and Glc monosaccharides units where the β form of the C1 atom of Gal is linked to the O4 atom of Glc through a β1,4 O-glycosidic linkage. So, Gal-β1,4-Glc provides an alternative name for this CBH. The glycosidic linkage is an important source of flexibility in oligosaccharides and a determining element in the biological function of the resulting oligosaccharide. The conformation of the glycosidic linkage is generally defined by the two dihedral angles, φ and ψ, (φ: $H1_{Gal}$-$C1_{Gal}$-$O1_{Gal}$-$C4_{Glc}$ and ψ: $C1_{Gal}$-$O1_{Gal}$-$C4_{Glc}$-$H4_{Glc}$, shown in Fig. 3b). A simpler, qualitative description is based on the relative position of the endocyclic oxygen O5 with respect to the linkage. If both oxygen atoms are on the same side of the linkage, the conformation is called *cis*, otherwise it is called *trans*. According to this notation, Fig. 3c presents lactose in a *cis* conformation.

The complexity grows tremendously when larger oligosaccharides are formed and the diversity, flexibility, and complexity (both chemical and structural) of oligosaccharides are spectacular. While proteins or DNA strands grow along one direction, oligosaccharides often feature branched, multidirectional structures, for example the one shown in Fig. 3d, which represents the mannose-rich glycan component of an N-linked glycopeptide. The number of different arrangements and conformations of an oligosaccharide chain can be orders of magnitude higher than the number of possible oligopeptides that can be built with the same number of amino acids. This is probably one reason why CBHs are given the role of carrying rich and subtle molecular information via the glycocode.

2.2 *Specific Aspects of Gas-Phase Spectroscopy of Carbohydrates*

The chemical structure of CBHs and, most critically, their high conformational flexibility and diversity, lead to specific difficulties for the interrogation of their conformational preferences by gas-phase spectroscopy.

To date, the most prolific source of experimental data for the investigation of the conformational preferences of neutral CBHs in the gas phase has been provided by vibrational spectroscopy in supersonic jets using the conformer selective double resonance laser spectroscopy IR–UV ion dip (IRID) and UV–IR and UV–UV hole burning (IRHB and UVHB) methods [32–59]. To apply these electronic spectroscopy-based double resonance methods to isolated CBHs, it has generally been necessary to attach an artificial chromophore (unlike the other biomolecular families – peptides, nucleotides, nitrogenous bases – CBHs do not include aromatic chromophores in their natural building blocks). Fortunately, for most CBH systems studied so far, the addition of the chromophore (generally a phenyl (Ph) group) at the anomeric OH-1 site does not significantly alter the conformational landscape – it is "conformationally benign." There is one *caveat*, however: it is obvious that the addition blocks interactions involving OH1 as an H-donor (but not O1 as an acceptor). However, in oligosaccharides this position is often occupied by glycosidic linkages and the interactions involving OH1 are rare. Furthermore, linking the chromophore to the CBH at the anomeric position has the advantage of blocking mutarotation, and thereby selectively defining the anomericity (α or β) of the system under study. For the most common monosaccharides, such as βDGlc, βDGal, and αDMan, compounds tagged with aromatic chromophores are commercially available. However, the need for tagged molecules can become a limiting factor when less "standard" systems are studied. It is then necessary to call for the expertise of glycochemists to obtain "synthesized to order" compounds. Carbohydrate chemistry is extremely challenging and glycochemists have had to develop sophisticated strategies to synthesize high purity samples of tagged CBHs in controlled anomeric forms in sufficient quantities to meet the requirements of gas-phase spectroscopists [57].

In the near IR (NIR) range, the vibrational spectra of CBHs are dominated by O–H stretch bands (and N–H stretch when amino- or N-acetylated monosaccharides are studied). Since these groups are direct actors in the non-covalent interactions that control the conformational landscapes of isolated CBHs, the properties (frequency, width, and intensity) of their NIR vibrational bands encode structural information. The frequency of the stretching mode of an H-bonded donor group is shifted towards low frequencies (red-shift) and the width and intensity of its IR band are increased. The magnitude of these spectral signatures of non-covalent interactions is related to the strength of the interaction. The difficulty in extracting this information comes from the high number of OH groups in the systems. Even for supersonically cooled CBHs of modest size, spectral congestion can become a limitation. On top of this difficulty, oligosaccharides are made of monosaccharide building blocks of masses in the 150–200 amu range, each containing between three and five groups with active bands in the NIR range. As a consequence, molecular and spectral complexities are often much higher for an n-saccharide than for an n-peptide.

Infrared MultiPhoton Dissociation (IRMPD) spectroscopy coupled to ElectroSpray Ionization (ESI) sources have also been applied to CBHs [60–66]. So far, the study of CBHs by ESI has focused on relatively simple mono- or disaccharides bound to metal atoms carrying the charge. This is because stabilizing the protonated forms of non-substituted carbohydrates with this ionization technique is difficult, although probably not impossible. However, it has recently been possible to observe the formation of a protonated monosaccharide (αMeGal-H^+) produced by UV photo-ionization of phenol bound to the CBH within a molecular complex [67].

Few biologically relevant monosaccharides have recently been studied by high resolution rotational microwave spectroscopy [31, 68–71]. Several groups have tried over the years but the high molecular density required in these experiments prevented any breakthrough. Vaporization techniques which had proved their efficiency for other biomolecules (oven sublimation, nanosecond laser ablation from single or multiple samples) were tested without success and, for a long time, only the spectra of decomposition products could be observed [72] – another illustration of the difficulty, caused by their high fragility, of dealing with CBHs in gas-phase spectroscopy experiments. To circumvent this difficulty and enable the investigation of CBH by microwave spectroscopy, it has been necessary to use picosecond UV laser to desorb the molecules from a pure solid sample of the CBH [31], while most vibrational spectroscopy experiments use IR or visible nanosecond evaporation laser and desorption samples where the molecule is mixed with graphite.

On the computational side, great strides have been made with access to supercomputers, allowing calculations with multiprocessors in ever faster computers to simulate larger systems more accurately. The methodology and the models used to explore the conformational landscapes of CBHs and interpret gas-phase spectroscopy experiments are the same as those used for the other biomolecules. Several CBH specific force fields (FFs) have been developed [73–80] but they have not been routinely used by gas-phase spectroscopists so far. One reason is that the FF exploration of the conformational landscape aims "only" at generating an exhaustive set of conformations before treating this set at a higher quantum chemical level. As long as the FF-based conformational search has been carefully conducted to guarantee that no conformation was omitted or "screened out," the outcome of the exploration is quite insensitive to the initial choice of the FF. Very recently 'carbohydrate specific' FF, GLYCAM06 [75] has been tested on isolated and micro-hydrated oligosaccharides, and compared to more general FFs (MMFFs [81] and OPLS [82]) for the first time in the context of gas-phase spectroscopy [57].

3 Isolated and Hydrated Neutral Carbohydrates

3.1 *Explicit Hydration of Carbohydrates*

3.1.1 Conformation of Hydrated Monosaccharides

The first investigations of isolated monosaccharides in the gas phase using vibrational spectroscopy did not indicate any basic general rules for their conformational preferences. Although adopting similar conformations, their relative energies (often in good correlation with the intensity of the associated signals in the experiments) were generally quite different [32, 33, 35]. The situation was spectacularly different when the monosaccharides interacted with a single water molecule [35]. All observed CBH-W_1 complexes favored the same type of conformation, with the water molecule inserting between two OH groups of the monosaccharide and replacing the weakest intramolecular OH–OH interaction by two strong intermolecular H-bonds. In this way, the water molecule played the role of a "conformational lock" by optimizing the length of the H-bonded chains, which were strengthened by inductive cooperativity effects [33, 35, 36, 40, 43, 45, 50–52, 54].

The effect of hydration on the conformational preference of monosaccharides has been further investigated for multiply micro-hydrated (W_n) systems, greatly helped by using heavy water (D_2O) instead of H_2O to reduce spectral congestion in the OH stretching region [52, 53]. Isotopic effects in vibrational spectra were first observed in the early 1920s [83, 84], and they have been used to help interpret complex data ever since. In the case of hydrated CBHs, they again proved very

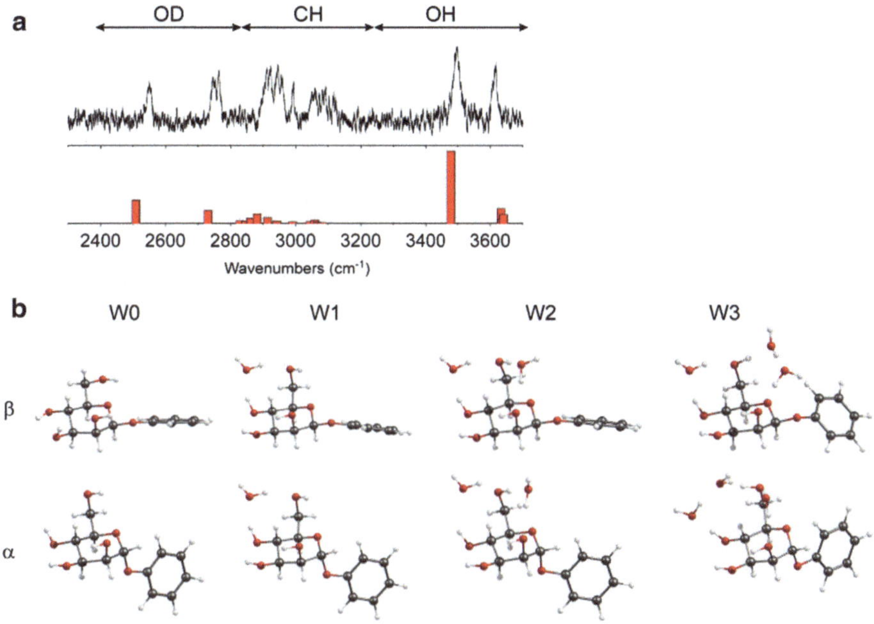

Fig. 4 (**a**) IRID spectrum of the singly hydrated complex αMan-O-Ph. The stretching bands of the deuterated water molecule are well separated from those of the monosaccharide. (**b**) Preferred conformations of isolated and micro-hydrated of the β (*top*) and α (*bottom*) anomers of Man-O-Ph

helpful by separating the vibrational bands of water, ODs, and those from the carbohydrates, OHs. This facilitated the interpretation of the IRID spectra, with the free or intramolecularly bound O–H stretches of the monosaccharides being expected in the 3,550–3,650 cm^{-1} region, the intermolecularly bound O–H stretches of the monosaccharide between 3,400 and 3,500 cm^{-1}, the free water O–D stretch around 2,750 cm^{-1}, and the bound water O–D below 2,600 cm^{-1}, as shown in Fig. 4a. The analysis of such IRID spectra with the simulated spectra predicted by quantum chemistry indicates that there is no H/D exchange between the heavy water and the monosaccharide.

The results established the propensity for stabilizing cooperative structures through multiple hydration and led to a generalization of the rules for conformational selectivity in hydrated monosaccharides [51, 52]. The preference of water molecules for cooperatively bonded conformations of the monosaccharide is such that it can promote conformations which are not favored for the isolated molecule. This is illustrated in the case of the different anomeric forms of Man-O-Ph in Fig. 4b, where the conformations of the isolated and sequentially hydrated monosaccharide, determined from the interpretation of gas-phase IRID spectra, are shown. The α anomer adopts a preferred cG-g+ cooperative conformation. It is unaffected by micro-hydration as water molecules can insert at different sites to increase the cooperative strengthening of the H-bond network. In the β anomer, the equatorial position of O1 "turns on" the OH2 → O1 interaction and the preferred conformation of bare βMan-O-Ph is then ccG+g-. However, single hydration is enough to perturb this preference and the water molecule selects the cG-g+ conformation to create the same fully cooperative network of H-bonds as in the α anomer [40].

Similar behavior is observed for Glc, Gal, Xyl, and Fuc, although for Xyl and Fuc it is slightly different [51], since these monosaccharides do not contain the exocyclic hydroxymethyl group involved in all the H-bonded networks of the other hydrated monosaccharides. Nevertheless, the propensity to favor cooperativity-enhanced H-bond networks is confirmed. In micro-hydrated GlcNAc, although the OH6 group is present, the N-acetyl group becomes the focal point of the bound water molecules. It features both H acceptor and H donor groups and its high flexibility permits fine adjustments to the geometry of the complexes to optimize the strength of the interaction. As a consequence, water molecules bind preferentially to the N-acetyl substituent, leaving the rest of the monosaccharide binding sites "dry" [43].

The corpus of data collected on hydrated monosaccharide clusters has led to the identification of five simple rules followed by all systems:

1. Cooperative structures where the longest chains of interacting OHs can be created are favored, and the first bound water molecule replaces the weakest $OH_{CBH} \rightarrow OH_{CBH}$ intramolecular interaction by two strong $OH_{CBH} \rightarrow OH_W \rightarrow OH_{CBH}$ intermolecular interactions.
2. When an N-acetyl group is present (GlcNAc, GalNAc), the water molecules bind to it.
3. In the absence of an N-acetyl group, the hydroxymethyl group (in Glc, Gal, and Man) is selected and the first water molecules preferentially insert between OH4 and OH6 when OH4 is equatorial or between OH6 and OH5 when OH4 is axial (following rule 1).
4. If neither of these groups are present (Xyl, Fuc), the water molecules bind to the CBH to obey as closely as possible the "cooperativity + weakest link" rule 1.
5. The monosaccharide OH groups "steer" successively bound water molecules on one side or another of the pyranose ring, reflecting hydrophobic and hydrophilic character of the faces on the pyranose ring.

The operation of these rules is illustrated in Fig. 5 for three typical examples: GlcNAc-W_3, Glc-W_3, and Xyl-W_2.

The computational exploration of the conformational landscapes of neutral hydrated monosaccharides is particularly challenging because the relative energies of the many possible conformational structures are very close and, in some cases, the associated vibrational spectra differ only very slightly. Singly hydrated βXyl-O-Ph is a typical example of these difficulties [50]. To improve the theoretical description of such systems, more sophisticated approaches than the standard DFT calculation have been applied. Vibrational anharmonicity has been accommodated ab initio, using Vibrational Self Consistent Field (VSCF) theory, and has reproduced very accurately the observed spectra [47, 50, 53]. The conformational interconversion of hydrated monosaccharides has been simulated using Ab Initio

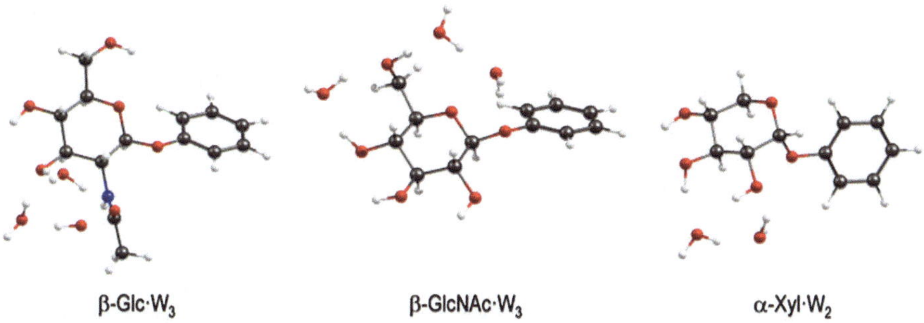

Fig. 5 Most stable structures of GlcNAc-W_3, Glc-W_3, and Xyl-W_2, three typical examples of the operation of the rules followed upon micro-hydration of monosaccharides

Molecular Dynamics (AIMD) [50]: at the low temperature attained in supersonic expansions, significant motion of the water molecule around the monosaccharide was prevented, but when the dynamics were simulated at 300 K it was found that the water molecule could "jump" from one binding site to another, on the picosecond timescale.

Micro-hydration studies of the different anomers of monosaccharides have enabled a novel means of exploring the anomeric effect – a text-book case of the effect of stereo-electronic interactions influencing molecular conformation [52]. In CBHs it leads to the preferred stabilization of electronegative substituents at C1 in the β orientation, although steric effects should favor the α orientation. Its origin is often pictured and understood in terms of electrostatic interactions between local dipoles at the endocyclic oxygen O5 and at the heteroatom atom bound to C1. Unambiguous experimental evidence of the core origins of the anomeric effect alone is difficult to identify, separately from other influences (steric effect, non-covalent and long range interactions within the molecule and with its environment, solvation). By eliminating some of these competing contributions, gas-phase spectroscopy offers a unique way of addressing the anomeric effect. Although the conformations and H-bond networks of the doubly hydrated clusters of both anomers of Man-O-Ph are very similar, significant differences can be observed in their IRID spectra (Fig. 6).

Monosaccharides exhibit an unusual preferential stabilization of pyranose rings containing an axial electronegative substituent at C1 carbon over the equatorial orientation expected, widely known as *the* anomeric effect. However, the underlying physical origins of this phenomenon are still not clear despite its identification in 1955 [85] and it is still subject to virulent scientific debate [49, 86, 87]. The differing spectral signatures were associated with subtle structural differences between the anomeric complexes which were consistent with the effect of 'lone pair repulsion' between O5 and the neighboring water molecule. A Natural Bond Orbital analysis gave an additional quantitative interpretation in terms of the variation of the *endo*-anomeric ($n(O5) \rightarrow \sigma^*(C1–O1)$) and *exo*-anomeric ($n(O1) \rightarrow \sigma^*(C1–O5)$) hyperconjugation effects. However, the observed vibrational

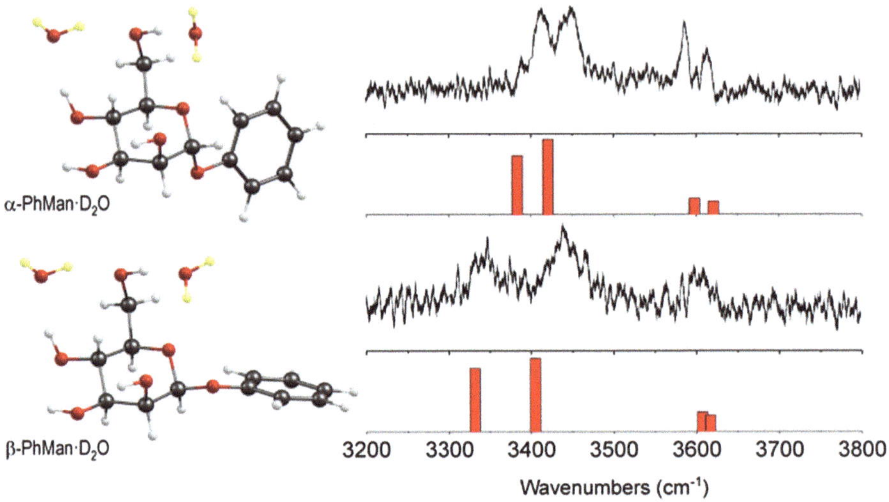

Fig. 6 Experimental IRID spectra, conformations, and calculated spectra of the doubly hydrated cluster of both anomers of Man-O-Phe in the OH region

signatures should not be considered as *direct* experimental probes of the intimate origin of *the*, as opposed to *an*, anomeric effect; the interpretation remains at the phenomenological stage and needs further theoretical exploration to be fully rationalized.

3.1.2 Isolated and Hydrated Disaccharides

Isolated in the gas phase, both lactose (Gal-β1,4-Glc) and cellobiose (Glc-β1,4-Glc, the repeated motif of cellulose), adopt a *cis* conformation about their O-glycosidic linkage [34, 42]. The experimental IRID spectra of these disaccharides, their assigned conformations, and their associated calculated vibrational spectra are shown in Fig. 7. The flexibility of the glycosidic linkage is dramatically reduced by strong cooperative interactions between the OH groups of the connected monosaccharides, supported by strong inter-ring interactions, bridging across the glycosidic linkage and rigidifying it. A fully cooperative chain with seven H-bonds can be created all around the lactose molecule while cellobiose has a slightly shorter chain of five H-bonds. The strength of these inter-ring H-bonds is reflected by the vibrational shifts of the donor OH groups involved. The enhanced red-shifts in the lactose spectrum indicate inter-ring interactions which are much stronger than those in cellobiose.

In crystalline or solvated phases, the situation is quite different from the gas phase: several studies have established that both lactose and cellobiose adopt a *trans* conformation about their glycosidic linkages [88–92]. Despite this, single hydration of the disaccharides (in the gas phase) does not affect their preference for

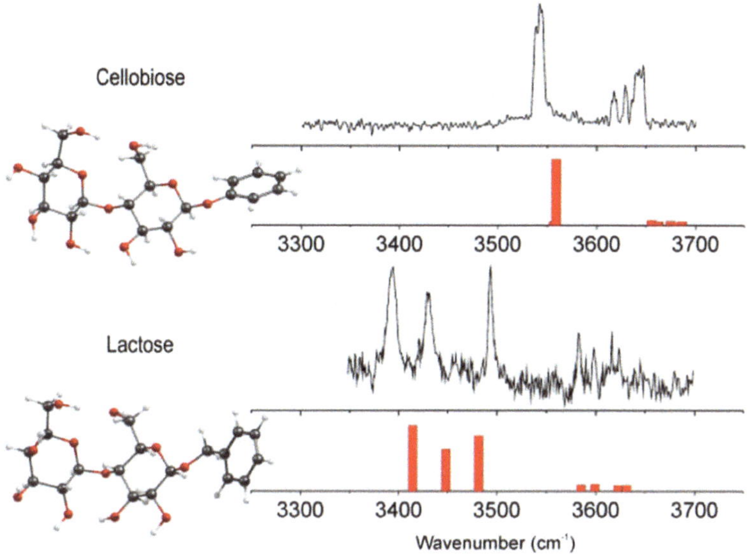

Fig. 7 Experimental and calculated vibrational spectra of the preferred *cis* conformations of the isolated Cellobiose and Lactose disaccharides

the *cis* conformation; see Fig. 8. Modification of the glycosidic linkage conformation by a water molecule would have to disrupt the inter-ring interactions. These interactions are very strong and the energy gain to replace them by intermolecular H-bonds with a single water molecule is not high enough to trigger the *cis* → *trans* transition. As a consequence, the bound water molecule follows the same trend as with monosaccharides. It inserts in binding sites where the OH groups of the disaccharide interact only weakly and it fortifies the cooperativity of the isolated disaccharides conformations by increasing the length of the cooperative H-bond network [42]. The resemblance with the behavior of water molecules binding to monosaccharides in the gas phase is striking. For both disaccharides, the water interacts with the hydroxymethyl groups, and in the case of cellobiose it inserts between the OH4 and OH6 groups of the terminal Glc monosaccharide (rule 3), and favors "all clockwise" (not observed for the non hydrated molecule) to follow the cooperativity driving force rule (rule 1).

The rigidity of the cellobiose hydrates has also been investigated by AIMD simulation [93]. At temperatures up to 300 K, the singly hydrated complex retains the *cis* conformation and the water molecule does not leave its preferred, lower energy binding site. For larger clusters (up to 25 water molecules) the *cis* conformation still remains energetically favored but acts as a "molecular surfactant," while the *trans* conformation is better "enveloped" by water. This may explain why fully solvated cellobiose adopts the *trans* conformation in an aqueous environment – extrapolation of the molecular dynamics findings suggests that it may take up to about 100 water molecules before this becomes the favored conformation.

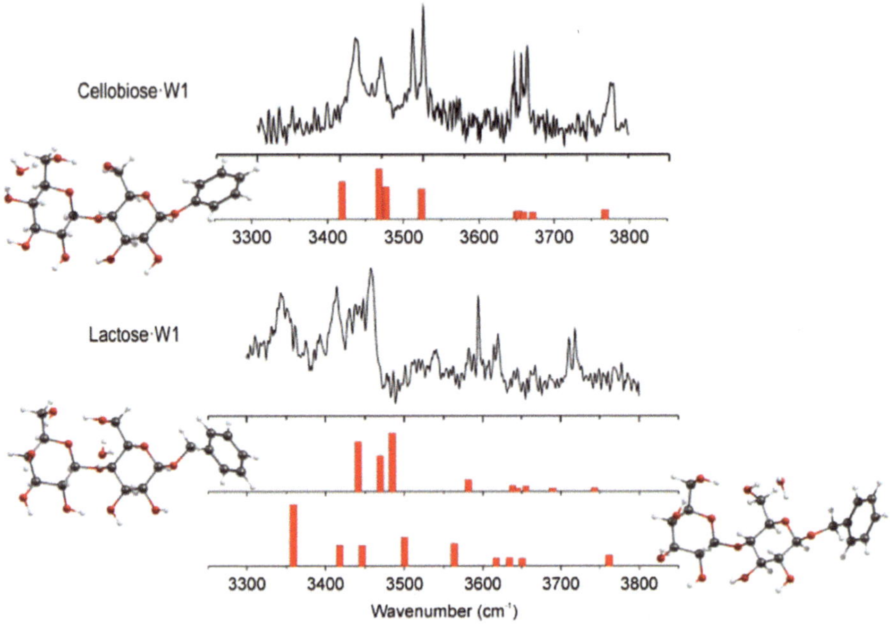

Fig. 8 Experimental and calculated spectra of the preferred *cis* conformations of singly hydrated Cellobiose and Lactose

The cases of lactose and cellobiose are not uncommon. For instance, the gas-phase structure of the Lewis^x antigen (a branched trisaccharide) [46] is different from the structure found in solution, which is stabilized by an unconventional CH $\rightarrow$ O interaction [94]. Experimental and theoretical gas-phase hydration studies on this system may help to understand how this rather odd interaction can become dominant in the solvated phase.

3.1.3 Binding Energy of Singly Hydrated Monosaccharides

The strength of the interaction between two biomolecular entities is closely related to the efficiency of their biological activities. If this strength is known for similar complexes composed of different molecules, it is possible to appreciate whether a molecule prefers to bind to a specific molecular motif or to another. The compared motifs can be either different molecules or different conformations of the same molecule. The determination of binding energy can then be a way to address quantitatively conformational selectivity and recognition, and complement the structural (qualitative) information derived from vibrational spectroscopic data.

In the gas phase, two color UV–UV experiments can allow an experimental determination of the ground state binding energy of bi-molecular complexes through the measurements of the ionization threshold of the bare chromophore

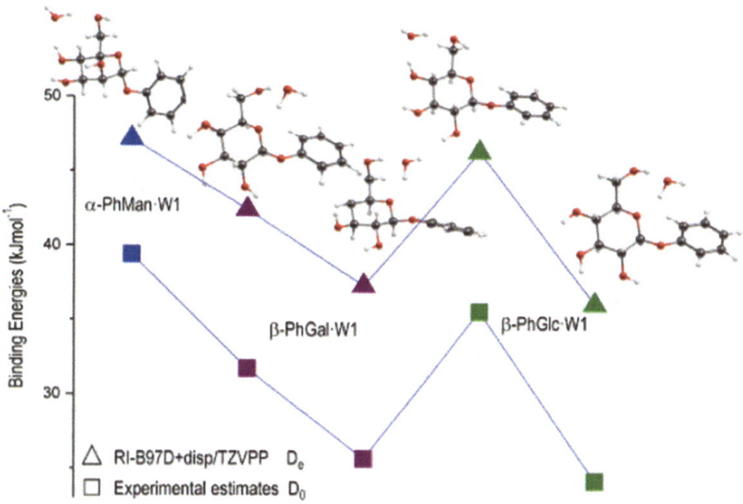

Fig. 9 Binding energies of different conformations of singly hydrated monosaccharides estimated from experimental measurements (*squares*) and calculated at the RI-B97-D/TZVPP level of theory (*triangles*)

and of the dissociation threshold of the cationic complex [95, 96]. Such measurements have been conducted on a series of singly hydrated monosaccharides [58]. The flexibility and conformational diversity of these systems in their neutral and ionic state prevents the measurement of adiabatic thresholds and a purely experimental, accurate determination of their binding energy is not possible. Nevertheless, considering the conformations of the most likely ionic dissociation products and taking into account the calculated relative energies of these conformations in the neutral ground state, it has been possible to propose reliable estimates of their binding energies, graphically represented in Fig. 9.

The experimental estimates of the zero-point averaged binding energies (D_0) have been compared with equilibrium values (D_e) calculated at increasing levels of theory and the correlation found to be very good for all methods taking into account dispersive interactions. In particular, the RI-B97D/TZVPP level of theory provided very good agreement with a considerable reduction in computational time.

As expected, the water molecule is more tightly bound in cooperative structures and preferentially inserts into binding sites where the intermolecular interaction strength can be optimized. Such considerations may seem obvious in the case of such "simple" complexes. However, they show that, even if the systems under study are not the best candidates to access the values of the binding energy from the experimental observations alone, it is possible to evaluate and compare them in different complexes – and conformations.

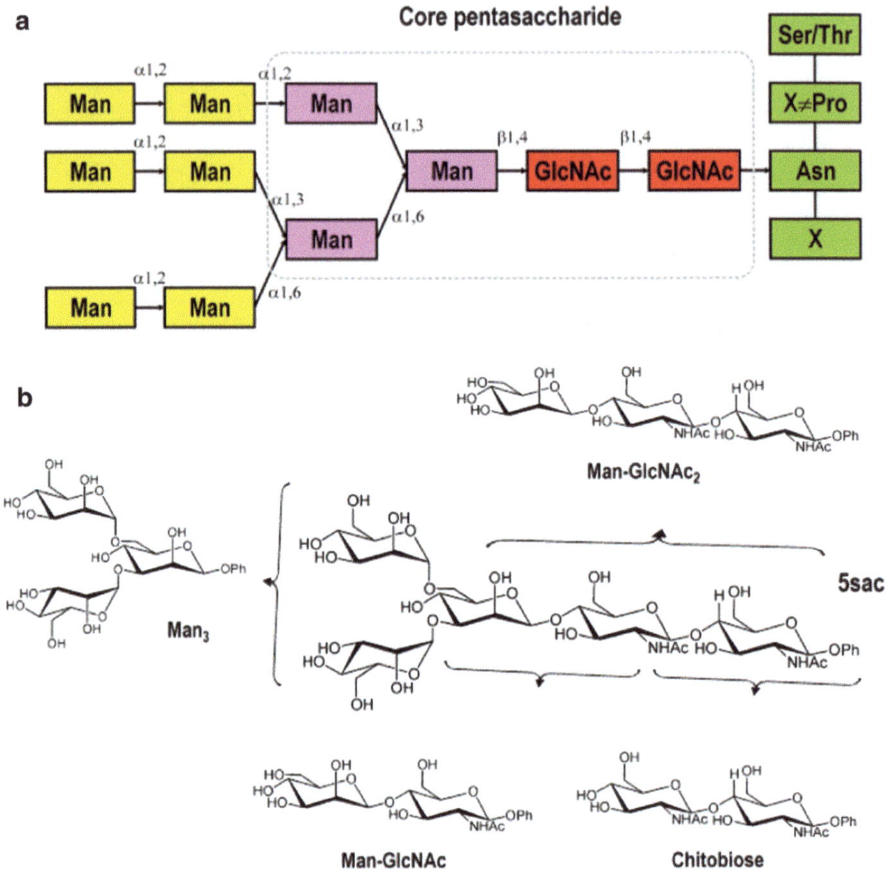

Fig. 10 (**a**) Schematic representation of the core pentasaccharide (*in the dashed box*) of *N*-glycans. The *green blocks* represent the peptide of the glycoprotein and the *yellow blocks* represent the oligosaccharide assembly that develops from *5sac*. (**b**) The "phenyl-tagged" core pentasaccharide (5sac) and the di- and trisaccharide sub-units selected for the gas-phase conformational investigation

3.2 Conformation of Oligosaccharides: Intrinsic Preference vs Solvent Promoted Conformations

The conserved core pentasaccharide unit, $Man_3GlcNAc_2$ (*5sac*), invariably lies at the protein-junction core of all oligosaccharides (glycans) attached to asparagine side-chains during the widespread and vital process of N-glycosylation of proteins. It is schematically represented in Fig. 10a. Although the ubiquity of this system suggests that it is central to the biological function of N-glycans, there is no clear consensus on the underlying physical reasons for the prevalence of this fundamental motif of glycobiology [9]. *5sac* features several intriguing structural properties, whose function has been questioned by glycochemists. What structural role is

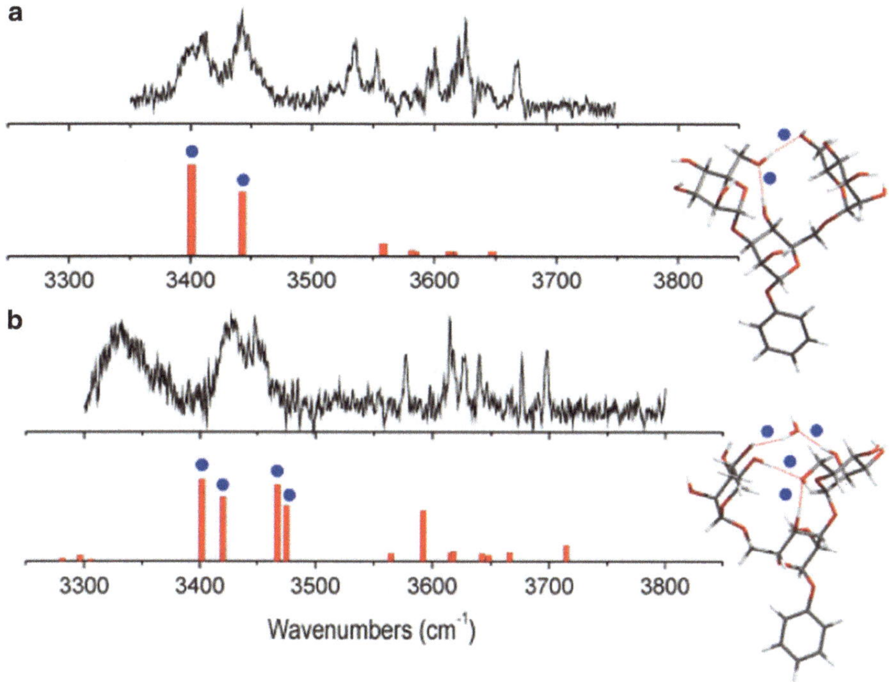

Fig. 11 Experimental and computed IR spectra of (**a**) Man_3 and (**b**) its singly hydrated complex. The *blue dots* indicate the strongly H-bonded OH groups in the structures and their corresponding vibrational bands in the computed spectra

played by the chitobiose (-GlcNAc-GlcNAc-) stem unit (shown in red in Fig. 10a)? Has the rather rare β-mannoside central linkage between this stem and the trimannoside unit (pink in Fig. 10a) been selected for a specific purpose? Do specific water pockets exist in the sub-units and in the whole assembly? Is the prevalence of the core pentasaccharide related to its structural properties and if so, are these intrinsic or influenced by the environment?

Following a bottom-up approach, a group of glycochemists and gas-phase spectroscopists undertook a long term joint project to explore, in the gas phase, the conformational structures of its various sub-units (presented in Fig. 10b) and the influence of controlled hydration on their conformational landscapes. This was initially prompted by evidence from NMR and molecular dynamics studies (in solution) that its potentially high flexibility was restrained in an aqueous environment by explicitly bound water molecules located at strategic 'binding sites' [11, 12].

The project began with an investigation of the spectroscopic signatures of the monosaccharide, αMan-O-Ph [35], which was then used to aid the interpretation of the spectra of the α1,3 and α1,6 dimannosides of the branched trimannoside unit, Man_3 [36]. An important stage was reached when the gas-phase conformations of the branched trisaccharide itself, and its singly hydrated complex, were determined from their IRID spectra, presented in Fig. 11 [41]. The stepwise, bottom-up approach was a posteriori validated since gas-phase conformations of the monosaccharide were retained in the di- and trisaccharides, despite variations in the contact partner, and the increased complexity of the system. Most importantly, the preferred conformation of Man_3 that it revealed was ideally suited to accept a water molecule linking the two (α1,3 and α1,6) arms of the branched trisaccharide, hence constraining any "wobbling" around their glycosidic linkages. It was an important result which could be paralleled to the interpretation of NMR data in solution [11, 12].

The key to understanding the structural properties of *5sac* turned out to involve the leading elements of the stem [57]. Well resolved IRID spectra of the isolated chitibiose ($GlcNAc_2$) and Man-GlcNac units, shown in Fig. 12, allowed their intrinsic conformations to be assigned. Chitibiose adopted two conformations which corresponded to its most stable, quasi iso-energetic, *trans* and *cis* conformers; in Man-GlcNac, again, the two most stable conformers (now both *cis*) were populated. In the trisaccharide unit, Man-$GlcNAc_2$, although the IRID spectrum featured three unresolved bands, the calculated spectrum of its most stable conformation, where both glycosidic linkages adopt a *cis* conformation, presented the best agreement with the observed spectrum.

In each of these observed conformers, torsional motion about their glycosidic linkages was blocked by inter-ring interactions; in the Mannose-containing systems, the interactions involved the Man OH6 group. However, this particular group could not come into play in the complete pentasaccharide since it was involved in the Man-α1,6-Man glycosidic linkage. When the Man-OH6 group (in Man-$GlcNAc_2$) was "switched off" by replacing it with a methyl group, it led to a drastic change in the conformation of the trisaccharide, shown in Fig. 13. Once the Man-OH6 $\rightarrow$ GlcNac-OH6 interaction was removed, the Man-GlcNac linkage was freed, allowing it to adopt an open and floppy *trans* conformation, supported by a single weak inter-ring interaction.

To date, hydrated complexes of the stem sub-units have not been observed experimentally, but the effect of hydration on the stem has been explored by quantum chemistry. In chitibiose, the *cis*/*trans* near degeneracy is lifted: single hydration favors the *trans* conformer with the water molecule inserting between the NH group of the terminal GlcNAc and the OH6 group of the other moiety. This NH $\rightarrow$ OH_w $\rightarrow$ OH6 link is extremely robust and leads to the stabilization of a rigid chitibiose *trans* conformation. In the most stable conformers of singly hydrated "methyl-capped" Man-GlcNac the *trans* conformation offers a binding site localized on the GlcNAc unit, where the water molecule can be accommodated without increasing the rigidity of the linkage.

The IRID spectrum of the complete pentasaccharide itself was, not surprisingly, too congested to be structurally informative [57]. It displayed a broad absorption extending from 3,100 to 3,700 cm^{-1}, composed of 14 O–H and 2 N–H unresolved stretching bands. The large size of the molecule (130 atoms, 986 amu) prevented a full exploration of its conformational landscape by quantum chemistry methods, without the guidance that a resolved IRID spectrum would have offered. As a consequence, its structural properties were explored using molecular mechanics, considering three situations: fully desolvated, explicitly hydrated (with three water

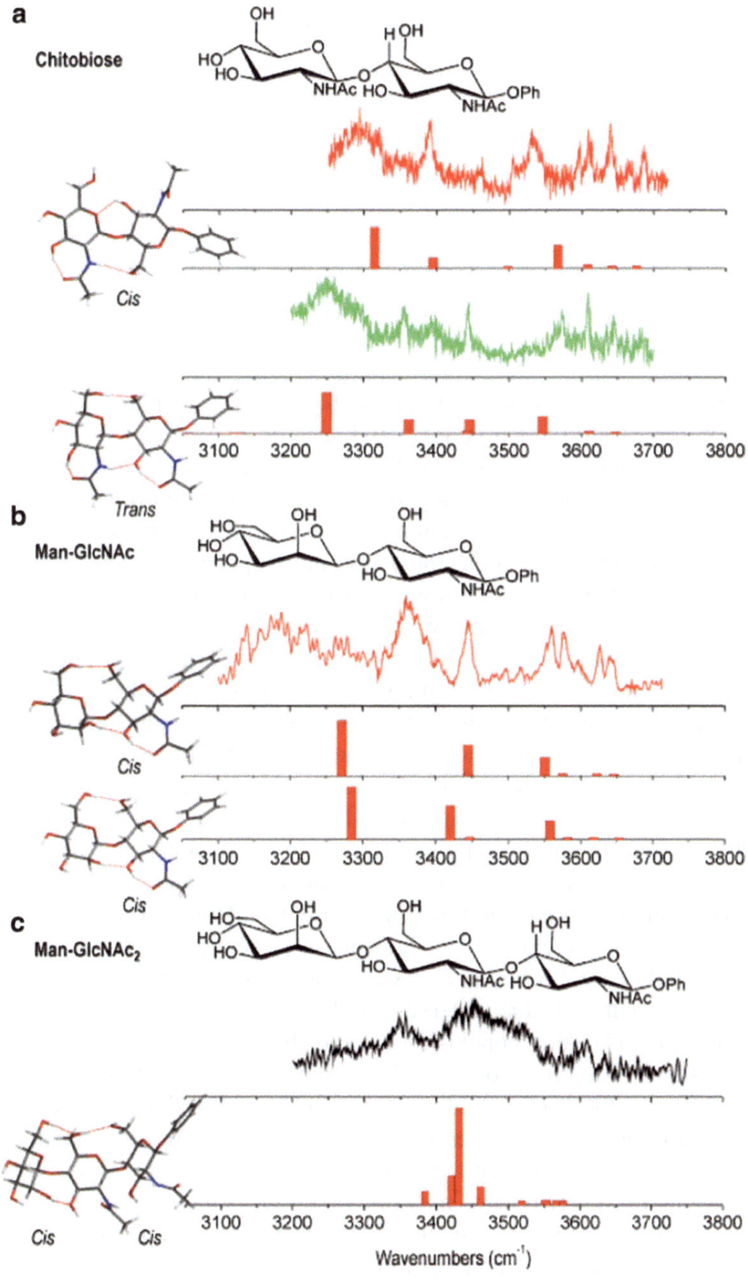

Fig. 12 (**a–c**) Gas-phase IRID spectrum, calculated spectra, and conformations of the stem sub-units of *5sac*

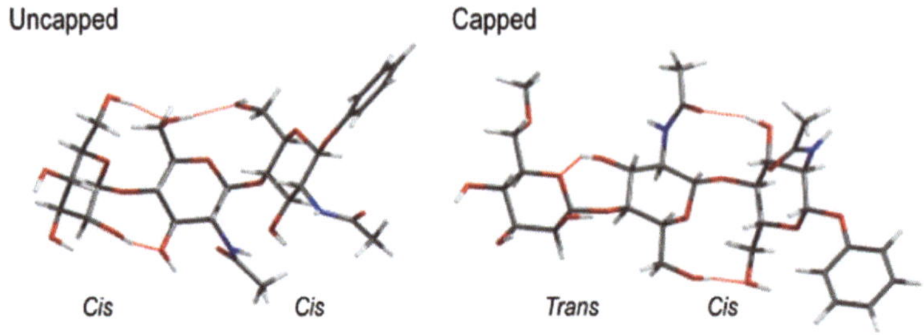

Fig. 13 Effect of capping the Man-OH6 group by a methyl group on the conformation of Man-GlcNAc-GlcNAc

molecules), and fully solvated in an implicit aqueous environment [57]. The OPLS2005 force field (which is not 'CBH specific') and GLYCAM06 (which is) were used and both led to qualitatively similar results. In the absence of solvent molecules, *5sac* could adopt many diverse "curled" conformations where the number of intramolecular interactions was maximized. The effect of explicit hydration (with only three molecules) was quite spectacular: the lowest energy structures unfolded, the *trans* conformation of the chitibiose stem was preferentially adopted, and the water molecules were often located in the vicinity of the strategic binding sites identified from the study of the sub-units (the water pocket of Man_3 and across the *trans* linkage of chitibiose). In the fully solvated case, the molecule adopted extended conformations with limited flexibility (indicated by a low dispersion of the maximum intramolecular distances distribution) and the *trans* conformation of chitibiose was retained. Representative low energy conformations of *5sac* for the three cases are shown in Fig. 14.

Such large molecular systems are extremely challenging and it took several years of a long-fought battle on the experimental, computational, and synthetic fronts to collect, understand, and interpret these data. However, this 'tour de force' led to a rewarding outcome – the gas-phase structural properties of the *5sac* sub-units could be established [57]. Extrapolating them in the context of *N*-linked glycoproteins, with all due reserves, and including the findings from other approaches [11–17], the conclusions could be summarized as follows. The chitibiose stem, in a *trans* conformation promoted by hydration, acts as a rigid "spacer" between the peptide and the saccharide manifold of the *N*-linked glycoproteins; while one of its N-acetyl groups (in the GlcNAc unit neighboring the peptide) is free to interact with the peptide chain to create an "anchor", the other participates in a water H-bonded network which locks the *trans* conformation. On the other hand, the singular Man-β1,3-GlcNac linkage remains flexible, allowing the subsequent glycan chain to adapt to its environment, while the flexibility of the Man_3 unit is hindered by a water-mediated bridge linking its two branches.

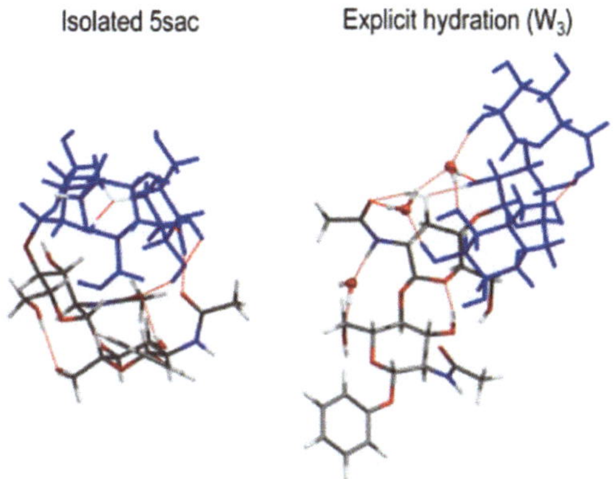

Fig. 14 Typical conformations predicted by the OPLS force field for the three solvation states: full desolvation (*left*), explicit hydration with three water molecules (*center*), and implicit solvation in an aqueous environment (*right*)

3.3 Microwave Spectroscopy of Monosaccharides

A chapter of this special issue is fully dedicated to microwave spectroscopy of biomolecules and contains a full section on CBHs. The observation of the pure rotational spectrum offers the advantage of not requiring additional aromatic chromophores for spectroscopic detection, so it may complement the vibrational data obtained from double resonance spectroscopy experiments, especially for the smallest CBHs.

The detection of intact monosaccharides by gas-phase microwave spectroscopy has been at the center of study of several specialists of this technique. The heating methods or IR-nanoseconds laser vaporization resulted in the observation of decomposition products only. The first publication reporting a microwave spectrum of an untagged sugar, ribose, made possible by UV picosecond laser vaporization, was only issued in 2012 [31]. Ribose is a pentose found as a five-membered ring, β-furanoside, configuration in biological compounds (RNA, ATP, etc.). The analysis of its rotational spectrum revealed a large conformational variability, with six structures very close in energy, all within ca. 4.6 kJ mol^{-1} (Fig. 15) and all forming a pyranose six-membered ring. Both anomeric forms were observed (β being the more stable) and quasi iso-energetic $^4C_1$ and $^1C_4$ ring configurations could be adopted. Notably, the furanose structures were not observed for ribose.

The Ribose study was soon followed by similar investigations of 2-deoxy-Ribose [71] Fructose [69], Xylose [68], Erythrose [97], and Glucose [98]. All observed monosaccharides (five or six carbon units) adopted the pyranose form while C4-sugars adopt furanose conformations as Erythrose [97]. These systems do not always follow the general trends observed in solution and in the crystalline phase where pyranose and furanose forms could be detected simultaneously. The absence of OH2 in 2-deoxy-ribose weakens the cooperative networks existing in ribose. Six conformations have been observed, four as β-anomers and two as

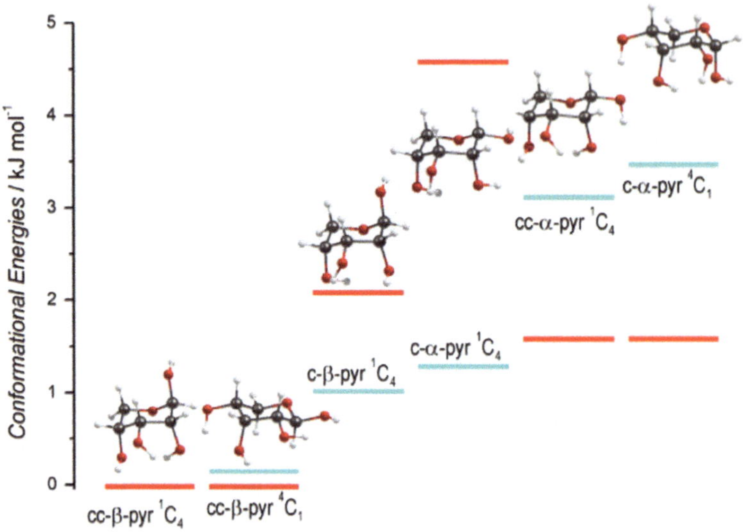

Fig. 15 The six detected conformations of free Ribose are six-membered pyranose rings, adopting $^{1}C_{4}$- or $^{4}C_{1}$-chair skeletons of both β- and α-anomers. The experimental conformational energies in *red* are estimations derived from the relative intensities of the microwave transitions. The *blue horizontal lines* represent the theoretical (MP2) ΔG_{298} predictions

α-anomers. Fructose [69, 70] was the first ketose to be observed in the gas phase and the situation is very different – a dominant conformation, the $^{2}C_{5}$ β-pyranoside (equivalent to the $^{1}C_{4}$ form of aldoses) plus another very weak $^{2}C_{5}$ β-pyranoside were observed. This dominant configuration corresponds to the most stable conformation, which is especially stabilized by five consecutive hydrogen bonds forming a cooperative network.

Microwave spectroscopy can also quantify experimentally the relative proportions of the observed species. However, for some of the detected monosaccharides the correlation between the experimental populations and the calculated relative energies is not always obvious. This might reflect incomplete conformational cooling in the supersonic expansion which does not lead to a relaxed population reflecting the equilibrium at low temperature (and the potential energy surface of the system). Alternatively, the pre-expansion population (governed by the free energy surface at the vaporization temperature) may have been frozen early in the expansion, or the observed population may simply reflect the initial distribution of structures prior to vaporization in the solid sample.

In addition, because of the very high resolution achieved, microwave spectroscopy can unambiguously assign rotational spectra for different isotopologues of the same molecule. In the studies of 2-deoxy-ribose and fructose, spectra of ^{13}C, ^{18}O, and ^{2}D substituted species, either in natural abundance (0.02–1.1%) or in enriched samples, could be observed. Since the isotopic spectral modifications are directly connected to the molecular moments of inertia, it is possible to determine experimentally the position of the isotopic substituents in the molecular structure without

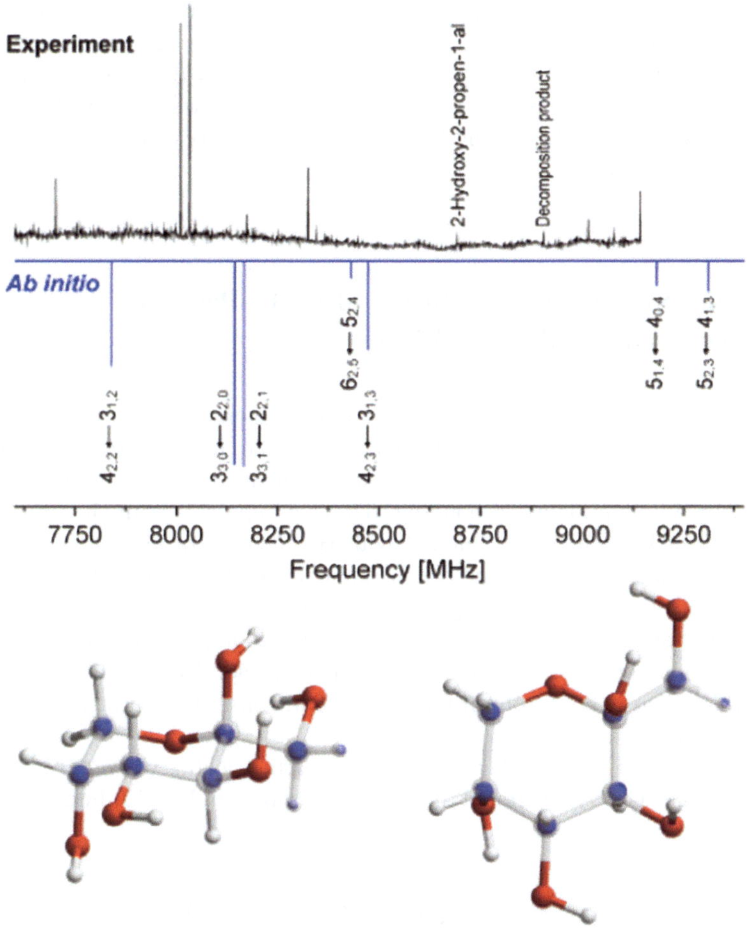

Fig. 16 Two views of the superposition of the position of atoms experimentally determined by the analysis of microwave spectra of several isotopomers of Fructose (blue spheres) and the corresponding calculated conformation, determined at the MP2 level

any computational information. This is spectacularly illustrated in Fig. 16 in the case of the observed dominant conformation of fructose.

The possibility of directly "measuring" molecular structures is an important advantage of microwave spectroscopy. In vibrational spectroscopy, isotopic substitution helps the interpretation of the spectra but the assigned structures are "only" the calculated ones which offer the best match between the calculated vibrational spectrum and the observed one. It is also difficult to determine relative abundances from electronic spectroscopy because the relative intensity of the observed electronic transitions of the chromophore can be affected by the dynamics of the excited state. Both techniques are complementary: vibrational spectroscopy can address the conformational preferences of large systems which microwave spectroscopy cannot deal with, and microwave spectroscopy can provide extremely valuable benchmark structures to validate the theoretical models used to describe the conformation of such flexible systems. Although microwave spectroscopy can only deal with relatively small systems at present, the advent of the microwave chirped pulse technique [99–101] pushes the current limits of its applicability to allow access to much larger molecular species.

4 Model Systems for the Study of CBH–Protein Interactions in the Gas Phase

When CBHs interact with proteins, the types of interaction can be quite diverse [27]. CBHs can be seen as "double sided molecular sticky tapes." They contain many groups (OH, NH, C=O) which can be engaged in strong and directional electrostatically-dominated H-bonding. These groups represent the "hydrophilic face" of the CBH which can interact with H-accepting or H-donating groups of the protein; water molecules and can also play a structural role. CBHs also have a "hydrophobic face" where CH groups can participate in longer range and less directional dispersion-dominated interactions, for example, non-covalent bonding CH $\rightarrow \pi$ interactions [102]. Dispersion is a universal, additive, and always attractive force which also plays a role in OH $\rightarrow \pi$ interactions and, more generally, in all non-covalent interactions.

CBHs are often found in "grooves" of the partner lectin, where all interactions can take place. This is illustrated in the cartoon shown in Fig. 17. Interactions with

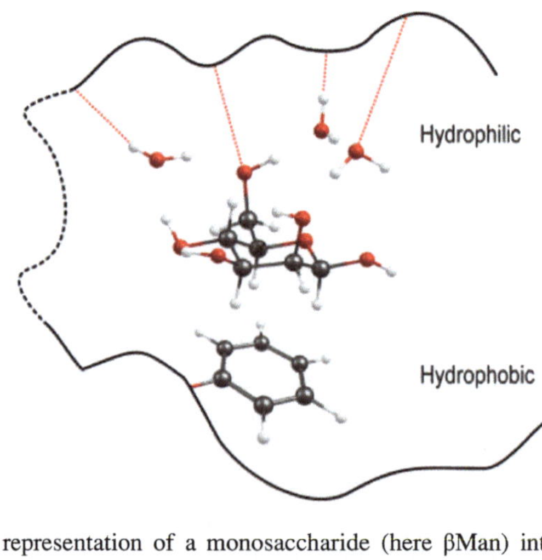

Fig. 17 Schematic representation of a monosaccharide (here βMan) interacting with a lectin (represented by the *curved line*). Dashed lines indicate H-bonds, directly between the lectin and the CBH, or indirectly via water molecules

ionic groups often occur too. In particular, charged alkali earth metals such as Ca^{2+} can promote bonding via very strong electrostatic interactions [103, 104].

These interactions are being investigated in a rapidly increasing range of monosaccharide complexes, via gas-phase spectroscopy [44, 48, 49, 59–63, 65, 66] with their partners judiciously selected to probe specific types of interaction and to determine the "minimum ingredients" involved in CBH–protein interactions.

4.1 Interaction with Aromatic Models

CBH-protein binding often involves interactions between the hydrophobic face of the CBH and aromatic amino acid residues. These have been modeled in the gas phase through studies of monosaccharides complexed with representative aromatic molecules [39, 44, 59]. The investigations can be challenging: it can be difficult to observe sufficient signals of the weakly bound complexes between the CBH (evaporated by laser desorption) and the aromatic compound seeding the carrier gas of the supersonic expansion. The theoretical part of the work is also challenging because many low-lying conformers may exist and levels of theory which include the dispersion interaction are essential. Fortunately, these have not prevented the observation of a series of monosaccharide complexes with toluene [39, 44] and the interpretation of their IRID spectra. Intuition – always suspect – might predict that one of the OH groups of the monosaccharide would be used to create an OH $\rightarrow \pi$ interaction. However, this would disrupt the intramolecular H-bond network of the monosaccharide and the best compromise seems to be to leave it unchanged. The bonding was dominated by CH $\rightarrow \pi$ interactions, leaving the OH groups of the monosaccharide free to interact intramolecularly, although a few OH $\rightarrow \pi$ bound complexes have been observed as well.

Similar studies on the complexes formed by *para*-hydroxy-toluene (*p*-Tol-OH, aka *p*-cresol) have also been conducted [59]. The addition of the phenolic OH group modified completely the structural selectivity of the observed complexes. The strong interaction between the phenolic OH group and the OH groups of the monosaccharide dominate and CH $\rightarrow \pi$ interactions only played a very minor role. This is illustrated in Fig. 18 for the βGlc-O-Me and βGal-O-Me complexes. The phenolic OH group inserts between neighboring OH groups on the monosaccharide to form cooperatively-enhanced H-bonded structures, following as closely as possible the basic micro-hydration rules. In the gas phase when H-bonding is possible, it clearly takes control over dispersive forces because of its much higher strength. This is not necessarily true in condensed media where many other competing effects (steric hindrance, dielectric effects, interactions with other bio- or solvent molecules) can come into play. It should be noted that although the dominant interaction is H-bonding, the inclusion of dispersion corrections in the structure calculation affects significantly the geometry of the complex (Fig. 18c).

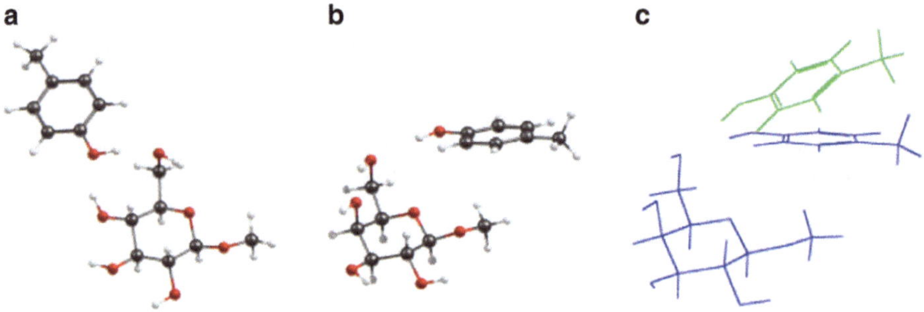

Fig. 18 Structures of the observed complexes of *p*-OH-Tol with (**a**) βGlc-O-Me and (**b**) βGal-O-Me. (**c**) The influence of dispersion correction on the calculated structure of the βGal-O-Me complex, showing the results of calculations using RI-B97 (*green*), which does not include a correction for dispersion, and RI-B97-D (*blue*), which does

4.2 Interaction with Peptide Models

As mentioned above, toluene and cresol can be considered as "poor man's" models for the interaction between the aromatic side-chain residues of phenylalanine (Phe) and tyrosine (Tyr). To get a little closer to 'reality' and to introduce possible interactions with a peptide chain model, subsequent investigations were extended to include complexes formed between monosaccharides such as Glc and Gal, and the capped peptides, AcPheNHMe [48, 49] and AcTyrNHMe [59]; see Fig. 19.

The conformational landscapes of both peptides are extremely similar. In short, they can adopt "open" conformations, with a fully extended peptide chain, and "folded" conformations where the NH groups of the chain are involved in intramolecular H-bonding: both types are observed in supersonic expansions of the free molecules [59, 105]. The monosaccharides can form complexes with either conformation, although with an apparent preference for the open one, which offers more readily available "contact points": all H-accepting and H-donating groups are free to interact with the CBH.

In the AcPheNHMe-CBH complexes, the monosaccharides interact exclusively with the peptide backbone, with their OH groups linked though H-bonds to NH and CO groups. Their structures keep the aromatic ring as remote as it can be from the monosaccharide. In the AcTyrNHMe complexes, however, the monosaccharides interact both with the peptide chain *and* with the Tyr phenolic OH. As a consequence, the aromatic ring is brought closer to the CBH and there is evidence of weak CH$\rightarrow\pi$ interactions, although they are hindered and constrained by the dominating H-bonds which govern the structure of the complexes. Those formed by βGal-O-Me and the two peptides are shown in Fig. 20, together with their observed and calculated vibrational spectra. Complexes of AcPheNHMe with the α anomers (of Gal-O-Me and Glc-O-Me) have also been observed and, as with the hydrated clusters of Man-O-Me, the peptide chain seems to be sensitive to the variations of local electron densities from one anomer to another [49].

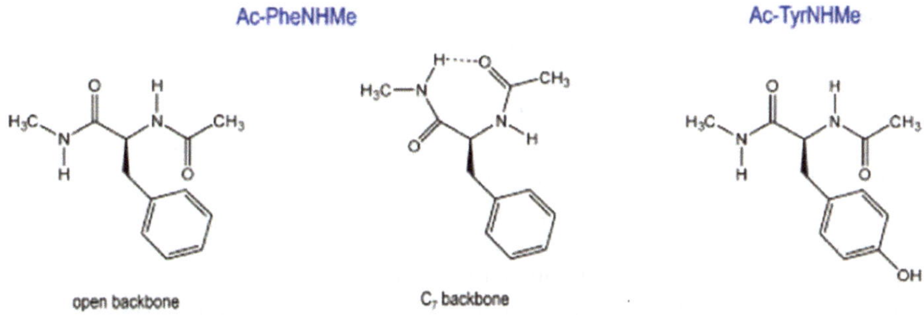

Fig. 19 The peptide models AcPheNHMe (*left*) and AcTyrNHMe (*right*). The open and folded conformations of AcPheNHMe are shown

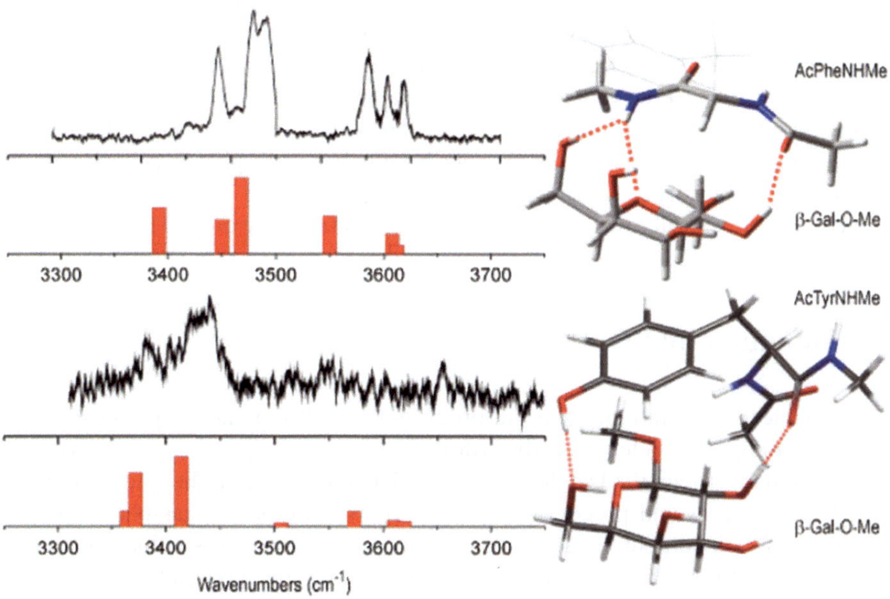

Fig. 20 Experimental IRID spectra, calculated spectra, and structures of the complexes formed by βGal-O-Me and two peptide models, AcPheNHMe (*top*) and AcTyrNHMe (*bottom*)

4.3 *Interaction with Charged Species*

Compared to the vast amount of experimental gas-phase spectroscopy of charged amino acids and peptides reported in the literature, investigations of charged CBHs, or of CBHs interacting with charged species (e.g., alkali metal ions [60–63, 65, 66]), both probed through IRMPD, are rare. Those that have been reported have focused mainly on establishing the vibrational spectroscopic signatures of the charged species and their potential exploitation to differentiate structures (anomers, epimers, isomers of disaccharides) which cannot be easily discriminated using mass spectrometry alone. The collection of acquired knowledge on these systems should be extremely helpful in improving sequencing techniques for glycans and oligosaccharides, a field far from the maturity reached by genomics and proteomics. In this context, promising IRMPD spectra of protonated and sodiated O-linked glycopeptides have recently been reported [106].

The structural information obtained from these early spectroscopic studies has not really provided insights into the structural basis of CBH-protein interactions and molecular recognition, although the field will surely expand soon and the potential of these high-throughput techniques holds much promise. The combination of optical gas-phase spectroscopy and mass spectrometry can be extremely informative. In particular, the coupling of ion mobility methods, such as ion drift [107] or Field Asymmetric Waveform Ion Mobility Spectrometry (FAIMS) [108] and optical spectroscopy [109], should enable conformer- or isomer-selective IRMPD spectroscopy, as shown by a recent IRMPD study by Maître et al. where different monosaccharides (Glc, Gal, Man and their anomers) have been separated by FAIMS before being probed by vibrational spectroscopy [110]. IRMPD experiments conducted very recently by Compagnon et al. have also brought the proof of principle of how mass spectrometry can benefit from gas-phase spectroscopy by identifying robust spectral signatures of phosphated and sulfated glucosamine which cannot be easily distinguished by mass spectrometry only [111].

5 Summary and Outlook

This chapter has summarized recent progress in experimental gas-phase spectroscopy of biologically relevant carbohydrates. Although very few groups have been involved in this topic, the last decade has shed a great deal of light onto a variety of important systems and processes in glycobiology. Combining gas-phase spectroscopic data with computational modeling and insights from crystalline and solvent phase studies (X-ray crystallography and NMR), it has been possible to interrogate several aspects of the fascinating structural properties of these molecules, such as conformational preferences, solvation effects, molecular recognition, and anomeric effects. Nonetheless, given the diversity of carbohydrates and of their biological functions, the field is very far from being exhausted. In particular, glycoconjugates – molecules formed by *chemical* bonding between CBHs and other biomolecules (glycopeptides, glycolipids) – have not yet been explored by gas-phase spectroscopy, in spite of their importance in biological processes. More thorough study of charged CBHs and the interaction of CBHs with charged species and reactive species (ions, acids, radicals) should also help to improve our understanding of glycochemistry (synthesis, hydrolysis, etc.). With the advent of new spectroscopic methods becoming more widely applied to the study of CBHs (microwave spectroscopy, IRMPD of ions), we can expect a bright future.

Despite this optimistic prospectus, two critical questions are worth asking.

5.1 *Should We Run the Race Against Molecular Size?*

As gas-phase biomolecular spectroscopy has evolved, great efforts have been made to increase the size of the systems under study, partly in view of approaching biological complexity, and also to try to answer the legitimate question – how far can we go and, starting from reductionist models, how easily can the experimental data be interpreted? Given the progress for carbohydrates described in this chapter, it is clear that one could answer "Well, pretty far!" The improvement of computational resources and the development of more efficient theoretical models without significant loss in interpretative capabilities have helped to push in this direction.

On the experimental side, the present state of the art is such that we can transfer virtually any molecular assembly into the gas phase, regardless of its mass, and observe its spectroscopic signals. However, spectral congestion can be an unbeatable limiting factor, particularly when H-bonded interactions lead to broad vibrational bands. The core pentasaccharide motif provides an illustration of this limitation and makes us wonder whether the limit of the approach is getting close.

We may have reached a point where, instead of trying to interrogate systematically ever-growing molecular system, we should focus on appropriate models of reasonable size, by gas-phase spectroscopists' standards, for which clear spectroscopic signatures can still be observed and, most importantly, understood. Isotopically-enriched species can help disentangle and interpret congested spectra of large molecules. This has been successfully applied to large peptide spectroscopy [112] and hydrated CBHs complexes.

Of course, several years ago, a rational gas-phase spectroscopist might have expressed reservations if asked whether it would be possible from gas-phase spectroscopy and quantum chemistry calculation to determine the conformation of a hydrated complex of a phenyl-tagged trisaccharide, such as Man_3 and Man_3-W_1 [40]. The point raised here may then be proved wrong sooner than expected, which would be very good news indeed!

5.2 *Can Size-Controlled Micro-Hydration be Extrapolated to Reflect Bulk Solvent Effects?*

Investigations of singly and multiply hydrated complexes of mono-, di-, and trisaccharides have revealed the effect of a few water molecules on their gas phase conformational choices. However, under the influence of bulk hydration they may behave quite differently, and reaching this limit may not be accessible to gas-phase spectroscopy experiments. In the case of cellobiose, molecular dynamics simulations suggest that it may take much more than a solvation shell (and maybe up to 100 water molecules) to stabilize the *trans* conformation of the glycosidic linkage observed in solution. Using laser desorption from frozen aqueous solution, it has proved possible to observe, in the gas phase, very large hydrated clusters

[113]. However, if the gas phase to bulk solvated phase transition is reached, is spectroscopy the right tool to provide structural information? This remains an open question. The quest for the structural properties of the core pentasaccharide in the gas phase revealed the key role that could be played by a *few* water molecules located at strategic sites of the molecule. They should not be considered as bulk solvent molecules but rather as "architectural" or "engineering" molecules. The biological role of these molecules, removed from the bulk solvent, is fundamental. Gas-phase spectroscopy provides a unique tool for locating the strategic water-binding sites ("water holes" or "water pockets") in the non-hydrated molecule and for exploring the specific influence of bound water molecules, in limited and controlled numbers, on biomolecular systems.

Finally, it should be mentioned that glycoscience now impacts on a very broad range of scientific communities, well beyond glycobiology [6]. For instance, in materials science, CBHs emerge as new components of bio-compatible, non-petroleum-based materials. CBHs are also extremely promising resources for sustainable and environmentally acceptable energy. These fields might offer new playgrounds and challenges for CBH gas-phase spectroscopy in the near future.

Acknowledgements Professor John P. Simons has pioneered the study of neutral CBHs in the gas phase. Several "generations" of students and post-docs have been lucky enough to spend some time in his group. We had the opportunity to contribute to very exciting studies and, most importantly, we found inspiration and confidence to pursue our own scientific interests. John has been especially important for the two of us for developing our own activities and we can never thank him enough for his support. We also want to thank all our friends from the "JPS sugar team" who contributed over the years. Francis Talbot, Rebeccah A. Jockush, Niel A. Macleod, Isabel Hunig, Cristina Stanca-Kaposta, Bo Liu, Timothy D. Vaden, Zheng Su, Nitzan Mayorkas, and Svemir Rudic.

References

1. Varki A (1993) Glycobiology 3(2):97–130
2. Berg JM, Tymoczko JL, Stryer L (2002) Carbohydrates. In: Biochemistry, 5th edn. W.H. Freeman, New York
3. Taylor ME, Drickamer K (2003) Introduction to glycobiology, 2nd edn. Oxford University Press, London, New York
4. Rudd PM, Dwek RA (2006) Curr Opin Struct Biol 16(5):559–560
5. Imperiali B (2012) J Am Chem Soc 134(43):17835–17839
6. A Roadmap for the Future (2012) National Research Council (US) committee on assessing the importance and impact of glycomics and glycosciences. National Academies Press, Washington. ISBN-13: 978-0-309-26083-1-ISBN-10: 0-309-26083-3
7. Simons JP, Jockusch RA, Carcabal P, Hung I, Kroemer RT, Macleod NA, Snoek LC (2005) Int Rev Phys Chem 24(3–4):489–531
8. Simons JP (2009) Mol Phys 107(23–24):2435–2458
9. Schwarz F, Aebi M (2011) Curr Opin Struct Biol 21(5):576–582
10. Arnold JN, Wormald MR, Sim RB, Rudd PM, Dwek RA (2007) Annu Rev Immunol 25:21–50
11. Wormald MR, Dwek RA (1999) Structure 7(7):R155–R160
12. Wormald MR, Petrescu AJ, Pao YL, Glithero A, Elliott T, Dwek RA (2002) Chem Rev 102(2):371–386

13. O'Connor SE, Imperiali B (1996) Chem Biol 3(10):803–812
14. O'Connor SE, Imperiali B (1998) Chem Biol 5(8):427–437
15. Weerapana E, Imperiali B (2006) Glycobiology 16(6):91r–101r
16. Woods RJ, Pathiaseril A, Wormald MR, Edge CJ, Dwek RA (1998) Eur J Biochem 258 (2):372–386
17. Bosques CJ, Tschampel SM, Woods RJ, Imperiali B (2004) J Am Chem Soc 126(27):8421–8425
18. Gabius HJ (2000) Naturwissenschaften 87(3):108–121
19. Gabius H-J, André S, Kaltner H, Siebert H-C (2002) Biochim Biophys Acta Gen Subj 1572 (2–3):165–177
20. Gabius H-J, André S, Jiménez-Barbero J, Romero A, Solís D (2011) Trends Biochem Sci 36 (6):298–313
21. Pilobello KT, Mahal LK (2007) Curr Opin Chem Biol 11(3):300–305
22. Feizi T, Chai W (2004) Nat Rev Mol Cell Biol 5(7):582–588
23. Davis BG (2000) Chem Ind, pp. 134–138
24. Shriver Z, Raguram S, Sasisekharan R (2004) Nat Rev Drug Discov 3(10):863–873
25. Ernst B, Magnani JL (2009) Nat Rev Drug Discov 8(8):661–677
26. Varki A, Cummings RD, Esko JD, Freeze HH, Stanley P, Bertozzi CR, Hart GW, Etzler ME (eds) (2009) Essential in glycobiology, 2nd edn. Cold Spring Harbor Laboratory Press, Cold Spring Harbor
27. Weis WI, Drickamer K (1996) Annu Rev Biochem 65(1):441–473
28. Bertozzi CR, Rabuka D (2009) Structural basis of glycan diversity. In: Varki A, Cummings RD, Esko JD et al (eds) Essentials of glycobiology, 2nd edn. Cold Spring Harbor Laboratory Press, Cold Spring Harbor
29. Breslow R (2012) J Am Chem Soc 134(16):6887–6892
30. Ruiz-Mirazo K, Briones C, de la Escosura A (2014) Chem Rev 114(1):285–366
31. Cocinero EJ, Lesarri A, Ecija P, Basterretxea FJ, Grabow J-U, Fernandez JA, Castano F (2012) Angew Chem Int Ed 51(13):3119–3124
32. Talbot FO, Simons JP (2002) Phys Chem Chem Phys 4(15):3562–3565
33. Jockusch RA, Kroemer RT, Talbot FO, Simons JP (2003) J Phys Chem A 107(49):10725–10732
34. Jockusch RA, Kroemer RT, Talbot FO, Snoek LC, Carcabal P, Simons JP, Havenith M, Bakker JM, Compagnon I, Meijer G, von Helden G (2004) J Am Chem Soc 126(18):5709–5714
35. Carcabal P, Jockusch RA, Hunig I, Snoek LC, Kroemer RT, Davis BG, Gamblin DP, Compagnon I, Oomens J, Simons JP (2005) J Am Chem Soc 127(32):11414–11425
36. Carcabal P, Hunig I, Gamblin DP, Liu B, Jockusch RA, Kroemer RT, Snoek LC, Fairbanks AJ, Davis BG, Simons JP (2006) J Am Chem Soc 128(6):1976–1981
37. Jockusch RA, Talbot FO, Rogers PS, Simone MI, Fleet GWJ, Simons JP (2006) J Am Chem Soc 128(51):16771–16777
38. Macleod NA, Johannessen C, Hecht L, Barron LD, Simons JP (2006) Int J Mass Spectrom 253(3):193–200
39. Screen J, Stanca-Kaposta EC, Gamblin DP, Liu B, Macleod NA, Snoek LC, Davis BG, Simons JP (2007) Angew Chem Int Ed 46(20):3644–3648
40. Cocinero EJ, Stanca-Kaposta EC, Scanlan EM, Gamblin DP, Davis BG, Simons JP (2008) Chem Eur J 14(29):8947–8955
41. Simons JP, Stanca-Kaposta EC, Cocinero EJ, Liu B, Davis BG, Gamblin DP, Kroemer RT (2008) Phys Scr 78(5), 058124
42. Cocinero EJ, Gamblin DP, Davis BG, Simons JP (2009) J Am Chem Soc 131(31):11117–11123
43. Simons JP, Davis BG, Cocinero EJ, Gamblin DP, Stanca-Kaposta EC (2009) Tetrahedron-Asymmetry 20(6–8):718–722
44. Su Z, Cocinero EJ, Stanca-Kaposta EC, Davis BG, Simons JP (2009) Chem Phys Lett 471(1–3):17–21

45. Drouin L, Stanca-Kaposta EC, Saundh P, Fairbanks AJ, Kemper S, Claridge TDW, Simons JP (2009) Chem Eur J 15(16):4057–4069
46. Su Z, Wagner B, Cocinero EJ, Ernst B, Simons JP (2009) Chem Phys Lett 477(4–6):365–368
47. Brauer B, Pincu M, Buch V, Bar I, Simons JP, Gerber RB (2011) J Phys Chem A 115 (23):5859–5872
48. Cocinero EJ, Carcabal P, Vaden TD, Davis BG, Simons JP (2011) J Am Chem Soc 133 (12):4548–4557
49. Cocinero EJ, Carcabal P, Vaden TD, Simons JP, Davis BG (2011) Nature 469(7328):76–U1400
50. Jin L, Simons JP, Gerber RB (2011) Chem Phys Lett 518:49–54
51. Mayorkas N, Rudic S, Cocinero EJ, Davis BG, Simons JP (2011) Phys Chem Chem Phys 13 (41):18671–18678
52. Mayorkas N, Rudic S, Davis BG, Simons JP (2011) Chem Sci 2(6):1128–1134
53. Pincu M, Cocinero EJ, Mayorkas N, Brauer B, Davis BG, Gerber RB, Simons JP (2011) J Phys Chem A 115(34):9498–9509
54. Jin L, Simons JP, Gerber RB (2012) J Phys Chem A 116(46):11088–11094
55. Sagar R, Rudic S, Gamblin DP, Scanlan EM, Vaden TD, Odell B, Claridge TDW, Simons JP, Davis BG (2012) Chem Sci 3(7):2307–2313
56. H-b X, Jin L, Rudic S, Simons JP, Gerber RB (2012) J Phys Chem B 116(16):4851–4859
57. Barry CS, Cocinero EJ, Carcabal P, Gamblin DP, Stanca-Kaposta EC, Remmert SM, Fernández Alonso MdC, Rudic S, Simons JP, Davis BG (2013) J Am Chem Soc 135 (45):16895–16903
58. Carcabal P, Cocinero EJ, Simons JP (2013) Chem Sci 4(4):1830–1836
59. Stanca-Kaposta EC, Carcabal P, Cocinero EJ, Hurtado P, Simons JP (2013) J Phys Chem B 117(27):8135–8142
60. Polfer NC, Valle JJ, Moore DT, Oomens J, Eyler JR, Bendiak B (2006) Anal Chem 78 (3):670–679
61. Stefan SE, Eyler JR (2009) Anal Chem 81(3):1224–1227
62. Cagmat EB, Szczepanski J, Pearson WL, Powell DH, Eyler JR, Polfer NC (2010) Phys Chem Chem Phys 12(14):3474–3479
63. Stefan SE, Eyler JR (2010) Int J Mass Spectrom 297(1–3):96–101
64. Brown DJ, Stefan SE, Berden G, Steill JD, Oomens J, Eyler JR, Bendiak B (2011) Carbohydr Res 346(15):2469–2481
65. Stefan SE, Ehsan M, Pearson WL, Aksenov A, Boginski V, Bendiak B, Eyler JR (2011) Anal Chem 83(22):8468–8476
66. Contreras CS, Polfer NC, Oomens J, Steill JD, Bendiak B, Eyler JR (2012) Int J Mass Spectrom 330:285–294
67. Rudic S, Xie H-b, Gerber RB, Simons JP (2012) Mol Phys 110(15–16):1609–1615
68. Pena I, Mata S, Martin A, Cabezas C, Daly AM, Alonso JL (2013) Phys Chem Chem Phys 15 (41):18243–18248
69. Cocinero EJ, Lesarri A, Ecija P, Cimas A, Davis BG, Basterretxea FJ, Fernandez JA, Castano F (2013) J Am Chem Soc 135(7):2845–2852
70. Bermudez C, Pena I, Cabezas C, Daly AM, Alonso JL (2013) Chemphyschem 14(5):893–895
71. Peña I, Cocinero EJ, Cabezas C, Lesarri A, Mata S, Écija P, Daly AM, Cimas Á, Bermúdez C, Basterretxea FJ, Blanco S, Fernández JA, López JC, Castaño F, Alonso JL (2013) Angew Chem Int Ed:n/a–n/a 52(45):11840–11845
72. Motiyenko RA, Alekseev EA, Dyubko SF, Lovas FJ (2006) J Mol Spectrosc 240(1):93–101
73. Foley BL, Tessier MB, Woods RJ (2012) Wiley Interdiscip Rev: Comput Mol Sci 2(4):652–697
74. Hansen HS, Hünenberger PH (2011) J Comput Chem 32(6):998–1032
75. Kirschner KN, Yongye AB, Tschampel SM, González-Outeiriño J, Daniels CR, Foley BL, Woods RJ (2008) J Comput Chem 29(4):622–655
76. Mackerell AD (2004) J Comput Chem 25(13):1584–1604
77. Damm W, Frontera A, Tirado–Rives J, Jorgensen WL (1997) J Comput Chem 18(16):1955–1970

78. Reiling S, Schlenkrich M, Brickmann J (1996) J Comput Chem 17(4):450–468
79. Woods RJ, Dwek RA, Edge CJ, Fraserreid B (1995) J Phys Chem 99(11):3832–3846
80. Sameera WMC, Pantazis DA (2012) J Chem Theory Comput 8(8):2630–2645
81. Halgren TA (1999) J Comput Chem 20(7):730–748
82. Kaminski GA, Friesner RA, Tirado-Rives J, Jorgensen WL (2001) J Phys Chem B 105 (28):6474–6487
83. Kratzcr A (1920) Z Phys 3:460
84. Loomis FW (1920) Astrophys J 62:248
85. Edward JT (1955) Chem Ind 1102–1104
86. Wang C, Ying F, Wu W, Mo Y (2011) J Am Chem Soc 133(34):13731–13736
87. Mo Y (2010) Nat Chem 2(8):666–671
88. Asensio JL, Jimenez-Barbero J (1995) Biopolymers 35(1):55–73
89. Cheetham NWH, Dasgupta P, Ball GE (2003) Carbohydr Res 338(9):955–962
90. Larsson EA, Staaf M, Söderman P, Höög C, Widmalm G (2004) J Phys Chem A 108 (18):3932–3937
91. Martín-Pastor M, Canales A, Corzana F, Asensio JL, Jiménez-Barbero J (2005) J Am Chem Soc 127(10):3589–3595
92. Olsson U, Serianni AS, Stenutz R (2008) J Phys Chem B 112(14):4447–4453
93. Pincu M, Gerber RB (2012) Chem Phys Lett 531:52–58
94. Zierke M, Smieško M, Rabbani S, Aeschbacher T, Cutting B, Allain FHT, Schubert M, Ernst B (2013) J Am Chem Soc 135(36):13464–13472
95. Mons M, Dimicoli I, Piuzzi F (2002) Int Rev Phys Chem 21(1):101–135
96. Mons M, Piuzzi F, Dimicoli I, Zehnacker A, Lahmani F (2000) Phys Chem Chem Phys 2 (22):5065–5070
97. Cabezas C, Pena I, Daly AM, Alonso JL (2013) Chem Commun 49(92):10826–10828
98. Alonso JL, Lozoya MA, Pena I, Lopez JC, Cabezas C, Mata S, Blanco S (2014) Chem Sci 5 (2):515–522
99. Brown GG, Dian BC, Douglass KO, Geyer SM, Shipman ST, Pate BH (2008) Rev Sci Instrum 79(5), 053103
100. Zaleski DP, Neill JL, Muckle MT, Seifert NA, Carroll PB, Weaver SLW, Pate BH (2012) J Mol Spectrosc 280:68–76
101. Neill JL, Harris BJ, Steber AL, Douglass KO, Plusquellic DF, Pate BH (2013) Opt Express 21 (17):19743–19749
102. Santana AG, Jiménez-Moreno E, Gómez AM, Corzana F, González C, Jiménez-Oses G, Jiménez-Barbero J, Asensio JL (2013) J Am Chem Soc 135(9):3347–3350
103. Haseley SR, Vermeer HJ, Kamerling JP, Vliegenthart JFG (2001) Proc Natl Acad Sci 98 (16):9419–9424
104. Carvalho de Souza A, Ganchev D, Snel ME, Eerden JJM, Vliegenthart JG, Kamerling J (2009) Glycoconj J 26(4):457–465
105. Gerhards M, Unterberg C, Gerlach A, Jansen A (2004) Phys Chem Chem Phys 6(10):2682–2690
106. Fukui K, Takahashi K (2012) Anal Chem 84(5):2188–2194
107. Pagel K, Harvey DJ (2013) Anal Chem 85(10):5138–5145
108. Brown LJ, Creaser CS (2013) Curr Anal Chem 9(2):192–198
109. Bellina B, Compagnon I, MacAleese L, Chirot F, Lemoine J, Maitre P, Broyer M, Antoine R, Kulesza A, Mitric R, Bonacic-Koutecky V, Dugourd P (2012) Phys Chem Chem Phys 14 (32):11433–11440
110. Hernandez O, Isenberg S, Steinmetz V, Glish G, Maitre P (2014) Submitted for publication
111. Schindler B, Joshi J, Allouche A-R, Simon D, Chambert S, Brites V, Gaigeot M-P, Compagnon I (2014) Phys Chem Chem Phys 16:22131–22138
112. Stearns JA, Seaiby C, Boyarkin OV, Rizzo TR (2009) Phys Chem Chem Phys 11(1):125–132
113. Shafizadeh N, Ha-Thi MH, Soep B, Gaveau MA, Piuzzi F, Pothier C (2011) J Chem Phys 135 (11):114303

Top Curr Chem (2015) 364: 335–402
DOI: 10.1007/128_2014_601

Published online: 27 February 2015

Microwave Spectroscopy of Biomolecular Building Blocks

José L. Alonso and Juan C. López

Abstract Microwave spectroscopy, considered as the most definitive gas phase structural probe, is able to distinguish between different conformational structures of a molecule, because they have unique spectroscopic constants and give rise to distinct individual rotational spectra.

Previously, application of this technique was limited to molecular specimens possessing appreciable vapor pressures, thus discarding the possibility of studying many other molecules of biological importance, in particular those with high melting points, which had a tendency to undergo thermal reactions, and ultimately degradation, upon heating.

Nowadays, the combination of laser ablation with Fourier transform microwave spectroscopy techniques, in supersonic jets, has enabled the gas-phase study of such systems. In this chapter, these techniques, including broadband spectroscopy, as well as results of their application into the study of the conformational panorama and structure of biomolecular building blocks, such as amino acids, nucleic bases, and monosaccharides, are briefly discussed, and with them, the tools for conformational assignation – rotational constants, nuclear quadrupole coupling interaction, and dipole moment.

Keywords Amino acids · Building blocks · Carbohydrates · Conformational analysis · Laser ablation · Microwave spectroscopy · Nitrogen bases

Contents

J.L. Alonso (✉) and J.C. López
Grupo de Espectroscopia Molecular (GEM), Edificio Quifima, Laboratorios de Espectroscopia y Bioespectroscopia, Unidad Asociada CSIC, Universidad de Valladolid, 47005 Valladolid, Spain
e-mail: jlalonso@qf.uva.es

1 Introduction

The study of the structure, function, reactivity, and properties of biomolecules covers a vast range of molecular sizes, from large macromolecules such as proteins, nucleic acids, lipids, or polysaccharides, down to small molecules such as amino acids, nucleic acid bases, neurotransmitters, or monosaccharides, to name but a few. Interest in this field of research has grown exponentially in the last few decades, across different areas, such as biology, medicine, chemistry, and physics, bringing with it a multitude of continuously developing experimental and theoretical methods. A major goal of this ongoing effort is to obtain information about the molecular structure of biomolecules and their related properties at the molecular level. Consequently, over the last few years, there has been a growing interest in the study of biomolecules in isolation conditions, that is, conditions matching those found in the gas phase [1], [2 and references therein], [3–6]. Gas-phase conditions are far from the condensed phase, in vivo, environment in which biomolecules work and biological reactions occur. Nevertheless, gas phase does provide ideal conditions to observe the intrinsic molecular properties, and thus it is a first step into the understanding of the molecular behavior in more complex media. For instance, in physiological media, many biological molecules are electrically charged species, because of processes of protonation (or deprotonation). A well-known example is amino acids, which exist as doubly-charged zwitterions ($^{+}NH_3$–CH(R)–COO^{-}) [7, 8] in either solution or crystal. Moreover, in the gas phase, perturbing agents, such as solvents or even intermolecular interaction, are absent and thus observation of neutral species becomes possible. Furthermore, most biochemical molecules are characterized by a great torsional flexibility, resulting in manifold conformational varieties relatively close in energy. Since studies in condensed phases are considerably affected by multiple intermolecular interactions, conformational preferences can be biased because of the surrounding environment and only gas-phase studies can reveal the inherent structural minima of bare molecules.

Despite these advantages, the number of biomolecules studied to date in the gas phase is still very small compared with the overwhelming number of condensed-phase studies. Supersonic jets [9, 10], in combination with different spectroscopic techniques, have been instrumental in providing conformational information on small biomolecules. Modern supersonic-jet laser spectroscopy [2–4, 9, 10] combined with double resonance techniques (such as hole burning UV–UV and ion-dip IR–UV) and mass detection (through resonance-enhanced multiphoton ionization – REMPI), giving electronic and vibrational information with mass and conformer selectivity, have pioneered this field. This kind of experiments must be interpreted with the aid of high level ab initio quantum chemical calculations, thus allowing assignment of discrete conformational structures within a reasonable degree of confidence. However, unfortunately, laser spectroscopy is hampered by the need for an absorbing chromophore possessing a sharp vibronic structure, such as an aromatic ring [2, 11].

Unlike the aforementioned techniques, microwave spectroscopy can be applied to any gaseous molecular system of moderate size, the only requirement being to have a nonzero electric dipole moment [12]. Microwave spectroscopy has long been considered to be a powerful method for precise determination of gas-phase structures, being the source of the majority of gas-phase structural data known nowadays [13, 14] (see, for example, the Landolt-Börnstein, New Series II series). The combination of Fourier transform microwave spectroscopy with supersonic jets has given rise to two different families of techniques: those using a narrow band molecular beam (MB-FTMW) [15–17] or, more recently developed, wide band chirped pulse FTMW spectroscopy (CP-FTMW [18] and IMPACT-FTMW [19]). By virtue of the extremely high resolution (sub-Doppler) and sensitivity of the FTMW techniques, all populated species in the jet, be they tautomers, conformers, or isotopomers, can be analyzed independently. Moreover, small hyperfine effects arising from electric or magnetic interactions, such as nuclear electric quadrupole coupling and nuclear spin-spin coupling, can be fully resolved, providing insight to the molecule electronic properties. Finally, the observation of tunneling doublings in the case of large amplitude motions, such as those arising from internal rotation or inversion, can give information on the intramolecular dynamics associated with these motions. These techniques have been widely used in the study of gaseous or easily vaporized compounds, weakly bounded complexes [20–22], and even short-lived species generated in situ by electric discharges [23–27]. In our laboratory it has been used to study axial/equatorial equilibria [28–31] in hydrogen-bonded complexes and weak C–H· · ·O bonds [32–35], and has been combined with electric discharges to generate and characterize new unstable compounds [36–39].

The vast majority of biomolecules have extremely low vapor pressures at room temperature and are thermally fragile. For that reason, complete series of molecules of biological interest, such as building blocks, have escaped microwave spectroscopy studies. Several laboratories have incorporated heatable nozzles [40–42] into their microwave spectrometers to overcome the problem, but, in many cases, thermal instability constituted an important drawback. The pioneering works of Brown [43] or Suenram [44] and co-workers on molecular building blocks were carried out using such heating methods. Laser ablation is an efficient method for vaporizing solid samples, which otherwise would decompose upon heating [45]. Laser-based vaporization techniques coupled with spectroscopic techniques [46–49] have been used for gas-phase studies of solid compounds. However, until recently, there have been few attempts to incorporate laser ablation into MB-FTMW spectrometers [50–55]. Suenram et al. [50] and Walker and Gerry [51] independently developed two laser ablation devices in combination with molecular-beam Fourier transform microwave spectroscopy (LA-MB-FTMW) to investigate the rotational spectra of metal oxides and halides. The first was also applied in the structural study of glycine in the gas phase, but it was reported that the classical heating method was more reliable. Thus the experiments devoted to organic solids [54, 55] were initially discontinued because of the poor experimental results. Over the last few years, several MB-FTMW spectrometers which implement laser ablation sources to vaporize molecules in the throat of the nozzle have been configured in our laboratories. This approach, laser-ablation molecular-beam Fourier transform microwave (LA-MB-FTMW) spectroscopy [56], was initially tested with several organic solids [57, 58]. With the initial design [59] and progressive improvements [60, 61 and references therein], relevant biomolecules have been in studied in order to identify and characterize their most stable conformations in the gas phase. This approach has been essentially followed in other laboratories [62]. More recently, laser ablation sources have been incorporated into broad-band CP-FTMW spectrometers [63, 64], extending the application of this method to a large number of thermally fragile systems with high melting points.

In the next sections, the experimental techniques, the procedure employed to characterize the different species observed, and the results obtained from the study of molecular building blocks such as amino acids, nucleic acid bases or sugars are discussed.

2 Experimental Techniques

2.1 Laser-Ablation Molecular-Beam Fourier Transform Microwave (LA-MB-FTMW) Spectroscopy

An extensive description of the experimental setup employed in our laboratory has been given elsewhere [56–61], but can be arranged into three main blocks: a laser ablation system, a Fabry–Pérot resonator, and an FTMW spectrometer, illustrated in Fig. 1. To accomplish vaporization efficiently, while retaining the advantages of the coaxial orientation of the resonator and supersonic jet axes, it was found necessary to modify the backside of the Fabry–Pérot resonator fixed mirror so that it can hold a specially designed laser ablation pulsed-jet nozzle. The samples are prepared as solid rods by pressing a fine powder of the pure compound with minimum quantities of a commercial binder. The rod is then placed vertically in the laser ablation nozzle, where a focused laser beam falls upon the sample laterally and the vaporized molecules are seeded in the supersonic jet expansion.

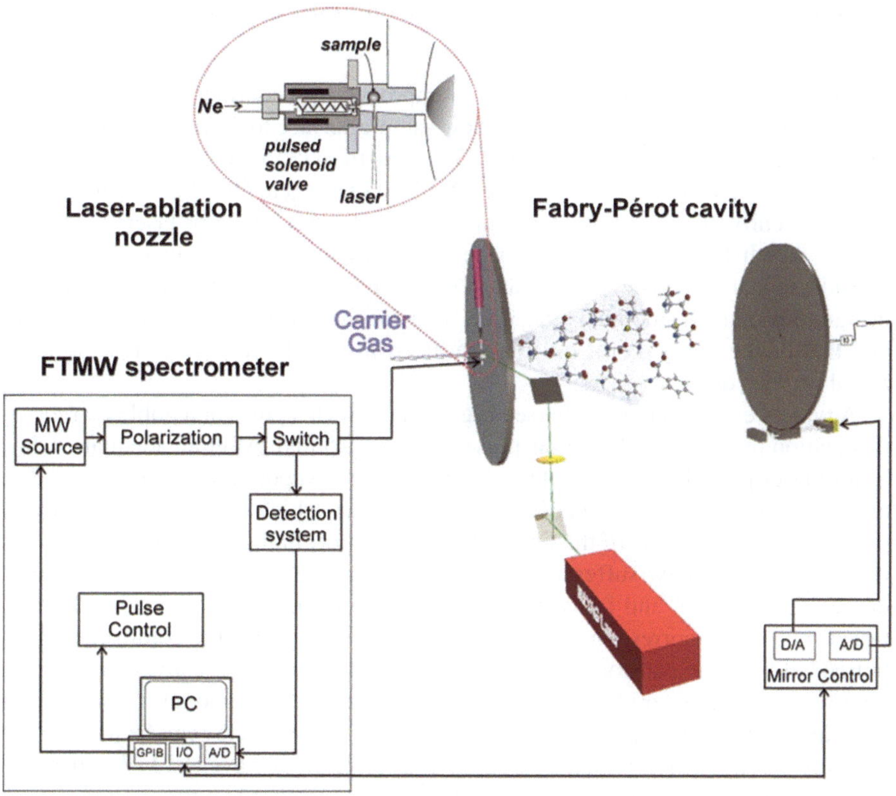

Fig. 1 Scheme of the LA-MB-FTMW spectrometer. (From [61])

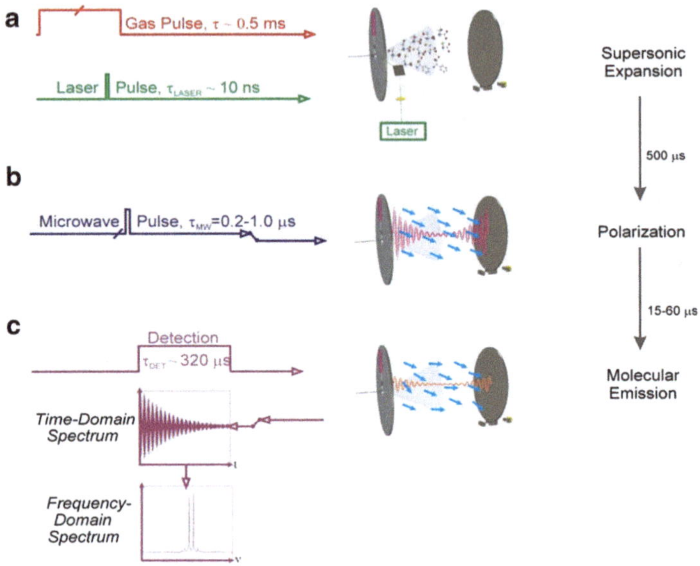

Fig. 2 Pulse sequence for a single experimental LA-MB-FTMW cycle. (From [61])

Reproducibility of the ablation process is improved by a continuous translation and rotation of the sample rod, exposing a new sample surface to each laser pulse. In our initial experimental setups, the second (532 nm), third (355 nm), or fourth (266 nm) harmonics of a nanosecond Q-switched Nd:YAG laser (ca. 50 mJ/pulse) have been used. Presently, the use of the harmonics of an Nd:YAG 20–35 ps lasers (7–15 mJ/pulse), which have been proved to be more efficient [see, for example, 65], is considered standard. The laser-ablation nozzle has been modified from its original design to provide a smoother transition between the aperture at the solenoid valve and that of the Fabry–Pérot mirror (see insert in Fig. 1), nowadays resembling a Laval nozzle [66].

A typical experimental cycle (see Fig. 2) starts with a pulse of a noble carrier gas (stagnation pressures 2–8 bar, gas pulse typically 0.5 ms). After an adequate delay, a short laser pulse hits the sample rod producing, the vaporization of the solid. The ablated molecules are then seeded in the carrier gas, which expands supersonically between the two mirrors of the Fabry–Pérot resonator. In the supersonic expansion, the seeded molecules suffer a strong cooling of the rotational and vibrational degrees of freedom, and individual conformers are frozen into the ground vibrational state of the corresponding potential energy well. Thus, the conformer distribution before the expansion may be preserved provided the interconversion barriers between conformers are high enough. Molecular collisions gradually disappear as the expansion proceeds, in such a way that the different species can be probed in a virtually isolated environment by Fourier transform microwave (FTMW) spectroscopy. A short microwave pulse (typically 0.3 μs) is subsequently applied, which produces a macroscopic polarization of the species in the jet. Once the excitation pulse ceases, the molecular emission signal (FID, free induction decay) is captured in the time domain. Its Fourier transformation to the frequency domain yields the molecular rotational transitions which appear as Doppler doublets because the supersonic jet travels parallel to the resonator axis. The molecular equilibrium frequencies are calculated as the arithmetic mean of the Doppler doublets, and are obtained with accuracy better than 3 kHz. A new experimental cycle can start once the vacuum cavity has been evacuated, and a repetition rate of 2 Hz is normally employed. For very weak signals, hundreds of cycles must be added coherently. To probe the jet at different microwave frequencies or conduct a frequency scan, the Fabry–Pérot resonator is tuned mechanically under computer control. The timing of the whole experiment, in particular the delay of the laser with respect to the molecular pulse, is crucial for an optimum signal. Laser pulse energy is also critical.

2.2 *Chirped-Pulse Fourier Transform Microwave Spectroscopy (CP-FTMW) Coupled to Laser Ablation (LA)*

A new approach of high-resolution rotational spectroscopy, providing a sensitive method for broadband detection (CP-FTMW), has recently been developed at the University of Virginia [18]. Such an instrument is based on the same principles as the original MB-FTMW instrument [15], but the burst of microwaves has been replaced by a microwave pulse which contains a fast linear sweep (chirp) over the entire frequency region being explored and is used to polarize the sample. This new approach increases the speed of the spectral acquisition, making the search of the different coexisting species in the jet much more efficient. In the last few years, a new alternative design of broadband Fourier transform microwave spectroscopy, called IMPACT (in-phase/quadrature-phase modulation passage acquired coherence technique) has been developed in our laboratory [19]. The aforementioned broadband techniques have been combined with a laser ablation source for the study of solid biomolecules [63, 64].

A schematic block diagram of the newly designed chirped-pulse Fourier transform microwave spectrometer CP-FTMW, combined with a *ps*-pulsed laser ablation system, is given in Fig. 3. The spectrometer, which uses the basic operation of the CP-FTMW instrument [18], is described elsewhere [63], with only the relevant details to this experiment being described here. It operates in the 6.0–18 GHz region. The solid sample, prepared as usual as a rod shape, was placed in a laser ablation nozzle, similar to that previously described [61] (**1** in Fig. 3) and vaporized using the second (532 nm) or third (355 nm) harmonics of a ps Nd:YAG laser (i.e.: Ekspla, 20 ps, 15 mJ/pulse) (**2** in Fig. 3). A motor controller (**3** in Fig. 3) allows a DC motor (Oriel Motor Mike 18074) (**4** in Fig. 3) to rotate and translate the rod up and down along the injection system to achieve the maximum exploitation of the

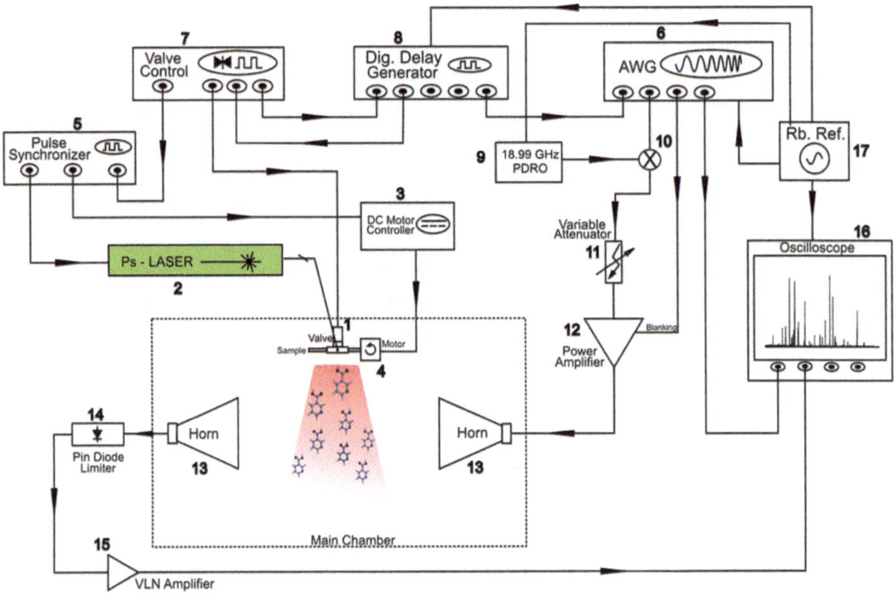

Fig. 3 Schematics of a laser ablation chirped-pulse Fourier transform microwave spectrometer. (From [63])

sample. So that good vacuum conditions are retained in the expansion chamber, and to reduce sample consumption, the repetition rate of the experiment is set to be 2 Hz, achieved via a shutter operating at said frequency, placed at the output of the laser and controlled via a pulse synchronizer (**5** in Fig. 3), used to adjust the repetition rate to 2 Hz from the standard rate of the laser (10 Hz). In the ablation nozzle, the sample molecules, once vaporized, are seeded in the flow of the carrier gas (Ne at backing pressure of 15 bar), being expanded adiabatically into the vacuum chamber, where they are probed by a microwave chirped pulse.

Briefly, the broadband microwave spectrometer works as follows. A 24 GS/s arbitrary waveform generator (**6** in Fig. 3) creates a fast chirp microwave pulse, covering the entire range to be explored, to polarize macroscopically the pulsed molecular beam sample created from the ablation nozzle by the valve driver (**7** in Fig. 3). A digital delay generator (**8** in Fig. 3) is used to trigger both the arbitrary waveform generator and the valve driver. The microwave pulse is upconverted by mixing it in a broadband mixer (**10** in Fig. 3) with an 18.99 GHz signal provided by a phase-locked dielectric resonator oscillator (PDRO) (**9** in Fig. 3). The upconverted signal is subsequently amplified by a 300 W traveling wave tube amplifier (**12** in Fig. 3). The power level necessary for the polarization of the molecular systems can be adjusted using a variable attenuator (**11** in Fig. 3). The amplified chirp pulse is broadcast across the vacuum chamber, where the jet expansion occurs, using one of the two standard horn antennas (**13** in Fig. 3) placed inside the chamber. These two horn antennas are separated by approximately

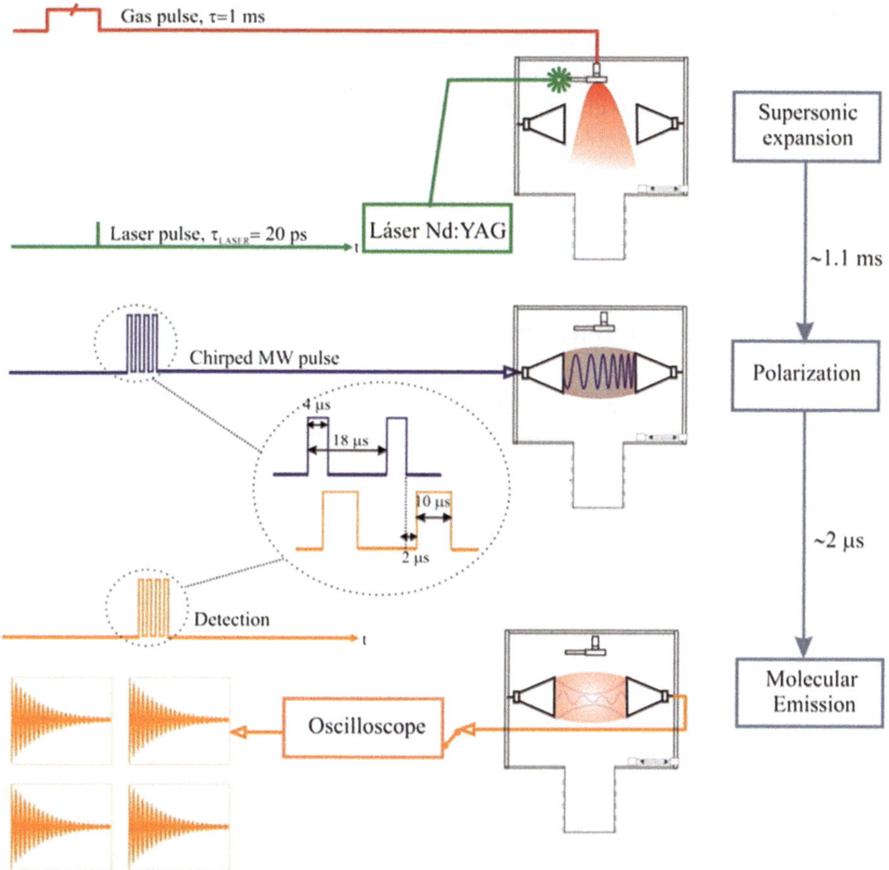

Fig. 4 Pulse sequence for a single experimental cycle including generation of a supersonic expansion and laser ablation, polarization and detection. (From [63])

60 cm. The second antenna is used to detect the free induction decay signal (FID) emitted by the sample as response to the microwave excitation. The FID is further amplified by a sensitive VLN amplifier (**14** in Fig. 3) which is protected from the high-power of the TWT amplifier by a pin diode limiter (**15** in Fig. 3). The amplified rotational free induction decay (FID) is recorded in the time domain by a digital oscilloscope (50 GS/s, 20 GHz hardware bandwidth) (**16** in Fig. 3) and Fourier transformed to the frequency domain. The phase reproducibility of the experiment is achieved by locking all frequency sources and the digital oscilloscope to a 10-MHz rubidium frequency standard oscillator (**17** in Fig. 3).

The operation sequence (see Fig. 4) starts with a molecular pulse of 1,000 μs duration which drives the carrier gas flow through the pulsed valve source. After an adequate delay (~850 μs), a laser pulse hits the solid and vaporizes the sample. To reduce sample consumption, four separate broadband rotational spectra are acquired in each injection cycle. The four individual broadband-chirped excitation pulses, of 4 μs width, are spaced by 18 μs. Then, 2 μs after each excitation pulse ceases, the rotational free induction decay is acquired for 10 μs. The internal pulse generator of the valve driver is used to create the digital pulses involved in the laser generation. Since the sample injection has a perpendicular arrangement with the microwave field, the transit time of the polarized molecular jet is quite short and linewidths of about 100 kHz full-width-half-maximum (FWHM) are achieved.

3 Tools in Conformational Analysis

The high flexibility of biomolecules produces the appearance of a significant number of low-energy conformers. Whereas covalent forces determine the molecular skeleton, conformational isomerism is also controlled by weaker nonbonded interactions within the molecule, especially hydrogen bonding. To treat this conformational problem and identify the different structures in the supersonic jet, a procedure represented in Fig. 5 has been followed in all the microwave studies. It is illustrated through its application to L-threonine [61] in this section and it was also described in our study of glutamic acid [67].

3.1 Model Calculations

To have an overall picture of the conformational landscape of L-threonine, theoretical predictions are used to find the most stable conformers on the potential energy surface. Only low-lying energy conformers are sufficiently populated in the supersonic jet to be observed in the rotational spectrum. Starting geometries for ab initio calculations are initially selected by considering all possible rotations around single bonds and identifying plausible intramolecular hydrogen bonds. To predict the most

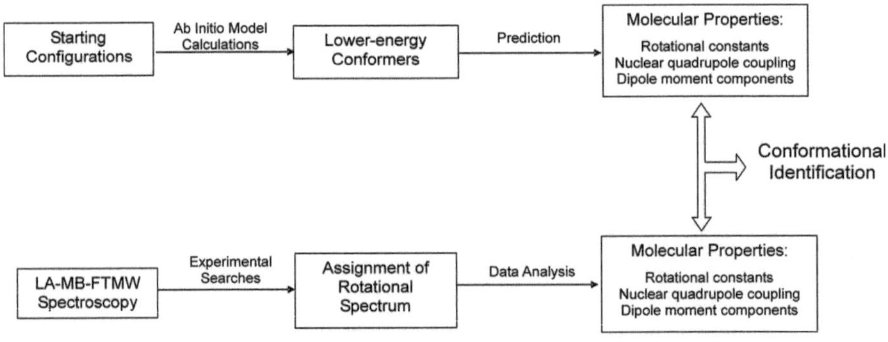

Fig. 5 Steps followed in the identification of the conformers of biomolecules. (From [61])

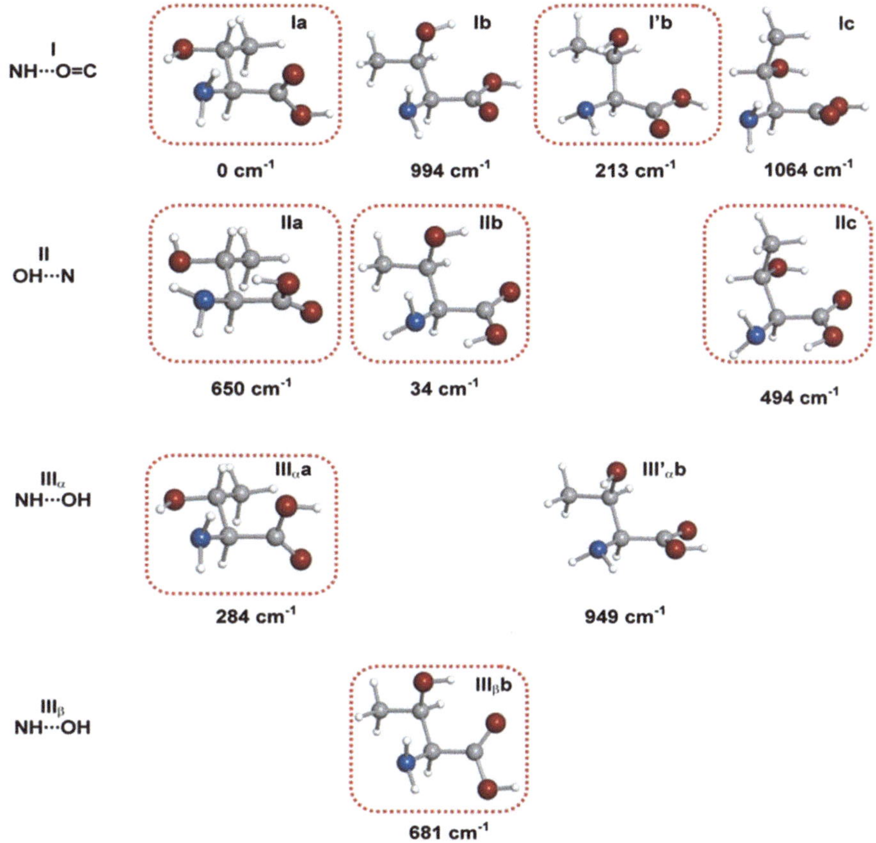

Fig. 6 Predicted low-energy conformers of threonine and energies relative (MP2/6-311++G(d,p)) to the global minimum in cm^{-1}. The detected conformers are encircled. (From [61])

stable forms, a series of structural optimizations is conducted for each of the starting configurations using the Gaussian suite of programs [68]. First, a cheap but computationally efficient calculation at the B3LYP/6-31G(d,p) level [69–73] is performed. In a second step, full geometry optimization calculations using second-order Møller–Plesset perturbation theory (MP) [74] and Pople's 6-311++G(d,p) [75] basis set are conducted. This level of theory has been found to behave satisfactorily in all the series of amino acids studied (see below). Finally, the ten conformers of Fig. 6 were predicted to be within 1,000 cm^{-1}. Conformers are classified as I, II, or III depending on the hydrogen bond established between the amino and carboxylic groups and as *a*, *b*, or *c* depending on the configuration adopted by the –CH(CH_3)OH side chain.

A distinguishing quality of rotational spectroscopy lies in its capability to generate very accurate spectroscopic parameters directly comparable with in vacuo ab initio predictions to provide unequivocal evidence of the conformers observed. Three sets of spectroscopic constants are relevant for the interpretation of the rotational spectra: rotational constants, nuclear quadrupole coupling constants, and electric dipole moment components. From the ab initio optimized structures of the lower-energy conformers of threonine (Fig. 6), those spectroscopic constants are calculated and listed in Table 1.

The rotational constants *A*, *B*, and *C*, which provide information on the mass distribution of the molecules [12, 76, 77], are compared with those obtained experimentally to identify the conformers present in the supersonic expansion. Such comparison is normally conclusive in the identification, but sometimes the difference in the values between conformers is not large enough to allow discrimination between them. This happens with some forms belonging to the same family in Table 1 (see, for example, Ia and IIa). In these cases, a different and independent way of identifying conformers based on the ubiquitous presence of ^{14}N in amino acids can be utilized. ^{14}N nuclei possess a nonzero quadrupole moment ($I = 1$) because of a non-spherical distribution of the nuclear charge which interacts with the electric field gradient created by the rest of the molecule at the site of those nuclei. This results in a nuclear hyperfine structure in the rotational spectrum [12, 76, 77]. FTMW spectroscopy provides the high resolution needed to resolve fully the various quadrupole hyperfine components (see Fig. 7). The associated experimentally-determined molecular properties are the quadrupole coupling constants χ_{gg} ($g = a,b,c$), referred to the principal inertial axes, which are directly related to the electronic environment of the quadrupolar nucleus referred to the principal inertial axes and strongly depend on the orientation of the amino group. For example, conformers Ia and IIa with similar predicted rotational constants have different orientations of the amino group (Fig. 7), which is reflected in the nuclear quadrupole coupling constants χ_{aa} and χ_{cc}, as can be seen by the predicted values in Table 1.

A third molecular property to be considered is the electric dipole moment. The selection rules and intensity of the rotational transitions of asymmetric tops depend on the dipole moment components along the principal inertial axes, that is, on μ_a, μ_b, and μ_c, which give rise to a-, b-, and c-type spectra, respectively. All conformers possess the same connectivity between the atoms but they differ in the orientation of the functional groups, and this necessarily produces diverse charge distributions reflected in different values of the dipole moment components, as can be seen in Table 1. The microwave power necessary for optimal polarization depends on the dipole moment component involved in a rotational transition. Hence, the difference in the values of the dipole moment components of conformers can be exploited to discriminate between specific conformers just by varying the polarization power. By itself, it cannot be used as a conclusive tool, but it can always corroborate the conformer identification achieved with the previously described molecular properties.

The electric dipole moment components are necessarily used in estimating relative conformational abundances, as is explained below. The predicted values of the electric dipole moment components of the threonine conformers are collected in Table 1.

Table 1 Ab initio spectroscopic constants for the low-energy conformers of threonine of Fig. 6. (From [61])

MP2/6-311++G(d,p)	Ia	IIa	$III_{\alpha}a$	Ib	I′b	IIb	$III'_{\alpha}b$	$III_{\beta}b$	Ic	IIc
A^{a} (MHz)	2,870	2,906	2,885	3,163	3,164	3,233	3,194	3,383	2,488	2,662
B (MHz)	1,608	1,668	1,578	1,542	1,521	1,543	1,498	1,488	1,838	1,793
C (MHz)	1,225	1,201	1,249	1,288	1,324	1,271	1,332	1,239	1,431	1,395
χ_{aa} (MHz)	−4.47	−0.44	−4.53	−2.82	−0.35	−3.72	0.57	−2.32	−2.32	−3.94
χ_{bb} (MHz)	2.76	2.67	2.85	1.89	2.94	1.88	2.96	−0.03	0.88	2.47
χ_{cc} (MHz)	1.71	−2.23	1.69	0.93	−2.59	1.84	−3.52	2.44	1.43	1.46
$\lvert\mu_a\rvert$ (D)	2.3	4.3	2.1	0.3	0.1	3.4	0.2	2.3	1.0	0.1
$\lvert\mu_b\rvert$ (D)	0.1	2.2	1.5	0.5	0.6	3.3	2.8	1.1	1.0	5.6
$\lvert\mu_c\rvert$ (D)	0.8	0.3	1.3	0.9	2.9	1.2	1.9	1.0	0.3	0.3

Source: From [61]

[a]A, B, and C are the rotational constants; χ_{aa}, χ_{bb}, and χ_{cc} are the diagonal elements of the ^{14}N nuclear quadrupole coupling tensor; μ_a, μ_b, and μ_c are the electric dipole moment components

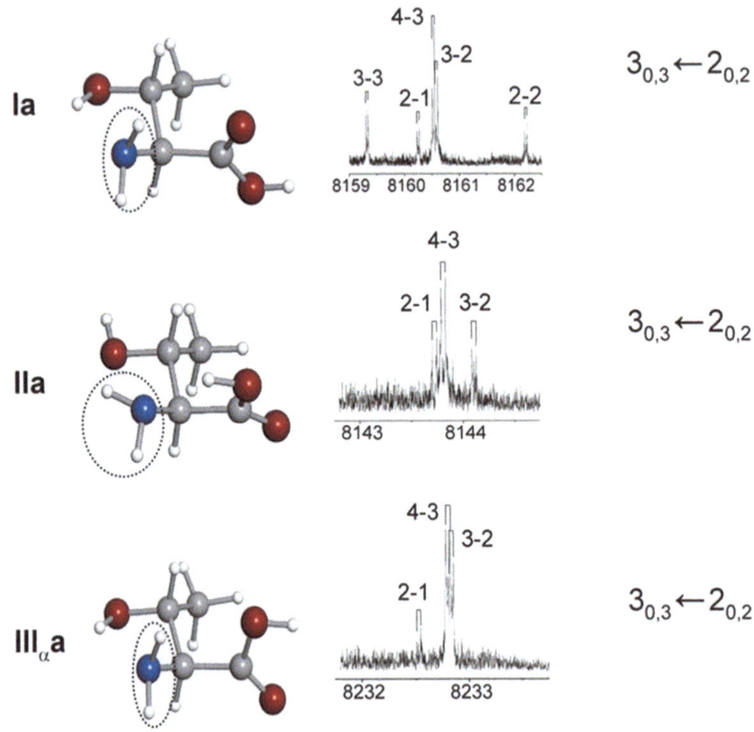

Fig. 7 Nuclear quadrupole hyperfine structure of the $3_{0,3} \leftarrow 2_{0,2}$ rotational transitions observed for rotamers M, N, and O, identified as IIa, Ia, and III$_\alpha$a, respectively. (From [61])

3.2 Analysis of Spectra and Conformer Identification

According to the values of the predicted electric dipole moment components and rotational constants listed in Table 1, the rotational spectra of most of the threonine conformers are dominated by μ_a-type, R-branch transitions forming groups of lines with characteristic patterns which appear at frequency intervals equivalent to B + C. Wide frequency scans with low power polarization conditions were conducted to search for such rotational transitions of conformers with relatively large μ_a. Several sets of R-branch lines corresponding to five different rotamers are labeled L, M, N, O, and P. Spectral searches, conducted to detect other sets of μ_a-type R-branch transitions, using high microwave power for polarization, were unsuccessful. Therefore, wide frequency scans were carried out to identify μ_b- and μ_c-type transitions. Finally, two more rotamers labeled Q and R were identified.

Apart from the instrumental Doppler doubling, all transitions are observed to be split into several close hyperfine components (see Fig. 7) arising from the ^{14}N nuclear quadrupole interaction described above, indicating that they arise from a molecule with a nitrogen nucleus. Successive predictions and new experimental measurements discarded or confirmed the initial assignment until a group of consistent rotational transitions, including all possible μ_a-, μ_b-, or μ_c-transitions, was collected for each conformer.

The observed transitions frequencies were analyzed with a Watson's semirigid rotor Hamiltonian ($H_R^{(A)}$) [12] supplemented with a nuclear quadrupole coupling term (H_Q): $H = H_R^{(A)} + H_Q$. The Hamiltonian is constructed in the coupled basis set $I + J = F$ and diagonalized. The A-reduced semirigid Watson Hamiltonian in the I^r representation is given by

$$H_R^{(A)} = AP_a^2 + BP_b^2 + CP_c^2 - \Delta_J P^4 - \Delta_{JK} P^2 P_a^2 - \Delta_K P_a^4 \\ -2\delta_J P^2 (P_b^2 - P_c^2) - \delta_K \left[P_a^2 (P_b^2 - P_c^2) + (P_b^2 - P_c^2) P_a^2\right], \tag{1}$$

where the coefficients A, B, and C represent the rotational constants and Δ_J, Δ_{JK}, Δ_K, δ_J, and δ_K are the quartic centrifugal distortion constants. Only Δ_J needed to be floated to obtain an rms deviation of the fit consistent with the estimated frequency accuracy. The term H_Q accounts for the interaction energy of the ^{14}N electric quadrupole moment (eQ) with the molecular electric field gradient ($q_{\alpha\beta} = \partial^2 V / \partial\alpha\partial\beta$; α, $\beta = a$, b, c) at the nitrogen nucleus. The determinable spectroscopic parameters are the elements of the nuclear quadrupole coupling tensor χ, linearly related to the electric field gradient by $\chi = -eQq$. Usually, only the diagonal elements of the tensor (χ_{aa}, χ_{bb}, χ_{cc}) are determined.

The experimental rotational constants and the χ_{aa}, χ_{bb}, and χ_{cc} nuclear quadrupole coupling constants for the rotamers L, M, N, O, P, Q, and R are collected in Table 2. A first look at the rotational constants of Table 2, and their comparison with the predicted constants of Table 1, allows us to classify easily rotamers as belonging to different families. Hence, rotamers M, N, and O belong to the "a" family, rotamers L, P, and Q belong to the "b" family, and, finally, rotamer R belongs to the "c" family. Conformers belonging to the "a" family have similar mass distributions: their rotational constants are very similar and they do not allow further discrimination. In these cases, as mentioned above, conclusive evidence comes from the values of ^{14}N quadrupole coupling constants because they are very sensitive to the orientation of the $-NH_2$ group with respect to the principal inertial axis system (see Table 1). The values of these constants, reflected in the hyperfine structure clearly identify rotamer M as IIa (see Fig. 6).

Rotamers N and O have very similar quadrupole coupling constants; comparison of their values with those predicted ab initio indicate that they are necessarily conformers Ia and III$_\alpha$a (see Tables 1 and 2). Because of the similar orientation of the amino group in these forms (see Fig. 6), they cannot be discriminated on the basis of the quadrupole constants. We can distinguish them from their selection rules and intensities of the observed transitions. The rotational spectrum of rotamer N shows strong μ_a-type transitions and fairly weak μ_c-type transitions, while form O presents strong μ_a-type transitions and medium-strength μ_b- and μ_c-type transitions. No μ_b-type transitions have been detected for conformer N. Considering the predicted dipole moment components of Table 1, these data are consistent with

Table 2 Experimental spectroscopic constants for the seven observed rotamers of threonine

Experimental	L	M	N	O	P	Q	R
A[a] (MHz)	3,379.841(14)[b]	2,912.6227(20)	2,872.77049(48)	2,889.93352(45)	3,232.4827(12)	3,148.59247(32)	2,670.72096(53)
B (MHz)	1,482.04984(21)	1,660.21807(34)	1,608.95699(26)	1,572.32152(50)	1,533.71801(32)	1,506.27679(37)	1,784.66894(60)
C (MHz)	1,237.59121(22)	1,189.31443(34)	1,211.39762(38)	1,241.83423(47)	1,267.88615(34)	1,316.33575(44)	1,383.75384(51)
κ	−0.77	−0.45	−0.52	−0.60	−0.73	−0.79	−0.38
χ_{aa} (MHz)	−2.201(14)	−0.544(11)	−4.1859(25)	−4.1529(32)	−3.4971(21)	−0.7403(21)	−3.7652(73)
χ_{bb} (MHz)	−0.157(50)	2.582(16)	2.6611(42)	2.5682(46)	1.7519(27)	2.8781(28)	2.4258(75)
χ_{cc} (MHz)	2.358(64)	−2.038(50)	1.5248(17)	1.5846(46)	1.7452(60)	−2.1378(70)	1.3394(20)

Source: From [61]

[a]A, B, and C are the rotational constants; χ_{aa}, χ_{bb}, and χ_{cc} are the diagonal elements of the ^{14}N nuclear quadrupole coupling tensor

[b]Standard error in parentheses in units of the last digit

the identification of rotamer N as conformer Ia and rotamer O as conformer $III_{\alpha}a$. The microwave power used to polarize optimally the rotational transitions is also in agreement with the predicted values for the dipole moment components of each conformer.

The rotamers L, P, and Q which belong to the "b" family, can be attributed to I′b, Ib, IIb, or $III_{\beta}b$ considering the values of the rotational constants. However, detailed comparison of the values of the ^{14}N quadrupole coupling constants conduct to the identification of rotamer L as conformer $III_{\beta}b$, P as IIb, and Q as I′b. In this case, the quadrupole coupling constants are essential for the discrimination of conformers.

Again, from the rotational constants alone it is only possible to match rotamer R to a conformer belonging to the "c" family, Ic or IIc. Consideration of the nuclear quadrupole coupling constants uniquely identifies form R as IIc.

The procedure illustrated here to analyze the spectrum of threonine and to identify the observed conformers has been followed throughout all the examples shown in the next sections.

4 Amino Acids

4.1 Introduction

An important subset of biologically relevant molecules is that of natural amino acids, the so-called "building blocks" of peptides and proteins. Amino acids have long been studied in solids [78–82] and in solution [83–85], where they are stabilized as doubly charged species or zwitterions ($R{-}CH(NH_3^+){-}COO{-}$). Gas-phase studies have the advantage of providing information on the neutral forms of amino acids ($R{-}CH(NH_2){-}COOH$, the canonical forms present in peptide chains) and on their inherent molecular properties free from the intermolecular interactions occurring in the condensed media. Furthermore, gas-phase data can be easily contrasted with theoretical models and used to refine the latter.

Figure 8 shows the 20 proteinogenic or coded α-amino acids ($NH_2{-}CH(R){-}COOH$) which can be classified according to the nature of their side chain (R). The α-amino acid backbone determines the primary sequence of a protein, but the nature of the side chains determines the protein properties [86–88]. For this reason, examination of the structural properties of amino acids with different side chains is important to the understanding of its functionality. Knowledge of the structures and conformational behavior of amino acids has been considerably expanded thanks to the extensive use of LA-MB-FTMW spectroscopy. Only the simplest of these amino acids, glycine [43, 44, 89–92] and alanine [93], were studied before by rotational spectroscopy, using heating methods. The amino acids encircled in a solid line in Fig. 9 have been studied with this technique in our laboratory and some of them are illustrated in this chapter.

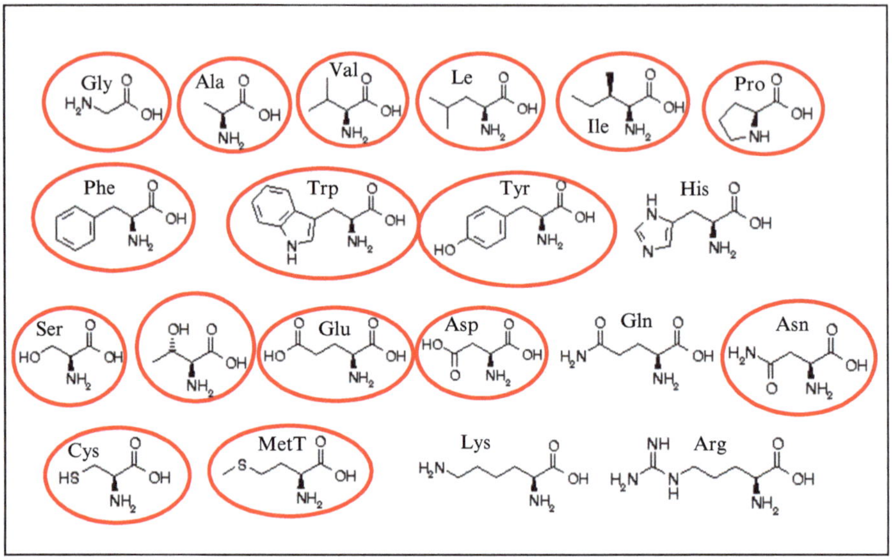

Fig. 8 Coded α-amino acids. The amino acids encircled have been studied by rotational spectroscopy

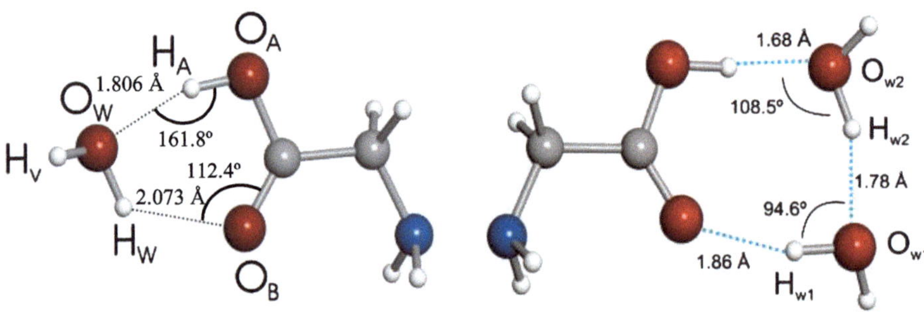

Fig. 9 Structure of the glycine-$(H_2O)_n$ complexes (From [100, 101])

The large flexibility of amino acids, which makes folding and unfolding of proteins possible, produces the appearance of a significant number of low-energy conformers. Whereas covalent forces determine the molecular skeleton, conformational isomerism is also controlled by weaker nonbonded interactions within the molecule, especially hydrogen bonding. Hydrogen bonding between the amino and carboxylic moieties is expected to control the configuration in non-polar side chain amino acids. The presence of a polar side chain brings about new sets of intramolecular interactions between the lateral side chain and the amino or acid groups, which greatly increase the number of low-energy forms. In β- or γ-amino acids, the balance of forces may change with the length of the chain between the amino and acid groups and, as shown below, new interacting forces may contribute to conformer stabilization. In the next sections, the conformational behavior of the different α-amino acids as observed from LA-MB-FTMW spectroscopy is described.

4.2 *Proteinogenic Amino Acids with Non-polar Side Chains*

The conformational behavior of α-amino acids with non-polar side chains is essentially determined by the three possible intramolecular hydrogen bonds established between the amino and carboxylic functional groups. Configuration I is stabilized by a hydrogen bond between the hydrogen atom of the amino group and the carbonyl oxygen of the carboxylic group (N–H···O=C), and a *cis* arrangement of the –COOH group. In configuration II, the hydrogen bond links the hydrogen atom of the hydroxyl group with the electronic lone pair at the nitrogen atom (N···H–O) while the –COOH group exhibits a *trans* arrangement. Configuration III bears a hydrogen bond between the amino group and the oxygen atom of the hydroxyl group (N–H···O–H) and presents a *cis*-COOH. On this basis, the conformational behavior of glycine, alanine, valine, leucine, and isoleucine are presented in the next sections.

4.2.1 Glycine and Its Hydrates

Glycine (R=H, m.p. = 240°C) is the smallest α-amino acid, and has been the object of different studies, so its structural properties in the gas phase are well understood. Following the simultaneous discovery by Brown et al. [43] and Suenram et al. [44] of glycine II, a subsequent search led to the detection of the weaker glycine I [89, 90]. Further work using jet expansions provided additional information on isotopic species, electric dipole moment, and ^{14}N nuclear quadrupole coupling hyperfine structure [92–93].[1] The rotational data are consistent with ground-state structures of C_s symmetry for glycines I and II. Recently, the rotational data have been extended into the millimeter-wave region [94] to improve radioastronomical searches.

These studies provided fundamental pieces of information about the dominant role of intramolecular hydrogen bonding between the polar moieties of the amino acid skeleton in molecular conformation, concluding that glycine rotamers are stabilized by intramolecular hydrogen bonding from either amine to carbonyl N–H···O=C (type I conformer) or hydroxyl to the amine nitrogen lone pair N···H–O (type II conformer). The third low-energy conformer III, with an intramolecular hydrogen bond N–H···O–H, could not be detected. The experimental evidence, including a recent reexamination of the glycine spectrum in our laboratory giving a conformational ratio N_I/N_{II}~6/1 (see footnote 1), concludes that conformer I is the global minimum, a result also supported by multiple theoretical studies [95–99].

[1] Investigation of the parent, $15NH_2CH_2COOH$ and $15NH_2CH_213COOH$ glycine species, partially reported as supplementary material of [100].

Hydration is known to trigger the transformation of amino acids from the neutral form observed in the gas phase to the charged zwitterion present in condensed media. The glycine–water complexes provide the simpler molecular models of the biologically important amino acid–water interaction, representing the initial steps of the hydration process. Although their generation is difficult in experiments where laser ablation is involved because of the formation of a hot plasma, the improvements in the LA-MBFTMW instruments by modifying the nozzles and incorporating picosecond (20–35 ps) laser technology [59, 61, 67] has rendered possible the observation of glycine microsolvates [100, 101]. In both cases, glycine–H_2O and glycine–$(H_2O)_2$, only conformer I was observed (see Fig. 9). As for bare glycine, conformer III was not detected.

On the basis of the excellent agreement between experimental and theoretical spectroscopic constants and the use of enriched samples of ^{15}N and $H_2{}^{18}O$, it was possible to determine the structure of the complexes shown in Fig. 9. The structures provide relevant chemical information on the nature of the hydrogen bond interactions which stabilize the adducts. Binding of one or two water molecules to glycine proceeds through the carboxylic group and gives rise to a closed ring structure in which the water bridges benefit from enhancing cooperative effects [102–104]. The structures of glycine–H_2O and glycine–$(H_2O)_2$ retain the preferred conformation I (NH···O=C, *cis*-COOH) of unsolvated glycine, so the water molecules in the complexes do not enter into competition with the intramolecular hydrogen bonds in glycine. Comparisons of the O–H···O hydrogen bond distances and angles show that the acidic character of the OH carboxylic group dominates the interactions between water and glycine.

4.2.2 Alanine

Alanine (R=CH_3) (m.p. = 315°C) was first studied in the gas phase by Godfrey et al. [93] from free-jet millimeter-wave absorption spectroscopy and observed two conformers of five possible low energy conformers. They were identified as I and II according to the type of their intramolecular hydrogen bonds. However, for conformer II, the torsion of the –COOH group gives rise to two close energy conformations. Because of the similar mass distribution of both structures, it was not possible to discriminate between them on the basis of the rotational constants alone. A later investigation of the rotational spectrum by high resolution LA-MB-FTMW spectroscopy [105] allowed one to resolve fully the ^{14}N quadrupole coupling hyperfine structure of the two observed conformers and confirmed their assignment as alanine I and IIa. The derived mean ratio $N_I/N_{IIa} = 3.7(5)$ confirmed that conformer I is the most stable species in the jet.

In addition, the rotational spectra of ten different isotopologues (one parent, one ^{15}N, three ^{13}C and five deuterated species) for the two most stable conformers of alanine were detected. The extensive isotopic data were analyzed to derive the substitution [106, 107] and effective structures for both I and IIa conformers. Unlike glycine, the amino acid skeleton in alanine is non-planar. Deuteration at the amino and

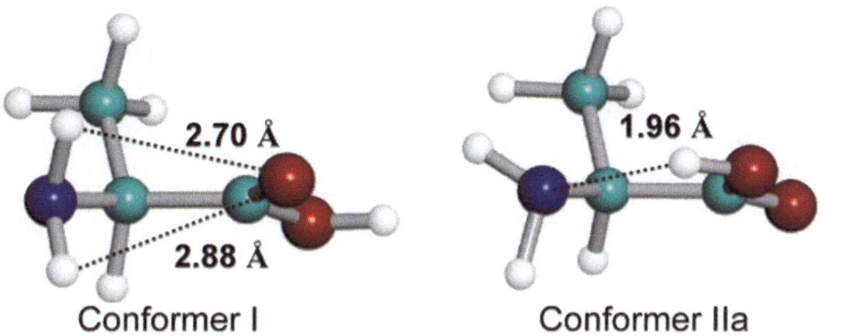

Fig. 10 The structures of conformers I and IIa of alanine determined in [105], showing the hydrogen bond distances

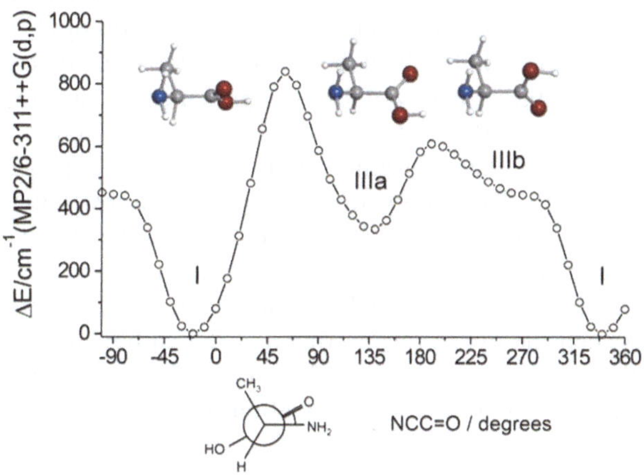

Fig. 11 Calculated MP2/6-311++G(d,p) energy profile of alanine along the torsional coordinate defined by dihedral angle ∠NCC=O. (From [105])

hydroxyl hydrogen atoms rendered information on intramolecular interactions: the data are consistent with the formation of an intramolecular hydrogen bond N···H–O in conformer IIa and with a non-symmetrically bifurcated hydrogen bond N–H···O=C established by the two amino hydrogen atoms in conformer I (see Fig. 10).

The absence of type III conformers which are predicted to fall in the group of lowest energy forms deserves some comment. This problem of "missing conformers" first observed in glycine was attributed by Godfrey et al. [93] to conformational relaxation, plausibly a III → I interconversion. It is accepted that this event takes place by collisions with the noble carrier gas in the adiabatic expansion. This phenomenon has been observed in systems involving only one degree of freedom, such as torsional isomerism [108–110] and axial/equatorial relaxation in hydrogen-bonded complexes [30, 111], when energy barriers are less than about 400 cm^{-1}. The ab initio calculation of a section of the potential energy surface for the interconversion III → I along the rotation of the –COOH group predicts a low barrier (see Fig. 11) [105], which is consistent with the relaxation hypothesis.

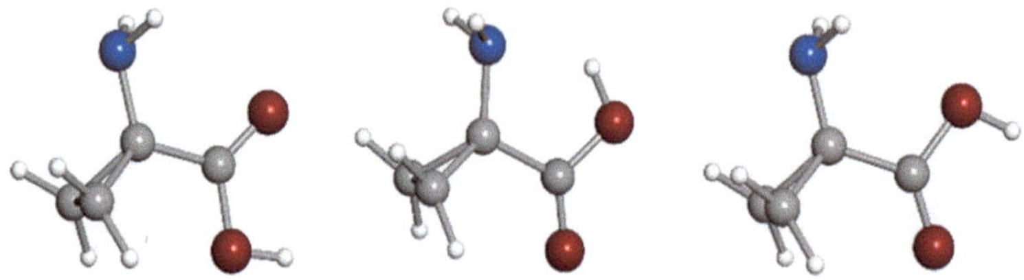

I II III

Fig. 12 Observed conformers for 1-aminocyclopropane carboxylic acid (Ac3c). (From [112])

The hypotheses of the III → I interconversion was confirmed by the study of the 1-aminocyclopropanecarboxylic acid (Ac3c) (R=cyclopropyl, m.p. = 231°C) [112], a natural α-amino acid in which the rotation of the –COOH group is hindered by the π-electron-donating capacity of the cyclopropane ring, able to conjugate with the COOH group [113, 114]. The LA-MB-FTMW investigation of Ac3c [30, 111] reveals the presence of the conformers I, II, and III of Fig. 12 with relative abundances I > III > II. The observation of the III conformer in Ac_3c is highly remarkable but not completely unexpected according to the above considerations. Relaxation from the III to the I form involves rotation around the C_α–C(O) bond and this conformational change should be highly restricted in Ac3c because of conjugation between the cyclopropane ring and the carboxylic acid. A predicted barrier of 2,000 cm^{-1} to interconversion from I and III conformers is high enough [108–110] to preclude conformational relaxation III → I.

4.2.3 Valine, Isoleucine and Leucine

Valine (R=isopropyl, m.p. = 295–300°C) [115], isoleucine (R=*sec*-butyl, m. p. = 288°C) [116], and leucine (R=iso-butyl, m.p. > 300°C) [117] complete the study of aminoacids with aliphatic non-polar side chains.

Because of the hydrophobic character of the side chain, these amino acids are usually involved in protein or enzyme construction but rarely in protein function [86–88]. The main issues in the study of such aliphatic α-amino acids are the increase of conformational possibilities from the multiple configurations associated to torsion about single bond in the lateral chain. Two conformers of types I and II shown in Fig. 13 were ultimately detected in the supersonic jet for valine, isoleucine, and leucine, and conclusively identified through comparison of the experimental rotational and ^{14}N nuclear quadrupole coupling constants with the predicted values ab initio, as described in Sect. 3.

In the three amino acids, the conformational preferences for the intramolecular interactions reproduced those of glycine and alanine, with the amino-to-carbonyl interaction (I) most stable. Interestingly, the two conformers observed for each amino acid present the same arrangement of the side chain (valine Ia and IIa, isoleucine Ia_1

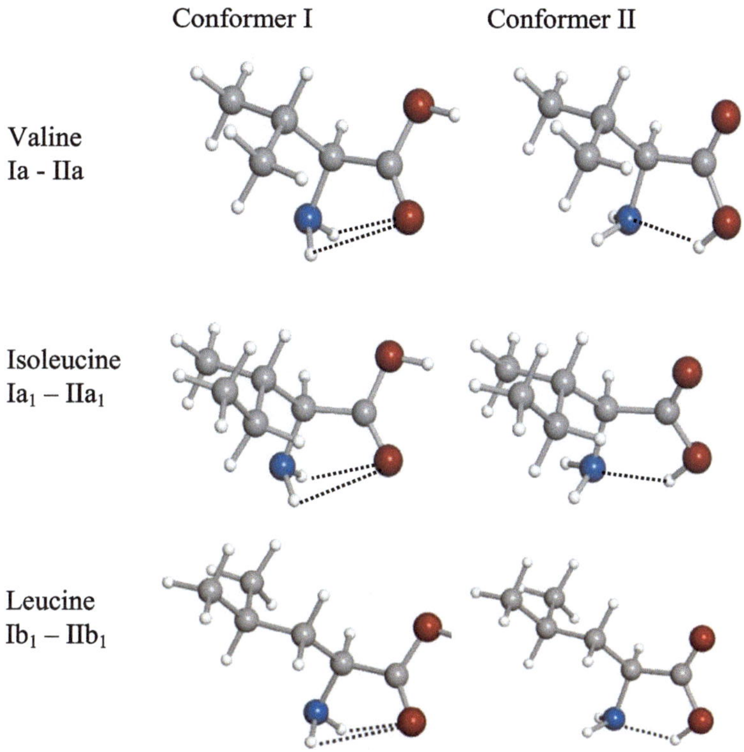

Fig. 13 The observed conformers for valine, leucine and isoleucine. (From [115–117])

and IIa_1, and leucine Ib_1 and IIb_1). This fact seems to indicate that the non-polar side chains do not interact significantly with the polar groups of the amino acid.

4.3 Imino Acids Proline and Hydroxyproline

Proline (m.p. = 228°C) is the only proteinogenic α-amino acid bearing a secondary amine in which the alpha carbon and the nitrogen atom are bound together to form a five-membered pyrrolidine ring. Proline has a unique role in protein formation because of its cyclic structure, acting at the end of α-helices or as structure disruptor in turns and loops [86–88]. Proline is also one of the major constituents of collagen, the most abundant protein in vertebrates, where it is accompanied by 4(*R*)-hydroxyproline (m.p. = 273°C), formed in a post-translational modification [118–124]. Collagen is a triple helix made of three super-coiled polyproline II-like chains with repetitive tripeptide sequences X-Y-Gly, where usually X=proline and Y = 4(*R*)-hydroxyproline. However, the diastereoisomer 4(*S*)-hydroxyproline (m.p. = 243°C)

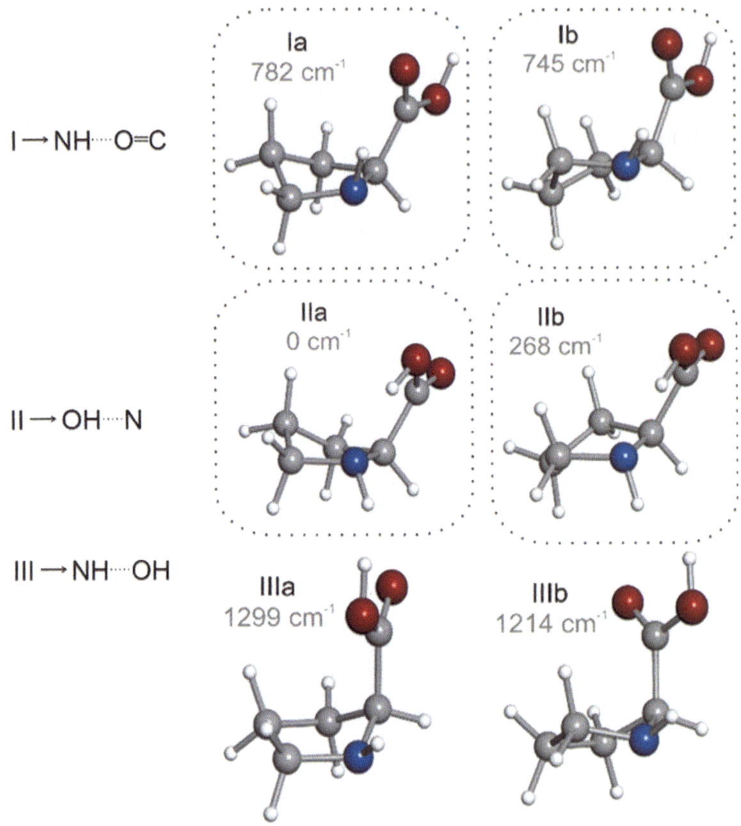

Fig. 14 Predicted low-energy conformers of proline and relative energies (MP2/6-311++G(d,p)) with respect to the global minimum. Conformers are labeled as *I*, *II*, or *III* depending on the hydrogen bond established between the amino group and carboxylic groups. Labels *a* or *b* indicate the configuration *endo* (*a*) or *exo* (*b*) adopted by the ring. The detected conformers are encircled. (From [59, 128])

inhibits the proper folding of the triple helix in either X or Y position [125, 126]. The fact that the diastereoisomer 4(*S*)-hydroxyproline produces a reverse destabilizing effect in the collagen triple helix makes a comparison of the three molecules interesting.

The low energy conformers of proline [127] are shown in Fig. 14 where they are labeled according to the hydrogen bond type I, II or III and the labels a or b to denote the C_γ-*endo* or C_γ-*exo* bent ring configurations, respectively. The first LA-MB-FTMW study on the rotational spectrum of an amino acid carried out in our laboratory was that of proline [59]. In this first study, only two conformers, *IIa* and *IIb*, were detected. Conformer IIa exhibits a *trans*-COOH arrangement, with the pyrrolidine ring adopting a C_γ-*endo* bent configuration. The distance between the nitrogen atom and the carboxylic hydrogen supports the existence of an N···H–O hydrogen bond, similar to those found in configuration II of aliphatic α-amino

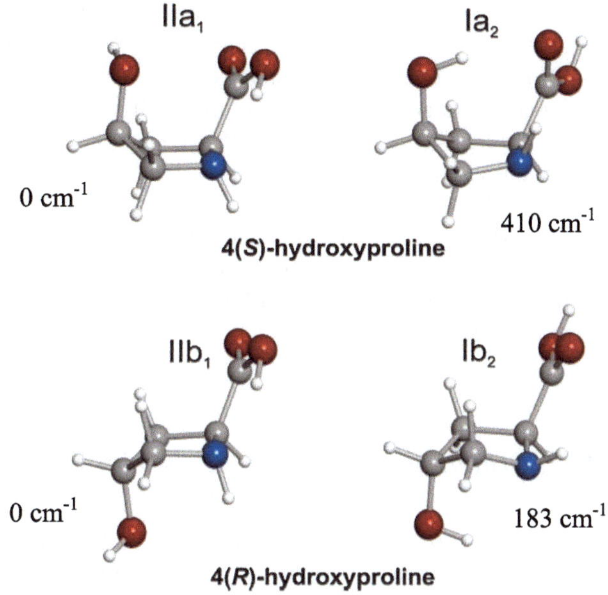

Fig. 15 The observed conformers of 4(*S*)-hydroxyproline and 4(*R*)-hydroxyproline (From [129])

acids. For conformer IIb, analysis of the rotational data yielded a similar N···H–O structure, but differing in a C_γ-*exo* ring. A later investigation taking advantage of the improvements in the laser ablation experiment [59] allowed the identification of two new conformers Ia and Ib [128]. These two new conformers are stabilized by interactions that link the hydrogen atom on the imino group to the oxygen atom of the carboxylic group (N–H···O=C) and differ in the pyrrolidine ring configurations *endo*-like (a) and *exo*-like (b). From the intensities of the lines, the population of the observed conformers was found to follow the order IIa > IIb > Ia ~ Ib, with the conformer IIa as the global minimum in good agreement with ab initio calculations. For the most abundant conformer, IIa, all five monosubstituted ^{13}C isotopomers could be observed in their natural abundance. This isotopic information, together with the data from enriched samples of ^{15}N and two deuterated species, was used to derive its effective structure. The preference for the N···H–O interaction in proline should be attributed to the geometrical constraints imposed by the pyrrolidine ring.

Two conformers were detected for each diastereoisomer of 4-hydroxyproline [129] (Fig. 15), which were labeled according to their intramolecular hydrogen bond pattern (II≡N···H–O; I≡N–H···O=C), ring puckering (a≡C_γ-*endo*; b≡C_γ-*exo*), and hydroxyl orientation (using arbitrary subscripts of the conformer type). In both 4-hydroxyprolines, the most stable conformer was found to exhibit an N···H–O hydrogen bond (4(*R*): IIb$_1$; 4(*S*): IIa$_1$), as occurred for proline. However, unlike proline, the second stable conformers (4(*R*): Ib$_2$; 4(*S*): Ia$_2$) displayed an N–H···O=C hydrogen bond. Significantly, the two forms detected in each case had the same ring puckering orientation (C_γ-*exo* for 4(*R*) and C_γ-*endo* in 4(*S*)). The propensity of 4(*R*)-hydroxyproline to give C_γ-*exo* puckerings would represent a form of preorganization of the collagen helix. Inspection of the most stable structures gives information on the intramolecular interactions causing this effect. Apart from easily recognizable hydrogen bond interactions, the arrangement of the 4-hydroxyl group in the most stable conformer IIa_1 of 4(*S*)-hydroxyproline suggests an n–π* interaction produced by a hyperconjugative delocalization of a non-bonding electron pair (*n*) of the oxygen atom in the hydroxyl group into the π* orbital at the carboxylic group carbon. This interaction is reminiscent of the preferred Bürgi–Dunitz trajectory for a nucleophilic addition to a carbonyl group [130, 131] because the predicted O···C=O distance of 2.97 Å and angle ∠O···C=O <108.5° are optimal for this approach. In this context, the possibility of further stereoelectronic contributions, as the so-called *gauche* effect [132, 133], were considered.

4.4 Proteinogenic Amino Acids with Polar Side Chain

The presence of polar functional groups in the side chains of α-amino acids is expected to increase dramatically the number of low-energy conformers. The functional group can establish additional interactions, which do not occur in other α-amino acids previously studied, and thus may affect the conformational preferences giving rise to a rich conformational space. The conformational behavior of serine [60], cysteine [134], threonine [61], aspartic acid [135], glutamic acid [67], and asparagine [136] have been revealed using LA-MB-FTMW spectroscopy. A summary of the results is collected in Table 3. The analysis of threonine

Table 3 The polar side chain (R) of α-amino acids with the number of observed conformers in each case

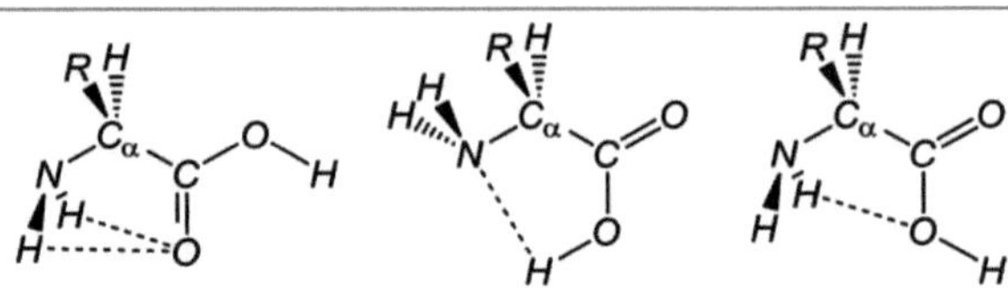

I (N-H···O=C) and cis-COOH

II (N···H-O)

III (N-H···O-H) and cis-COOH

R	Amino acid	Observed conformers
$-CH_2-OH$	Serine	7
$-CH_2-SH$	Cysteine	6
$-CH(OH)-CH_3$	Threonine	7
$-CH_2-COOH$	Aspartic acid	6
$-CH_2-CH_2-COOH$	Glutamic acid	5
$-CH_2-CONH_2$	Asparagine	(1)

Source: From [136]

was described in detail in Sect. 3. The results for the other are presented in the next pages.

4.4.1 Serine, Cysteine, and Threonine

Serine [60] [R=(CH_2OH), m.p. = 240°C] is the simplest proteinogenic amino acid with a polar group (R=CH_2OH) in its side chain. The hydroxyl group can interact as proton donor to the amino or carboxyl groups, or as proton acceptor through the non-bonding electron pair at its oxygen atom. Ab initio calculations predicted a significant number of conformers within 1,000 cm^{-1} (see Fig. 16). Up to seven rotamers were observed in the rotational spectrum following the procedure described in Sect. 3. The experiment also provided information on the relative stability of the serine conformers which follow the order Ia > IIb > I'b > IIc > III_βb ≈ III_βc ≈ IIa, Ia, with an N–H· · ·O=C hydrogen bond, being the global minimum. All observed conformers are stabilized by a network of

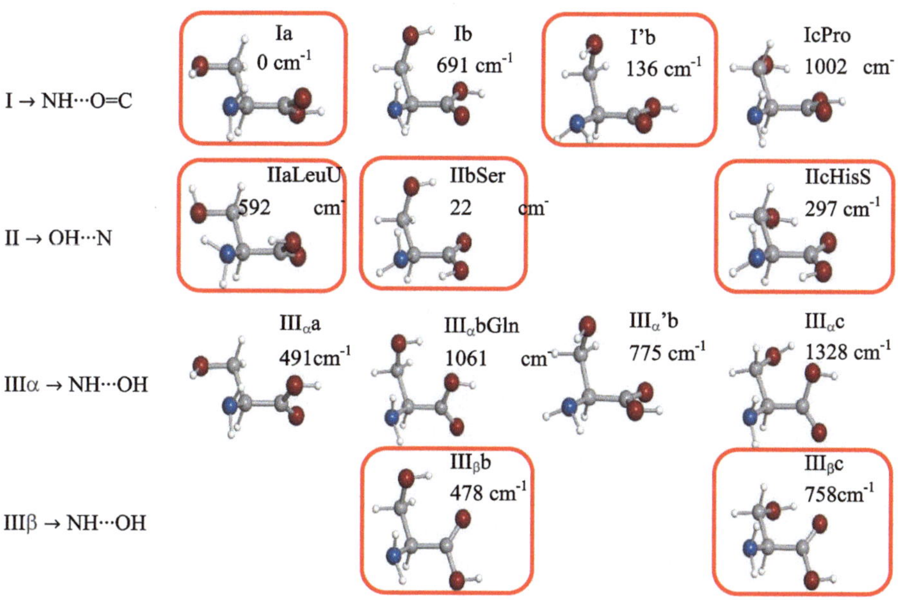

Fig. 16 The predicted low energy conformers of serine and relative energies (MP4/6-311++G(d, p)) with respect to the global minimum (in cm^{-1}). *I*, *II*, or *III* refer to the intramolecular hydrogen bonds between the –NH_2 and –COOH groups. *a* corresponds to a + *sc* configuration for the –OH and –NH_2 groups viewed along the C_β–C_α bond according to IUPAC terminology, *b* corresponds to -*sc* and *c* to *ap* configurations. The prime label indicates a down orientation of the –NH_2 along with an O–H· · ·N intramolecular interaction with the side chain. In conformers *III* an α or β subscript indicates the H atom of the amino group bonded to the hydroxyl or carboxylic groups. The detected conformers are encircled. (From [60])

hydrogen bonds established between the different polar groups. Conformers Ia and I'b bear, in addition to the N–H···O=C amino acid backbone interaction, an O–H···N bond between the hydroxyl group in the side chain and the lone pair of the nitrogen atom. The side chain hydroxyl group interacts in a different way in each of the type II conformers. In IIa, the –OH acts as a proton acceptor of one of the amino hydrogen atoms. In conformer IIc, the –OH interacts as a proton donor to the carbonyl oxygen of the –COOH group, and in conformer IIb the –OH acts both as proton donor, to the –COOH group, and as proton acceptor of one of the $–NH_2$ group hydrogen atoms. The amphoteric character of the hydroxyl group is also manifested in conformer III_βb, where it interacts with both amino and carboxylic groups. In conformer III_βc, the hydroxyl group only interacts with the carbonyl oxygen of the carboxylic group.

The observation of III_βb and III_βc conformers in serine is in direct contrast with the behavior observed in non-polar side chain amino acids. As explained above, in those cases the non-observation of type III conformers is attributed to the collisional relaxation from III to the most stable I conformers. For serine, conformers III_βb and III_βc are predicted to be more stable than the corresponding Ib and Ic forms. A low barrier of ~140 cm^{-1} was predicted for the interconversion from conformer Ib to conformer III_βb. A similar energy profile was found for the interconversion from Ic to III_βc conformer. The microwave work on serine [60] concluded that the I > III preference of α-amino acids with non-polar side chains no longer holds for the "b" and "c" side chain arrangements because of the cumulative effects of the intramolecular hydrogen bonds established between all polar groups. The non-detection of conformer III_αa or conformer III'_αb can be explained by a relaxation to the corresponding more stable type I conformers (see Fig. 11). In those cases, the side-chain hydroxyl group can only establish an O–H···N hydrogen bond with the amino group. The difference in stability is completely attributable to the interaction between the $–NH_2$ and –COOH groups, and the I > III preference of α-amino acids with non-polar side chains is maintained.

Similar behavior to that of serine was found for cysteine [134], [R=CH_2SH, m.p. = 300°C], a natural amino acid which bears an –SH group in the side chain instead of the –OH group. The microwave spectra reveals six rotamers identified as conformers Ia, IIa, Ib, IIb, III_βb, and III_βc. In cysteine, conformer Ib has been observed instead of I'b observed in serine, while conformed IIc observed in serine has not been observed for cysteine. The main differences between both molecules come from the marked less acidic character of the –SH group compared to the –OH group.

Threonine (R=–CH(CH_3)OH, m.p. = 256°C) [61] is related to serine as it only differs on the C_γ methyl group. Figure 7 shows the seven low energy conformers detected for threonine. While form III_βb observed in serine [60] and cysteine [134] was also observed in threonine, conformer III_βc was not detected. This can be attributed to the raising of the energy of this conformer caused by steric interaction of the C_γ methyl group with the –COOH group. Conformer III_αa has been observed, in contrast to serine and cysteine, where this conformer relaxes in the supersonic expansion to the Ia form. The only explanation in this case is that the steric interactions between the C_γ methyl group and the –COOH group would increase the barrier of interconversion between those forms, predicted to be ~1,000 cm^{-1} in threonine [61] which cannot be surmounted via collisions with the carrier gas.

4.4.2 Aspartic and Glutamic

Aspartic acid (R=–CH_2–COOH, m.p. >300°C) is a natural amino acid with a carboxyl group in the side chain. Ab initio searches in the potential energy surface predict up to 14 low energy conformers to lie within 800 cm^{-1} [135]. Investigation of the microwave spectrum of laser ablated aspartic acid yields the identification of six conformers (see Fig. 17). The relative population ratios follow the order Ia-I > Ib-I ≈ IIa-I > Ia-II ≈ IIb-I > III_βb-I. Conformer Ia-I is the most abundant conformer in the molecular beam, in poor agreement with the ab initio calculations which predict Ia-I to be ca. 300 cm^{-1} above the global minimum. This fact can be attributed to relaxation processes between conformers [108–110, 135].

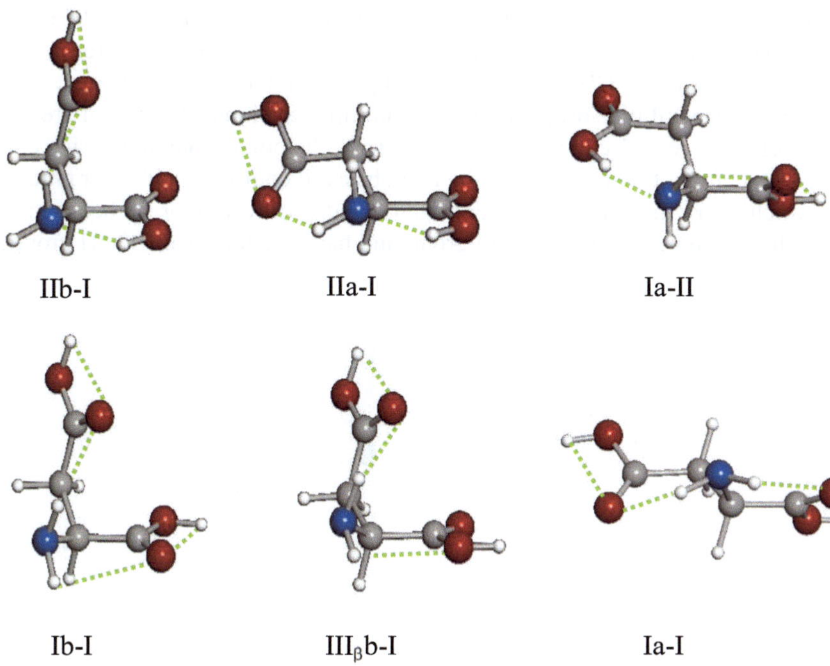

Fig. 17 The observed conformers of aspartic acid. Labeling is the same as used in serine, cysteine or threonine. An additional label indicating the hydrogen bond type (*I*, *II*, or *III*) between the –NH_2 and the β-COOH groups is used to avoid ambiguities. (From [135])

It was confirmed again that the presence of a nitrogen nucleus makes the ^{14}N quadrupole coupling constants a unique tool to establish the nature of the intramolecular hydrogen bonding. All the conformers detected are stabilized through a network of intramolecular hydrogen bonds, in which all functional groups participate (see Fig. 19). The six conformers observed belong to the "a" or "b" families, the β-COOH group being synclinal to the amino group, which allows these two functional groups to interact. This indicates that the β-COOH prefers to interact with the amino group rather than with the COOH group in α. The $-NH_2$ group acts as a bridge connecting the –COOH groups in the hydrogen bond network. Furthermore, all observed conformers except one are of type I, indicating that the β-COOH acts essentially as proton acceptor rather that as proton donor.

Aspartic acid follows the trends observed by all but one of the aliphatic amino acids studied: type I conformers are more populated than type II conformers. The most populated form of aspartic acid is conformer Ia, which is stabilized by two N–H···O hydrogen bonds established between the amino group and each of the carboxylic groups, both of them in a *cis* configuration. The amino group disposition in Ia-I gives rise to considerably shorter N–H···O hydrogen bonds than in other conformers, which is usually associated with greater hydrogen bond strength and could explain the stability of Ia-I. In the case of aspartic acid, the observation of the conformer III_{β}b-I can be ascribed to the high potential barrier (around 1,000 cm^{-1}), as predicted from ab initio calculations [135] for relaxation to the Ib-I form.

Glutamic acid (R=$-CH_2-CH_2-COOH$, m.p. = 205°C) is a clear example of the great torsional flexibility originating from multiple torsional degrees of freedom because of its large side chain. As in aspartic acid, the polar group in the side chain is a –COOH group. Figure 18 shows the five observed conformers [67] for glutamic acid which show intramolecular hydrogen bonds of type I or II between the carboxylic group in α and the amino group and have the terminal COOH group in

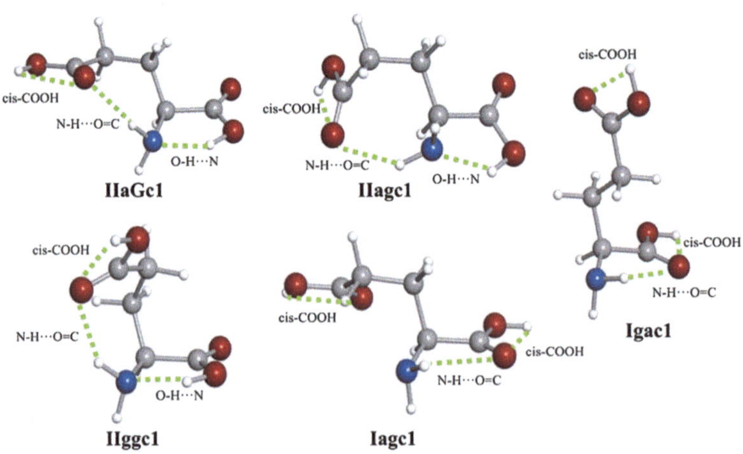

Fig. 18 The five observed conformers of glutamic acid showing their intramolecular interactions. (From [67])

a *cis* configuration. No type III conformers have been observed in this case. Conformer Iagc1 presents a type I hydrogen bond, i.e., an N–H···O=C hydrogen bond between the $-NH_2$ and the α-COOH which adopts a *cis* arrangement. Glutamic acid thus follows the behavior displayed by the overwhelming majority of the aliphatic α-amino acids studied so far with a type I conformer as global minimum. Conformer Igac1 is also stabilized by an N–H···O=C hydrogen bond between the carboxylic group in alpha (in a *cis* disposition) and the amino group. Similar to the most abundant conformer Iagc1, it displays an extended backbone configuration where the acid group in the side chain does not establish any additional interactions with the other polar groups in the amino acid. The other conformers identified, IIggc1, IIagc1, and IIaGc1, present a similar intramolecular hydrogen bond network. As for aspartic acid, the interactions of the terminal –COOH group with the amino acid backbone are dominated by the type I interaction NH···O in which the terminal –COOH acts as proton acceptor.

The γ-COOH group of Iagc1 and Igac1 does not establish hydrogen bonds with any of the other functional groups. This is in sharp contrast with what was observed for other α-amino acids with polar side chains, described above, where all conformers flaunted interactions between all functional groups. For example, the most abundant conformer of aspartic acid [135], the polar amino acid more closely related to glutamic acid, was stabilized by two N–H···O=C hydrogen bonds between the amino group and the carboxylic groups in alpha and beta. In comparison with aspartic acid, glutamic acid has a longer side chain which can adopt a larger number of dispositions. For some of these dispositions the γ-COOH group is unable to establish any interactions with the other polar groups in the amino acid. These conformations, with less intramolecular interactions, are entropically favored and Iagc1 is the most populated conformer in the supersonic jet.

Type III conformers have only been observed in those α-amino acids where rotation of the α-COOH group is hindered or in those for which the I/III relative energies are reversed. Glutamic acid has a polar side chain $-CH_2-CH_2-COOH$ with the same functional group as that of aspartic acid ($-CH_2-COOH$) but possessing just one more methylene group. However, this apparently small increase in chain length produces a strong impact on the conformational preferences.

4.4.3 Asparagine: Conformational Locking

Asparagine ($R{=}-CH_2-CONH_2$, m.p. = 236°C) is similar to aspartic acid but with an amide group, $-CONH_2$, in the lateral chain instead of a –COOH group. In this case, only one structure has been identified (see Fig. 19) in the jet cooled rotational spectrum [136]. This is in sharp contrast to the multiconformational behavior observed in other proteinogenic amino acids with polar side chains. This conformational locking to a single conformer IIa is caused by a network of three cooperative intramolecular hydrogen bonds, $N_\alpha \cdots H{-}O_\alpha$, $N_\alpha{-}H \cdots O_\beta$, and $N_\beta{-}H \cdots O_\alpha$, forming the intramolecular sequence shown in Fig. 19. Such arrangement of functional groups into hydrogen bond networks leads to the phenomenon of

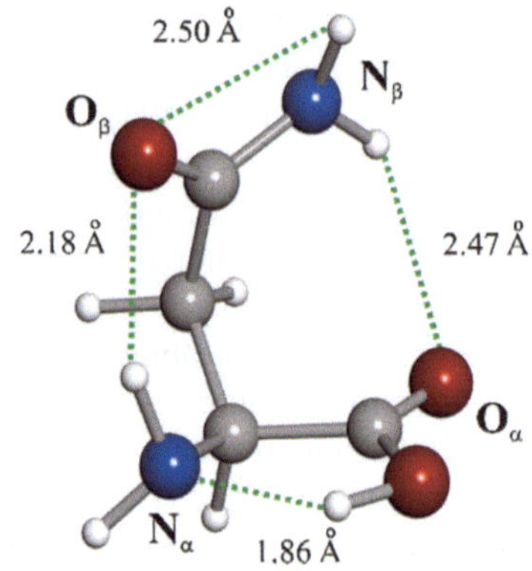

Fig. 19 Observed conformer, atom labeling for asparagine and hydrogen bond distances taken from the ab initio structure (From [136])

cooperativity [102–104] under which the strength of individual hydrogen bond interactions is notably enhanced. In addition, the simultaneous formation of $N_β\cdots H–O_α$ and $N_α–H\cdots O_β$ may contribute to a further stabilization of the amide group by favoring the resonance form which confers a double bond character to the C–N amide bond.

To conclude, in serine (R=–CH_2–OH) [60], threonine (R=CH(OH)–CH_3) [61], cysteine (R=–CH_2–SH) [134], and aspartic acid (R=–CH_2–COOH) [135], the side chain has been found to change the conformational preferences with respect to aliphatic α-amino acids without polar side chains. The number of possible conformations increased and conformers with type III hydrogen bonds between the α-COOH and –NH_2 groups were observed and sometimes found to be more stable than their analogues with type I hydrogen bonds. The longer side chains of glutamic acid (R=–CH_2–CH_2–COOH) [67] confer larger flexibility and multiply the conformational possibilities, but also changes the influence of the interactions involving the γ-COOH group and the other polar groups in the amino acid. The increased length of the side chain in glutamic acid makes entropic contributions arising from the presence of intramolecular interactions more significant than in other polar amino acids. This is reflected in the fact that the most populated conformer of glutamic acid is an extended conformer where the side chain polar group does not interact with the –NH_2 or α-COOH groups. The rich conformational behavior of these amino acids contrasts with that of asparagine [136], which collapses to only one conformer, highly stabilized by a network of intramolecular hydrogen bonds, involving all polar groups in the molecule.

4.5 Amino Acids with Aromatic Side Chain

As already explained in the introduction of this review, the pioneering studies of the proteinogenic amino acids in the gas phase were carried out using laser spectroscopy techniques. However, since these techniques can be applied only to molecules

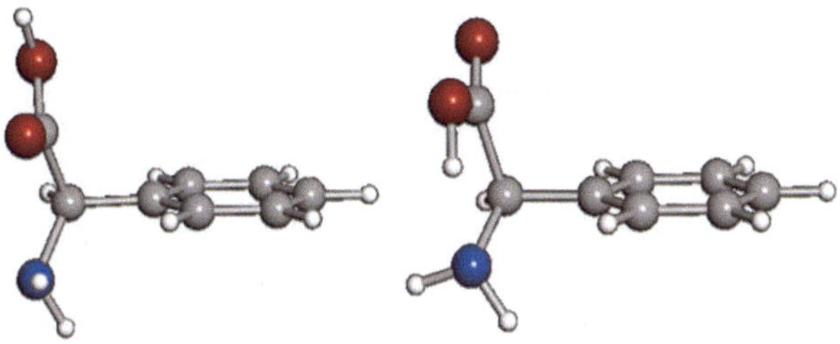

Fig. 20 The two observed conformers, I (*left*) and II (*right*), of the unnatural amino acid phenylglycine in the gas phase (From [152])

bearing chromophore groups, their application to proteinogenic amino acids has been limited to phenylalanine (R=benzyl) [11, 137–147], tyrosine (R=*p*-hydroxybenzyl) [3, 11, 138, 148], and tryptophan (R=3-methylene indolyl) [149–151] which have been the subject of a large number of investigations. While the studies of non-aromatic amino acids by microwave spectroscopy using laser ablation techniques has been successful, the same techniques applied to the aromatic ones have been shown to have problems related to photofragmentation/ionization of these molecules taking place during the laser ablation process.

4.5.1 Phenylglycine and Phenylalanine

Phenylglycine (R=–C_6H_5, m.p. = 290°C) is the simplest aromatic α-amino acid, with the phenyl group directly attached to the α-carbon. It was the first aromatic amino acid studied by LA-MB-FTMW and thus provided a suitable test for the application of laser ablation chromophoric systems [152]. The analysis of the rotational spectrum leads to the identification of two conformers shown in Fig. 20. The most abundant conformer Ia exhibits a hydrogen bond interaction N–H···O=C and a *cis*-COOH arrangement reminiscent of glycine I. The second amino group hydrogen atom points to the phenyl ring, indicative of an N–H···π hydrogen bond interaction. The conformer II presents an N···H–O interaction with a *trans*-COOH configuration. The population ratio derived for the phenylglycine forms in the supersonic jet is $N_{Ia}/N_{II} \sim 4$, which demonstrates the predominance of the type I conformer, as occurs in all aliphatic α-amino acids. The amino acid skeleton of phenylglycine reproduces the primary conformational preferences of aliphatic amino acids.

The conformational landscape of phenylalanine (R=CH_2–C_6H_5, m.p. = 270–275°C) has been widely investigated [11, 137–147]. Six conformational species were identified using laser induced fluorescence LIF, hole burning UV–UV, and ion dip IR–UV spectroscopy coupled with ab initio calculations [11, 137–141]. Lee et al. [141] carried out a definitive identification of the conformers of phenylalanine, based upon comparisons between the partially resolved ultraviolet band contours and that simulated by ab initio computations. The study of the rotational spectrum of phenylalanine by LA-MB-FTMW [153] showed rather weak spectra of only two conformers, IIa and IIb (see Fig. 21). Both conformers exhibit a *trans* configuration in the COOH group, being stabilized by O–H···N and N–H···π

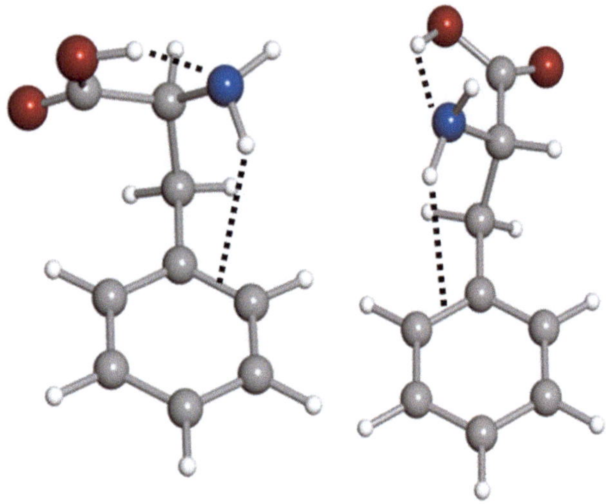

Fig. 21 The two observed conformers, IIa (*left*) and IIb (*right*), of phenylalanine (From [153])

intramolecular hydrogen bonds. The low intensity of the observed spectra and the non-observation of the conformers, detected with other techniques, can be caused by ionization and photofragmentation processes during laser ablation taking place with different rates for the different conformers. Kim and co-workers [145–147], from their measurements of the ionization energies (IE) of the low-energy conformers of phenylalanine, affirm that the observed conformers IIa and IIb have higher IEs than the other conformers.

4.6 Non-proteinogenic Amino Acids

Rotational studies extended to non-coded α-amino acids are also of biochemical relevance [154]. The effects on the conformational behavior of enlarging the amino acid backbone chain have been analyzed on β-alanine [155, 156] and γ-amino butyric acid (GABA) [157]. Both are neurotransmitters which bind to the same sites as glycine [158–160]. These are also the simplest β-amino and γ-amino acids and so are the natural starting point to analyze the conformational panorama of this type of amino acids. Other studies include α-aminobutyric acid [161], the *N*-alkylated species sarcosine [162], *N,N*-dimethylglycine [163], and taurine [164]. Brief results on β-alanine and GABA are presented.

4.6.1 β-Alanine

The rotational spectrum of β-alanine (NH_2–CH_2–CH_2–COOH, m.p. = 202°C) was studied by Godfrey et al. [155] using heating methods of vaporization.

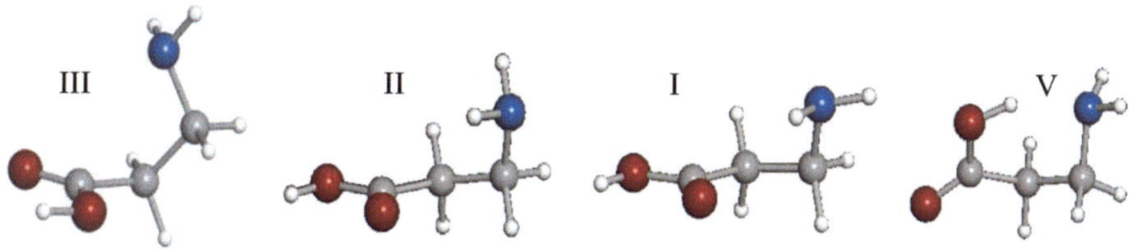

Fig. 22 Observed conformers of β-alanine. (From [156])

Two conformers (I and V shown in Fig. 22) were identified, stabilized by N–H···O=C and N···H–O intramolecular bonds. An LA-MB-FTMW study [156] led to the characterization of two new conformers, II and III. Full resolution of the ^{14}N quadrupole coupling structure was invaluable to distinguish between the most abundant conformers I and II, which exhibit an N–H···O=C hydrogen bond with different orientation of the amino group. Conformer III is not stabilized by a hydrogen bond, but by an n–π* interaction between the electron lone pair at the nitrogen atom and the π* orbital of the –COOH carbonyl group which produces electronic delocalization by hyperconjugation to the π* orbital [130, 131]. Such interaction is described in more detail below.

4.6.2 γ-Aminobutyric Acid (GABA)

In γ-aminobutyric acid (GABA) (NH_2–CH_2–CH_2–CH_2–COOH, m.p. = 204°C) the five hindered rotations around the single bonds generate a plethora of conformational species. An overall picture of the conformational landscape obtained from theoretical predictions [157] confirms the richness of GABA: up to 30 feasible conformers shown in Fig. 23 were localized on the ab initio potential energy surface with relative energies below 900 cm^{-1}. Thorough analysis of the rotational spectra finally led to the assignment of nine different rotamers of GABA [157] encircled in Fig. 23. A close look at the detected conformers indicates that in *GG2*, *aG1*, *aa1*, and *ga1*, the two polar groups are far apart and no intramolecular interactions are apparent, apart from a stabilizing *cis*-COOH functional group interaction. In folded configurations, non-covalent interactions can be established between the two polar groups. Conformers *gG2* and *GG1* are stabilized by type II, O–H···N, and conformer *GG3* by type I N–H···O=C intramolecular hydrogen bonds similar to those observed in non-polar aliphatic α-amino. As occur for the III form of β-alanine, conformers gG1 and gG3 of GABA show an arrangement of the amino and carboxyl groups which resembles the Bürgi–Dunitz trajectory [130, 131] which describes the most favorable approach of a nucleophile nitrogen to a carbonyl group carbon in an addition reaction. This is a signature of the existence of an n → π* interaction arising from the hyperconjugative delocalization of the non-bonding

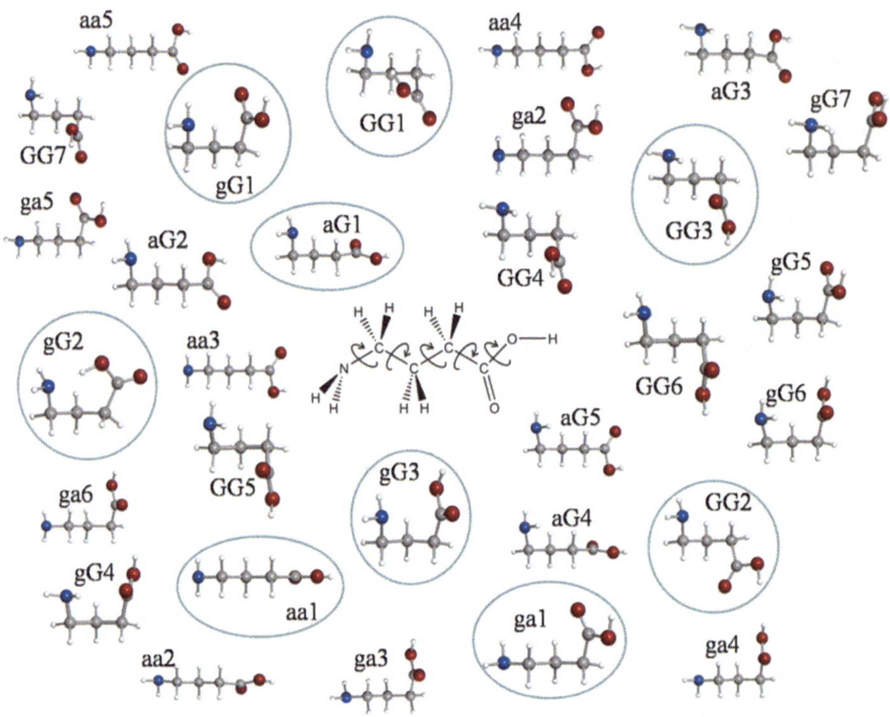

Fig. 23 Predicted low energy conformers of GABA. The nine observed conformers are encircled. (From [157])

electron pair of the nitrogen atom to the π* orbital at the carbonyl group (see Fig. 24). In the Bürgi–Dunitz trajectory, the approach path of the nucleophile (N:) is described to lie in the plane bisecting the ∠R–C–R′ angle with an angle α of about 105°±5° for sort N to C distances [130, 131]. The ab initio structures confirm that gG1 and gG3 conformers of GABA show an optimal geometrical arrangement for this interaction.

The relative populations of the GABA conformers *GG2* > *aG1* > *gG1* > *aa1* > *ga1* can be taken as proof of the coexistence of GABA conformers having intramolecular interactions (folded) with those free of them (extended). Conformers of GABA free of intermolecular interactions are the most abundant. Intramolecular interactions contribute to decrease entropy and to increase the Gibbs energy, thus diminishing number density. The works on β-alanine [156] and GABA [157] show that an increment in the number of methylene groups between the amino and carboxylic groups gives rise to the existence of stabilizing interactions, such as n → π*, different from the hydrogen bond. In addition, the presence of low energy extended conformers free of intramolecular interactions becomes significant.

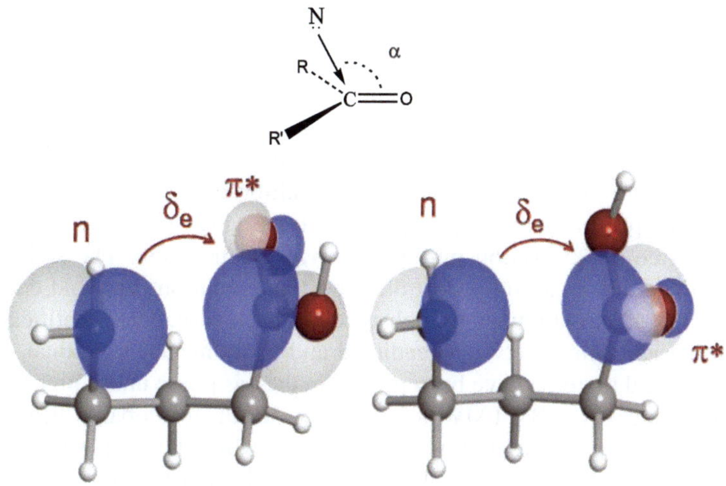

Fig. 24 Scheme of the Bürgi-Dunitz trajectory of addition of a nucleophile N to a carbonyl carbon atom and representation of the n→π* interaction in conformers gG1(a) and gG3(c) of GABA. (From [157])

5 Nitrogen Bases

Much effort has been devoted to the identification of preferred tautomers of nucleobases since the structure of nucleic acid and its base pairs were first reported [165]. The best experimental approach to address the structural preferences of nucleobases is to place them under isolation conditions in the gas phase, cooled in a supersonic expansion. Under these conditions, the various tautomers/conformers can coexist and are not affected by the bulk effects of their native environments, which normally mask their intrinsic molecular properties. The main restriction to the gas-phase study of these building blocks is the difficulty in their vaporization owing to their high melting points (ranging from 316°C for guanine to 365°C for adenine) and associated low vapor pressures. The success of LA-MB-FTMW experiments to the study of coded amino acids prompted their application to nucleic acids uracil [166], thymine [167], guanine [168], and cytosine [169], as well as the monohydrates of uracil and thymine [170]. This technique gives a precise interpretation of the structure and relative energies of the different forms of nucleic acid bases to solve apparent discrepancies between the previous studies. The rotational constants, which are the main tool to identify the different forms of a biomolecule, had a minor role when trying to discern between the different tautomers of a nucleobase. It is the quadrupole coupling hyperfine structure caused by the presence of ^{14}N nuclei which constitute authentically the fingerprints of every tautomer. In the next sections, the results on nucleobases are discussed.

5.1 *Uracil, Thymine, and Their Monohydrates*

Uracil and thymine have similar structures and only differ in the presence of a methyl group in thymine (5-methyluracil). These can exist in various tautomeric forms which differ by the position of the hydrogen atoms, which may be bound to either nitrogen or oxygen atoms (keto-enolic equilibrium). The first observations of the rotational spectra of uracil [171] and thymine [172] were made using Stark modulation free-jet absorption millimeter-wave spectroscopy. The solid samples were vaporized under carefully controlled heating conditions to avoid decomposition. Only the diketo tautomer was observed in each case. The identification was based on the agreement between predicted and experimental rotational constants value. Uracil and thymine have been probed in the gas phase under high-resolution conditions using LA-MB-FTMW spectroscopy. Both nucleobases bear two ^{14}N atoms with nonzero quadrupole moments ($I = 1$), which interact with the electric field gradient at the nucleus, resulting in a complicated hyperfine structure. As mentioned in Sect. 3, the experimental values of the quadrupole coupling constants provide an independent approach to identify the tautomeric forms.

In the case of thymine, the rotational spectrum was complicated not only by the hyperfine structure of two ^{14}N atoms but also by a further doubling arising from the coupling of the internal rotation of the methyl group to the overall rotation (see Fig. 25). The resulting structure was completely resolved and analysis of the spectrum yielded the quadrupole coupling constants for both ^{14}N nuclei and the barrier to internal rotation of the methyl group, $V_3 = 1.502(9)$ kcal/mol obtained from the A–E doublet (see Fig. 25).

For uracil, the intensity of the observed spectrum led to the observation of the spectra of the ^{15}N(1)-^{14}N(3) and ^{14}N(1)-^{15}N(3) isotopomers in natural abundance. In a subsequent step, a ^{15}N-^{15}N enriched sample was used to observe the

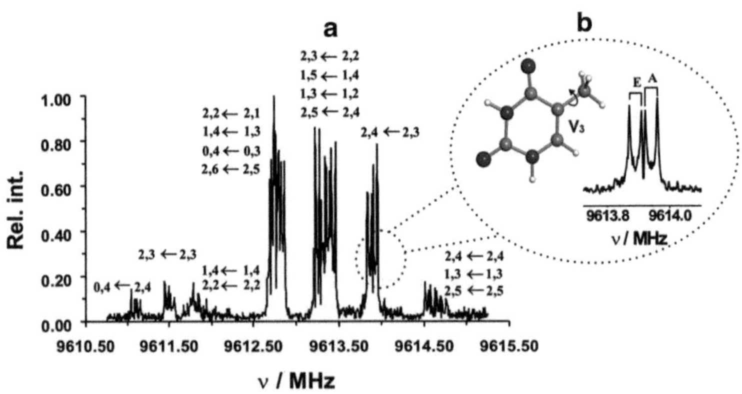

Fig. 25 (**a**) The $4_{1,4}$-$3_{0,3}$ rotational transition of thymine showing the ^{14}N quadrupole components labeled with the quantum numbers $I',F' \leftarrow I'',F''$. (**b**) Detail of the $I,F = 2,4 \leftarrow 2,3$ quadrupole component, which is split into four lines because of the internal rotation (A–E doublet) and Doppler effects (⊓). (From [167])

spectra of all ^{13}C and ^{18}O monosubstituted isotopologues. The inertial defect ($\Delta_c = I_c - I_a - I_b = \Sigma_i m_i c_i^2$) measures the mass extension out of the *ab* inertial plane and is close to zero for planar molecules. The values of parameter for the observed isotopologues range from −0.129 to −0.134 uÅ^2, allowing one to establish the planarity of uracyl. The substitution and equilibrium structures of this molecule [166] are shown in Table 4. Recently, the computational composite scheme [173] has been applied to the first study of the rotational spectrum of 2-thiouracil. The joint experimental–computational study allowed the determination of the accurate molecular structure and spectroscopic properties of this nucleobase.

The uracil–water and thymine–water complexes provide the simplest molecular models of the interactions between biologically important nitrogen bases and water. These monohydrates have been the subject of many theoretical studies [174 and references therein, 175, 176] and experimental studies [177–179]. In the microwave work [170] by LA-MBFTMW spectroscopy, only one conformer of each complex has been observed. Investigation of the structure of the adducts from the rotational constants of the different isotopologues shows that the observed conformers correspond to the most stable forms in which water closes a cycle with the nucleic acid bases forming $N{-}H_{NB}\cdots O_w$ and $H_W\cdots O{=}C_2$ hydrogen bond (see Fig. 26). Both adducts present similar hydrogen bond structures which are also comparable to those observed for related complexes, such as formamide–water [180–182] *N*-methylformamide–water [183], or 2-pyridone–water [184], as could be expected given the similar natures of the hydrogen bonds in all these systems.

5.2 Guanine

Theoretical calculations predict the existence of four low-energy forms [185–187]: keto N7H, keto N9H, and enol N9H *cis* and *trans* (Fig. 27). The features observed from UV laser spectroscopy of guanine [188–193], IR techniques in He nanodroplets experiments [194], and electron diffraction study [195] led to some controversy. The four more stable forms were first identified by de Vries and co-workers in the R2PI spectrum [188, 191]. However, in successive works by Mons and co-workers [189, 192, 193] and Seefeld et al. [190] it was concluded that the spectrum is dominated by the less stable tautomers N7H enol and the two keto imine tautomers. This was attributed to the occurrence of a fast non-radiative relaxation of the excited states of the N7H keto, N9H keto, and N9H enol *trans* which prevents their observation in the R2PI spectrum. On the other hand, the fourth most stable form was identified by IR techniques in He nanodroplet experiments [194], while an electron diffraction study confirmed the existence of N9H keto form [195].

The difficulties in detecting conclusively the most stable tautomers of guanine showed the need for a gas-phase high sensitivity structural probe such as microwave spectroscopy insensitive to excited-state dynamics. However, guanine has the drawback of having five ^{14}N atoms in its structure so their quadrupole coupling

Table 4 Atom coordinates in the principal inertial axis system and substitution (r_s) structure for the diketo tautomer of uracil. Distances are given in Å and angles in degrees

	a	b	c[a]
N_1	1.1884(10)[b]	−1.0115(12)	0
C_2	1.1975(10)	0.375(3)	0
N_3	−0.091(13)	0.9640(12)	0
C_4	−1.2925(9)	0.295(4)	0
C_5	−1.1992(10)	−1.1535(10)	0
C_6	0.05(2)	−1.7396(7)	0
O_7	2.2323(5)	1.0189(12)	0
O_8	−2.3273(5)	0.9432(13)	0
$r(N_1–C_2)$	1.386(5)[c]	$\angle N_1C_6C_5$	122.3(6)
$r(N_3–C_4)$	1.38(2)	$\angle C_2N_1C_6$	123.0(11)
$r(C_4–C_5)$	1.451(4)	$\angle C_5C_4N_3$	115.4(16)
$r(C_5–C_6)$	1.379(4)	$\angle C_6C_5C_4$	118.8(12)
$r(C_6–N_1)$	1.352(14)	$\angle N_1C_2O_7$	122.3(8)
$r(C_2–O_7)$	1.219(4)	$\angle C_5C_4O_8$	118.8(7)
$r(C_4–O_8)$	1.22(2)		

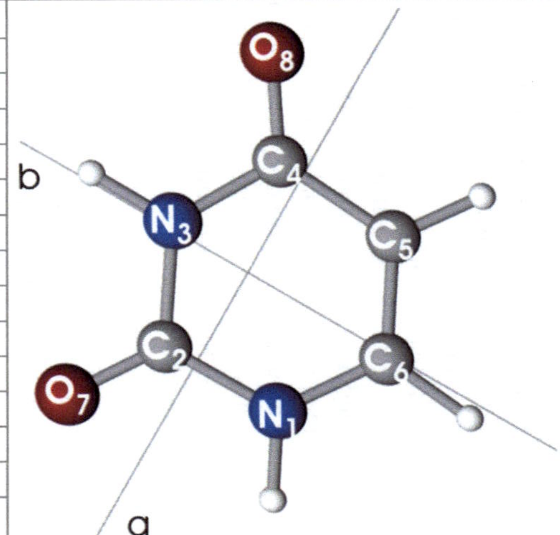

Source: From [166]

[a]Assumed value for all atoms, since uracil is a planar molecule lying on the ab inertial plane

[b]Derived errors in parentheses in units of the last digit

[c]Derived errors in parentheses in units of the last digit

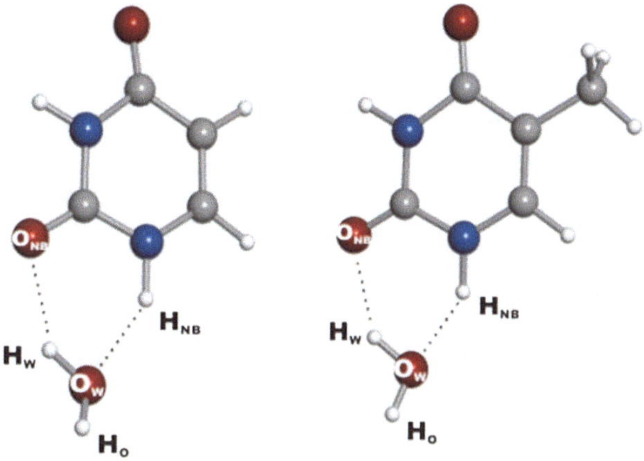

Fig. 26 The uracil-water (*left*) and thymine water (*right*) observed complexes. (From [170])

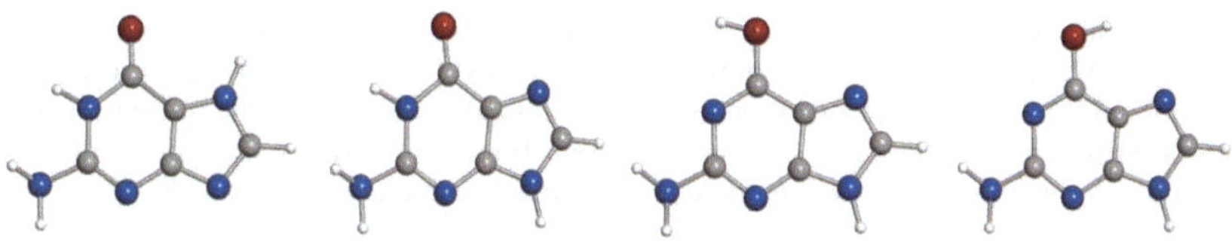

Fig. 27 The four observed conformers of guanine. (From [168])

hyperfine structure gives very complex patterns in the rotational spectra. With this background the LA-MB-FTMW rotational spectra was investigated and four different rotamers were recognized [168]. All observed transitions were split into many components, confirming that they belong to guanine. No attempt was made to assign the quadrupole hyperfine components, and the rotational frequencies were measured as the intensity-weighted mean of the line clusters. The values of the rotational constants and the electric dipole moment components were used to identify the observed rotamers as the four most stable forms of guanine in Fig. 27. The post-expansion abundances measured from relative intensity measurements point to a higher stability of the N9H and N7H keto forms. Finally, the values of the inertial defect, ranging from 0.48 to 0.68 uÅ^2, show that the tautomers of guanine are slightly non-planar.

5.3 *Cytosine*

The molecular system of cytosine (CY) is more complex than that of guanine. Figure 28a shows the five most stable species according to theoretical calculations

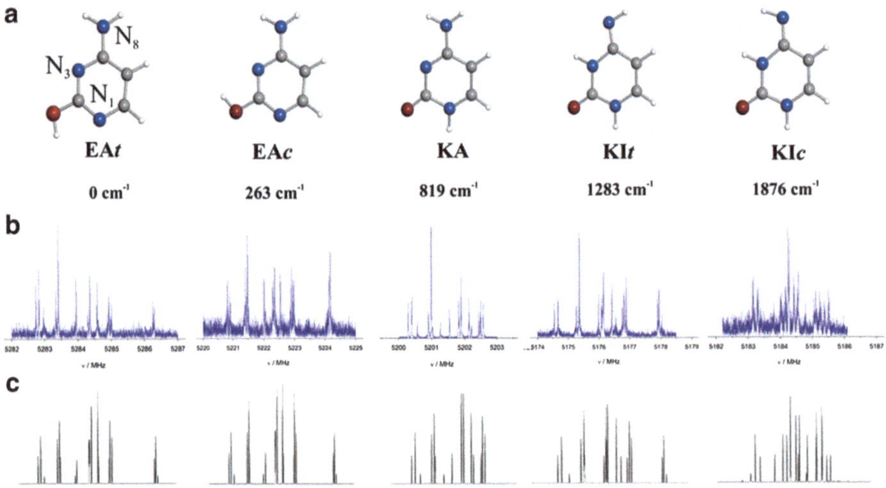

Fig. 28 (**a**) The five more stable species of cytosine: enol-amino *trans* (EA*t*), enol-amino *cis* (EA*c*), keto-amino (KA), keto-imino *trans* (KI*t*), keto-imino *cis* (KI*c*) given in order of stability according to MP2/6-311++G(d,p) ab initio calculations. (**b**) LA-MB-FTMW spectra for the $1_{1,1}$-$0_{0,0}$ rotational transition of the five species. (**c**) Theoretical simulation of the nuclear quadrupole hyperfine structure for the $1_{1,1}$-$0_{0,0}$ rotational transition. The differences among the various patterns act as fingerprints for tautomeric/conformational assignment. (From [169])

[169]: enol–amino *trans* (EAt), enol–amino *cis* (EAc), keto–amino (KA), keto–imino *trans* (KIt), and keto–imino *cis* (KIc). Different experiments have been conducted to reveal the tautomerism of cytosine. The infrared spectra in inert gas matrices were interpreted in terms of a mixture of KA and EA [196]. IR laser spectroscopy in helium nano-droplets [197] characterized the EAt, EAc, and KA species. Two features observed in the vibronic spectra were attributed to the KA and EA forms [198]. The electron diffraction pattern [199, 200] was interpreted in terms of a conformation mixture dominated by the EA forms. In very recent experiments on Ar-matrix, photoisomerization processes [201, 202] were interpreted in terms of the coexistence of various tautomers of cytosine. The free jet millimeter-wave absorption spectra [203] of the three detected species were tentatively assigned to KA, Eat, and KI forms. Five rotamers were observed in the high resolution rotational spectrum of cytosine investigated by LA-MB-FTMW spectroscopy [169]. Rotational transitions exhibit a very complex hyperfine structure caused by the presence of three ^{14}N nuclei. Analysis of this hyperfine structure yields the nuclear quadrupole coupling constants, which are extremely sensitive to the electronic distribution around the quadrupolar nuclei $^{14}N_1$, $^{14}N_3$, and $^{14}N_8$. Comparison between experimental and predicted spectroscopic constants leads to a conclusive identification of the detected rotamers.

Species EAt and EAc have very similar values for the three ^{14}N nuclei, being the differences between the two conformers only caused by the different orientation of the hydroxyl group. In those conformers, the positive values of χ_{cc} for $^{14}N_1$ and $^{14}N_3$, in the range 1.0–1.8 MHz, indicate that these are pyridinic nitrogen atoms, while the negative values for $^{14}N_8$ indicate this is an amino nitrogen atom. A conclusive discrimination of both species was achieved by using the trend of the change on the rotational constants. In tautomer KA, the χ_{gg} ($g = a$, b, c) values associated with atoms $^{14}N_3$ and $^{14}N_8$ do not change very much with respect to the previous tautomers, but those for atom $^{14}N_1$ change radically, indicating that it is an imino N atom. In passing to conformers KI, the quadrupole coupling constants associated with atom $^{14}N_3$ change and the corresponding value of χ_{cc} reveals that it is now an imino N atom. Finally, χ_{cc} of $^{14}N_8$ of KIt and KIc are "chemically" different from all other χ_{cc} values and were finally identified as being part of a C=N–H group (see Table 5). ^{14}N nuclear quadrupole patterns make it possible to obtain the spectral signatures for each individual tautomer in the complex sample and thus act as a sort of fingerprint (Fig. 28b, c). The relative intensity measurements indicate that the EA forms are more abundant in the gas phase than the canonical KA form. The values of the inertial defects show that all species are effectively planar.

Table 5 Spectroscopic constants for the observed tautomers and conformers of cytosine

	EA*t*	EA*c*	KA	KI*t*	KI*c*
A/MHz	3,951.85325(32)[a]	3,889.46510(38)	3,871.54618(31)	3,848.18174(41)	3,861.2966(12)
B/MHz	2,008.95802(12)	2,026.31804(12)	2,024.97804(11)	2,026.31068(31)	2,011.41032(62)
C/MHz	1,332.47228(08)	1,332.86951(10)	1,330.33627(08)	1,327.99167(10)	1,323.19999(22)
N_1 χ_{aa}/MHz	−2.6373(13)	−2.8007(18)	1.6211(19)	1.8518(69)	1.898(23)
χ_{bb}/MHz	1.1672(28)	1.0340(27)	1.4772(34)	2.0545(40)	2.104(28)
χ_{cc}/MHz	1.4701(28)	1.7667(27)	−3.0983(34)	−3.9063(40)	−4.002(28)
N_3 χ_{aa}/MHz	2.2619(20)	2.2371(23)	2.5217(12)	2.1383(33)	2.105(20)
χ_{bb}/MHz	−3.6570(22)	−3.3890(25)	−3.5140(16)	1.6064(42)	1.764(28)
χ_{cc}/MHz	1.3951(22)	1.1519(25)	0.9923(16)	−3.7448(42)	−3.870(28)
N_8 χ_{aa}/MHz	2.2167(17)	2.2237(17)	2.1802(17)	1.8033(87)	−2.091(15)
χ_{bb}/MHz	1.9511(20)	1.9832(20)	1.8429(26)	−3.1572(58)	0.940(14)
χ_{cc}/MHz	−4.1678(20)	−4.2069(20)	−4.0231(26)	1.3539(58)	1.151(14)
Δ^b_c/uÅ^2	−0.1676(3)	−0.1767(4)	−0.2212(3)	−0.1789(5)	−0.2023(13)
σ^c/kHz	1.1	1.4	1.1	1.3	3.9
N[d]	71	74	84	54	45

Source: From [169]
[a]Errors in parenthesis are expressed in units of the last digit
[b]$\Delta_c = I_c - I_a - I_b$ is the inertial defect. Conversion factor: 505,379.1 MHz uÅ^2
[c]Standard deviation of the fit
[d]Number of fitted lines

6 Monosaccharides

Carbohydrates are one of the most versatile biochemical building blocks, widely acting in energetic, structural, or recognition processes [204, 205]. The importance of its structure has been the driving force behind the development of methods for elucidating the shape of their building blocks, monosaccharides. Thus, it comes as no surprise that 3D structures and relative stability of conformers of monosaccharides continue to be an area of great research interest [204, 205]. The subtle variation in hydroxyl arrangement is thought to account for differences in chemical and physical properties of the sugars. This is also relevant to distinguish between different conformers. Additionally, monosaccharides are also of interest in the field of astrophysics. The availability of rotational data has been the main bottleneck for examining the presence of these building blocks in the interstellar medium (ISM) [206]. Based on the rotational spectra identification, the simplest C_2 sugar of glycolaldehyde [207, 208] has been identified, but has yet to detect the C_3 sugar of glyceraldehyde [209].

The experimental results obtained in condensed phases [210–217] seem to indicate that a subtle balance between intrinsic and environmental effects governs the conformational preferences of monosaccharides; the structure and relative stability of isolated sugars are different from their counterparts in solution. To separate these contributions, it is crucial to obtain data on the isolated monosaccharides in the gas phase. This highlights the importance of generating sugars in isolated conditions, free from the influence of environmental effects to determine its intrinsic conformational properties relevant to understand its biological activity [218]. In the particular case of biomolecular building blocks, the group of Prof. Simons in Oxford, one of the pioneers in the field of laser spectroscopy, has heavily contributed to the study of carbohydrates [219, 220 and references therein] as can be seen in the chapter dedicated to this subject in this book [221].

Presently, Fourier transform microwave spectroscopy techniques in supersonic jets, combined with laser ablation techniques [61–63], can bring intact monosaccharides into the gas phase for structural investigation. The low-temperature environment of a supersonic expansion provides the ideal medium for preparing individual conformers of sugars in virtual isolation conditions, ready to be interrogated by a short burst of microwave radiation. To date, rotational investigations of monosaccharides have been carried out for C_4 sugars [222], C_5 sugars [62, 223, 224], and C_6 sugars [225–227]. All factors contributing to stabilization of the observed species are given in the next sections.

6.1 *C_4 Sugars: D-Erythrose*

D-Erythrose ($C_4H_8O_4$, see Fig. 29) may be present in linear or cyclic furanose forms. Aqueous solution NMR studies [228, 229] have shown that furanose forms

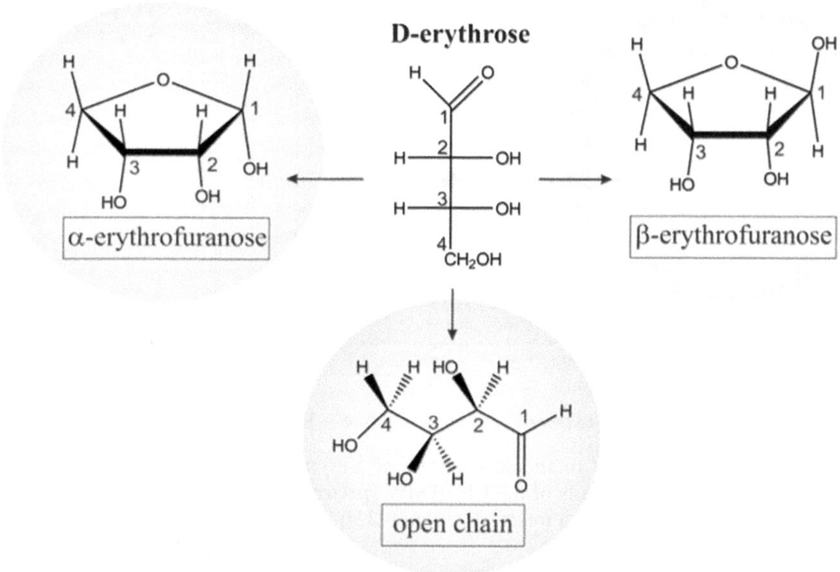

Fig. 29 Fisher projection of D-erythrose (*center*). Haworth projections of the α and β anomers and sketch of the open chain form. (From [222])

(25% and 65% of α and β furanose forms, respectively) are in equilibrium with an appreciable amount of acyclic forms. The five-membered ring structures contain an asymmetric carbon C_1 which leads to the appearance of two stereochemical α and β anomeric species, according to the position of the anomeric OH group (see Fig. 32). Puckering of the ring gives rise to envelope (E) and twist (T) configurations which are interconvertible by rotation of single bonds. This adds complexity, as each of these furanose rings may give rise to many conformers because of the relative arrangement of the OH groups.

Erythrose is a syrup at room temperature and is thermally decomposed using conventional heating methods, so it has been vaporized using laser ablation of solid NaCl doped with D-erythrose [222]. In the experimental procedure, some drops of D-erythrose were mixed with finely powdered NaCl and a small amount of a commercial binder. The laser vaporized products were probed by CP-FTMW spectroscopy [62]. The broadband spectrum (see Fig. 30) shows, apart from the strong NaCl rotational transitions, additional weak lines (inset in Fig. 30) which were attributed to two rotamers A and B of D-erythrose. The derived experimental rotational constants were contrasted with those from ab initio calculations on the lowest lying conformations [229]. Rotamer A was unequivocally identified as conformer α-^{2}E-cc (predicted as global minimum) and rotamer B as conformer β-$^1T_2$-cc. Conformer α-^{2}E-cc (Fig. 31) has the three hydroxyl groups on the same side of the furanose ring forming a cyclic cooperative intramolecular hydrogen

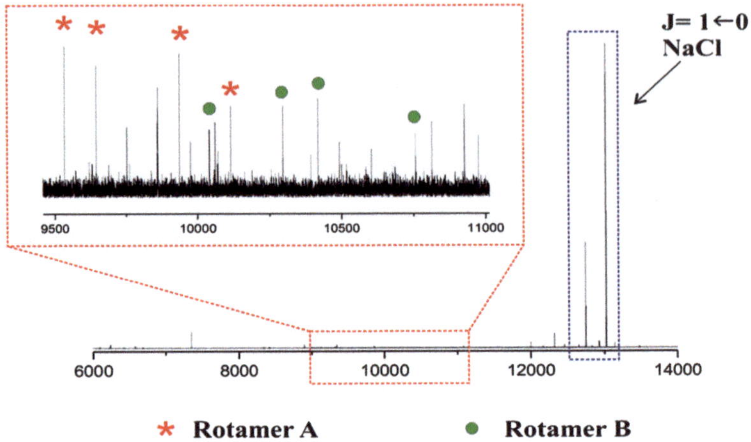

Fig. 30 Broadband CP-FTMW rotational spectrum of D-erythrose and NaCl in the 6–14 GHz frequency region. *Top inset*: details of the CP-FTMW spectrum showing the feature ascribed to several transitions for both detected rotamers. (From [222])

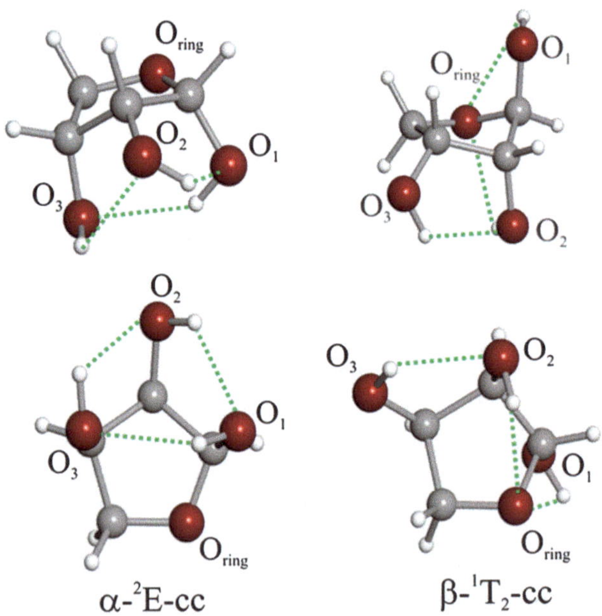

Fig. 31 The three-dimensional structures of the two observed conformers of D-erythrose showing the intramolecular hydrogen bond networks. The notation used to label the conformers includes the symbols α and β to denote the anomer type, E and T with lower and upper subscripts indicate the ring puckerings, and the symbols "c" or "cc" to indicate the clockwise or counterclockwise configuration of the adjacent OH bonds, respectively. (From [222])

bond network, $OH_1 \cdots OH_3 \cdots OH_2 \cdots OH_1$, with a counterclockwise arrangement. Such cyclic network, explains the stability associated with this conformer.

Conformer β-$^1T_2$-cc has two hydroxyl groups on the same side of the furanose ring. Hence, $OH_3 \cdots OH_2 \cdots O_{ring}$ and $OH_1 \cdots O_{ring}$ hydrogen bond motifs together with the axial position of the hydroxyl group OH_1 (anomeric effect) are the main stabilizing factors for this conformer.

6.2 2-Deoxy-D-Ribose and Ribose

2-Deoxy-D-ribose (2DR, $C_5H_{10}O_4$) (Fig. 32a) is an important naturally occurring monosaccharide, present on nucleotides' structures, known as the building blocks of DNA [230]. In DNA, 2DR is present in the furanose (five-membered) ring form, whereas in aqueous solution it is present as five- and six-membered rings species, with the latter being dominant [231, 232]. In the six-membered ring, the C_1 carbon atom is an asymmetric centre, yielding two possible stereochemical α and β anomeric species (Fig. 32b). In aqueous solution, 2DR primarily exists as a mixture of nearly equal amounts of α- and β-pyranose forms, present in their low energy chair conformations, $^4C_1$ and $^1C_4$ (Fig. 32c) [210, 233–235]. Former experiments to determine the conformation of monosaccharides based on X-ray and NMR

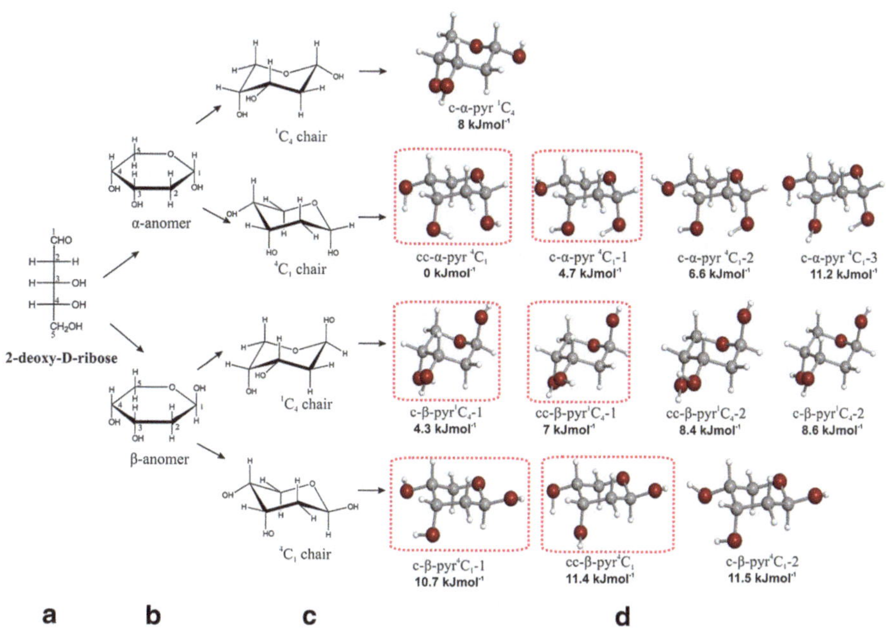

Fig. 32 (**a**) Fisher projection of 2-deoxy-D-ribose. (**b**) Haworth projections of α and β anomers. (**c**) $^1C_4$ and $^4C_1$ chair conformations. (**d**) Predicted conformers within 12 kJ mol^{-1} from MP2(full)/6-311++G(d,p) ab initio computations; the observed conformers are encircled. (From [224])

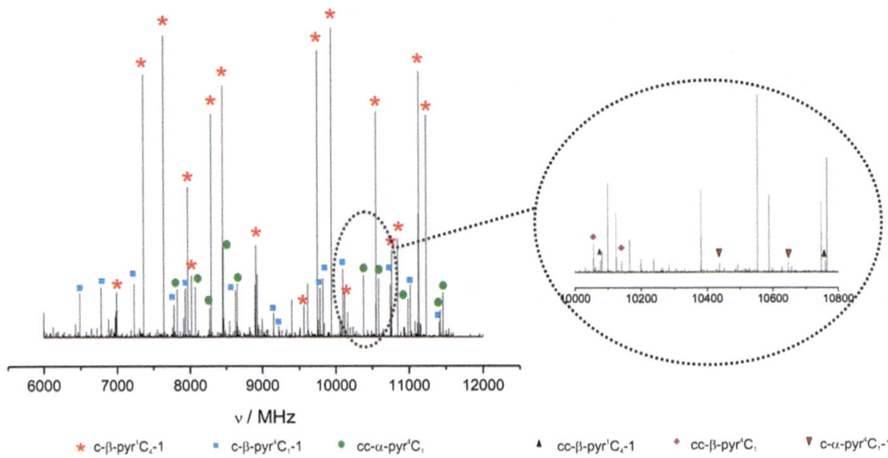

Fig. 33 Broadband microwave spectrum of 2-deoxy-D-ribose. (From [224])

measurements [233, 234, 236, 237] are influenced by environmental effects associated with the solvent or crystal lattice. An IR spectrum of 2DR isolated in an inert matrix [238] has been interpreted by summing the modeled spectra for several α and β conformers.

The conformational panorama of isolated 2-deoxy-D-ribose (m.p. = 89–90°C) has been recently unveiled [224] using CP-FTMW spectroscopy in conjunction with a picosecond laser ablation LA source. The broadband spectra of Fig. 33 allowed the assignment of six different rotameric species labeled I to VI. The rotational constant values were found to be consistent with those predicted ab initio for the conformers shown in Fig. 33d. In addition, spectral measurements have been extended to all five monosubstituted ^{13}C species and the endocyclic ^{18}O species in their natural abundance (~1.1% and ~0.2%) for the most abundant c-β-pyr-$^1C_4$-1 conformer using laser ablation combined with MB-FTMW. The isotopic information was used to derive its structure [224]. The population ratios for α and β conformers estimated from transition intensities indicate that 2DR exists in the gas phase as a mixture of approximately 10% of α- and 90% of β-pyranose forms, thus displaying the dominant β-$^1C_4$ pyranose form, as found in the previous X-ray crystalline study [236]. No evidence has been found of either α/β-furanoses or any linear forms in gaseous 2DR (Fig. 33).

The detected conformers of 2DR, depicted in Fig. 34, can be rationalized in terms of factors that may contribute to their stabilization. The two observed α conformers, cc-α-pyr-$^4C_1$ and c-α-pyr-$^4C_1$-1, are stabilized by anomeric effects; they have a $^4C_1$ ring configuration, thus leading the anomeric OH group towards the axial position. The hydroxy groups of both conformers are located at the same side of the ring, and are able to form chains of hydrogen bonds, which, in turn, are strongly reinforced by sigma hydrogen bond cooperativity [102–104]. The most abundant α form cc-α-pyr-$^4C_1$ presents a counterclockwise arrangement of the OH

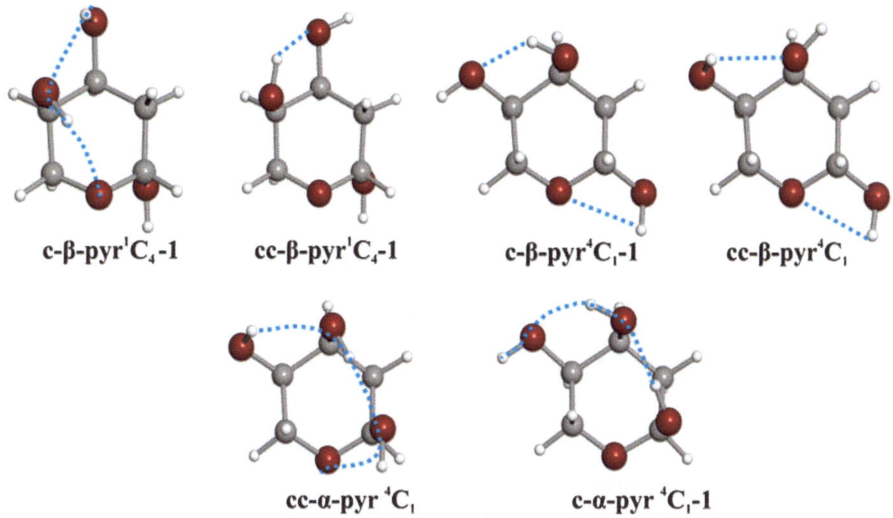

Fig. 34 The six observed conformers of 2-deoxy-D-ribose showing the intramolecular hydrogen bond arrangements. (From [224])

groups with a chain of three hydrogen bonds $O_{(4)}H\cdots O_{(3)}H\cdots O_{(1)}H\cdots O_{ring}$, while the less abundant c-α-pyr-$^4C_1$-1 shows a chain of two $O_{(1)}H\cdots O_{(3)}H\cdots O_{(4)}H$. The anomeric effect in the most abundant β form c-β-pyr-$^1C_4$-1 is reinforced by the intramolecular hydrogen bond network $O_{(3)}H\cdots O_{(4)}H\cdots O_{ring}$. Conformers c-β-pyr-$^4C_1$-1 and cc-β-pyr-$^4C_1$, with the anomeric hydroxy group in equatorial position, are stabilized by two non-cooperative intramolecular hydrogen bonds.

Similar to 2DR, ribose ($C_5H_5O_5$) is one of the most important monosaccharides since it constitutes a subunit of the backbone of RNA. NMR studies have shown that ribose in solution is a mixture of α- and β-pyranose and α- and β-furanose forms, the β-pyranose form being predominant. The recently settled crystal structures have shown that the α- and β-pyranose forms are present in the solid phase [239–243]. The structure in the gas phase has been experimentally investigated using a laser ablation molecular beam Fourier transform microwave spectroscopy (LA-MBFTMW) technique [62]. The high resolution rotational spectrum has provided structural information on a total of six rotamers of ribose, three belonging to the α-pyranose forms and other three to the β-pyranose forms. Recently, D-ribose (m.p. 95°C) has been submitted to a laser ablation broadband (CP-FTMW) spectroscopic study and eight conformers (two new α-pyranose forms) have been identified. A broadband section of the spectra is shown in Fig. 35 and the detected conformers depicted in Fig. 36.

Compared to ribose, the absence of the hydroxy group at C_2 in 2-deoxyribose limits the possibility of forming hydrogen bonds and in practice leads to weakening of the cooperative hydrogen-bond network, altering the relative abundances. For example, the most stable α-pyranose form c-α-pyr-$^1C_4$ of ribose has not been

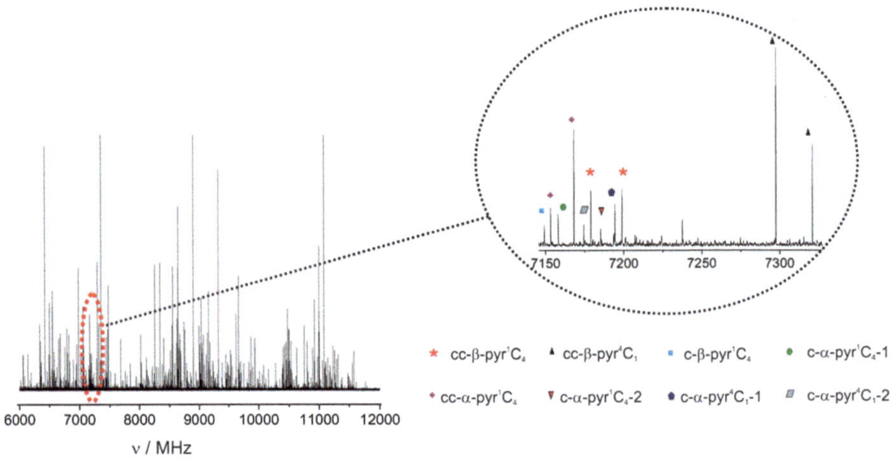

Fig. 35 Broadband spectrum of D-ribose

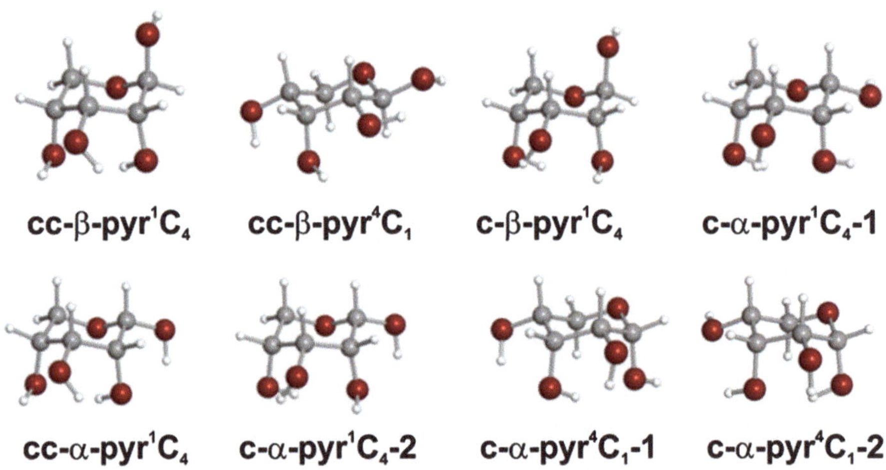

Fig. 36 The eight detected conformers of D-ribose, adopting $^1C_4$ or $^4C_1$ -chair structures

detected in 2DR. The absence of an $O_{(2)}H$ group reverses the arrangement of the OH groups in the most stable β-pyranose forms (from clockwise in c-β-pyr $^1C_4$ of 2DR to counterclockwise in ribose cc-β-pyr $^1C_4$) to maximizes the number of hydrogen bonds (two in cc-orientation vs one in the clockwise arrangement). The evidence collected so far supports that pyranose forms of ribose and deoxyribose are more stable both in gas phase and solution, so the biological pathway to the insertion of furanose forms of ribose and deoxyribose in RNA or DNA cannot be merely attributed to a preference for the furanose forms in the physiological medium.

6.3 D-Glucose and D-Xylose

D-Glucose ($C_6H_{12}O_6$, see Fig. 37a) is one archetypical monosaccharide representing the major building block for many carbohydrate systems [244–247]. Similar to other six-carbon monosaccharides, it may exist in either linear or cyclic, pyranose or furanose, forms. In aqueous solution, NMR studies have shown that the pyranose form is dominant [210, 212, 213]. The cyclization leads to the occurrence of two anomeric species, α and β, according to the position of the OH group (see Fig. 37b). It is commonly believed that the α anomer is more stable than the β anomer because of the stereoelectronic anomeric effect [248, 249]. However, when D-glucose is dissolved in water, the α and β anomers are present in a 40:60 ratio in the $^4C_1$ ring conformation (Fig. 37c). The observed abundance of the β anomer in water could only be explained by taking into consideration strong solvation effects, which overcome the preference for the α anomer [250–253].

The glucopyranose's hydroxymethyl group configuration must be considered in the conformational analysis of D-glucose in terms of three staggered conformers, designated G+, G−, and T (see Fig. 37d), associated with the C6–O6–C5–O5 torsional angle, which assumes values of *ca.* 60°, −60° or 180°, respectively. Experimental observations in both the solid phase [211, 214–217, 254] and solution [212, 213, 233] display approximately equal populations of G+ and G− conformers, with an almost complete absence of the T conformer. This propensity in glucopyranosides to adopt gauche conformations is known as gauche effect [255, 256 and references therein]. Finally, the structural analysis of D-glucose requires the consideration of intramolecular hydrogen bond networks involving adjacent OH groups. The orientation of the hydroxyl groups is relevant to distinguish between the different conformers of D-glucose and is thought to account for

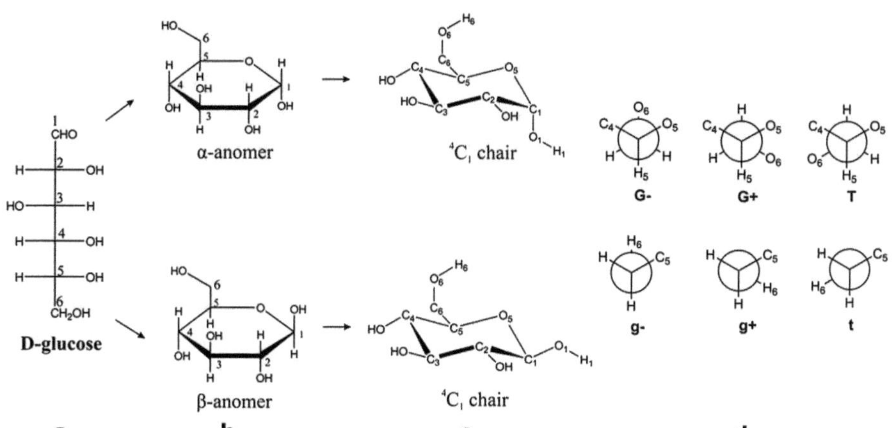

Fig. 37 (**a**) Fisher projection of D-glucose. (**b**) Haworth projections of α and β anomers. (**c**) $^4C_1$ chair conformations. (**d**) Newman projections of the plausible conformations of the hydroxymethyl group around C_5–C_6 and C_6–O_6 bonds. (From [225])

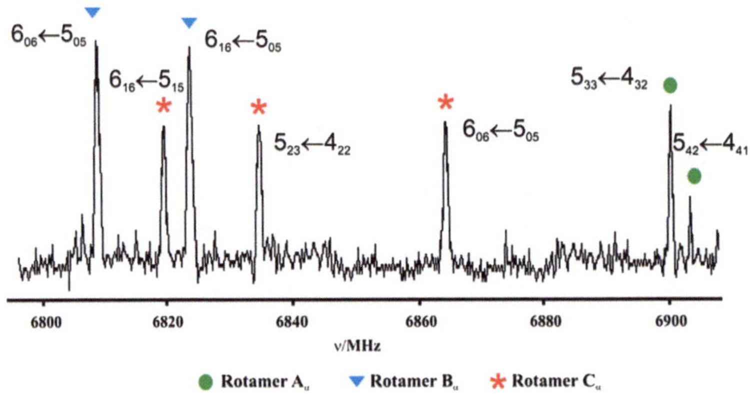

Fig. 38 A section of the LA-MB-FTMW spectrum of α-D-glucopyranose showing the rotational transitions for three of the four observed rotamers. (From [224])

differences in chemical and physical properties of this biologically relevant biomolecule [244–247].

Apart from the number of theoretical studies on D-glucopyranose [257, 258], only one vibrational spectroscopic study of α-D-glucopyranose isolated in Ar matrix has been reported [259]. Laser spectroscopy through UV–UV and IR–UV double-resonance techniques has contributed to the description of the conformations of some β-phenylglucopyranosides and their hydrates [219, 220, 260, 261] but these studies are limited to vibrational resolution and the structural conclusions are not totally transferable to D-glucose because of the electronic chromophore at the anomeric position.

The gas-phase structures of α- and β-D-glucopyranose (m.p = 153°C and 157°C, respectively) have been examined using LA-MB-FTMW spectroscopy [225] (see Fig. 38). After completing a wide frequency scan, it became possible to assign the rotational spectra of four different rotamers of α-D-glucopyranose Their identification was based on the agreement between the experimental and theoretical values of the rotational constants and their trends of variation upon subtle structural changes. In the same way, the agreement between the electric dipole moment s and the observed rotational selection rules allow one to confirm the assignments. Relative populations of the identified conformers (see Fig. 39) G-g+/cc/g+ : G+g−/cc/g+ : Tg+/cc/g+ : G-g+/cl/g− = 1:0.9(2):0.5(1):0.4(2), estimated by relative intensity measurements of rotational transitions, were found to be in reasonable agreement with those calculated from the ab initio Gibbs free energies of 1:0.90:0.30:0.26. Gauche $^4C_1$ glucopyranose forms with a counterclockwise arrangement of OH groups dominating the conformational panorama of α-D-glucopyranose. The four observed conformers of α-glucopyranose, depicted in Fig. 42, are stabilized by anomeric effect; they have a $^4C_1$ ring configuration with the anomeric OH group towards the axial position. The hydroxyl groups located at equatorial positions are able to form chains of hydrogen bonds, strongly reinforced by sigma-hydrogen bond cooperativity [102–104]. The most abundant α conformers, G-g+/cc/t and G

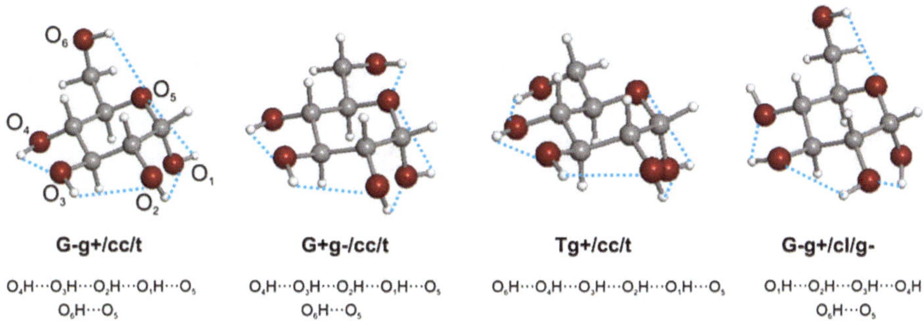

Fig. 39 The four observed conformers of α-D-glucopyranose showing the intramolecular hydrogen bond networks. The symbol in capital letters, *G*+, *G*−, or *T*, describe the torsion angle O_6–C_6–C_5–O_5 (see Fig. 1) of about 60°, −60° or 180°, respectively, which describe the configuration of the hydroxymethyl group. The lower case symbol *g*+, *g*−, or *t*, describe in the same way the torsion angle H_6–O_6–C_6–C_5 (see Fig. 37d). These symbols are followed by a slash and the symbol cl or cc describing, respectively, the clockwise (*cl*) or counterclockwise (*cc*) arrangement of the cooperative network of intramolecular hydrogen bonds. Then, after the slash, the last symbol, *g*+, *g*− or *t*, gives the value of the torsion angle H_1–O_1–C_1–C_2 which describes the orientation of the anomeric hydroxyl group hydrogen atom. (From [225])

+g−/cc/t, present a counterclockwise arrangement of the OH groups with a chain of four cooperative hydrogen bonds $O_4H\cdots O_3H\cdots O_2H\cdots O_1H\cdots O_5$ and additional $O_6H\cdots O_5$ interaction with G− or G+ configurations of hydroxymethyl side chains, respectively. The least abundant G-g+/cl/g− conformer presents a clockwise arrangement of three cooperative hydrogen bonds $O_1H\cdots O_2H\cdots O_3H\cdots O_4H$ and one non-cooperative $O_6H\cdots O_5$. The $O_1H\cdots O_5$ interaction does not take place in the clockwise oriented network explaining its low abundance. Gauche $^4C_1$ glucopyranose forms with a counterclockwise arrangement of OH groups dominate the conformational panorama of α-D-glucopyranose.

In β-D-glucopyranose, rotational spectra revealed the presence of three conformers G-g+/cc/t, G+g−/cc/t and Tg+/cc/t with relative abundances 0.9(2):1:0.2(1), respectively. The two most abundant β conformers G-g+/cc/t and β-G+g−/cc/t shown in Fig. 40, exhibit the same conformational shape as observed in α forms with the obvious differences in the anomeric OH group.

The observation of conformers with an *anti* orientation of the dihedral angle (O_6–C_6–C_5–O_5) in α- and β-D-glucopyranose constitutes a remarkable fact. Numerous experimental studies on α- and β-glucopyranosides, both in solid [211, 214–217, 254] and solution phases [262–264], have shown that the dihedral angle (O_6–C_6–C_5–O_5) displays a preference for G− and G+ gauche configurations, which has been attributed to the gauche effect [255]. This feature was exemplified in a statistical analysis of X-ray structures of glucopyranosyl derivatives [265], yielding a rotamer population of 40:0:60 (G+/T/G−). In contrast to previous results, our gas-phase experiment revealed the existence of *trans* configurations in α-Tg+/cc/t and β-Tg+/cc/t conformers. In agreement with ab initio calculations, these forms have a higher energy and are less abundant in the jet. Both conformers exhibit a

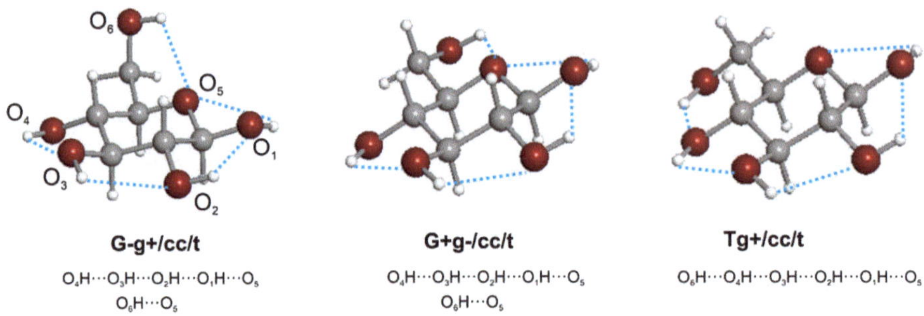

Fig. 40 The three observed conformers of β-D-glucopyranose showing the intramolecular hydrogen bond networks. (From [225])

chain of five cooperative hydrogen bonds $O_6H\cdots O_4H\cdots O_3H\cdots O_2H\cdots O_1H\cdots O_5$ oriented counterclockwise involving the hydroxymethyl group. Therefore, the structure and relative stability of isolated α- and β-D-glucopyranose are different from their counterparts in condensed phases.

D-Xylose ($C_5H_{10}O_5$) is the aldopentose analogue to the aldohexose D-glucose, lacking the hydroxymethyl group. Xylose exists predominantly in the pyranose form in the condensed phase [210, 233, 234, 266], and it is believed to maintain this structure in the gas phase [267]. The pyranose structures have two enantiomers designated as α and β depending on the OH group position at the chiral C_1. Crystalline samples of D-xylose (m.p. = 162°C) have been vaporized by laser ablation and probed by CP-FTMW [222] using a new parabolic reflector system [268]. The recorded broadband spectrum, shown in Fig. 41, allows identification of the conformers cc-α-$^4C_1$ and c-α-$^4C_1$ (see Fig. 42). The population ratio estimated from the relative intensity measurements cc-α-$^4C_1$: c-α-$^4C_1 \approx 1{:}0.03$ is in good agreement with the theoretical predictions. The isotopic information corresponding to the five monosubstituted ^{13}C species of the most abundant conformer cc-α-$^4C_1$ was used to derive its structure [222]. Despite the sensitivity reached in the experiment, no traces belonging to β-pyranose forms have been detected. This is in accordance with the fact that the rotational spectra of laser ablated crystalline D-xylose should reflect the α form found in the crystalline sample [266]. The interconversion between α and β anomers is usually a solvent-mediated reaction and would not occur that easily during the laser ablation process or in the gas phase [269].

The conformers of α-D-xylose, depicted in Fig. 42, present the most favorable chair configuration $^4C_1$, with the largest substituent in equatorial position and the anomeric hydroxyl group in axial position (anomeric effect) [248, 249]. They correlate with the corresponding G-g+/cc/g+ and G-g+/cl/g− of α-D-glucopyranose. Both conformers show arrangements of hydroxyl groups into intramolecular H-bond networks which can lead to a phenomenon known as cooperativity [102–104]. Under cooperativity, directionally arranged H-bonds that form an H-bond network can increase the strength of an individual H-bond donor or acceptor.

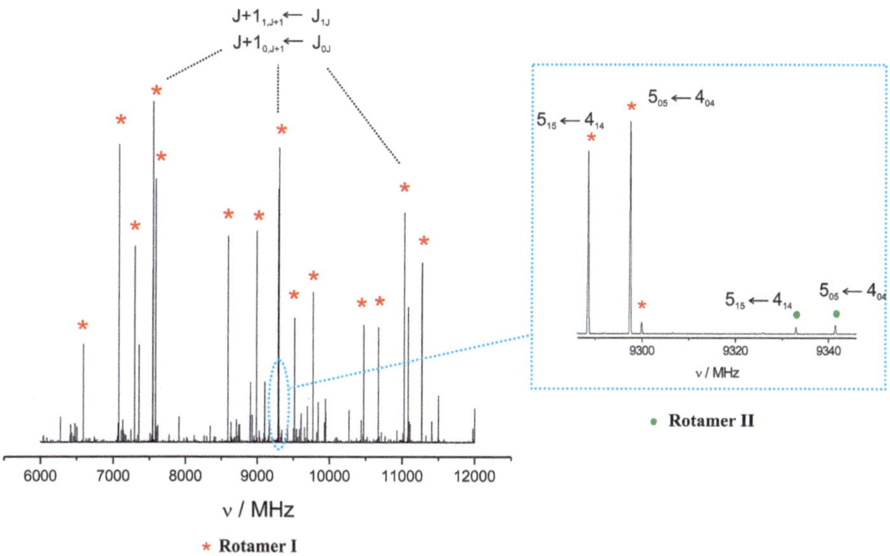

Fig. 41 The broadband rotational spectrum of D-xylose showing the intense rotational transitions for rotamer I. The *inset* shows the characteristic μ_a-R-branch progressions for rotamer I and rotamer II. (From [223])

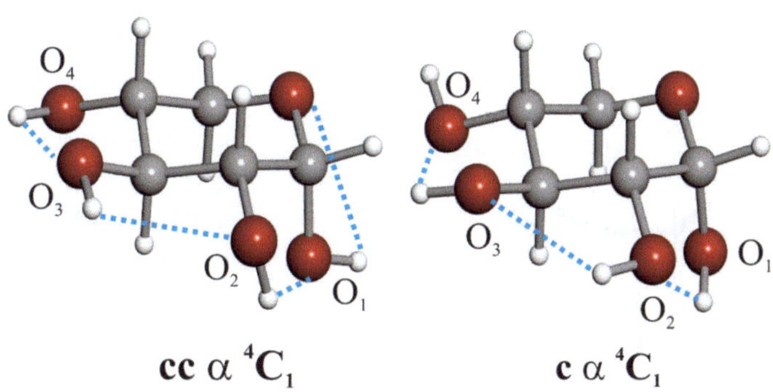

Fig. 42 The three-dimensional structures of the two observed conformers of α-D-xylose showing the intramolecular hydrogen bond arrangements. (From [223])

The most abundant conformer cc-α-$^4C_1$ exhibits a counterclockwise arrangement with a chain of four hydrogen bonds $O_{(4)}H\cdots O_{(3)}H\cdots O_{(2)}H\cdots O_{(1)}H\cdots O_{ring}$. The less abundant c-α-$^4C_1$ species presents a three intramolecular hydrogen bond network $O_{(1)}H\cdots O_{(2)}H\cdots O_{(3)}H\cdots O_{(4)}H$ orientated clockwise. The $O_{(1)}H\cdots O_{ring}$ hydrogen bond found in conformer cc-α-$^4C_1$ might be the cause of the over-stabilization of this species. These result demonstrate the pivotal role that intramolecular hydrogen-bonding network plays in the conformational behavior of free monosaccharides.

Gas phase structures of phenyl α- and β-D-xylopyranoside have been investigated by UV–UV and IR–UV laser spectroscopic techniques coupled with theoretical calculations [270]. The authors hypothesized that the substitution of the anomeric OH group for a phenoxy group compatible with the UV excitation scheme has little effect in the conformational behavior. Present results clearly show that the substitution of the anomeric OH group by the chromophore phenoxy affects the intramolecular hydrogen bond network and, consequently, the conformational behavior and the structural conclusions are transferable to D-xylose. Indeed, the related c-α-$^4C_1$ conformer has not been observed in the phenyl derivative.

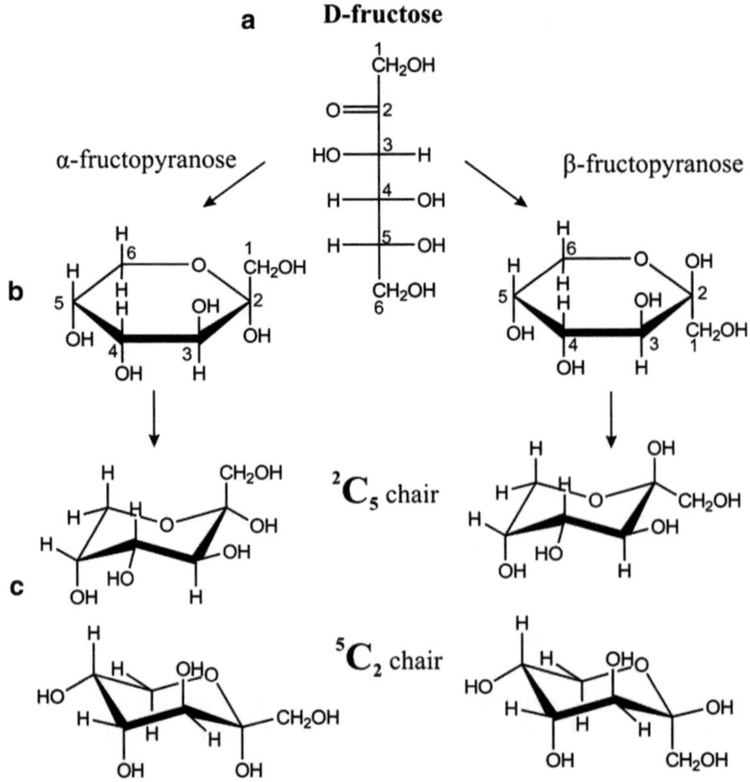

Fig. 43 (**a**) Fisher projection of D-fructose. (**b**) Haworth projections of α and β anomers. (**c**) $^2C_5$ and $^5C_2$ chair configurations. (From [226])

6.4 D-Fructose

D-Fructose ($C_6H_{12}O_6$) is a six-carbon polyhydroxyketone (Fig. 43a). Although ketohexoses such as fructose can exhibit a linear form, D-fructose rapidly cyclizes in aqueous solution to form mixtures of pyranose and furanose forms [246, 247]. The cyclization reaction converts C_2 in a chiral carbon, yielding two enantiomers designated α and β (Fig. 43b). For D-fructose, the equilibrium concentrations in water are around 82% of pyranose forms and 12% of furanose forms [271]. However, in its crystalline form, the unique species found is the β-D-fructopyranose [272, 273]. Besides other higher energy forms, pyranoses preferably adopt a rigid chair backbone with the conformations $^5C_2$ and $^2C_5$ shown in Fig. 43c.

The most stable structures of D-fructose in isolated conditions of gas phase have been unveiled by CP-FTMW [226] spectroscopy, bringing crystalline D-fructopyranose (m.p.120°C) into the gas phase by laser ablation. Once the lines from known photofragmentation species are removed from the broadband spectrum (see Fig. 44), rotational spectra of two rotamers I and II could be identified. The match between the experimental and calculated ab initio values of the rotational constants leads to the irrefutable identification of rotamer I as conformer cc β $^2C_5$ g− and rotamer II as conformer cc β $^2C_5$ t. Both conformers (see Fig. 45) are present in the most favorable chair configuration $^2C_5$, with the largest substituent in an equatorial position and the anomeric hydroxyl group ($OH_{(2)}$) in an axial position (anomeric effect) [248, 249]. It is stabilized by a five cooperative intramolecular hydrogen bond network (Fig. 2) $OH_{(5)}\cdots OH_{(4)}\cdots OH_{(3)}\cdots OH_{(2)}\cdots OH_{(1)}\cdots O_{(ring)}$, with a counterclockwise arrangement of the OH groups. This hydrogen bond cooperative interaction is a form of intramolecular solvation which reinforces the

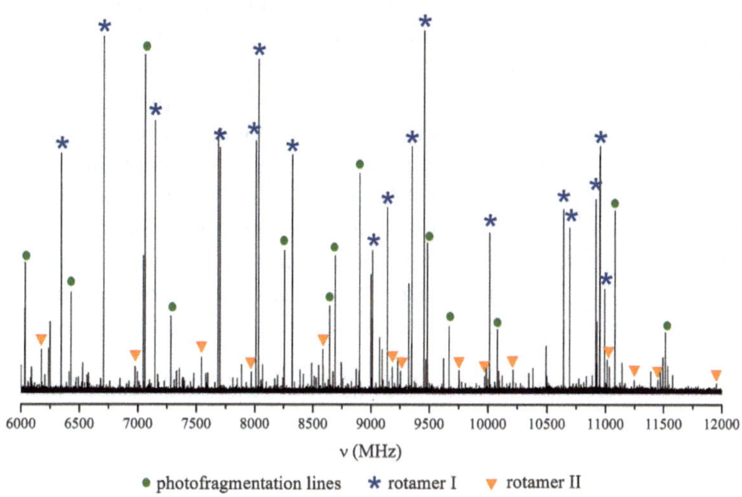

Fig. 44 Broadband microwave spectrum of D-fructose. (From [226])

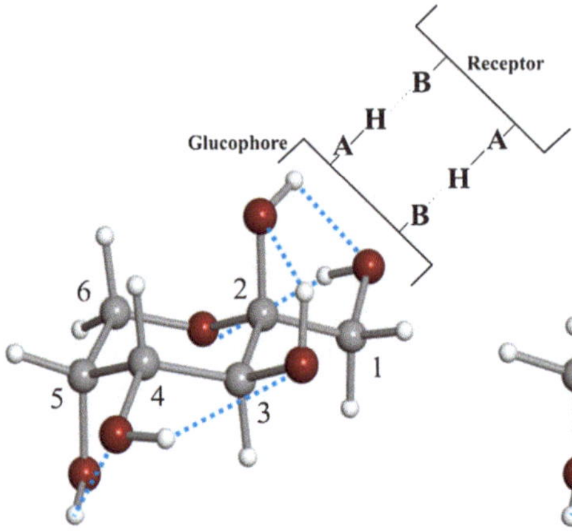

Fig. 45 The three-dimensional structures of the two observed conformers of β-D-fructopyranose showing the intramolecular hydrogen-bonding networks and the glucophore unit. (From [226])

stability. Conformer cc β $^2C_5$ t shows a chain of three cooperative $OH_{(5)}\cdots OH_{(4)}\cdots OH_{(3)}\cdots OH_{(2)}$ and one non-cooperative $OH_{(1)}\cdots O_{(ring)}$ hydrogen bonds. The estimated population ratio β $^2C_5$ g− : cc β $^2C_5$ t as 1: 0.02 is in excellent accordance with the computed energies; conformer cc β $^2C_5$ t is predicted 700 cm^{-1} above the global minimum cc β $^2C_5$ g−. The most abundant conformer cc β $^2C_5$ g− has also been examined by LA-MB-FTMW spectroscopy [227] and its structure has been determined from the rotational spectra of parent and monosubstituted isotopic species.

D-Fructose is the sweetest naturally occurring carbohydrate. Its sweetness in solution is directly related to the D-fructopyranose proportion [274–279]. A simple rationalization of the structure–sweetness relationship is based on the existence of a basic structural unit formed by proton donating A-H and proton accepting B electronegative groups [274–279]. The concept of a tripartite glucophore AH-B ("sweetness triangle") has its merit as a unifying criterion proved useful in rationalizing the sweetness in diverse classes of compounds. The assignment of the AH-B tripartite in detected conformers in Fig. 45 is very complicated as each OH group can function as AH and/or B. After examination of the structure of cc β $^2C_5$ g− conformer, one can state that $OH_{(1)}$ and $O_{(2)}$ can be considered as the most likely AH-B glucophore causing the sweet response via interaction with a complementary hydrogen bond donor and acceptor in the taste receptor. Hence, the most abundant conformer cc β $^2C_5$ g− might be responsible for the sweetness of D-fructose.

7 Summary and Outlook

The main objective of this chapter is to summarize the advances attained in the knowledge of the structure of isolated biomolecular building blocks by the combination of the experimental methods that combine laser ablation with Fourier transform microwave spectroscopy techniques in supersonic jets.

The results described confirm that these methods are formidable tools to investigate the conformational landscape of solid biomolecules in the gas phase.

The high accuracy of frequency measurements, the resolution achieved, and the supersonic expansion cooling are clear advantages for the study of the complex conformational behavior of such molecules. They allow unambiguous discrimination between conformers and provide rich information about the intermolecular forces at play. The molecular properties extracted from the analysis of the rotational spectrum, such as the rotational and quadrupole coupling constants, can be directly compared with those predicted ab initio to achieve a conclusive identification of the most abundant conformers in the supersonic expansion. ^{14}N quadrupole coupling interaction has been shown to be an invaluable alternative tool to identify the nature and most stable forms of these compounds. The characteristic hyperfine structure pattern produced by this coupling has been found to be a fingerprint for each observed species. Moreover, rotational data provide a benchmark against which quantum theory calculations can be checked.

Thanks to the LA-MB-FTMW technique, knowledge of the structural properties of non-aromatic neutral amino acids in the gas phase, unavailable through other techniques, has expanded considerably. The investigation of individual neutral amino acids in the gas phase reveals a complex and subtle network of forces that condition the conformations adopted by amino acids. Conformational stabilization is dictated, in all cases, by hydrogen bonding. The presence of a polar side chain has been found to play an influential role in all α-amino acids studied, reflected in the increased number of low-energy conformers, with non-polar side-chain amino acids exhibiting a limited conformational variety. The hydrogen bond interaction of the side chain with the other polar groups, in the amino acid backbone, may reverse the relative stabilities of conformers or even increase the interconversion barriers. Interestingly, in asparagine, a conformational locking occurs because of the interplay of several hydrogen bond interactions. The increase of the amino acid backbone chain in the series of α-, β-, and γ-amino acids also has important consequences for conformational behavior, because of the presence of other stabilization forces such as n–π* interactions.

The LA-MB-FTMW technique has also contributed to extending our knowledge on the tautomer equilibria in important nucleobases such as guanine or cytosine. Again, ^{14}N quadrupole coupling interaction has been shown to be instrumental in identifying the nature of the different nitrogen atoms and thus to establish unambiguously the most stable tautomers/conformers of these compounds. Moreover, characteristic hyperfine structure patterns produced by this coupling have been found to be fingerprints for each observed species.

The future perspectives of this research area depend heavily on the application of CP-FTMW spectrometers, which have changed the scope of rotational spectroscopy in recent years. Chirped-pulse Fourier-transform microwave spectroscopy, combined with laser ablation, opens a new era in the investigation of isolated biomolecules, as can be shown by its application to carbohydrates as discussed in this chapter. The broad frequency and large dynamic range make it possible to extend the range of detectable conformers to less stable forms and to access structural determinations in molecular systems of increasingly larger sizes from measurements of heavy atom (^{13}C, ^{15}N, ^{18}O) isotopes detected in natural abundance. This opens promising perspectives for the structural determination of key systems, and the study of heavier systems such as new amino acids, sugars, dipeptides or tripeptides, glycosides, and other biomolecules of increasing complexity.

References

1. Schermann JP (2008) Spectroscopy and modelling of biomolecular building blocks. Elsevier, Amsterdam
2. Robertson EG, Simons JP (2001) Getting into shape: conformational and supramolecular landscapes in small biomolecules and their hydrated clusters. Phys Chem Chem Phys 3:1
3. Zwier TS (2001) Laser spectroscopy of jet-cooled biomolecules and their water-containing clusters: water bridges and molecular conformation. J Phys Chem A 105:8827
4. Weinkauf RP, Schermann J-P, de Vries MS, Kleinermanns K (2002) Molecular physics of building blocks of life under isolated or defined conditions. Eur Phys J D 20:309
5. Simons JP (2004) Bioactive molecules in the gas phase. Phys Chem Chem Phys 6:E7
6. Simons JP (2007) Spectroscopic probes of molecular recognition. Phys Chem Chem Phys 32:4429–4608
7. Kistenmacher TJ, Rand GA, Marsh RE (1974) Refinements of the crystal structures of DL-serine and anhydrous L-serine. Acta Crystallogr B 30:2573
8. Görbitz C, Dalhus HB (1996) Redetermination of L-leucine at 120K. Acta Crystallogr C 52:1754
9. Levy DH (1980) Laser spectroscopy of cold gas-phase molecules. Annu Rev Phys Chem 31:197
10. Scoles G (ed) (1988) Atomic and molecular beam methods, vol 1. Oxford University Press, New York/Oxford
11. Martinez SJ, Alfano JC, Levy DH (1992) J Mol Spectrosc 156:421
12. Gordy W, Cook RL (1984) Microwave molecular spectra. Wiley, New York
13. Structure data of free polyatomic molecules, vol 7 (1976), vol 15 (1987), vol 21 (1992), vol 23 (1995), vol 25A (1998), vol 25B (1999), vol 25C (2000), vol 25D (2003), vol 28A (2006), vol 28B (2006), vol 28C (2007), vol 28D (2007). Springer, Berlin
14. Graner G, Hirota E, Iijima T, Kuchitsu K, Ramsay DA, Vogt J, Vogt N (1998) Structure data of free polyatomic molecules. Basic data. Springer, Berlin, 214 pp
15. Balle TJ, Flygare WH (1981) Rev Sci Instrum 52:33
16. Grabow J-U, Stahl W (1990) Z Naturforsch A 45:1043
17. Andresen U, Dreizler H, Grabow JU, Stahl W (1990) Rev Sci Instrum 61:3694
18. Brown GG, Dian BC, Douglass KO, Geyer SM, Shipman ST, Pate BH (2008) Rev Sci Instrum 79:053103

19. Grabow J-U, Mata S, Alonso JL, Peña I, Blanco S, López JC, Cabezas C (2011) Phys Chem Chem Phys 13:21063
20. Leopold KR, Fraser GT, Novick SE, Klemperer W (1994) Chem Rev 94:1807
21. Bauder A (1996) In: Fausto R (ed) Low temperature molecular spectroscopy. Kluwer, The Netherlands, pp 271–289
22. Legon AC, Millen DJ (1986) Chem Rev 86:635
23. Endo Y, Kohguchi H, Ohshima Y (1994) Faraday Discuss 97:341
24. McCarthy MC, Thaddeus P (2001) Chem Soc Rev 30:177
25. Sutter DH, Dreizler H (2001) Z Naturforsch A Phys Sci 56:425
26. Gerry MCL (1992) Chem Phys Lett 188:213
27. Brupbacher B, Brupbacher T (1999) J Chem Phys 111:6300
28. Antolínez S, López JC, Alonso JL (1999) Angew Chem Int Ed 38:1772
29. Sanz ME, López JC, Alonso JL (1999) Chem Eur J 5:3293
30. Sanz ME, Lesarri A, López JC, Alonso JL (2001) Angew Chem Int Ed 40:935
31. Antolínez S, López JC, Alonso JL (2001) ChemPhysChem 2:114
32. Alonso JL, Antolínez S, Blanco S, Lesarri A, López JC, Caminati W (2004) J Am Chem Soc 126:3244
33. Caminati W, López JC, Alonso JL, Grabow J-U (2005) Angew Chem Int Ed 44:2
34. Ottaviani P, Caminati W, Favero LB, Blanco S, López JC, Alonso JL (2006) Chem Eur J 12:915
35. López JC, Caminati W, Alonso JL (2006) Angew Chem Int Ed 45:290
36. Blanco S, Sanz ME, Mata S, Lesarri A, López JC, Dreizler H, Alonso JL (2003) Chem Phys Lett 375:355
37. Blanco S, Sanz ME, Lesarri A, López JC, Alonso JL (2004) Chem Phys Lett 397:379
38. Blanco S, López JC, Sanz ME, Lesarri A, Dreizler H, Alonso JL (2004) J Mol Spectrosc 227:202
39. Sanz ME, Alonso JL, Blanco S, Lesarri A, López JC (2004) Astrophys J 621:L157
40. Suenram RD, Lovas FJ, Fraser GT (1988) J Mol Spectrosc 127:472
41. Welzel A, Stahl W (1999) Phys Chem Chem Phys 1:5109
42. Blanco S, López JC, Alonso JL, Ottaviani P, Caminati W (2003) J Chem Phys 119:880
43. Brown RD, Godfrey PD, Storey JWV, Bassez MPJ (1978) Chem Soc Chem Commun 547–548
44. Suenram RD, Lovas FJ (1978) J Mol Spectrosc 72:372
45. Levis RJ (1994) Annu Rev Phys Chem 45:483
46. Simard B, Mitchell SA, Humphries MR, Hackett PA (1988) J Mol Spectrosc 129:186
47. Cable JR, Tubergen MJ, Levy DH (1989) J Am Chem Soc 111:9032
48. Barnes M, Fraser MM, Hajigeorgiou PG, Merer AJ, Rosner SD (1995) J Mol Spectrosc 170:449
49. Elam JW, Levy DH (1998) J Phys Chem B 102:8113
50. Suenram RD, Lovas FJ, Fraser GT, Matsumura K (1990) J Chem Phys 92:4724
51. Walker KA, Gerry MCL (1997) J Mol Spectrosc 182:178
52. Oshima Y, Endo Y (1993) Chem Phys Lett 213:95
53. Low RJ, Varberg TD, Sonnely JP, Auty AR, Howard BJ, Brown JM (1993) J Mol Spectrosc 161:499
54. Kretschmer U, Consalvo D, Knaack A, Schade W, Stahl W, Dreizler H (1996) Mol Phys 87:1159
55. Lovas FJ, Kawashima Y, Grabow J-U, Suenram RD, Fraser GT, Hirota E (1995) Astrophys J Lett 455:L201
56. Lesarri A, Mata S, López JC, Alonso JL (2003) Rev Sci Instrum 74:4799
57. Antolinez S, Lesarri A, Mata S, Blanco S, López JC, Alonso JL (2002) J Mol Struct 612:125
58. Lesarri A, Mata S, Blanco S, López JC, Alonso JL (2004) J Chem Phys 120:6191
59. Lesarri A, Mata S, Cocinero EJ, Blanco S, López JC, Alonso JL (2002) Angew Chem Int Ed 41:4673

60. Blanco S, Sanz ME, López JC, Alonso JL (2007) Proc Natl Acad Sci U S A 104:20183
61. Alonso JL, Pérez C, Sanz ME, López JC, Blanco S (2009) Seven structures of threonine in the gas phase: a LA-MB-FTMW study. Phys Chem Chem Phys 11:617
62. Cocinero EJ, Lesarri A, Ecija P, Basterretxea FJ, Grabow J-U, Fernández JA, Castaño F (2012) Angew Chem Int Ed 51:3119–3124
63. Mata S, Peña I, Cabezas C, López JC, Alonso JL (2012) J Mol Spectrosc 280:91–96
64. Peña I, Daly AM, Cabezas C, Mata S, Bermúdez C, Niño A, López JC, Grabow J-U, Alonso JL (2013) J Phys Chem Lett 4:65–69
65. Cabezas C, Alonso JL, López JC, Mata S (2012) Angew Chem Int Ed 51:1375–1378
66. Atkinson DB, Smith MA (1995) Rev Sci Instrum 66:4434
67. Peña I, Sanz ME, López JC, Alonso JL (2012) Preferred conformers of proteinogenic glutamic acid. J Am Chem Soc 134:2305
68. Frisch MJ, Trucks GW, Schlegel HB, Scuseria GE, Robb MA, Cheeseman JR, Scalmani G, Barone V, Mennucci B, Petersson GA, Nakatsuji H, Caricato M, Li X, Hratchian HP, Izmaylov AF, Bloino J, Zheng G, Sonnenberg JL, Hada M, Ehara M, Toyota K, Fukuda R, Hasegawa J, Ishida M, Nakajima T, Honda Y, Kitao O, Nakai H, Vreven T, Montgomery JA Jr, Peralta JE, Ogliaro F, Bearpark M, Heyd JJ, Brothers E, Kudin KN, Staroverov VN, Kobayashi R, Normand J, Raghavachari K, Rendell A, Burant JC, Iyengar SS, Tomasi J, Cossi M, Rega N, Millam JM, Klene M, Knox JE, Cross JB, Bakken V, Adamo C, Jaramillo J, Gomperts R, Stratmann RE, Yazyev O, Austin AJ, Cammi R, Pomelli C, Ochterski JW, Martin RL, Morokuma K, Zakrzewski VG, Voth GA, Salvador P, Dannenberg JJ, Dapprich S, Daniels AD, Farkas Ö, Foresman JB, Ortiz JV, Cioslowski J, Fox DJ (2009) Gaussian 09, revision B.01. Gaussian, Wallingford
69. Becke AD (1993) J Chem Phys 98:5648–5652
70. Lee C, Yang W, Parr RG (1988) Phys Rev B 37:785–789
71. Vosko SH, Wilk L, Nusair M (1980) Can J Phys 58:1200–1211
72. Stephens PJ, Devlin FJ, Chabalowski CF, Frisch MJ (1994) J Phys Chem 98:11623–11627
73. Ditchfield R, Hehre WJ, Pople JA (1971) Self-consistent molecular orbital methods. 9. Extended Gaussian-type basis for molecular-orbital studies of organic molecules. J Chem Phys 54:724
74. Møller C, Plesset MS (1934) Phys Rev 46:618
75. Raghavachari K, Binkley JS, Seeger R, Pople JA (1980) Self-consistent molecular orbital methods. 20. Basis set for correlated wave-functions. J Chem Phys 72:650–654
76. Townes CH, Schawlow AL (1975) Microwave spectroscopy. Dover, New York
77. Kroto HW (1975) Molecular rotation spectra. Wiley, London
78. Albrecht G, Corey RB (1939) J Am Chem Soc 61:1087
79. Marsh RE (1958) Acta Crystallogr 11:654
80. Levy HA, Corey RB (1941) J Am Chem Soc 63:2095
81. Donohue J (1950) J Am Chem Soc 72:949
82. Hirokawa S (1955) Acta Crystallogr 8:637
83. Gaffney JS, Pierce RC, Friedman L (1977) J Am Chem Soc 99:4293
84. Edsall JT, Blanchard MH (1933) J Am Chem Soc 55:2337
85. Ellzy MW, Jensen JO, Hameka HF, Kay JG (2003) Spectrochim Acta A 59:2619
86. Meister A (1965) Biochemistry of the amino acids, vol 2, 2nd edn. Academic, New York
87. Barrett GC (ed) (1985) Chemistry and biochemistry of the amino acids. Chapman and Hall, Glasgow
88. Barrett GC (1998) Amino acids and peptides. Cambridge University Press, Cambridge
89. Schafer L, Sellers HL, Lovas FJ, Suenram RDJ (1980) Am Chem Soc 102:6566–6568
90. Suenram RD, Lovas FJ (1980) J Am Chem Soc 102:7180
91. Godfrey PD, Brown RD (1995) J Am Chem Soc 117:2019
92. McGlone SJ, Elmes PS, Brown RD, Godfrey PD (1999) J Mol Struct 485:225
93. Godfrey PD, Firth S, Hatherley LD, Brown RD, Pierlot AP (1993) J Am Chem Soc 115:9687

94. Ilyushin VV, Alekseev EA, Dyubko SF, Motiyenko RA, Lovas FJ (2005) Millimeter wave spectrum of glycine. J Mol Spectrosc 231:15
95. Császár AG (1992) J Am Chem Soc 114:9568
96. Hu C-H, Shen M, Schaefer HF (1993) J Am Chem Soc 115:2923
97. Barone V, Adamo C, Lejl F (1995) J Chem Phys 102:364
98. Császar AG (1995) J Mol Struct 346:141
99. Bludský O, Chocholoušová J, Vacek J, Huisken F, Hobza P (2000) J Chem Phys 113:4629
100. Alonso JL, Cocinero EJ, Lesarri A, Sanz ME, López JC (2006) The glycine-water complex. Angew Chem Int Ed 45:3471
101. Alonso JL, Peña I, Sanz ME, Vaquero V, Mata S, Cabezas C, López JC (2013) Chem Commun 49:3443
102. Saenger W (1979) Nature 279:343
103. Jeffrey GA (1997) An introduction to hydrogen bonding. Oxford University Press, Oxford
104. Jeffrey GA, Saenger W (1991) Hydrogen bonding in biological structures. Springer, New York
105. Blanco S, Lesarri A, López JC, Alonso JL (2004) J Am Chem Soc 126:11675
106. Kraitchman J (1953) Am J Phys 21:17
107. Van Eijck BP (1982) J Mol Spectrosc 91:348
108. Ruoff RS, Klots TD, Emilsson T, Gutowsky HS (1990) J Chem Phys 93:3142
109. Florio GM, Christie RA, Jordan KD, Zwier T (2002) J Am Chem Soc 124:10236
110. Zwier T (2006) J Phys Chem A 110:4133
111. Sanz ME, Lopez JC, Alonso JL (2002) Chem Eur J 8:4265
112. Jimenez AI, Vaquero V, Cabezas C, Lopez JC, Cativiela C, Alonso JL (2011) The singular gas-phase structure of 1-aminocyclopropanecarboxylic acid (Ac(3)c). J Am Chem Soc 133:1062
113. de Meijere A (1979) Angew Chem Int Ed Engl 18:809–826
114. Rademacher P (2003) Chem Rev 103:933–975
115. Lesarri A, Cocinero EJ, López JC, Alonso JL (2004) Angew Chem Int Ed 43:605
116. Lesarri A, Sánchez R, Cocinero EJ, López JC, Alonso JL (2005) J Am Chem Soc 127:12952
117. Cocinero EJ, Lesarri A, Grabow J-U, López JC, Alonso JL (2007) ChemPhysChem 8:599
118. Ramachandran GN, Reddi AH (eds) (1976) Biochemistry of collagen. Plenum, New York
119. Nimmi ME (ed) (1988) Collagen, vol 1–4. CRC, Boca Raton
120. Adams E, Frank L (1980) Annu Rev Biochem 49:1005
121. Ramachandran GN, Kartha G (1955) Nature 176:593
122. Rich A, Crick FHC (1961) J Mol Biol 3:483
123. Fraser RD, McRae TP, Suzuki E (1979) J Mol Biol 129:463
124. Beck K, Brodsky J (1998) Struct Biol 122:17
125. Rosenbloom J, Prockop DJ (1971) J Biol Chem 246:1549
126. Inouye K, Sakakibara S, Prockop D (1976) J Biochim Biophys Acta 420:133
127. Czinki E, Császár AG (2003) Chem Eur J 9:1008
128. Mata S, Vaquero V, Cabezas C, Peña I, Pérez C, López JC, Alonso JL (2009) Observation of two new conformers of neutral proline. Phys Chem Chem Phys 11:4141
129. Lesarri A, Cocinero EJ, López JC, Alonso JL (2005) The shape of 4(S)- and 4(R)-hydroxyproline in gas phase. J Am Chem Soc 127:2572
130. Bürgi HB, Dunitz JD, Shefter E (1973) J Am Chem Soc 95:5065
131. Bürgi HB, Dunitz JD, Lehn JM, Wipff G (1974) Tetrahedron 30:1563
132. Deslongchamps P (1983) Stereoelectronic effects in organic chemistry. Pergamon, New York
133. Tatcher GRJ (1993) The anomeric effect and associated stereoelectronic effects. American Chemical Society, Washington, DC
134. Sanz ME, Blanco S, López JC, Alonso JL (2008) Rotational probes of six conformers of neutral cysteine. Angew Chem Int Ed 47:6216
135. Sanz ME, Lopez JC, Alonso JL (2010) Six conformers of neutral aspartic acid identified in the gas phase. Phys Chem Chem Phys 12:3573

136. Cabezas C, Varela M, Peña I, Mata S, López JC, Alonso JL (2012) The conformational locking of asparagine. RSC Chem Commun 48:5934
137. Snoek LC, Robertson EG, Kroemer RT, Simons JP (2000) Conformational landscapes in amino acids: infrared and ultraviolet ion-dip spectroscopy of phenylalanine in the gas phase. Chem Phys Lett 321:49
138. Lee KT, Sung J, Lee KJ, Kim SK, Park YD (2002) J Chem Phys 116:8251
139. Lee KT, Sung J, Lee KJ, Park YD, Kim SK (2002) Angew Chem Int Ed 41:4114
140. Lee KT, Sung J, Lee KJ, Kim SK, Park YD (2003) Chem Phys Lett 368:262
141. Lee YH, Jung J, Kim B, Butz P, Snoek LC, Kroemer RT, Simons JP (2004) J Phys Chem A 108:69
142. Lindinger A, Toennies JP, Vilesov AF (1999) J Chem Phys 110:3
143. Cohen R, Brauer B, Nir E, Grace L, de Vries MS (2000) J Phys Chem A 104:6351
144. Grace LI, Cohen R, Dunn TM, Lubman DM, de Vries MS (2002) J Mol Spectrosc 215:204
145. Kaczor A, Reva ID, Proniewicz LM, Fausto R (2006) J Phys Chem A 110:2360
146. von Helden G, Compagnon I, Blom MN, Frankowski M, Erlekam U, Oomens J, Brauer B, Gerbercd RB, Meijera G (2008) Phys Chem Chem Phys 10:1248
147. Huang Z, Yu W, Lin Z (2006) J Mol Struct THEOCHEM 758:195
148. Inokuchi Y, Kobayashi Y, Ito T, Ebata T (2007) J Phys Chem A 111:3209
149. Rizzo TR, Park YD, Peteanu LA, Levy DH (1986) J Chem Phys 84:2534
150. Snoek LC, Kroener RT, Hockridge MR, Simons JP (2001) Phys Chem Chem Phys 3:1819
151. Bakker JM, Aleese LM, Meijer G, von Helden G (2003) Phys Rev Lett 91:203003
152. Sanz ME, Cortijo V, Caminati W, Lopez JC, Alonso JL (2006) The conformers of phenylglycine. Chem Eur J 12:2564
153. Pérez C, Mata S, Blanco S, López JC, Alonso JL (2011) J Phys Chem A 115:9653
154. Wang L, Schultz PG (2005) Angew Chem Int Ed 44:34
155. McGlone SJ, Godfrey PD (1995) J Am Chem Soc 117:1043
156. Sanz ME, Lesarri A, Peña I, Vaquero V, Cortijo V, López JC, Alonso JL (2006) The shape of beta-alanine. J Am Chem Soc 128:3812
157. Blanco S, López JC, Mata S, Alonso JL (2010) The conformations of γ-aminobutyric acid (GABA): the role of n→π* interaction. Angew Chem Int Ed 49:9187
158. Choquet D, Korn H (1988) Neurosci Lett 84:329
159. Wu FS, Gibbs TT, Farb DH (1993) Eur J Pharmacol 246:239
160. Horikoshi T, Asanuma A, Yanagisawa K, Anzai K, Goto S (1988) Mol Brain Res 4:97
161. Cocinero EJ, Lesarri A, Sanz ME, López JC, Alonso JL (2006) Conformations of β-aminobutyric acid in the gas phase. ChemPhysChem 7:1481
162. Cocinero EJ, Villanueva P, Lesarri A, Sanz ME, Blanco S, Mata S, López JC, Alonso JL (2007) The shape of neutral sarcosine in gas phase. Chem Phys Lett 435:336
163. Lesarri A, Cocinero EJ, López JC, Alonso JL (2005) Gas-phase structure of N, N-dimethylglycine. ChemPhysChem 6:1559
164. Cortijo V, Sanz ME, López JC, Alonso JL (2009) J Phys Chem A 113:14681–14683
165. Watson D, Crick FH (1953) Molecular structure of nucleic acids: a structure for deoxyribose nucleic acid. Nature 1953(17):737–738
166. Vaquero V, Sanz ME, López JC, Alonso JL (2007) The structure of uracil: a laser ablation rotational study. J Phys Chem A 111:3443
167. López JC, Peña I, Sanz ME, Alonso JL (2007) Probing thymine with laser ablation molecular beam Fourier transform microwave spectroscopy. J Chem Phys 126:191103
168. Alonso JL, Peña I, López JC, Vaquero V (2009) Rotational spectral signatures of four tautomers of guanine. Angew Chem Int Ed 48:6141
169. Alonso JL, Vaquero V, Peña I, López JC, Mata S, Caminati W (2013) All five forms of cytosine revealed in the gas phase. Angew Chem 125:3287
170. López JC, Alonso JL, Peña I, Vaquero V (2010) Hydrogen bonding and structure of uracil–water and thymine–water complexes. Phys Chem Chem Phys 12:14128
171. Brown RD, Godfrey PD, McNaughton D, Pierlot AP (1988) J Am Chem Soc 110:2329

172. Brown RD, Godfrey PD, McNaughton D, Pierlot AP (1989) J Chem Soc Chem Commun 37–38
173. Puzzarini C, Barone V (2011) Phys Chem Chem Phys 13:7158
174. Kabeláč M, Hobza P (2008) Phys Chem Chem Phys 9:903
175. Yoshikawa A, Matsika S (2008) Chem Phys 347:393
176. Wang FF, Zhao DX, Yang ZZ (2009) Chem Phys 360:141
177. Casaes RN, Paul JB, McLaughlin RP, Saykally RJ, van Mourik T (2004) J Phys Chem A 108:10989
178. Graindourze M, Smets J, Zeegers-Huyskens T, Maes G (1990) J Mol Struct 222:345
179. Choi MY, Miller RE (2005) Phys Chem Chem Phys 7:3565
180. Blanco S, López JC, Lesarri A, Alonso JL (2006) J Am Chem Soc 128:1211
181. Lovas FJ, Suenram RD, Fraser GT, Gillies CW, Zozom JJ (1988) Chem Phys 88:722
182. Fraser GT, Suenram RD, Lovas FJ (1988) J Mol Struct 189:165
183. Caminati W, López JC, Blanco S, Mata S, Alonso JL (2010) How water links to cis and trans peptidic groups: the rotational spectrumof N-methylformamide–water. Phys Chem Chem Phys 12:10230
184. Maris A, Ottaviani P, Caminati W (2002) Chem Phys Lett 360:155
185. Hanus M, Ryjacek F, Kubar T, Bogdan TV, Trygubenko SA, Hobza P (2003) J Am Chem Soc 125:7678
186. Elshakre M (2005) Int J Quant Chem 104:1–15
187. Chen H, Li S (2006) J Phys Chem A 110:12360
188. Nir E, Janzen CH, Imhof P, Kleinermanns K, de Vries MS (2001) J Chem Phys 115:4604
189. Mons M, Dimicoli I, Piuzzi F, Tardivel B, Elhanine M (2002) J Phys Chem A 106:5088
190. Seefeld K, Brause R, Häber T, Kleinermanns K (2007) J Phys Chem A 111:6217
191. Nir E, Grace L, Brauer B, de Vries MS (1999) J Am Chem Soc 121:4896
192. Chin W, Mons M, Piuzzi F, Tardivel B, Gorb IL, Leszczynski J (2004) J Phys Chem A 108:8237
193. Mons M, Piuzzi F, Dimicoli I, Gorb L, Leszczynski J (2006) J Phys Chem A 110:10921
194. Choi MY, Miller RE (2006) J Am Chem Soc 128:7320
195. Gahlmann A, Park ST, Zewail AH (2009) J Am Chem Soc 131:2806
196. Szczesniak M, Szczepaniak K, Kwiatkowski JS, KuBulat K, Person WB (1988) J Am Chem Soc 110:8319
197. Dong F, Miller RE (2002) Science 298:1227
198. Nir E, Müller M, Grace LI, de Vries MS (2002) Chem Phys Lett 355:59
199. Shorokhov D (2000). Ph.D. dissertation, University of Oslo
200. Feyer V, Plekan O, Richter R, Coreno M, de Simone M, Prince KC, Trofimov AB, Zaytseva IL, Schirmer J (2010) J Phys Chem A 114:10270
201. Lapinski L, Nowak MJ, Reva I, Rostkowska H, Fausto R (2010) Phys Chem Chem Phys 12:9615–9618
202. Lapinski L, Reva I, Nowak MJ, Fausto R (2011) Phys Chem Chem Phys 13:9676–9684
203. Brown RD, Godfrey PD, McNaughton D, Pierlot AP (1989) J Am Chem Soc 111:2308
204. Colins P, Ferrier R (1995) Monosacccharides: their chemistry and their roles in natural products. Wiley, New York
205. Pigman W, Horton D (1972) The carbohydrates: chemistry and biochemistry. Academic, New York
206. Herbst E, van Dishoeck EF (2009) Annu Rev Astron Astrophys 47:427
207. Hollis JM, Lovas FJ, Jewell PR (2000) Astrophys J 540:L107
208. Hollis JM, Vogel SN, Snyder LE, Jeweel PR, Lovas FJ (2001) Astrophys J 554:L81
209. Lovas FJ, Suenram RD, Plusquellic DF, Møllendal H (2003) J Mol Spectrosc 222:263
210. Angyal SJ (1969) Angew Chem Int Ed 8:157–226
211. Sponsler OL, Dore WH (1931) J Am Chem Soc 53:1639
212. Mackie W, Perlin AS (1966) Can J Chem 44:2039
213. Perlin AS (1966) Can J Chem 44:539

214. McDonald TRR, Beevers CA (1952) Acta Crystallogr 5:654
215. Ferrier WG (1960) Acta Crystallogr 13:678
216. Ferrier WG (1963) Acta Crystallogr 16:1023
217. Brown GM, Levy HA (1965) Science 147:1038
218. (2002) See for example the special issue of Chem. Rev. (issue 102, 2002) dedicated to glycosylation
219. Simons JP, Jockusch RA, Carçabal P, Hung I, Kroemer RT, Mcleod NA, Snoeck LC (2005) Int Rev Phys Chem 24:489–531
220. Simons JP (2009) Mol Phys 107:2435–2458
221. Cocinero EJ, Carçabal P (2014) Carbohydrates. Top Curr Chem. doi:10.1007/128_2014_596
222. Cabezas C, Peña I, Daly AM, Alonso JL (2013) Chem Commun 49:10826
223. Peña I, Mata S, Martín A, Cabezas C, Daly AM, Alonso JL (2013) Phys Chem Chem Phys 15:18243
224. Peña I, Cocinero EJ, Cabezas C, Lesarri A, Mata S, Ecija P, Daly AM, Cimas A, Bermúdez C, Basterretxea FJ, Blanco S, Fernández JA, López JC, Castaño F, Alonso JL (2013) Angew Chem Int Ed 52:11840
225. Alonso JL, Lozoya MA, Peña I, López JC, Cabezas C, Mata S, Blanco S (2014) Chem Sci 5 (2):515
226. Bermúdez C, Peña I, Cabezas C, Daly AM, Alonso JL (2013) ChemPhysChem 14:893–895
227. Cocinero EJ, Lesarri A, Écija P, Cimas A, Davis BJ, Basterretxea FJ, Fernández JA, Castaño F (2013) J Am Chem Soc 135:2845
228. Risley JM, Van Etten RL (1982) Biochemistry 21:6360
229. Serianni AS, Pierce J, Huang SG, Barker R (1982) J Am Chem Soc 104:4037
230. Azofra LM, Alkorta I, Elguero J, Popelier PLA (2012) Carbohydr Res 358:96
231. Watson JD, Crick FHC (1953) Nature 171:737–738
232. Saenger W (1984) Principles of nucleic acid structure. Springer, New York, pp 1–556
233. Rudrum M, Shaw DF (1965) J Chem Soc 52–57
234. Lemieux RU, Stevens JD (1966) Can J Chem 44:249
235. Cortes SJ, Mega TL, Van Etten RL (1991) J Org Chem 56:943–947
236. Furberg S (1960) Acta Chem Scand 14:1357–1363
237. Breitmaier E, Hollstein U (1976) Org Magn Reson 8:573
238. Nikolaenko TY, Bulavin LA, Govorun DN (2011) J Appl Spectrosc 78:751–754
239. Furberg S, Hordvik A, Taulbol K (1956) Acta Chem Scand 10:135
240. Sisak D, McCusker LB, Zandomeneghi G, Meier BH, Bläser D, Boese R, Schweizer WB, Gylmour R, Dunitz JD (2010) Angew Chem 122:4605
241. Sisak D, McCusker LB, Zandomeneghi G, Meier BH, Bläser D, Boese R, Schweizer WB, Gylmour R, Dunitz JD (2010) Angew Chem Int Ed 49:4503
242. Saenger W (2010) Angew Chem 122:6633
243. Saenger W (2010) Angew Chem Int Ed 49:6487
244. Lemieux RU (1990) Explorations with sugars: how sweet it was. American Chemical Society, Washington, DC
245. Collins P, Ferrier R (1995) Monosaccharides: their chemistry and their roles in natural products. Wiley, New York
246. Lehninger AL (2004) Principles of biochemistry, 4th edn. W. H. Freeman, New York
247. Sinnot ML (2007) Carbohydrate chemistry and biochemistry. RSC, Cambridge
248. Juaristi E, Cuevas G (1992) Tetrahedron 48:5019–5087
249. Perrin CL, Armstrong KB, Fabian MA (1994) J Am Chem Soc 116:715–722
250. Cramer CJ, Truhlar DG (1993) J Am Chem Soc 115:5745–5753
251. Corchado JC, Sánchez ML, Aguilar MA (2004) J Am Chem Soc 126:7311–7319
252. Wladkowski BD, Chenoweth SA, Jones KE, Brown JW, Phys J (1998) J Phys Chem A 102:5086–5092
253. Barrows SE, Storer JW, Cramer CJ, French AD, Truhlar DG (1998) J Comput Chem 19:1111

254. Hough E, Neidle S, Rogers D, Troughton PGH (1973) Acta Crystallogr Sect B: Struct Crystallogr Cryst Chem 29:365
255. Wolfe S (1972) Acc Chem Res 5:102–111
256. Kirschner KN, Woods RJ (2001) Proc Natl Acad Sci U S A 98:10541–10545
257. Mihura N, Taniguchi T, Monde K, Nishimura SI (2006) Chem Phys Lett 419:326
258. Hoffmann M, Rychelewski J (2001) J Am Chem Soc 123:2308–2316
259. Kovács A, Ivanov AY (2009) J Phys Chem B 113:2151
260. Çarçabal P, Jockusch RA, Hünig I, Snoek LC, Kroemer RT, Davis BG, Gamblin DP, Compagnon I, Oomens J, Simons JP (2005) J Am Chem Soc 127:11414–11425
261. Çarçabal P, Cocinero EJ, Simons JP (2013) Chem Sci 4:1830–1836
262. Ohrui H, Nishida Y, Watanabe M, Hori H, Meguro H (1985) Tetrahedron Lett 26:3251–3254
263. Nishida Y, Ohrui H, Meguro H (1984) Tetrahedron Lett 25:575–1578
264. Bock K, Duus JØ (1994) J Carbohydr Chem 13:513–543
265. Marchessault RH, Pérez S (1979) Biopolymers 18:2369–2374
266. Hordvik A (1971) Acta Chem Scand 25:2175
267. Guler LP, Yu Y-Q, Kenttämaa HI (2002) J Phys Chem A 106:6754
268. Jahn MK, Dewald DA, Wachsmuth D, Grabow J-U, Mehrotra SC (2012) J Mol Spectrosc 280:54
269. Finch P (1999) Carbohydrates: structures, syntheses and dynamics. Kluwer, The Netherlands
270. Hünig I, Painter AJ, Jockusch RA, Çarçabal P, Marzluff EM, Snoek LC, Gamblin DP, Davis BG, Simons JP (2005) Phys Chem Chem Phys 7:2474–2480
271. Doddrell D, Allerhan A (1971) J Am Chem Soc 93:2779–2781
272. Kanters JA, Roelofsen G, Alblas BP, Meinders I (1977) Acta Crystallogr B 33:665–672
273. Takagi S, Jeffrey GA (1977) Acta Crystallogr B 33:3510–3515
274. Shallenberger RS, Acree TE (1967) Nature 216:480–482
275. Kier LB (1972) J Pharm Sci 61:1394–1397
276. van der Heijden A (1997) Pure Appl Chem 69:667–674
277. Woods RJ, Szarek WA, Smith VH (1990) J Am Chem Soc 112:4732–4741
278. Shallenberger RS (1978) Pure Appl Chem 50:1409–1420
279. Deutsch EW, Hansch C (1966) Nature 211:75

Index